Up and Running with Bluebeam Revu 2019

Deepak Maini
National Technical Manager - Named Accounts
Cadgroup Australia

Technical Editor
April McCall
AEC Application Specialist / Certified Instructor / BIM Modeler - Kelar Pacific
www.kelarpacific.com

Cover Image

Karl De Wet
Director, Draftech Developments (MEP Solutions)
https://www.draftechdevelopments.com.au/

ISBN: 9798612567578

Dedication

To all Architects, Engineers, and Designers who create innovative products and make this world a better place to live

To my mum, Late Mrs. Veena Maini, who taught me the most important lessons of my life of working hard and being patient

To my dad, Mr. Manmohan Lal Maini who has always supported me unconditionally in my endeavors

To my wife Drishti and my son Vansh whose motivation, support, and inspiration made this textbook possible

Foreword

This comprehensive textbook covers the vast array of markup, measurement, and stamp tools available in Bluebeam Revu. In addition, the text provides an in-depth insight into effective quantity take-off processes that utilize the specialized tools in Revu, allowing the readers to quantify their projects using the native PDFs developed during planning and design. Where readers need further options in the cloud, Studio Projects and Sessions are covered in detail and provide a clear and concise guide on how multiple project stakeholders can collaborate in realtime through the PDF file format.

This textbook will provide a valuable resource to project teams and practitioners alike through Deepak's flair for making the complex appear simple and clear. Deepak has provided a fantastic overview of the multitude of dynamic tools available in Bluebeam Revu to provide a firm foundation for project teams to enable resilient digital workflows to make their delivery process efficient and effective.

Russell Bunn
Regional CAD Leader - Transport

Acknowledgments

I would like to thank the following people for helping me throughout the process of writing this textbook and their continued support:

April McCall
AEC Application Specialist / Certified Instructor / BIM Modeler - Kelar Pacific
www.kelarpacific.com

Karl De Wet
Director, Draftech Developments (MEP Solutions)
www.draftechdevelopments.com.au

Kelly Furtado
Regional Director, APAC, Bluebeam, Inc.

Lilian Magallanes
Program Manager, Community Development, Bluebeam Inc.

About the Author

Deepak Maini (Sydney, Australia) is a qualified Mechanical Engineer with more than 20 years of experience working in the design industry. He is a Bluebeam Certified Instructor and a Bluebeam Certified Consultant and has authored the ***"Up and Running with Autodesk Navisworks"***, ***"Autodesk Navisworks for BIM/VDC Managers"***, and ***"Up and Running with Autodesk Advance Steel"*** series of books. He is currently working as the National Technical Manager - Named Accounts with Cadgroup Australia, a Bluebeam Platinum Partner.

Deepak is a regular speaker at various conferences around the world and was the **Top Speaker** at the Bluebeam XCON 2019 conference in Washington DC. He was also awarded the "**Top Autodesk University Speaker**" two years in a row in 2018 and 2017 in the Instructional Demo category. Deepak is also one of the "**Top Rated Speakers**" at various BILT conferences in ANZ and Asia.

Outside his full-time work, Deepak is a Guest Lecturer at the University of Technology Sydney (UTS) and the University of New South Wales (UNSW), two of the leading universities in Australia.

Deepak's Contact Details

Email: *deepak@deepakmaini.com*
Website: *https://www.deepakmaini.com/*

Accessing Tutorial Files

The author has provided all the files required to complete the tutorials in this textbook. To download these files:

1. Visit https://www.deepakmaini.com/Revu/Revu.htm

2. Click on cover page of the book whose tutorial files you want to download.

3. On the top right of the page, click on **ACCESSING TOC/TUTORIAL FILES**.

4. Click on the **Tutorial Files** link.

Free Teaching Resources for Faculty

The author has provided the following free teaching resources for the faculty:

1. Video of every tutorial in the textbook.
2. PowerPoint Slides of all chapters in the textbook.
3. Teacher's Guide with answers to the end of chapter **Class Test Questions**.
4. Help in designing the course curriculum.

To access these resources, please contact the author at **deepak@deepakmaini.com**.

Accessing Videos of the Tutorials in this Textbook

The author provides complimentary access to the videos of all tutorials in this textbook. To access these videos, please Email your proof of purchase to the author at the following Email address:

deepak@deepakmaini.com.

Preface

Welcome to Up and Running with Bluebeam Revu 2019.

This is a comprehensive textbook consisting of twelve chapters for the Architecture, Engineering, Construction, and Operations (AECO) industry covering markup, measurement, and stamp tools of Bluebeam Revu Standard. The process of Quantity Takeoff using specialized tools in Revu Standard is also discussed in detail in both Imperial and Metric units. This will equip the readers to takeoff accurate quantities in their real world projects. This book also covers Bluebeam Studio Projects and Sessions in detail, helping users learn how to get multiple stakeholders to review and markup PDF files together in realtime.

There are two projects at the end of the book that cover the Quantity Takeoff process from the HVAC and Electrical files in both Imperial and Metric units. The concept covered in these projects can also be used to takeoff quantities from the other services, such as Plumbing and Fire.

The chapters in this textbook start with a detailed description of the Revu tools and concepts. These are then followed by the hands-on tutorials. Every section of the tutorials starts with a brief description of what you will be doing in that section. This will help you to understand why and not just how you have to do certain things.

Real-world projects have been carefully selected to discuss the tools and concepts in the tutorials of every chapter. You will be able to find various similarities between the files used in this textbook and your current projects. This will allow you to apply the concepts learned in this textbook to your day-to-day work.

*I have also added the "**What I Do**" sections in most chapters. In these sections, I have discussed the approach I take while working with Revu. You will also find a number of "**Notes**" and "**Tips**" that discuss additional utilities of various concepts.*

I hope you find learning the software using this textbook an enriching experience and are able to apply the concepts in real-world situations.

If you have any feedback about this textbook, please feel free to write to me at the following email address:

deepak@deepakmaini.com

TABLE OF CONTENTS

Chapter 3 - Working with the Markup Tools - II

Chapter 4 - Working with the Markup Tools - III

Chapter 5 - Creating and Managing Custom Tool Sets

Chapter 6 - Working with the Measurement Tools - I

Chapter 7 - Working with the Measurement Tools - II

Chapter 8 - The Markups List

Chapter 9 - Document Management and Hyperlinks

Chapter 10 - Searching and Comparing PDFs and Inserting Images

Chapter 11 - Working with Stamps

Chapter 12 - Working with Bluebeam Studio

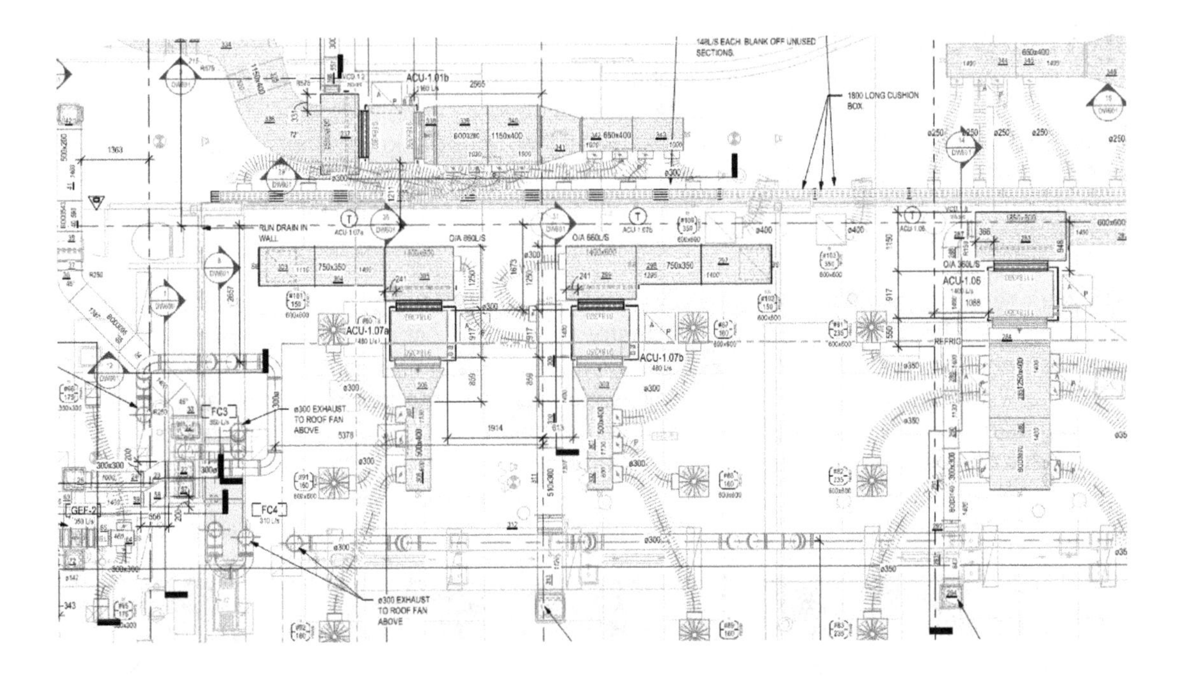

Chapter 1 - Introduction to Revu

The objectives of this chapter are to:

- √ *Introduce you to Revu*
- √ *Explain various Revu flavors*
- √ *Introduce you to Revu interface*
- √ *Explain the concept of profiles and teach how to create and manage profiles*
- √ *Explain the process of opening PDF files using the **Open** tool*
- √ *Explain the process of navigating through the PDF files*

INTRODUCTION TO REVU

Revu is purpose-built software for the Architecture, Engineering, Construction, and Operations (AECO) industry that allows you to create, markup, review, organize, manage, and collaborate using Portable Document Format (PDF) files. Revu facilitates design and construction workflows and streamlines office and field processes by creating a smart set of documents to be shared with various project stakeholders. It is important to note here that the key is PDF files as this is the file format that Revu works with.

The exponential growth in Building Information Modeling (BIM) industry has seen a number of 3D format files being used in the AEC industry. However, most of the design review, markup, and drawings exchange processes still use PDF files. One of the main reasons for this is that PDF is a widely recognized format that allows intellectual property to be maintained with the authors of the data. It is important to note that the smart set of documents created or marked up in Revu can be opened in any other recognized PDF viewer.

Figure 1 shows the zoomed in view of a sample Revu PDF file with some markups added.

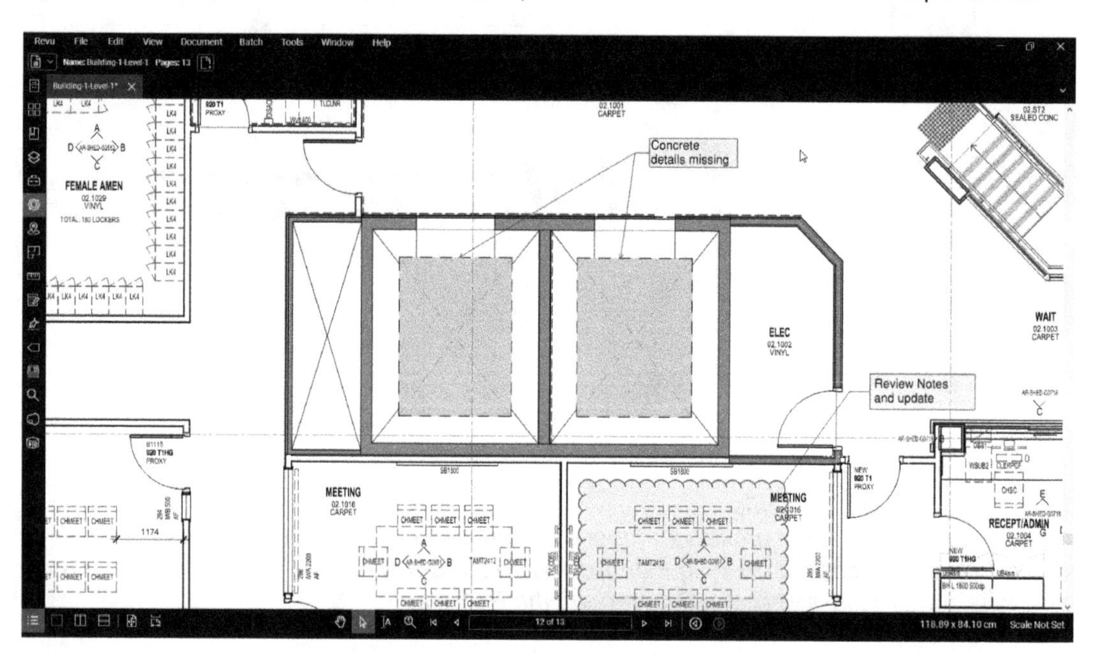

Figure 1 A sample Revu PDF with markups

REVU FLAVORS

The following are the flavors of Revu for Windows and iPad:

Revu Standard

This is the base level flavor of Revu that provides the essential tools for PDF editing, marking-up, and collaborating using smart sets of tools and workflows. Revu Standard is generally used by the General and Speciality Contractors, Estimators, Superintendents, Supervisors, and Owners in the AEC industry.

Tip: Using the tools available in Revu Standard, you can also create 2D and 3D PDF files. You will learn more about this in later chapters of this book.

Revu CAD

This flavor or Revu includes all the functionality of Revu Standard, along with the functionality of creating 2D or 3D PDF files from various CAD/BIM authoring tools. Therefore, Revu CAD is specifically meant for CAD users who require plug-ins within their CAD/BIM tools to create single or multiple PDF files. The Revu CAD plug-ins to create 2D PDF files are available for AutoCAD, Autodesk Revit, Autodesk Navisworks, SolidWorks, and SketchUp. Additionally, you can create 3D PDF files using the Revu plug-in inside Autodesk Revit, Autodesk Navisworks, SolidWorks, and Sketchup. This flavor of Revu is generally used by Architects, Engineers, or general CAD/BIM users.

Revu eXtreme

This flavor or Revu includes all the functionality of Revu CAD, along with the functionality to automate and batch process various repetitive tasks. These tasks include batch hyperlink multiple documents, digitally sign a batch of multiple documents, and so on. Some of the additional functionality of Revu eXtreme includes allowing automatic creation of forms, running Optical Character Recognition (OCR) to convert scanned documents into a document with searchable text, using JavaScript to create custom intuitive stamps, live linking of quantity take-off to Microsoft Excel, and so on. This flavor of Revu is generally used by any of the Revu Standard or Revu CAD users that require the additional functionality of Revu eXtreme.

Revu for iPad

This flavor or Revu is used to review, markup, and collaborate using PDF files on an iPad. Similar to Revu for Mac, this flavor of Revu includes some of the tools or Revu Standard, but is not as robust as Revu Standard.

Note: All flavors of Revu include Bluebeam Studio, the flagship module of Revu that allows document management and real-time collaboration of PDF files on the Cloud.

*Tip: All flavors of Revu 2019 can be run in the **View Mode** as a free viewer. As a result, Bluebeam discontinued the free Vu viewer. However, you can still download Vu 2017 or older version from the Bluebeam website.*

REVU INTERFACE

Figure 2 shows the default interface of Revu 2019. Various components in this interface are discussed below.

Menu Bar

The **Menu Bar** is available at the top of the window and is used to invoke various menus. You can click on any of these menus to activate Revu tools.

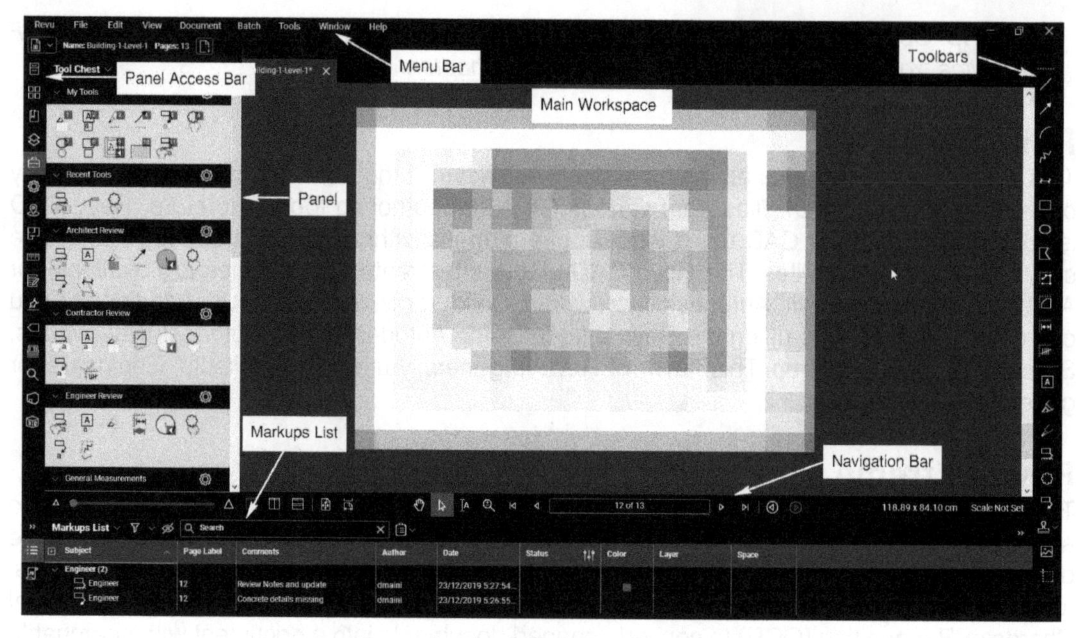

***Figure 2** The default Revu interface*

Panel Access Bar

The **Panel Access Bar** is available on the left of the window and is used to invoke various panels. These panels can be docked on the left or right of the window or kept floating, depending on the way you want the Revu interface to look like. Clicking on any of the panel buttons from the **Panel Access Bar** will display that particular panel. You can click on the same button to hide that panel.

Panels

As mentioned above, clicking on any button in the **Panel Access Bar** will display that particular panel. Figure 2 above shows the **Tool Chest** panel turned on.

Navigation Bar

The **Navigation Bar** is available at the bottom of the window. This bar provides a number of tools to navigate through the PDF file. You can split the view vertically or horizontally using the tools available on the left of this bar. The sheet size of the active PDF file is listed on the right of this bar, along with the measurement scale that you have set. Note that if there is no measurement scale defined, it will show **Scale Not Set** on the right side of this bar.

Markup List

The **Markup List** is available below the **Navigation Bar** and is used to display all the markups added to the active PDF file. Please note that if the active PDF file has multiple pages, the **Markup List** will show the markups added to all the pages.

Toolbar

In the default Revu interface, there are some toolbars displayed on the right of the window. These toolbars provide you tools to review and markup PDF files.

***Tip:** You can right-click on the toolbars to display the list of additional toolbars. Clicking on any other toolbar in that list will display that toolbar also in the Revu interface.*

Main Workspace

The center of the Revu window is the **Main Workspace** where the PDF file is displayed.

CREATING AND MANAGING PROFILES

Profiles in Revu are used to organize and save the visibility and location of various interface components in order to assist you in completing your tasks efficiently. Revu also allows you to customize the interface by resizing and changing the location of its components to better suit your design review workflows. For example, you can turn off the toolbars and dock the most commonly used panel on the right of the window. Once customized, the settings can be saved as a new profile and then shared with other machines running Revu.

The program comes with some default profiles. These can be accessed by clicking on **Revu > Profiles** on the top left of the window, as shown in Figure 3.

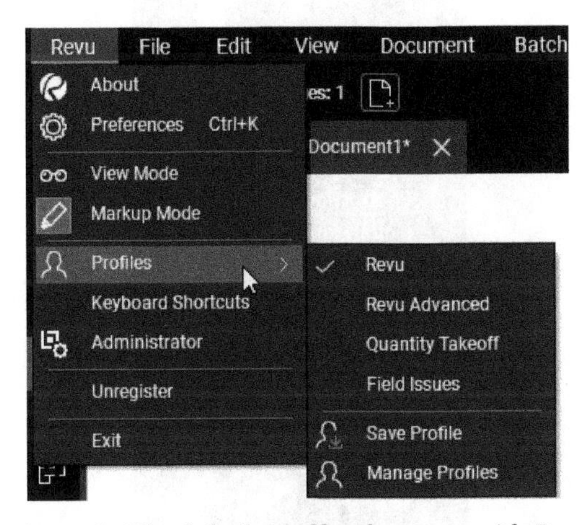

***Figure 3** The default profiles that come with Revu*

Procedures for Creating and Managing Profiles

The following are various procedures for creating and managing profiles.

Procedure for Changing Profiles

The following is the procedure for changing profiles.

1. From the top left of the Revu window, click **Revu > Profiles > Revu Advanced**; the Revu interface will be changed and a number of toolbars will be displayed below the **Menu Bar**.

2. From the top left of the Revu window, click **Revu > Profiles > Quantity Takeoff**; the Revu interface will be changed and the **Measurements** panel will be displayed on the left side of the **Main Workspace**. Also, a panel called the **Markup List** will be displayed at the bottom.

3. From the top left of the Revu window, click **Revu > Profiles > Revu**; the default Revu interface will be restored.

Procedure for Creating a Custom Profile

The following is the procedure for creating a custom profile.

1. From the top left of the Revu window, click **Revu > Profiles > Revu** to ensure the default Revu interface is restored.

2. Right-click on one of the toolbars on the right of the **Main Workspace**; a shortcut menu with the names of all the toolbars will be displayed, as shown in Figure 4.

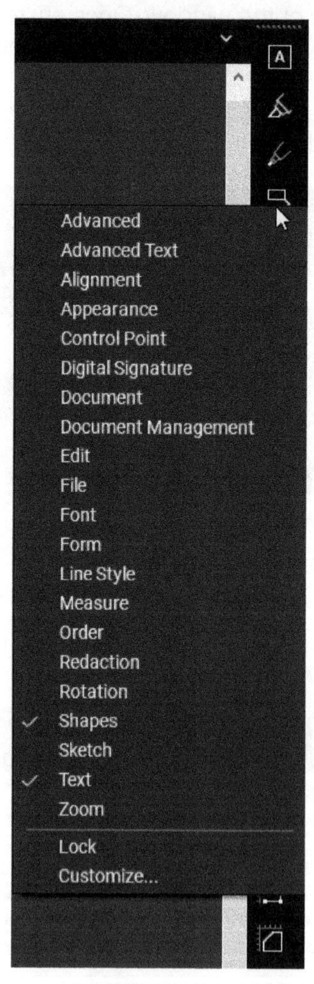

Figure 4 *The shortcut menu displayed on right-clicking on one of the toolbars*

The toolbars with a check mark on the left of their names are the ones that are turned on. You will now turn these toolbars off by clicking on their names.

3. Click on the name of the **Shapes** toolbar in the shortcut menu; the visibility of this toolbar will be turned off.

4. Repeat steps 2 and 3 above to turn off the visibility of all other toolbars.

5. From the **Panel Access Bar** on the left of the **Main Workspace**, right-click on the **Settings** panel and click **Attach > Right**, as shown in Figure 5; the **Settings** panel will be docked on the right of the **Main Workspace**.

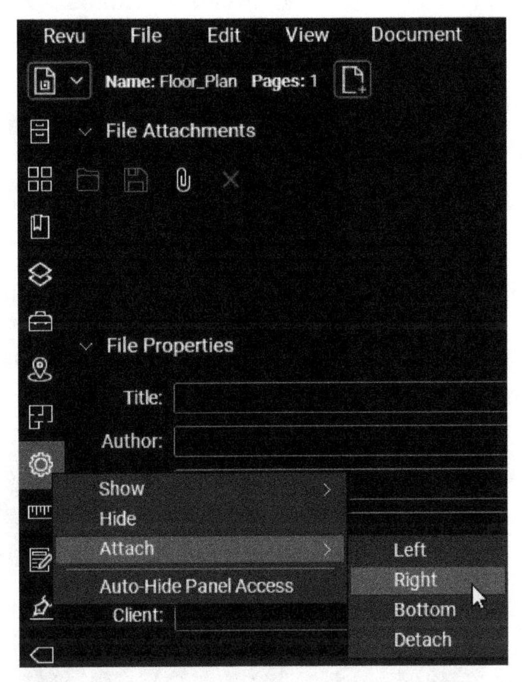

*Figure 5 Docking the **Settings** panel on the right of the main workspace*

***Note:** You can resize the width of the docked panels by hovering the cursor over their edge that aligns with the **Main Workspace**. Once the double-sided arrow is displayed, drag it to resize the panel.*

6. Reduce the width of the **Settings** panel by dragging its left edge.

7. From the **Panel Access Bar** on the left, click on the **Tool Chest** panel to turn it on.

8. Reduce the width of the **Tool Chest** panel by dragging its right edge.

9. From the top left of the window, click **Revu > Profiles > Manage Profiles**; the **Manage Profiles** dialog box will be displayed.

10. In the dialog box, click **Add**; the **Add Profile** dialog box will be displayed.

11. Enter **Revu Training** as the name of the profile and click **OK** in the dialog box; you will be returned to the **Manage Profiles** dialog box with the new profile listed in it.

12. Click **OK** in the **Manage Profiles** dialog box; the new profile will be created and saved.

13. From the top left of the Revu window, click **Revu > Profiles > Revu**; the default Revu profile will be activated and the toolbars are redisplayed on the right of the **Main Workspace**.

14. From the top left of the Revu window, click **Revu > Profiles > Revu Training**; the custom profile that you created will be restored.

Procedure for Modifying an Existing Profile

The following is the procedure for modifying an existing profile.

1. From the top left of the Revu window, click **Revu > Profiles** and activate the profile that you want to modify.

2. Make changes to the Revu interface components, their sizes, or their locations.

3. From the top left of the Revu window, click **Revu > Profiles > Save Profile**, as shown in Figure 6; the profile will be saved with the new changes.

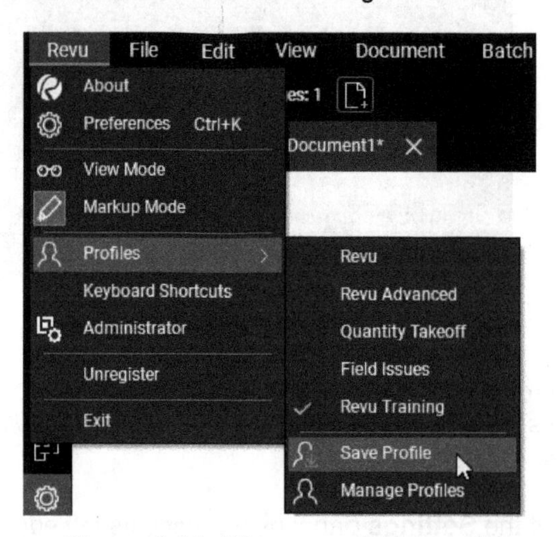

Figure 6 *Modifying an existing profile*

OPENING FILES IN REVU

Revu provides you with a number of methods to open files. These methods are discussed next.

Opening Files using the Open Tool

Menu: File > Open
Keyboard Shortcut: CTRL + O

Open Ctrl+O

This method is used to open an existing PDF file. Additionally, this tool can be used to open a non-PDF format file, such as an image file or a 3D format file. It is important to note that if you open an image format file, it will be converted into PDF format during the open process. The 3D format files will be converted into 3D PDFs that you can spin around and section to review the design. When you invoke this tool, the **Open** dialog box will be displayed, as shown in Figure 7.

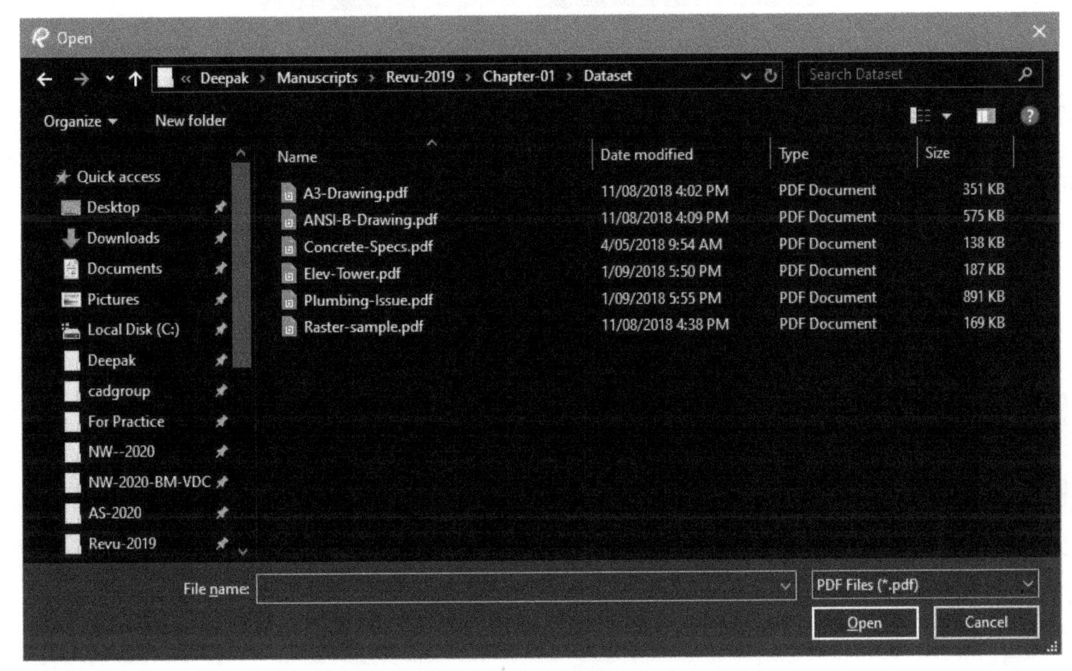

Figure 7 The ***Open*** *dialog box to open files*

By default, the drop-down list on the lower right of this dialog box shows **PDF Files (*.pdf)**. You can browse to the location of the PDF file you need to open and then double-click on it to open.

To convert an image file into a PDF format and open it, click on the drop-down list on the bottom right of the **Open** dialog box and select **All Image Files**, as shown in Figure 8, and then browse and double-click on the image file.

Figure 8 Opening an image format file in Revu

On doing so, the image file will be converted into a PDF file and opened. Note that you will have to save the newly created PDF file manually to ensure you can open the PDF file of the image next time.

To convert a 3D format file **(WORKS ONLY FOR REVU CAD AND eXTREME)** into a 3D PDF and open it, click on the drop-down list on the bottom right of the **Open** dialog box and select **3D Files (*.ifc,*.u3d)**, as shown in Figure 9. Then browse and double-click on the 3D file.

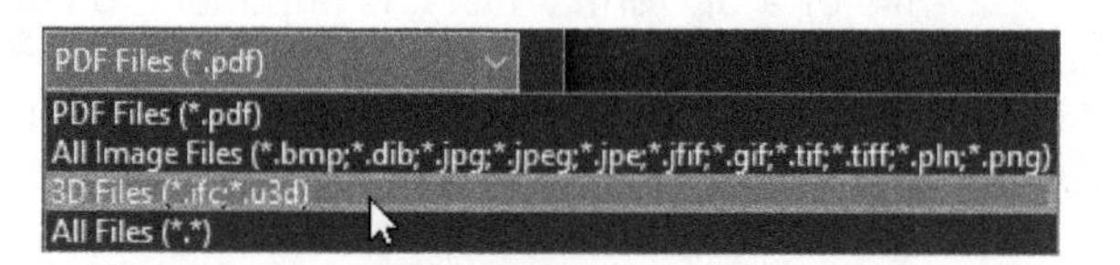

Figure 9 Opening a 3D format file in Revu

On doing so, the process of converting the 3D file into a 3D PDF file will start. Once that process is completed, the **New 3D PDF** dialog box will be displayed, as shown in Figure 10. The options in this dialog box are discussed next.

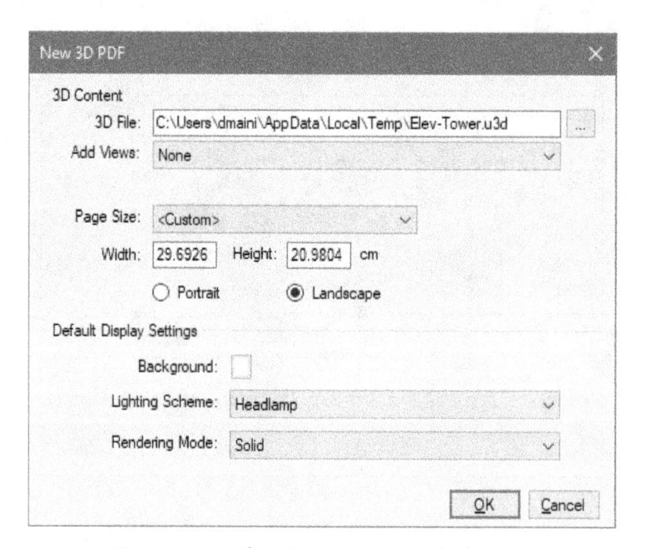

*Figure 10 The **New 3D PDF** dialog box*

3D File

This field displays the location of the .u3d file that gets automatically created to display the 3D content in the 3D PDF file. This location is generally the **Temp** folder of the current Windows user profile. You can change this default location using the **[...]** button available on the right of this list.

Add Views

This drop-down list is used to add predefined orthogonal and perspective views to the 3D PDF file. If you select an option to add any of these views, there will be a toolbar displayed at the top in the 3D PDF file that shows the list of the views you can activate, as shown in Figure 11.

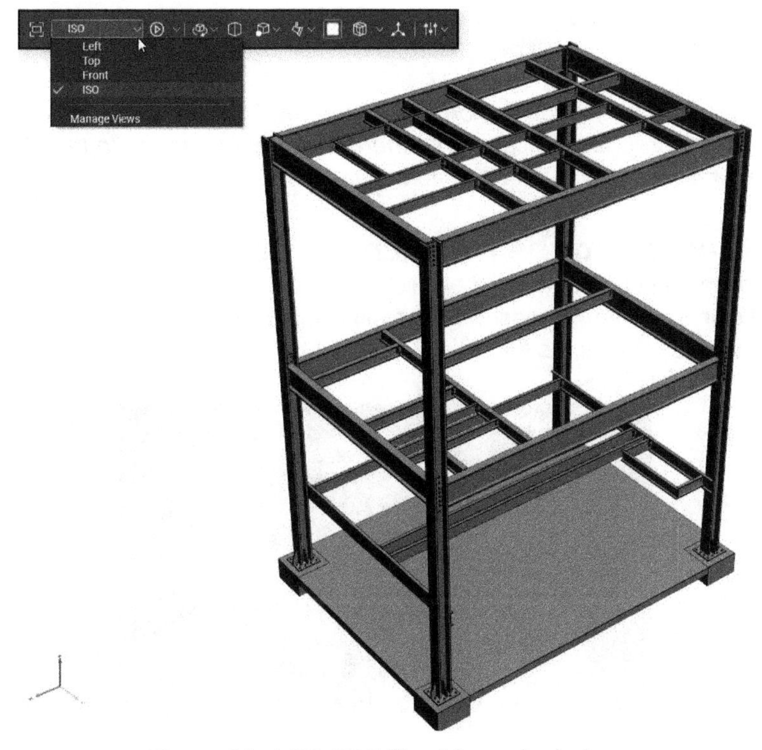

Figure 11 A 3D PDF file with standard views

Paper Size

This list is used to select the default paper size of the PDF file. You can select from one of the standard sizes or enter a custom size in the **Width** and **Height** edit boxes available below this list.

Portrait/Landscape

These radio buttons allow you to specify the orientation of the 3D PDF file.

Default Display Settings Area

The options available in this area are discussed next.

Background

This button is used to define the default background color of the 3D PDF file. Figure 11 above shows the 3D PDF file with a White background.

Lighting Scheme

This list is used to define the default lighting scheme for displaying the objects in the 3D PDF file. Note that this scheme can be changed anytime using the toolbar at the top in the 3D PDF file.

Rendering Mode

This list is used to define the default rendering mode for displaying the objects in the 3D PDF file. This mode can also be changed using the toolbar at the top in the 3D PDF file.

Once you select the desired options in the **New 3D PDF** dialog box, click **OK**; the dialog box will be closed and the 3D PDF file will be opened in Revu.

***Tip**: The asterisk (*) displayed on the right of the PDF file name in the **File Tab** above the **Main Workspace** indicates that there are changes made to the file that have not been saved yet. Saving the PDF file will make the asterisk disappear.*

Opening Files using the File Access Panel

The **File Access** panel can be invoked by clicking on the **File Access** button at the top in the **Panel Access Bar**. The **File Access** toolbar available at the top left in this panel allows you to access files using four modes, as shown in Figure 12.

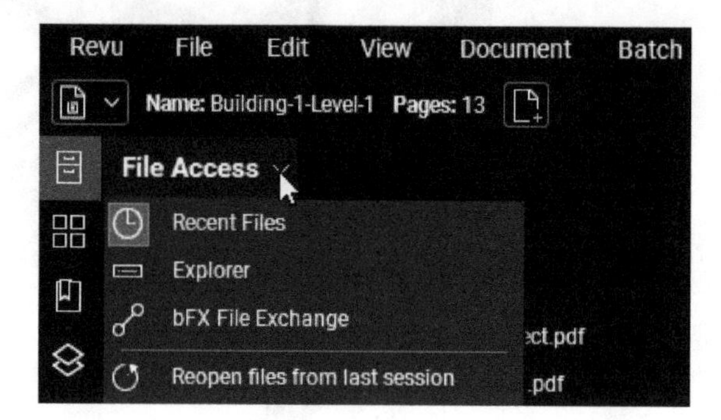

***Figure 12** The **File Access** toolbar shown the four modes to open files*

These modes are discussed next.

Recent Files Mode

This mode provides quick access to the recently used files, as shown in Figure 13. This mode also lets you pin the regularly used files so they do not get dropped off the recent files list. By default, these files are sorted by the most recently used date. There are various other ways to sort the recent files in this mode. The following are the procedures for changing the sort types of the recent files and also for pinning files.

***Tip**: Holding down the CTRL key and clicking on a recent file opens it in the background, with the currently opened PDF file still shown in the Revu **Main Workspace**.*

Procedure for Changing the Sorting of Recent Files

By default, the most recently used files are listed at the top in the **Recent Files** mode. The following is the procedure for changing this sort type.

1. Click on the **Calendar** icon on the right of the **File Access** toolbar, as shown in Figure 14.

2. Select **By Folder**; the **Calendar** icon changes to the **Folder** icon and various folders are listed in the **File Access** panel. These folders are sorted using the drive letter and the folder name and contain the recently used files.

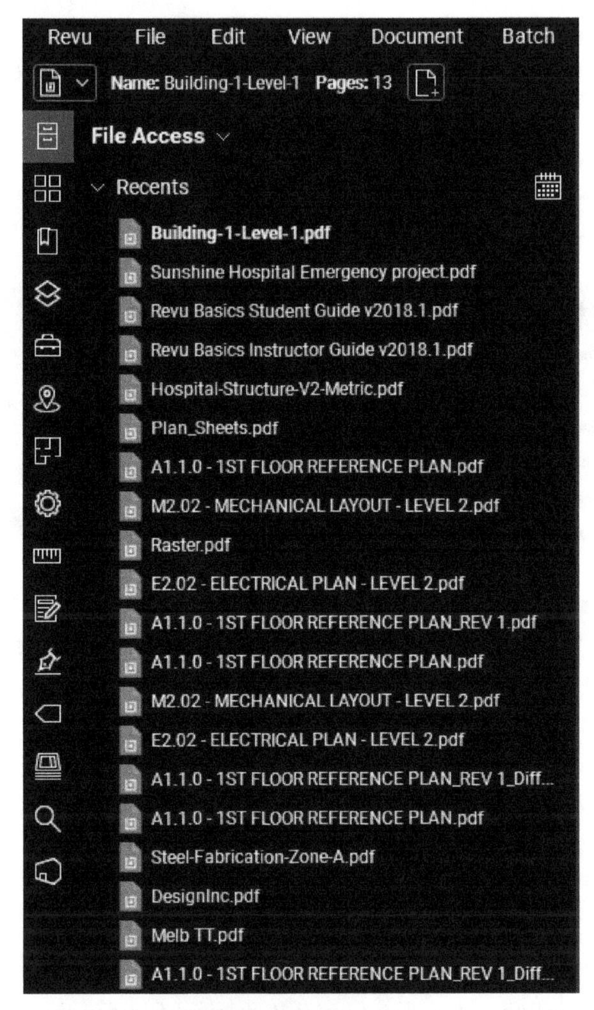

*Figure 13 The **File Access** panel showing recent files*

Figure 14 Changing the sort type for the recent files

3. Click on the **By Folder** icon and select **Most Accessed** to list the files by the number of times the files have been accessed, with the most commonly accessed file listed at the top.

4. Click on the **Most Accessed** icon and select **Access History** to group files based on their access history.

Procedure for Pinning Files

Revu allows you to prevent the files from getting dropped from the recent files list by pinning them. The following is the procedure for doing that.

1. Hover the cursor over a recent file to show the **Pin** icon on the right of its name.

2. Click on that **Pin** icon and select **Pin File** from the shortcut menu, as shown in Figure 15.

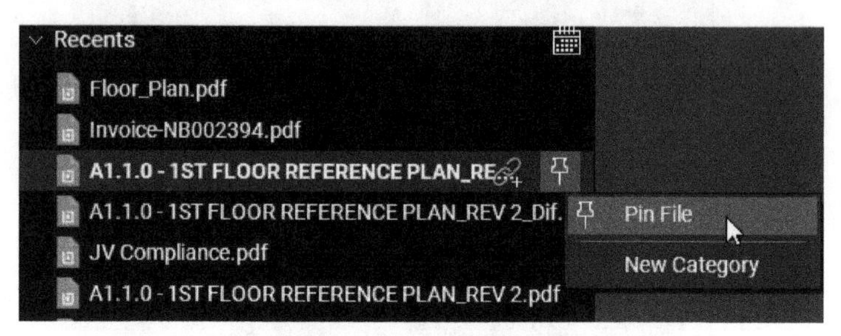

Figure 15 *Pinning a file*

On doing so, that file will be displayed under the name of its original folder in the **Pinned** section, as shown in Figure 16.

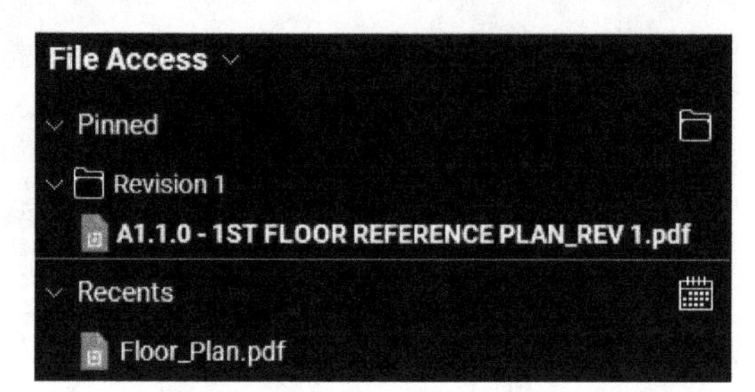

Figure 16 *A pinned file listed inside its original folder in the* ***Pinned*** *section*

3. To create a new category to pin the file under instead of the name of its original folder, hover the cursor over the file and click the **Pin** icon.

4. Select **New Category** from the shortcut icon. Then enter the name of the new category in the **Pinned Category** dialog box and click **OK**; the file will be listed under the category name you defined.

Explorer Mode

This mode is used to access the local or network drives on your computer to open files. When you change the **File Access** panel to the **Explorer** mode, a toolbar is displayed at the top, as shown in Figure 17.

*Figure 17 The toolbar displayed in the **Explorer** mode*

The drop-down list at the top in this toolbar allows you to access the desktop of your computer, local or network drives on the computer, or the **Documents** folder. The **Select File Filter** menu below this list allows you to change the files of type from **PDF Files** to documents or all files. Remember that if you open any non-PDF format file, it will be converted into a PDF file during the open process and you will be prompted to specify the location to save that PDF file.

This mode also allows you to pin the folder of the frequently used files. The following is the procedure for doing that.

1. Browse to the folder that has the files that you frequently use. Make sure you are inside the folder and you can see the content of the folder.

2. Click the **Pin Folder** icon on the toolbar, as shown in Figure 18; the folder will be pinned and will be available in the drop-down list above this icon.

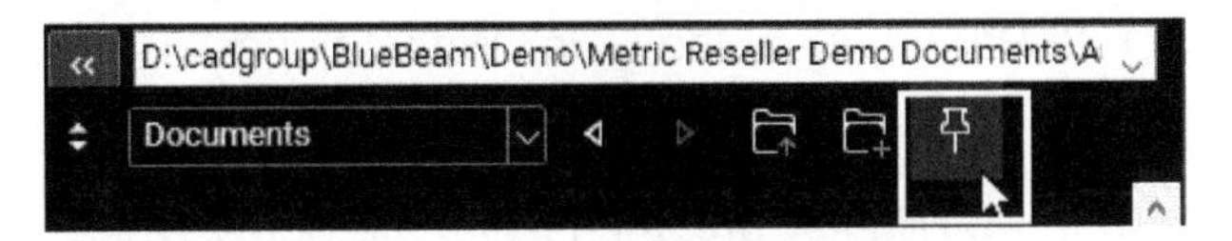

Figure 18 Pinning the folder

***Note**: Using the **Recent Files** or **Explorer** mode, you can also hyperlink a PDF file at a location in the currently opened PDF file. This will be discussed in later chapters.*

bFX File Exchange Mode

This mode is used to track and manage PDF markups that utilize Bluebeam's proprietary .bFX exchange format.

Reopen files from last session Mode

Clicking on this option allows to you reopen all the files that were open in the last Revu session before you closed the program. Each file is opened at the exact same location in the file that you were reviewing before closing the Revu session. Note that in the case of multi-page PDF files, the file will be reopened on the same page that was active before closing the Revu session.

NAVIGATING THROUGH THE PDF FILES

Revu provides you with a number of tools to navigate through the PDF files. These tools are available on the **Navigation Bar**, which is displayed below the **Main Workspace**. These tools can be used to split the view in the **Main Workspace**, zoom or pan the PDF file, select markups or text, go to additional pages of the PDF file, or go to previous or next views. In addition to using these tools to navigate through the PDF file, you can also use the Scroll Wheel on the mouse to zoom, pan, or scroll through the PDF file. Remember that the behavior of the Scroll Wheel on the mouse changes depending on the size of the PDF sheet. To make it convenient for you to understand the sheet size, Revu displays it on the right side of the **Navigation Bar**, as shown in Figure 19.

Figure 19** The **Navigation Bar

If you open a PDF file that has a drawing format sheet of size ANSI B or larger (for Imperial) or A3 or larger (for Metric), the Scroll Wheel on the mouse will allow you to zoom in and out of the sheet. However, if you open a PDF file of a document, the Scroll Wheel on the mouse will allow you to scroll through the pages of the PDF file.

***Tip**: Irrespective of the page size of the PDF file you opened, holding down the Scroll Wheel on the mouse will allow you to pan in the PDF file.*

This default Scroll Wheel behavior can be overridden by using the **One Full Page** or **Scrolling Pages** button available on the left side of the **Navigation Bar**. For example, with a drawing format PDF file opened, when you click the **Scrolling Pages** button, the Scroll Wheel behavior changes to scrolling through multiple pages. Similarly, with a document format PDF file opened, when you click the **One Full Page** button, the Scroll Wheel behavior changes to zooming in and out of the page.

***Tip**: Holding down the CTRL key and then using the Scroll Wheel on the mouse also reverses the default behavior.*

If you have calibrated the PDF file or sheet with a measurement scale, it will be displayed on the right side of the **Navigation Bar**. However, if there is no scale calibrated yet, the **Navigation Bar** will show a message **Scale Not Set**, as shown in Figure 19.

VECTOR PDF VS RASTER PDF

Depending on the method that was used to create the PDF file, it can be a vector PDF or a raster PDF. A vector PDF is the one that is generated directly from a native software, such as a CAD or BIM software or a Microsoft Office application. These PDF files have lines that the markup or measurement tools in Revu can snap to or text that can be searched. The raster PDFs are the ones that are generated from images or scanned documents. For example, a drawing that was printed on a paper and then scanned as a PDF will be a raster PDF. The content in the raster PDFs is formed from a series of dots, called pixels. As a result, if you zoom close to the content of a raster PDF file, it will appear blurry. Remember that the markup or measurement tools in Revu will not be able to snap to the lines in a raster PDF file. Also, the text in those PDFs will not be searchable.

Figure 20 shows a zoomed in view of a vector PDF file. Notice the content of the PDF is sharp and crisp. Figure 21 shows the same zoomed in view of a raster PDF. Notice that the content of this PDF is blurry.

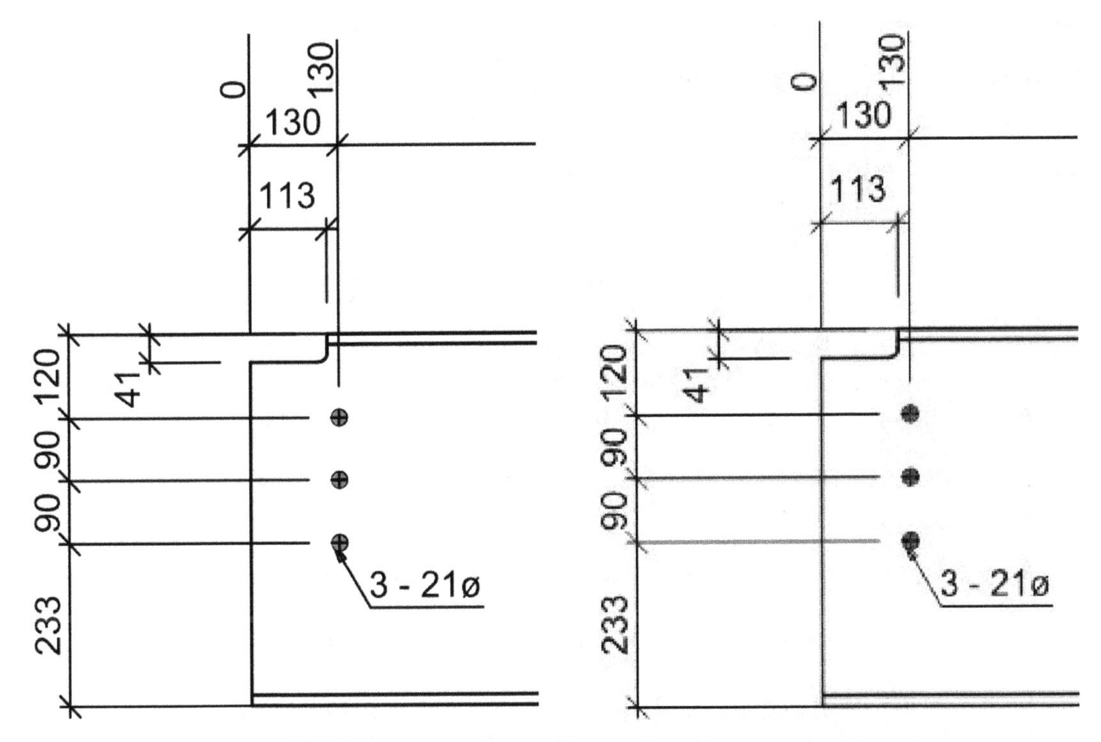

Figure 20 Zoomed in view of a vector PDF

Figure 21 Zoomed in view of a raster PDF

DARK THEME VS LIGHT THEME

Revu 2019 allows you to set the theme to the **Dark** or **Light** theme, depending on your preference. This can be done using the **Preferences** dialog box. The following is the procedure for setting the Revu theme.

1. From the top right of the window, click **Revu > Preferences**; the **Preferences** dialog box is displayed.

2. From the left side of the dialog box, select **General**, if not already selected.

3. On the right side of the dialog box, select the preferred theme from the **Theme** list.

4. Click **OK** in the dialog box to close it and change the theme to the one you selected.

Hands-on Tutorial	*In this tutorial, you will complete the following tasks:* *1. Open a PDF file using the **Open** dialog box.* *2. Customize the Revu interface and create a new profile.* *3. Use the **File Access** panel to pin files and folders.* *4. Navigate through PDF files.* *5. Open an IFC file and convert it into a 3D PDF file.* *6. Open an image file and covert it into a PDF format.* *7. Use the **File Access** panel to open files from the last Revu session.*

Section 1: Using the Open Dialog box to Open a PDF File

In this section, you will use the **Open** dialog box to open the **ANSI-B-Drawing.pdf** or the **A3-Drawing.pdf** file, depending on whether you use Imperial or Metric units.

1. Start Revu 2019. By default, Revu opens in a profile called **Revu**. To ensure that is the case on your machine as well, you will first restore that profile.

2. From the top left, click **Revu > Profile > Revu**; the interface resets to the default **Revu** profile.

3. From the top left in the Revu window, click **File** > **Open**; the **Open** dialog box is displayed.

4. Browse to the **Tutorial Files > C01** folder and open the appropriate file *(refer to Page vii of the book to download the Tutorial Files, if you have not downloaded them yet)*:

   For Imperial units - **ANSI-B-Drawing.pdf** file
   For Metric units - **A3-Drawing.pdf** file

   The Revu interface looks similar to the one shown in Figure 22.

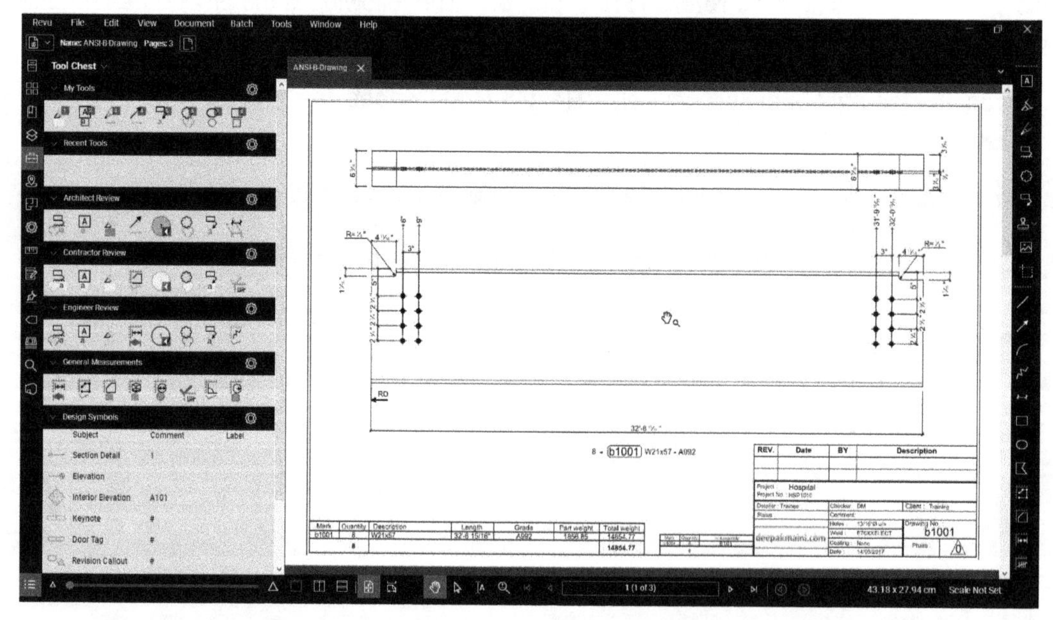

***Figure 22** The default Revu interface with the **ANSI-B-Drawing.pdf** file opened*

By default when you open Revu, the **Pan** tool is active and the cursor is displayed as a **Hand** cursor, as shown in Figure 22. You will now change the cursor to the **Select** cursor. This can be done using the **Select** tool on the **Navigation Bar** displayed below the **Main Workspace** or by pressing the V key on the keyboard.

5. Press the **V** key on the keyboard; the cursor changes to the select cursor.

Section 2: Customizing the Revu Interface and Creating a New Profile

In this section, you will customize the Revu interface by turning off all the toolbars. You will then dock the **Properties** panel on the right side of the **Main Workspace**. You will also turn on the visibility of the **Status Bar**. Finally, you will create a new profile to save the interface customization.

The default **Revu** profile has some toolbars displayed on the right of the **Main Workspace**. You will first turn these toolbars off.

1. Right-click on one of the toolbars displayed on the right of the **Main Workspace**; the shortcut menu with the names of all toolbars is displayed. Note that the toolbars that are currently turned on are displayed with a check mark on the left of their names, as shown in Figure 23.

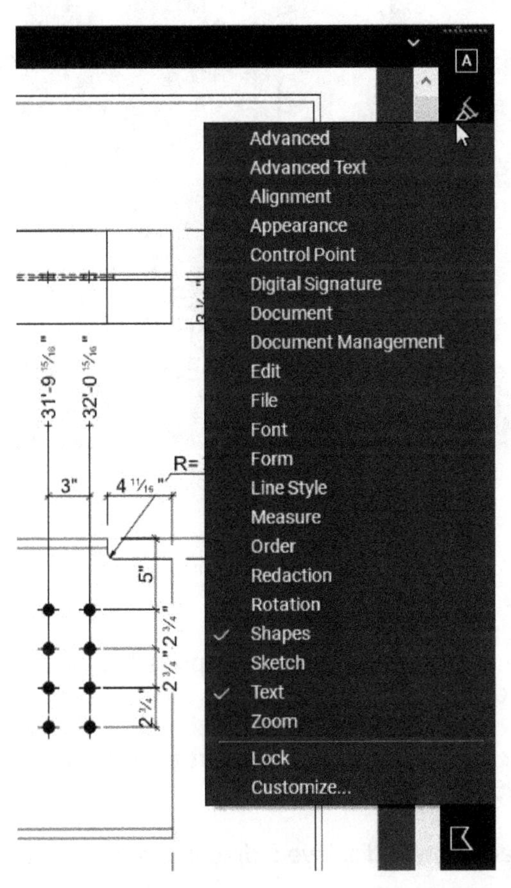

Figure 23 Right-clicking on a toolbar to display the shortcut menu

2. Click on the **Shapes** toolbar name; this toolbar is turned off.

   The **Text** toolbar is still displayed. You will now turn it off.

3. Right-click on the **Text** toolbar and click on its name from the shortcut menu to turn it off.

   Next, you will dock the **Properties** panel on the right of the **Main Workspace**. This panel is displayed in the **Panel Access Bar** on the left of the **Main Workspace**.

4. From the **Panel Access Bar**, right-click on the **Properties** panel button and click **Attach > Right**, as shown in Figure 24; this panel is attached to the right of the **Main Workspace**.

*Figure 24 Attaching the **Properties** panel on the right of the main workspace*

Although the **Properties** panel is attached to the right of the **Main Workspace**, you need to stretch it to ensure it is displayed.

5. If the **Properties** panel is not automatically displayed, click on its icon on the right of the **Main Workspace** to display it.

6. Hover the cursor over the left edge of the **Properties** panel to display a double-sided arrow, as shown in Figure 25.

7. Once the double-sided arrow is displayed, drag it to resize the panel, refer to Figure 25 for the size of this panel.

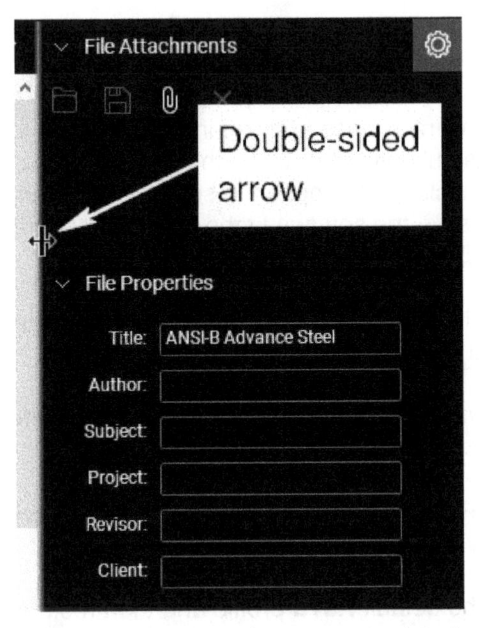

*Figure 25 Resizing the **Properties** panel*

You will now create a new profile to save these settings.

8. From the top left, click **Revu > Profiles** > **Manage Profiles**; the **Manage Profiles** dialog box is displayed.
9. In the dialog box, click **Add**; the **Add Profile** dialog box is displayed.
10. Enter **Revu Training**, as the name of the new profile, as shown in Figure 26.

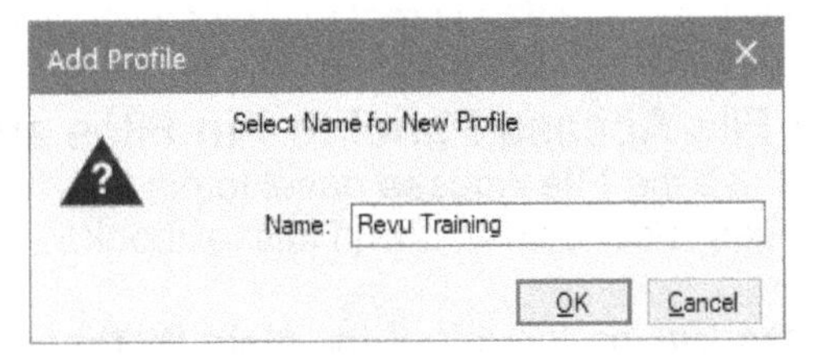

Figure 26 Creating the new profile

11. Click **OK** in the **Add Profile** dialog box; the new profile is created and is listed in the **Manage Profiles** dialog box.

***Tip:** You can click on the name of the profile in the **Manage Profiles** dialog box and click **Export** to export the profile and the dependencies in a **.bpx** format file. You can then double-click on this file on any other machine to import this profile and all the related settings. You will learn about the dependencies in later chapters.*

12. Click **OK** in the **Manage Profiles** dialog box; the custom Revu interface is now saved in a profile and the newly created profile is now the active profile.

    One of the things you were supposed to do in this profile is turn on the **Status Bar**. This bar provides various useful buttons, such as the ones to snap on to the content or markups on the PDF file. It also includes the option to set the measurement scale for the sheet. You will now turn it on and save the settings in the profile to ensure it is updated to include the visibility of the **Status Bar** as well.

13. From the **Menu Bar**, click **Tools** > **Toolbars** > **Status Bar**; the **Status Bar** is displayed at the bottom in the Revu window.

***Tip**: Instead of using the **Tools** menu, you can also use the Function 8 (F8) key on the keyboard to toggle the **Status Bar** on or off.*

14. From the top left, click **Revu** > **Profiles** > **Save Profile**; the **Revu Training** profile is updated and now includes the visibility of the **Status Bar**.

    You will now activate the default **Revu** profile and then restore the custom **Revu Training** profile to see how the interface changes.

15. From the top left, click **Revu** > **Profiles** > **Revu**; the interface changes to the default profile and the **Properties** panel is no longer docked on the right of the **Main Workspace**. Also, notice that the **Status Bar** is not displayed in this profile.

16. From the top left, click **Revu** > **Profiles** > **Revu Training**; the custom profile you created is restored and the **Properties** panel is docked on the right of the **Main Workspace**. Also, the **Status Bar** is displayed.

***Note**: The panel that was active on the left of the **Main Workspace** when you created the custom profile is the panel that is restored when you activate your custom profile.*

Section 3: Using the File Access Panel to Pin Files and Folders

In this section, you will use the **File Access** panel to pin files from the **C01** folder. You will then pin the **Tutorial Files** folder provided with this textbook.

1. From the **Panel Access Bar** on the left of the **Main Workspace**, click the **File Access** button, which is the first button; the **File Access** panel is displayed.

   By default, the **File Access** panel is displayed in the **Recent Files** mode where the files that were opened recently in Revu are displayed. You will first pin the file that you have currently opened.

2. In the **File Access** panel hover the cursor over the **ANSI-B-Drawing.pdf** file or the **A3-Drawing.pdf** file, depending on the one that you have currently opened; the link and pin icons are displayed on the right of its name.

3. Click on the pin icon and select **Pin File** from the shortcut menu, as shown in Figure 27; the selected file is pinned in the **Pinned** category at the top in the **File Access** panel.

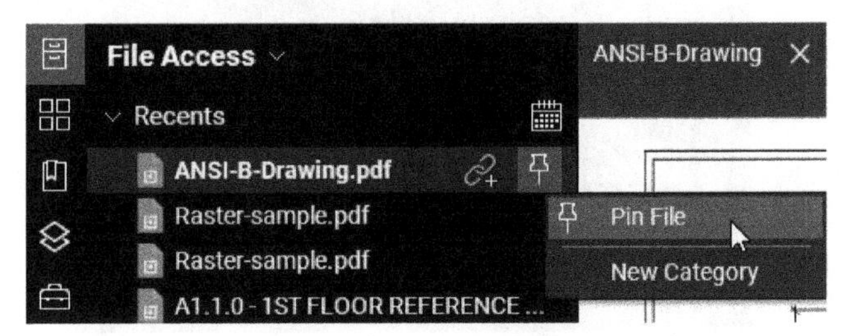

*Figure 27 Pinning the **ANSI-B-Drawing.pdf** file*

Note that this file is pinned under the **C01** category. This is the name of the folder in which this file is saved. Next, you will pin another file from the **Explorer** mode of the **File Access** panel.

4. From the **File Access** toolbar at the top in the **File Access** panel, click **Explorer**, as shown in Figure 28; the display in the **File Access** panel becomes similar to the Windows Explorer display. Also, a toolbar is displayed at the top in the **File Access** panel.

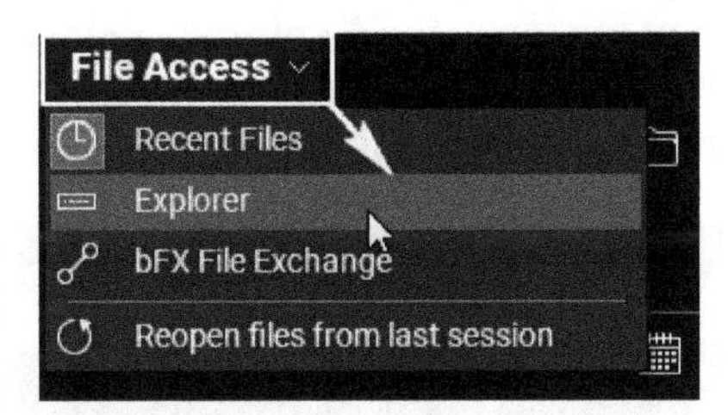

*Figure 28 Invoking the **Explorer** mode*

By default, the drop-down list at the top in the **Explorer** mode displays the path to the user's **Document** folder. You will now click on this drop-down list and browse to the folder where you have saved the **Tutorial Files** folder that was provided with this textbook.

5. Click on the drop-down list at the top in the **Explorer** mode, as shown in Figure 29, and select **Computer**.

6. Browse to the folder where you have saved the **Tutorial Files > C01** folder that came with this textbook; the content of that folder is displayed, as shown in Figure 29.

***Tip:** In Figure 29, only the PDF files from the **C01** folder are displayed. This is because the **Select File Filter** menu displays **PDF Files**. If you change this to **All Files**, you will notice that this folder also has an IFC file and a JPG file. It is important to note that you can only pin PDF files and not the other format files.*

Notice the horizontal pin icon on the right of the file you have currently opened. This informs you that this file is already pinned. You will now pin the **Concrete-Specs.pdf** file.

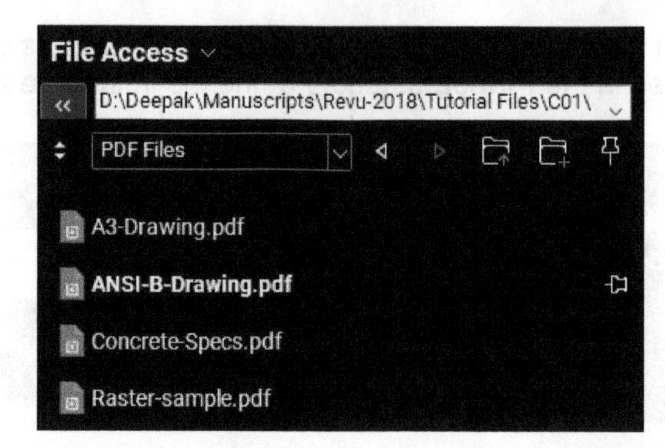

*Figure 29 The PDF files in the **C01** folder displayed*

7. Hover the cursor over the **Concrete-Specs.pdf** file; the link and pin icons are displayed.

8. Click the pin icon and select **Pin File** from the shortcut menu; the pin icon on the right of this file changes to horizontal, informing you that this file is pinned.

   You will now pin the **C01** folder.

9. From the toolbar displayed at the top in the **Explorer** mode, click the **Pin Folder** button, as shown in Figure 30; the folder is pinned. Note that if you cannot see the **Pin Folder** icon then you will have to resize the **File Access** panel and make it wider.

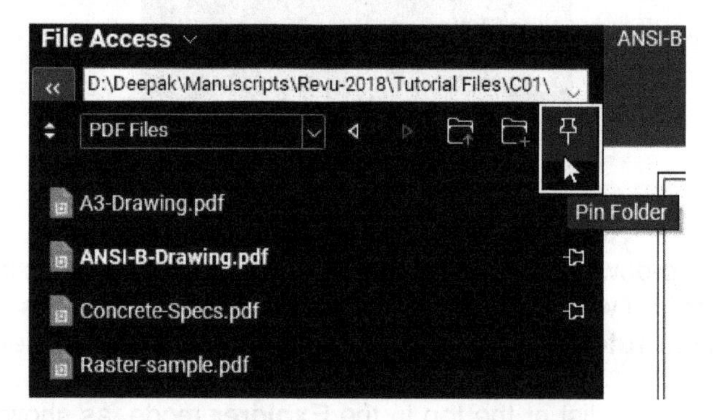

*Figure 30 Pinning the **C01** folder*

   The pinned folders are displayed in the drop-down list at the top in the **Explorer** mode. You will now click on this drop-down list to see how the pinned folder appears.

10. Click on the drop-down list displayed at the top in the **Explorer** mode; the **C01** folder, along with its path is displayed at the bottom in this drop-down list, as shown in Figure 31.

    Next, you will pin the **Tutorial Files** folder for easy access to the files in the later chapters.

11. From the toolbar displayed at the top in the **Explorer** mode, click the **Up One Level** button; all the subfolders in the **Tutorial Files** folders are displayed.

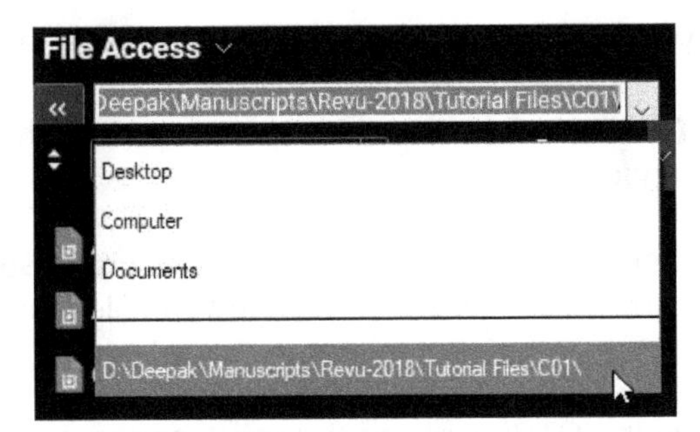

*Figure 31 The pinned **C01** folder displayed in the drop-down list*

12. From the toolbar displayed at the top in the **Explorer** mode, click the **Pin Folder** button; the **Tutorial Files** folder is pinned.

    Next, you will restore to the **Recent Files** mode in the **File Access** panel.

13. From the **Toolbar** at the top in the **File Access** panel, click **Recent Files**; the recent files are now listed in the **File Access** panel.

    Notice that in the **Pinned > C01** category, there are two files listed now. The file that you have currently opened is displayed in bold.

Section 4: Navigating through the PDF Files

In this section, you will use the Scroll Wheel on the mouse to navigate through the currently opened PDF file. You will also open the **Concrete-Specs.pdf** file and notice how the Scroll Wheel behavior changes.

Note that it is assumed that you have the **ANSI-B-Drawing.pdf** or **A3-Drawing.pdf** file currently opened in Revu.

1. Move the cursor near the bottom right of the title block where it displays **Drawing No.**

   Revu is a really smart program and it automatically changes the behavior of the Scroll Wheel on the mouse to zoom or scroll, depending on whether you have a drawing format file or a document format file opened. In this case, you currently have a drawing format file opened. Therefore, the Scroll Wheel on the mouse will allow you to zoom in or out of the drawing.

2. Scroll forward using the Scroll Wheel on the mouse; the display zooms in the area.

3. Hold down the Scroll Wheel on the mouse and drag to the left; the drawing is panned to the left.

4. Scroll back using the Scroll Wheel on the mouse, the display zooms out of the PDF.

5. Double-click the Scroll Wheel on the mouse; the display zooms to the extents of the PDF.

6. Similarly, try zooming in and out of other areas of the PDF.

7. Double-click the Scroll Wheel on the mouse; the display zooms to the extents of the PDF.

   Next, you will open the **Concrete-Specs.pdf** file and notice how the behavior of the Scroll Wheel on the mouse changes.

8. From the **File Access** panel, click on **Pinned > C01 > Concrete-Specs.pdf** file; this file is opened in the Revu window. Also, notice that there are two file tabs displayed at the top of the **Main Workspace**.

*Tip: If you hold down the CTRL key and then click on the name of the PDF file in the **File Access** panel, that file is opened in the background in the Revu window and the currently opened file is still active.*

   The file that you have opened is a multi-sheet PDF file of a document. You will now notice how the scroll behavior of the Scroll Wheel is different in this file.

9. Scroll back using the Scroll Wheel on the mouse; the PDF scrolls down through various pages of this multi-sheet PDF file.

10. Scroll forward using the Scroll Wheel on the mouse; the PDF scrolls up through various pages of this multi-sheet PDF file.

    Because this is a document format PDF file, Revu automatically changes the scroll behavior of the Scroll Wheel to scroll through the pages of this document.

11. From the file tab at the top, click on the **ANSI-B-Drawing** or **A3-Drawing** file to restore the drawing PDF.

12. Scroll back and forward using the Scroll Wheel on the mouse; the PDF zooms in and out.

    As you can see, depending on whether you are in a drawing format file or a document format file, Revu automatically changes the scroll behavior.

13. Double-click the Scroll Wheel on the mouse; the display zooms to the extents of the drawing.

    You will now override this scroll behavior by holding the CTRL key.

14. Still in the drawing format file, hold down the CTRL key and scroll back once; the second sheet of the drawing file is displayed.

    Notice that the **Navigation Bar** below the **Main Workspace** shows **2(2 of 3)** as the page number.

15. Again, hold down the CTRL key and scroll back once more; the third sheet of the drawing file is displayed and the **Navigation Bar** shows **3 (3 of 3)** as the page number.

16. Release the CTRL key and scroll back and forward; the original zoom in and out behavior is restored.

17. Hold down the CTRL key and scroll forward twice to display the first sheet of the drawing file. The **Navigation Bar** should show **1 (1 of 3)** as the page number.

    You will now switch to the document PDF file and hold down the CTRL key to zoom in and out of the text in that file.

18. From the file tab at the top, click on the **Concrete-Specs** file to activate this PDF.

19. Hold down the CTRL key and scroll back; the display zooms out in the PDF.

20. Hold down the CTRL key and scroll forward; the display zooms in.

21. Release and CTRL key and then scroll; the default behavior of scrolling through the pages of the PDF file is restored.

***Note**: If you want to permanently change the behavior of the Scroll Wheel on the mouse in the current document, you can use the **One Full Page** button or the **Scrolling Pages** button available on the left side in the **Navigation Bar**. If the **One Full Page** button is turned on, the scroll behavior changes to zoom in and out, irrespective of which file you have currently opened. If the **Scrolling Pages** button is turned on, the scroll behavior changes to scrolling through various sheets of the multi-sheet PDF file, irrespective of which file is currently opened.*

You will leave these two files open in the Revu session.

Section 5: Opening an IFC File and Converting it into a 3D PDF File (Works only in Revu CAD and Revu eXtreme Starting Version 2019)

In this section, you will use the **Open** dialog box to open an IFC file and convert it into a 3D PDF file. Note that although this options appears in Revu Standard, but it only works in Revu CAD and Revu eXtreme starting version 2019.

1. From the **Menu Bar** on the top left, click **File > Open**; the **Open** dialog box is displayed.

2. Browse to the **Tutorial Files > C01** folder, if not already in that folder.

   By default, only the PDF files are displayed in this folder. This is because the drop-down list on the bottom right of the **Open** dialog box shows **PDF Files (*.pdf)**.

3. Click on the drop-down list on the bottom right of the **Open** dialog box and select **3D Files (*.ifc,*.u3d)**; the **Elev-Tower.ifc** file is displayed in the dialog box.

4. Double-click on the **Elev-Tower.ifc** file; the IFC converter starts to convert the IFC file into a 3D PDF file.

   Once the conversion finishes, the **New 3D PDF** dialog box is displayed.

5. From the **Add Views** drop-down list in the **New 3D PDF** dialog box, select **Perspective: Left, Top, Front, Iso**, as shown in Figure 32.

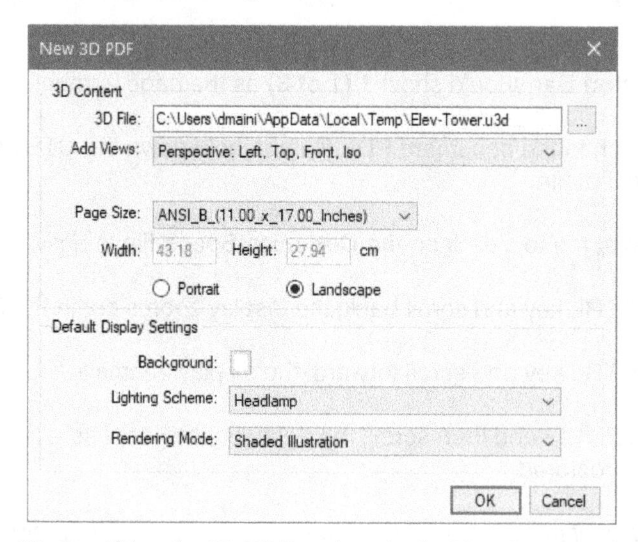

*Figure 32 Specifying the 3D PDF settings in the **New 3D PDF** dialog box*

6. From the **Page Size** drop-down list, select **ANSI_B_(11.00_x_17.00_Inches)** or **ISO_A3_(297.00_ x_420.00_MM)**, depending on whether you work with the Imperial or Metric paper sizes.

7. Click the **Landscape** radio button, as shown in Figure 32, to ensure the 3D PDF is created with the landscape orientation.

8. From the **Rendering Mode** drop-down list, select **Shaded Illustration**, as shown in Figure 32.

9. Click **OK** in the dialog box; the 3D PDF file is created and displayed, as shown in Figure 33.

   Notice the toolbar displayed at the top left of the 3D PDF file. You can use the tools in this toolbar to navigate and manipulate this PDF file. You will learn more about these tools in later chapters.

10. Press and hold down the Left Mouse Button and drag it to orbit the view in the 3D PDF file.

11. Scroll back and forward using the Scroll Wheel on the mouse to zoom in and out of the view in the 3D PDF file.

12. From the **Views** drop-down list in the toolbar at the top left of the 3D PDF file, click **Front**, as shown in Figure 34; the perspective front view of the model is displayed.

13. From the same list, click **Iso**; the perspective isometric view of the model is displayed.

    Next, you will create a custom isometric view to show the model in the 3D PDF file and then save that custom view.

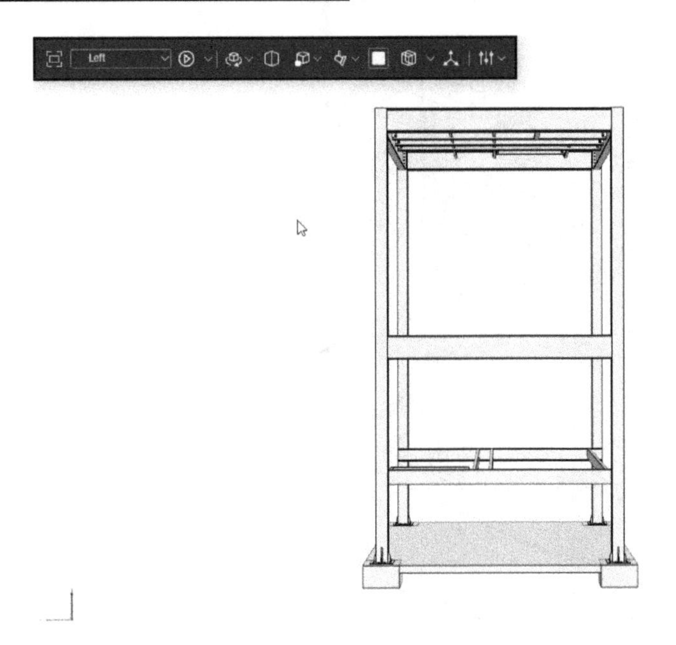

Figure 33 The 3D PDF file opened in Revu

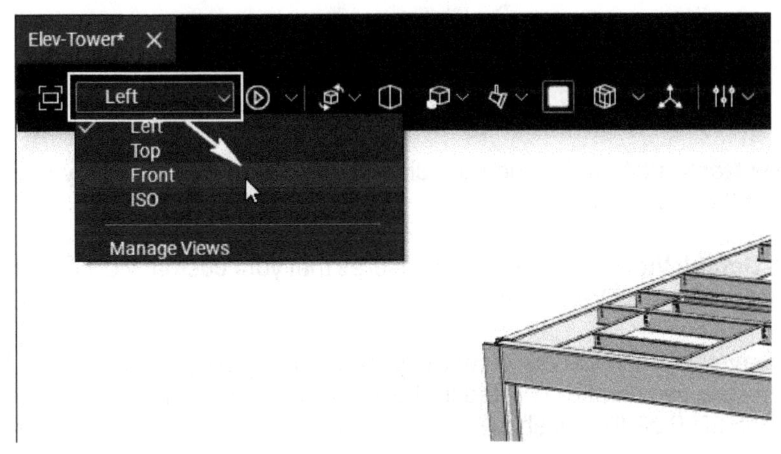

Figure 34 Changing the view in the 3D PDF file

14. Press and hold down the Left Mouse Button and orbit the model, as shown in Figure 35.

15. From the **Views** drop-down list in the toolbar at the top left of the 3D PDF file, click **Manage Views**; the **3D View Properties** dialog box is displayed.

16. Click the + button on the left of the **Views** area; a new view with the default name of **New View** is added to the list.

Figure 35 Orbiting the model in the 3D PDF file

17. In the **View Name** edit box on the right side of this dialog box, enter **My Custom View** as the name of the new view.

18. Select the **Default View** check box. This ensures that your custom view is activated as soon as you click the **Default View** button.

19. Click the **Save** button in the **3D View Properties** dialog box; the dialog box is closed and the custom view you created is listed in the **Views** drop-down list, which is available at the top left of the 3D PDF file, as shown in Figure 36.

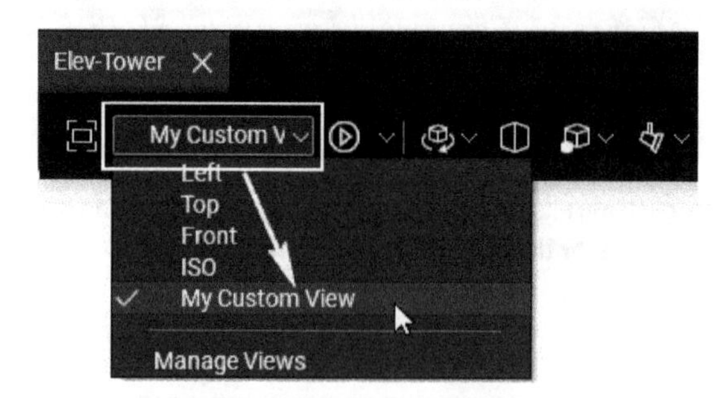

*Figure 36 The custom view listed in the **Views** list*

20. Press and hold down the Left Mouse Button and drag to orbit the model.

21. Click on the **Default View** button on the left of the **Views** drop-down list to restore your custom view. 

***Note**: You will learn about the other tools available in the 3D PDF toolbar, such as the **Sectioning** tools, in later chapters.*

Next, you will save this 3D PDF file.

22. From the **Menu Bar**, click **File > Save**; the **Save As** dialog box is displayed.

    The default name of this PDF file is displayed as **Elev-Tower.pdf**. This is because this file gets its name from the IFC file that was used to create it.

23. Browse to the **Tutorial Files > C01** folder, if not already active.

24. Click the **Save** button; the 3D PDF file is saved.

***Tip**: The asterisk (*) displayed on the right of the name of the file in the tab above the **Main Workspace** indicates that there are changes made in the PDF file that have not been saved.*

You will leave this PDF file also opened in the Revu session.

Section 6: Opening an Image File and Converting it into a PDF File

In this section, you will use the **Open** dialog box to open an image file and convert it into a PDF file.

1. From the **Menu Bar** on the top left, click **File > Open**; the **Open** dialog box is displayed.

2. Browse to the **Tutorial Files > C01** folder.

   By default, only the PDF files are displayed in this folder. This is because the drop-down list on the bottom right of the **Open** dialog box shows **PDF Files (*.pdf)**.

3. Click on the drop-down list on the bottom right of the **Open** dialog box and select **All Image Files**; the **Clash.JPG** file is displayed in the dialog box.

4. Double-click on the **Clash.JPG** file; the image file is converted into a PDF file and opened in the Revu interface, as shown in Figure 37.

5. From the **Menu Bar**, click **File > Save**; the **Save As** dialog box is displayed.

   The default name of this PDF file is displayed as **Clash.pdf**. This is because this file gets its name from the image file that was used to create it.

6. Browse to the **Tutorial Files > C01** folder, if not already active.

Figure 37 The image file converted into a PDF file

7. Click the **Save** button; the PDF file is saved.

Section 7: Opening Files from the Last Revu Session

In this section, you will close Revu and then restart it. You will then use the **File Access** panel to reopen the PDF files that were opened in the last Revu session.

1. From the files tab at the top of the **Main Workspace**, click **Concrete-Specs** to activate this PDF file.

2. Scroll to page 5 of this PDF file.

3. From the files tab at the top of the **Main Workspace**, click **ANSI-B-Drawing** or **A3-Drawing** to activate this PDF file.

4. Navigate to the bottom right of this PDF file where the drawing number is displayed.

   You will now close the Revu session.

5. Close the Revu session by clicking on the **X** button displayed on the top right of this window.

6. Restart Revu.

7. Click on the **File Access** toolbar at the top of this tab and select **Reopen files from last session**, as shown in Figure 38; all the files that were opened in the last Revu session are opened.

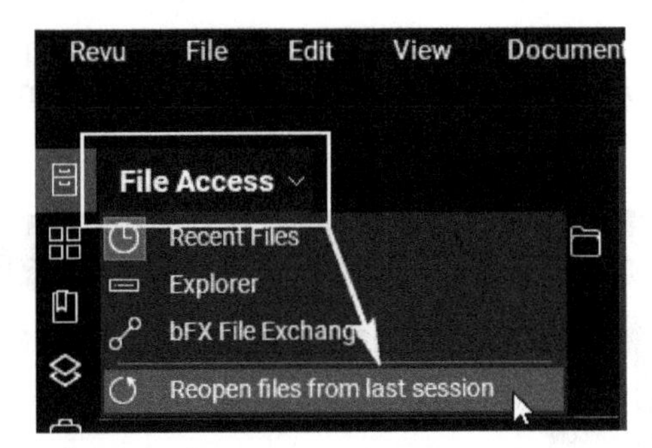

Figure 38 *Opening files from the last Revu session*

Notice that the drawing PDF file opens zoomed in at the lower right corner. This is the area where you zoomed in before you closed the last Revu session.

8. Double-click the Scroll Wheel button on the mouse to zoom to the extents of this PDF file.

9. From the files tab at the top of the **Main Workspace**, click **Concrete-Specs** to activate this PDF file; the PDF file opens on page 5, which is the page that was active when you closed the previous Revu session.

   These steps show that when you use the **Reopen files from last session** option, the files open in the same view that was active when you closed the Revu session.

10. Scroll to the first page of this PDF file.

Section 8: Opening a Raster PDF File

In this section, you will use the **Explorer** mode of the **File Access** panel to open the raster PDF file saved in the **C01** folder. You will then navigate through this PDF file to see how the content of this file appears.

1. From the **File Access** toolbar at the top in the **File Access** panel, click **Explorer**, as shown in Figure 39; the display in the **File Access** panel becomes similar to the Windows Explorer display. Also, as mentioned earlier, a toolbar is displayed at the top in the **File Access** panel.

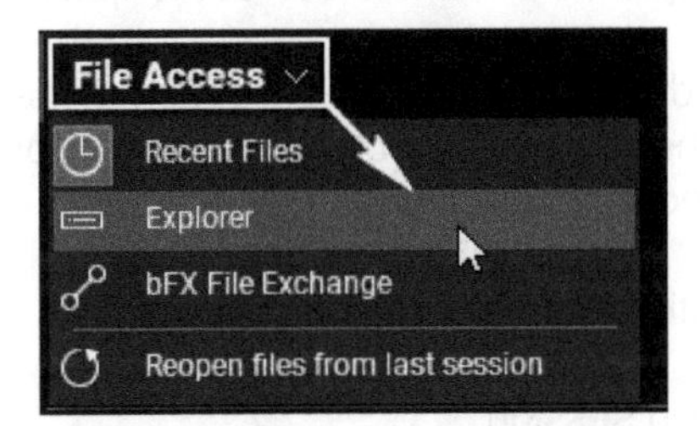

Figure 39 *Changing the **File Access** panel to the **Explorer** mode*

The drop-down list in the toolbar displays the two folders that you pinned in the earlier sections of this tutorial. You will now invoke the **C01** folder from this drop-down list.

2. Click on the drop-down list available at the top in the **Explorer** mode and click on the link of the **C01** folder, as shown in Figure 40; the PDF files available in this folder are listed in the **File Access** panel.

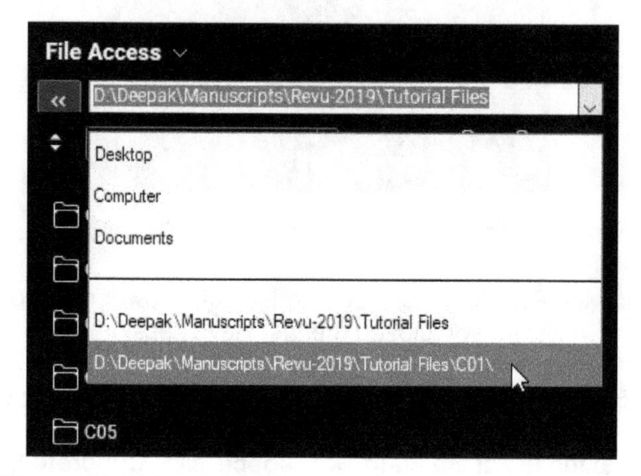

***Figure 40** Using the **Explorer** mode to access the **C01** folder*

3. Double-click on the **Raster-sample.pdf** file; this file is opened in the Revu window.

4. Using the Scroll Wheel button on the mouse, navigate through this PDF file and notice the quality of the content.

   As you zoom closer to the content of this PDF file, you will notice that the text and content in this PDF file become blurry. This is because this PDF file was created from an image. As a result, there is no vector data in this PDF file.

5. Switch to the **ANSI-B-Drawing** or the **A3-Drawing** file and notice the quality of the content in that PDF file. The content in this PDF file appears sharp and crisp as this is a vector PDF file generated from CAD software.

6. Return to the **Raster-sample** file and then look at the quality of this PDF file again.

   You can clearly notice the difference between the quality of the two PDF files. The raster PDF file has poor quality and the content gets blurry when you zoom in. However, the content in the vector PDF file does not get blurry when you zoom in.

7. Zoom to the extents in both raster and vector PDF files.

   You will now close the Revu session.

8. Close the Revu session by clicking on the **X** button displayed on the top right of this window.

Skill Evaluation

Evaluate your skills to see how many questions you can answer correctly. The answers to these questions are given at the end of the book.

1. Revu only lets you open PDF files. (True/False)

2. You can pin files or folders in Revu. (True/False)

3. Revu does not allow you to create a custom profile. (True/False)

4. While working with 3D PDF files, you can create and save custom views. (True/False)

5. The panels in Revu can only be docked on the left of the **Main Workspace**. (True/False)

6. Which panel allows you to pin files and folders?

   (A) **Properties** (B) **Section**
   (C) **Open** (D) **File Access**

7. Which type of file cannot be opened using the **Open** dialog box?

   (A) .DWG (B) .PDF
   (C) .JPG (D) .IFC

8. To save the custom interface settings in Revu, you need to save the settings in a:

   (A) Profile (B) Shape
   (C) File (D) Section

9. Which key is used to override the default behavior of the Scroll Wheel on the mouse?

   (A) ENTER (B) TAB
   (C) CTRL (D) SHIFT

10. What is the default profile in which Revu opens?

    (A) Revu (B) Test
    (C) Bluebeam (D) My Profile

Class Test Questions

Answer the following questions:

1. Explain briefly the process of overriding the behavior of the Scroll Wheel on the mouse.
2. How will you create a 3D PDF file in Revu using an IFC file?
3. Explain briefly how will you create a custom profile?
4. Explain briefly the process of docking a panel on the right of the **Main Workspace**.
5. What is the process of converting an image file into a PDF file?

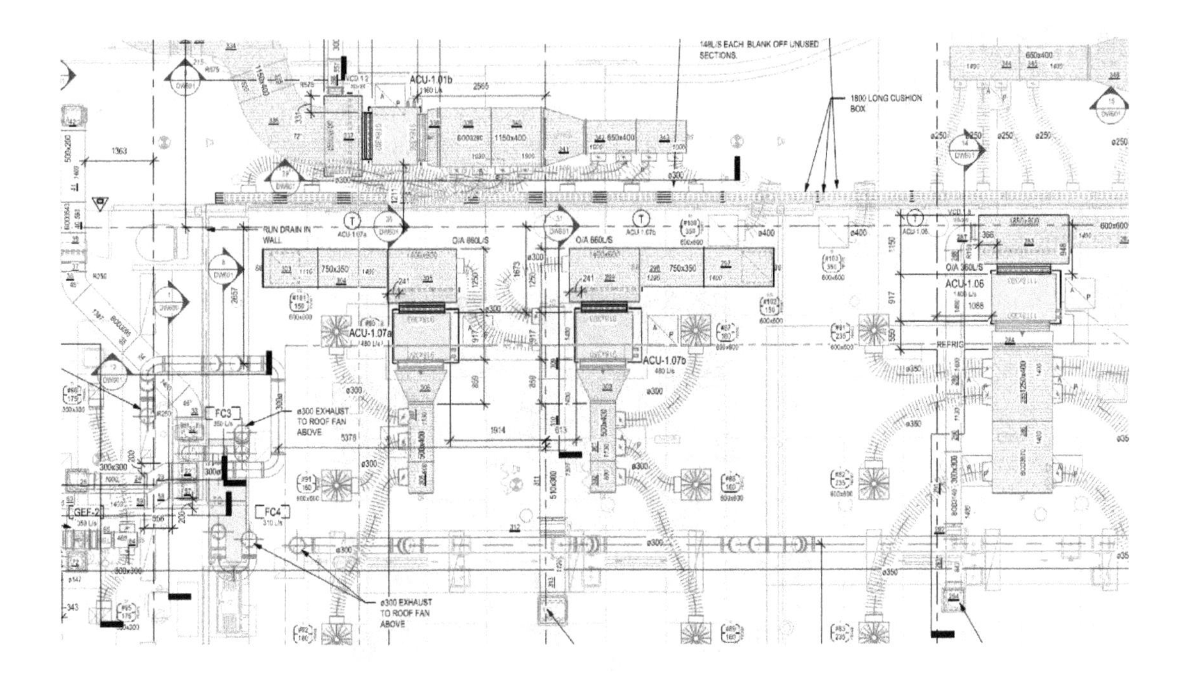

Chapter 2 - Working with the Markup Tools - I

The objectives of this chapter are to:

- √ *Explain various types of text markup tools*
- √ *Teach how to customize markup tools*
- √ *Explain how to match properties of the markups*
- √ *Teach you how to work with the **Recent Tools** tool set*
- √ *Explain various types of pen markup tools*

MARKUP TOOLS

One of the main reasons Revu is an extremely popular program in the Architecture, Engineering, and Construction (AEC) industry is because of the specialized markup tools that it provides. These tools can also be customized to your project requirements and saved for future use.

Figure 1 shows the **Tools** menu with various markup tools. In this chapter, you will learn about the various text and pen markup tools. You will also learn how to customize these tools.

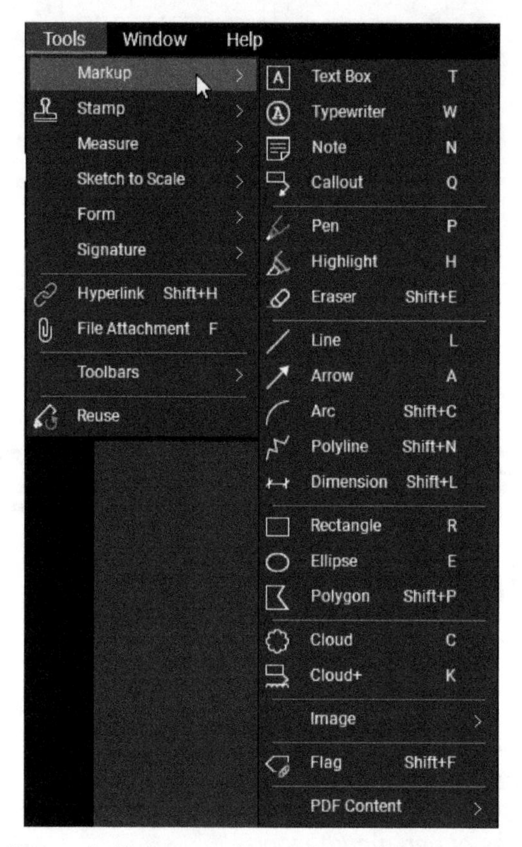

Figure 1 *Various markup tools available in Revu*

***Tip**: Revu is one of the most user-friendly software available in the industry. Whenever you invoke any tool, Revu displays the prompt sequence informing what you need to do on the left side of the **Status Bar**, which is available below the **Navigation Bar**. The **Status Bar** can be turned on and off using the F8 key.*

WORKING WITH THE TEXT MARKUP TOOLS

Revu provides four text markup tools: **Text Box**, **Typewriter**, **Note**, and **Callout**. All these tools are discussed next.

The Text Box Markup

Menu: Tools > Markup > Text Box
Keyboard Shortcut: T

This tool allows you to write a text enclosed inside a box. When you invoke this tool, you will be prompted to select a region to place the text box. Note that the prompt sequence is displayed on the left side of the **Status Bar**. After you specify the two opposite corners of the text box, the real-time text editor will be displayed in which you can write the text.

Once you have finished writing the text, you can press the ESC key or click anywhere outside the text box to exit the tool. On doing so, the text will be written. Also, the text box markup will be highlighted and the control handles displayed, as shown in Figure 2. These handles can be used to resize or rotate the text box.

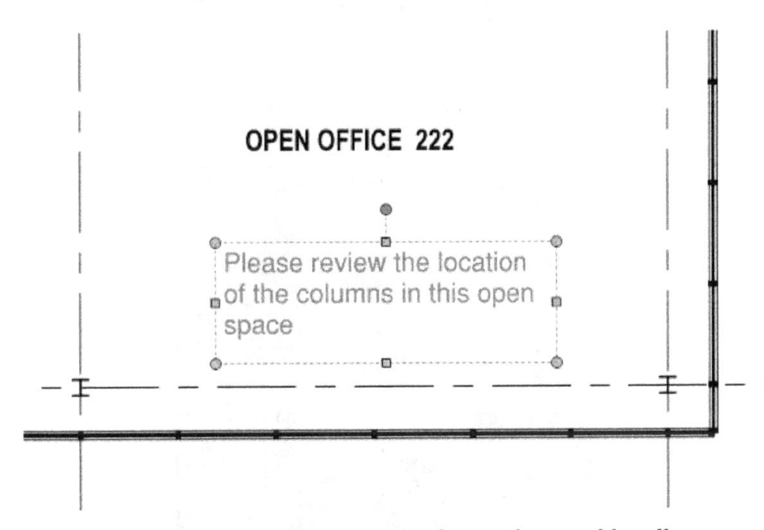

Figure 2 *A default text box markup with control handles*

The default text box markup uses some standard settings, such as Red color, 12 point font size, and so on. All these settings can be customized using the **Properties** panel or using the **Properties** toolbar displayed below the menu bar when you invoke the **Text Box** tool.

The following are various procedures to customize the text box markup.

Procedure for Modifying the Appearance of the Text Box Markup

The following is the procedure for modifying the appearance of the text box markup.

1. Click on the text box markup to ensure it is selected; the **Properties** panel will display the appearance properties of the text box markup, as shown in Figure 3.

2. Click on the **Color** swatch and select the required color for the text box. Note that this will not change the text color.

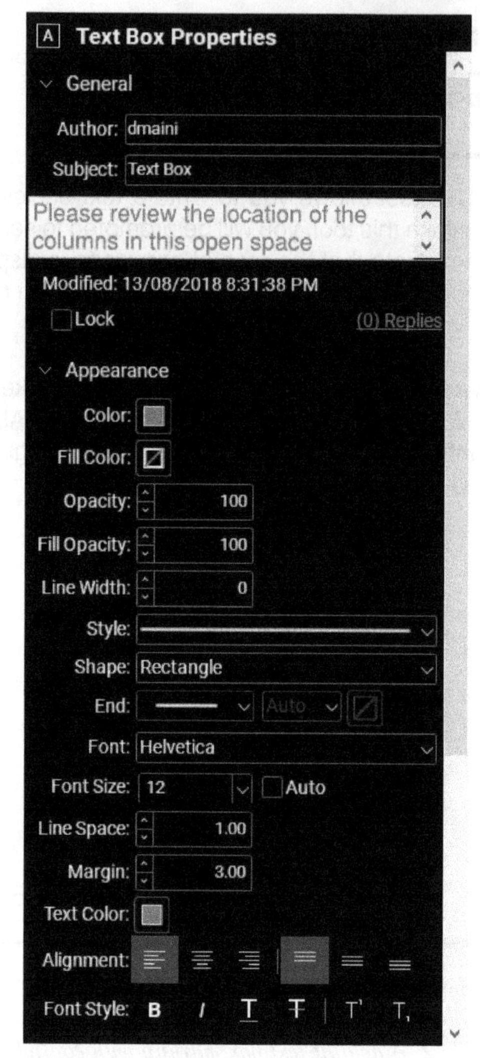

*Figure 3 The **Properties** panel displaying the appearance properties of the text box markup*

3. Click on the **Fill Color** swatch to specify the fill color of the text box.

4. In the **Opacity** edit box, enter the value for the text opacity. This value can vary from 0 to 100.

5. In the **Fill Opacity** edit box, enter the value for the opacity of the fill color of the text box. Figure 4 shows a text box with the fill opacity value as 100 and Figure 5 shows a text box with the fill opacity as 50. Notice that in Figure 5, the content of the PDF behind the text box is visible.

6. In the **Line Width** edit box, enter the line width of the text box.

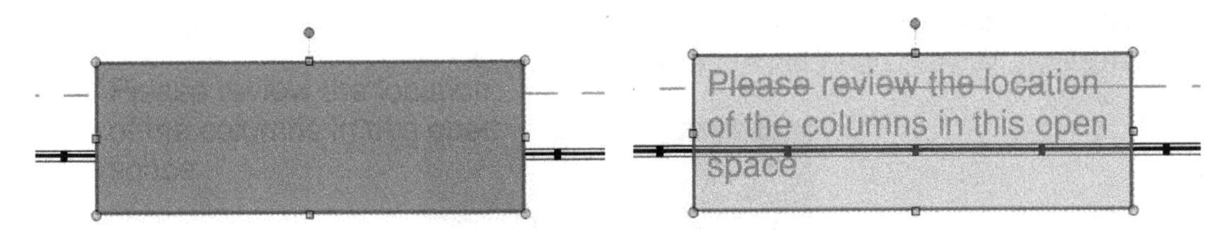

Figure 4 Text box with fill opacity value of 100   *Figure 4 Text box with fill opacity value of 50*

7. From the **Style** drop-down list, select the line style of the text box.

8. From the **Shape** drop-down list, select the shape of the text box. By default, the shape is set to **Rectangle**. As a result, a rectangular box is created around the text. You can change this text box to circle or triangle by selecting their respective options from this drop-down list.

9. From the **Font** drop-down list, select the required font.

*Note: The fonts in the **Font** drop-down list are displayed in three sections. The first section shows the fonts used in the currently selected markup. The second section shows the list of standard fonts supported by all PDF viewers. The third section shows all other available fonts. Note that if you use the fonts from the third section, they will have to be embedded in the PDF file for viewing in other PDF viewers.*

10. In the **Font Size** edit box, enter the size of the text font.

11. Select the **Auto** check box if you want to automatically modify the font height so the text fits the width of the text box.

12. In the **Line Spacing** edit box, enter the line spacing between the multiple lines of text.

13. In the **Margin** edit box, enter the value for the space between the border of the text box and the text.

14. Click on the **Text Color** swatch and select the color of the text.

15. From the **Alignment** area, select the button for the required text alignment.

16. From the **Font Style** area, select the required option to make the font bold, italic, underline, strikethrough, or add superscript or subscript to the text.

Procedure for Updating the Size of the Text Box around the Text

When you define two opposite corners of the text box, that size may not be the right size for the content of the text you write. Therefore, you can autosize the text box to match the content of the text. The following is the procedure for doing this.

1. Click on the text box markup to ensure it is selected.

2. Right-click to display the shortcut menu and select **Autosize Text Box**, as shown in Figure 6. Alternatively, you can press the ALT+Z key on the keyboard.

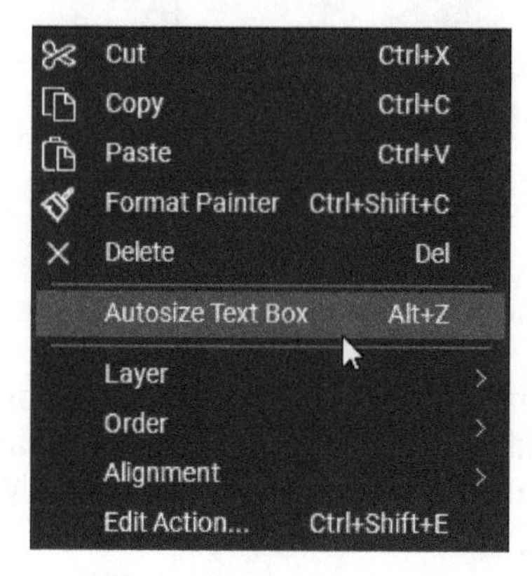

Figure 6 Autosizing the text box

Procedure for Autosizing Text Box in Revu Preferences

If you do not want to autosize the text box every time you add a text box markup, you can set the option to do it automatically in the Revu preferences. The following is the procedure for doing this.

1. From the top left, click **Revu > Preferences**; the **Preferences** dialog box will be displayed.

2. From the left pane of the dialog box, select **Tools**; the **Markup** option is selected in the right pane of the dialog box, as shown in Figure 7.

3. From the right pane of the dialog box, select the **Autosize Text Box and Callout Markups** check box, as shown in Figure 7.

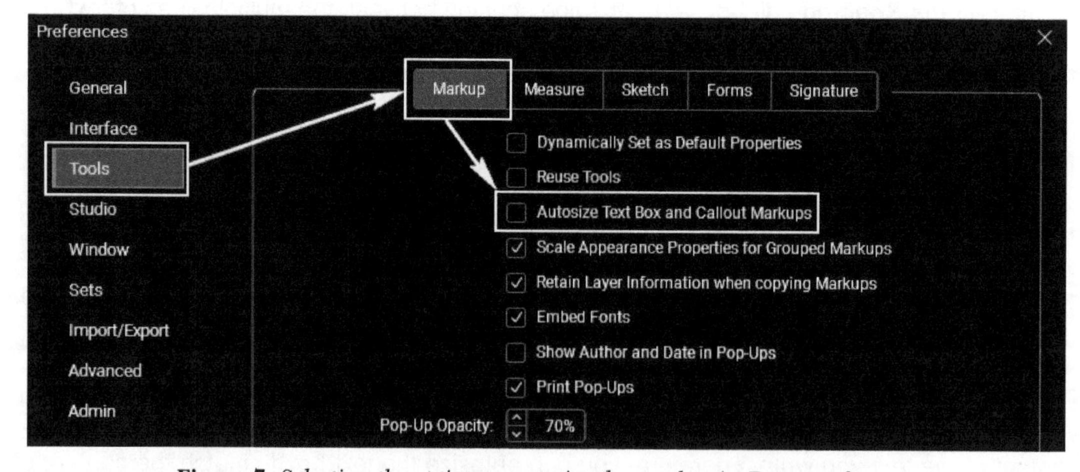

Figure 7 Selecting the option to autosize the text box in Revu preferences

4. Close the **Preferences** dialog box.

Procedure for Rotating the Text Box Markup

The default orientation of the text box markup is horizontal. However, if required, you can rotate this text to any required angle. The procedure for doing this is discussed next.

1. Click on the text box markup; the **Properties** panel will display the properties of the markup.

2. Scroll down to the **Layout** section in the **Properties** panel.

3. Enter the rotation angle in the **Rotation** spinner, as shown in Figure 8.

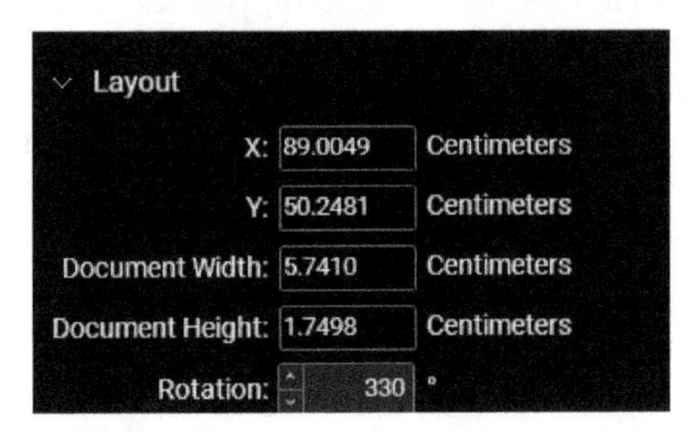

Figure 8 The ***Layout*** *section of the* ***Properties*** *panel showing the* ***Rotation*** *spinner*

4. Alternatively, you can drag the rotate handle that is displayed above the top center control handle of the text box. Note that while dragging, by default this handle will snap at 15-degrees increments. However, if you hold down the SHIFT key, the handle will snap at 1-degree increments.

The Typewriter Markup

Menu: Tools > Markup > Typewriter
Keyboard Shortcut: W

This tool also allows you to add a text markup without the need of specifying a box. When you invoke this tool, you will be prompted to select the location to place the typewritten text. As soon as you click at a location, the real-time text editor will be displayed and you will be able to write the text. Once you have finished typing the text, press the ESC to exit the tool.

Because this tool does not require a text box to be created, it is most suited to enter details when you are filling a PDF form. You can click on any location on the form and start typing the details you need to enter.

Similar to the text box markup, you can also customize the typewriter markup using the options in the **Properties** panel. These options were discussed earlier.

The Note Markup

Menu: Tools > Markup > Note
Keyboard Shortcut: N

If you need to write large paragraphs of notes and you do not have enough space on the document, then you can use the **Note** markup tool. This tool allows you to enter the required text in a pop-up window that can be hidden and displayed as the **Note** icon when you do not need to refer to the content of the note. But when you need to refer to the content, you can double-click on the **Note** icon and the text pop-up window will be displayed showing the content. Figure 9 shows the **Note** pop-up window with the text displayed. To close the pop-up window, click on the **X** icon on the top right of this window. On doing so, only the **Note** icon will be displayed in the document.

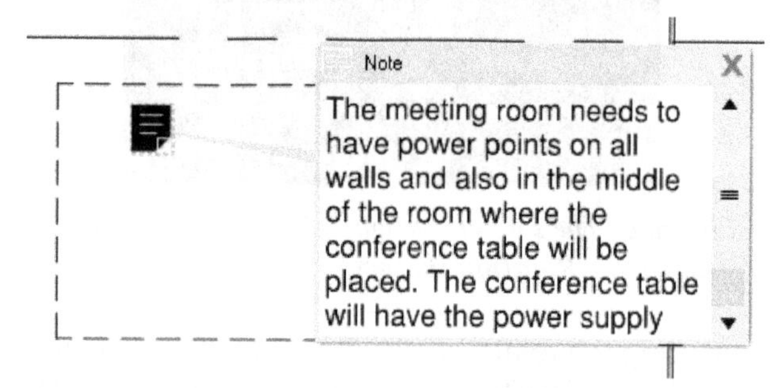

Figure 9 The note markup with the ***Note*** *icon and the text pop-up window*

Similar to the other markup tools, you can use the **Properties** panel to modify the appearance of the **Note** icon also. The other icon types that are available are **Cross**, **Help**, **Insert**, **Key**, and **New Paragraph**.

The Callout Markup

Menu: Tools > Markup > Callout
Keyboard Shortcut: Q

This markup adds a text box with a leader line pointing to a specified location. If required, you can also add multiple leader lines to the same text box. When you invoke this tool, you first need to specify the start of the arrowhead. Once you specify that, you can then specify the location of the text box. On doing so, the real-time text editor will be displayed that allows you to write the text. After you are finished writing the text, press the ESC key or click anywhere outside the text box to exit the tool. Once the markup is added, you can use the control points on the text box and the leader to resize the two components independently. To move the text box, press and drag it to the new location. On doing so, the leader line will be automatically adjusted.

Similar to the other markup, you can also customize this markup using the options in the **Properties** panel. Figure 10 shows a customized callout markup.

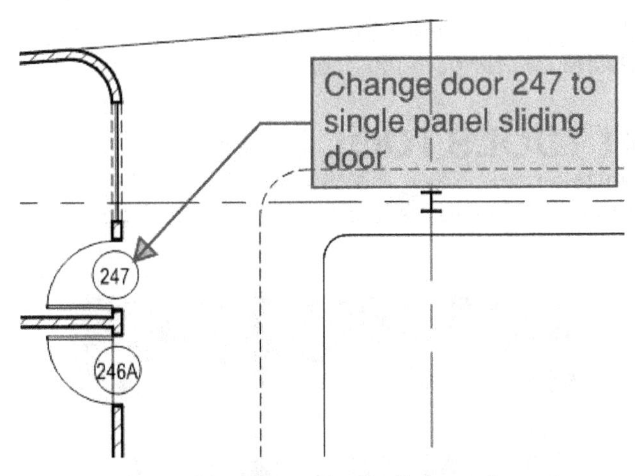

Figure 10 A customized callout markup

Procedure for Creating Additional Leader Lines on the Callout Markup

As mentioned earlier, you can create additional leader lines on the callout markup so that the same text can point to multiple locations. The following is the procedure for doing that.

1. Click on the callout markup to select it.
2. Right-click on the text box and select **Add Leader** from the shortcut menu; another leader line will start from the landing of the first leader line.
3. Point the arrowhead of the additional leader line to the required location. Figure 11 shows a callout markup with multiple leader lines.

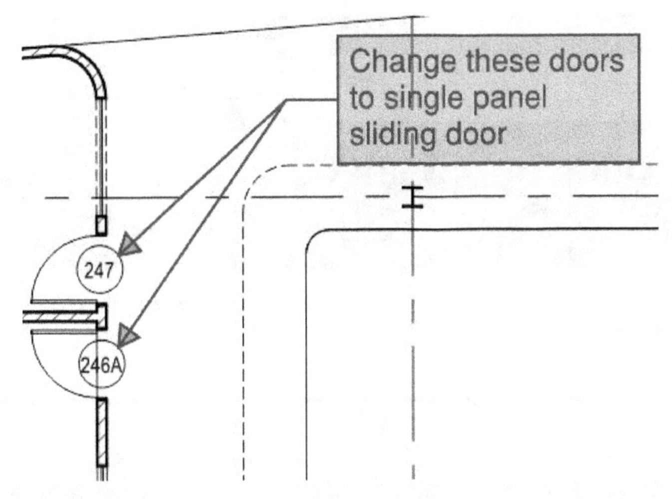

Figure 11 A callout markup with multiple leader lines

***Note:** To delete any leader line from the callout markup, right-click on that leader line and select **Delete Leader** from the shortcut menu.*

THE RECENT TOOLS TOOL SET

Whenever you customize a markup tool, the custom version of that tool is automatically saved in the **Recent Tools** tool set. This tool set is available on the **Tool Chest** panel, as shown in Figure 12.

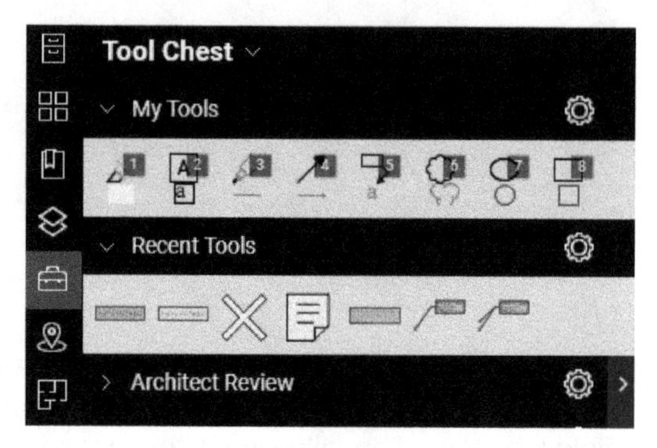

***Figure 12** The **Recent Tools** tool set showing the recent tools that were used to markup the PDF file*

As you keep using tools in Revu, they will be added to this **Recent Tools** tool set. However, it is important to note that by default, only the last twelve tools are saved in this tool set. To change this value, click on the cogwheel symbol on the top right of the **Recent Tools** tool set and select the desired number of maximum recent tools to keep, as shown in Figure 13.

***Figure 13** Changing the value of the number of recent tools to keep in the **Recent Tools** tool set*

What I do

*It is important to note here that when you close the Revu session and start again, the **Recent Tools** tool set is wiped out clean and all the custom tools from this tool set are removed. To ensure that the custom version of my tools are permanently saved, I add them to my own custom tool set. You will learn about creating custom tool sets in later chapters.*

The tools saved in the tools sets can function in one of the following two modes.

Drawing Mode

In this mode, the new markup is added as the exact same copy of the original markup that was customized. It is similar to copying and pasting the markup from one location to the other. If the tool is currently in the **Drawing** mode in the tool set, it appears as the smaller image of the markup you added, as shown in Figure 14.

Figure 14 The ***Recent Tools*** *tool set showing the tools in the* ***Drawing*** *mode*

Properties Mode

To switch to the **Properties** mode, double-click on the custom tool button in the tool set. This mode is used when the content of the mark you want to add is different from the original markup, but you want to retain the **Appearance** properties that you customized for the original markup. If the tool is currently in the **Properties** mode in the tool set, it appears as the image of the markup tool, as shown in Figure 15.

Figure 15 The ***Recent Tools*** *tool set showing the tools in the* ***Properties*** *mode*

Tip: *The* ***Recent Tools*** *tool set can be configured to automatically save the custom tools in the* ***Properties*** *mode or the* ***Drawing*** *mode. This can be done by clicking on the cogwheel icon on the top right of this tool set and selecting or clearing the* ***Properties Mode*** *option. If this option is turned on, the next time you customize a tool, it will be automatically saved in the* ***Properties*** *mode in the* ***Recent Tools*** *tool set.*

Note: *You will learn more about the* ***Tool Chest*** *and other tool sets in detail in later chapters.*

MATCHING PROPERTIES OF THE MARKUPS

Once you modify the **Appearance** properties of a markup, those properties can be used as the source properties and can then be quickly copied to the other similar markups using the **Format Painter** tool. For example, you can select a text box markup with custom properties as the source and copy those properties to the callout markup. To match properties, right-click on the existing markup with the custom properties that you want to use as the source properties and then select **Format Painter** from the shortcut menu, as shown in Figure 16.

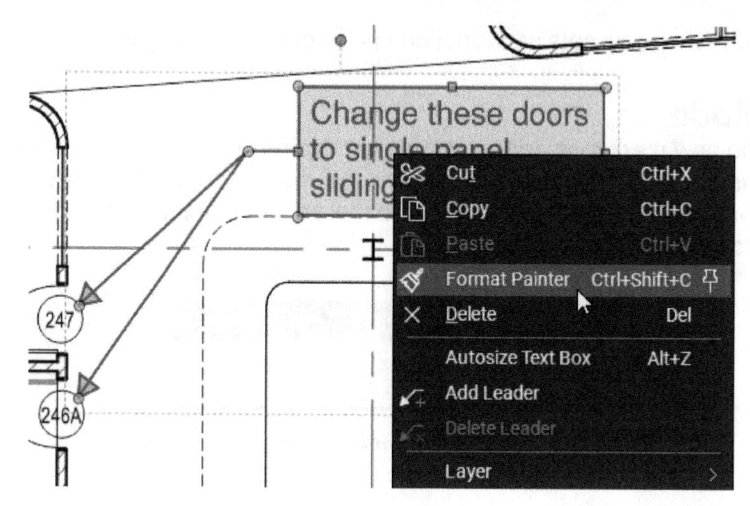

Figure 16 *Invoking the* ***Format Painter*** *tool to match properties between markups*

Next, hover the cursor over the markup that needs to be updated; the cursor will change to the paintbrush cursor. Once that happens, click on the markup; the **Appearance** properties of this markup will be matched with those of the source markup. You can continue selecting multiple markups to match their properties with those of the source markup.

WORKING WITH THE PEN MARKUP TOOLS

Revu provides three pen markup tools: **Pen**, **Highlight**, and **Eraser**. These tools are discussed next.

The Pen Markup

Menu: Tools > Markup > Pen
Keyboard Shortcut: P

This tool allows you to create a free form shape by pressing and holding down the Left Mouse Button and dragging the cursor. When you invoke this tool, the cursor changes to the pen cursor. You can then press and hold down the Left Mouse Button and drag the cursor to draw the free form shape. As soon as you release the Left Mouse Button, the free form shape will end. You can continue drawing as many disjoint free form shapes as part of the same **Pen** tool. However, it is important to note that all the disjoint free form shapes created by releasing the Left Mouse Button are considered as separate markups. As a result, you can modify their properties independently.

Using the **Properties** panel > **Appearance** area, you can modify the color, opacity, and line width of the pen markup. Similar to the other markups, you can use the control handles of the pen markup also to change its shape and size. The top center control handle can be used to change the rotation of this markup.

The Highlight Markup

Menu: Tools > Markup > Highlight
Keyboard Shortcut: H

This tool allows you to highlight the content of the PDF file by dragging the cursor on top of it. When you invoke this tool, the cursor changes to the highlighter icon. This tool has the following two modes.

Pen Mode

This is the default mode is used to create free form highlights in the PDF file. You can press and hold down the Left Mouse Button and drag to highlight the linework in the PDF file. Note that if you press and hold down the SHIFT key before dragging the mouse, you can draw straight line highlights. Figure 17 shows a free form and a straight line highlight.

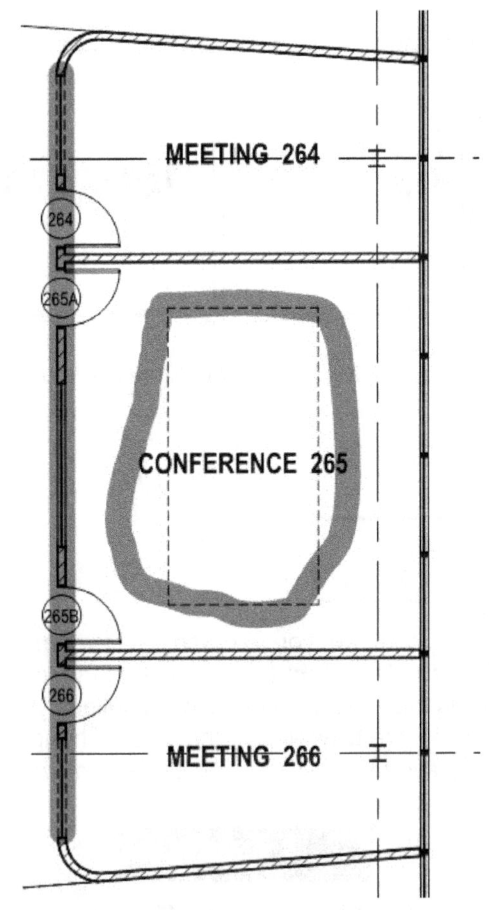

Figure 17 *A free form and a straight line highlight*

Text Mode

This mode is used to highlight the text in the vector PDF file. As soon as you hover the highlighter cursor over a text in the PDF file, the cursor changes to the **Text** mode. You can then press and hold down the Left Mouse Button and drag the cursor over a text to highlight it. While highlighting the text, it is recommended to zoom close to it to avoid highlighting any unwanted text. Figure 18 shows some text highlighted in the PDF file.

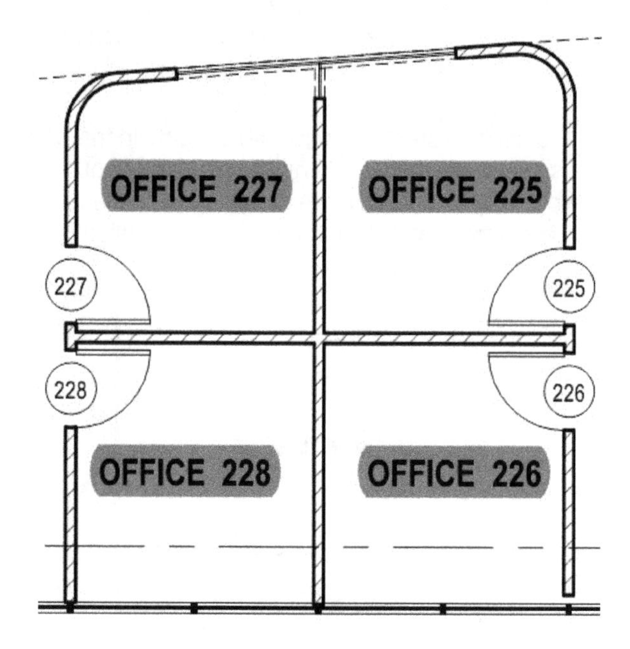

Figure 18 *Text highlighted in the PDF file*

The Eraser Tool

Menu: Tools > Markup > Eraser
Keyboard Shortcut: SHIFT + E

This tool allows you to erase the free form markups created using the **Pen** and **Highlight** tools. Note that the text highlight cannot be erased using this tool. To delete a text highlight, you need to select it and press the Delete key. When you invoke the **Eraser** tool, the cursor changes to the eraser cursor. Also, the **Eraser Hover Bar** is displayed below the menu bar, as shown in Figure 19.

Figure 19 *The* ***Eraser Hover Bar***

By default, the **Stroke Eraser** button is selected from the **Eraser Hover Bar**. As a result, any stroke of the free form pen shapes or the free form highlights you click on will be erased. This means that if the pen or highlight markup was used to create multiple end-connected segments, only the segment on which you click will be erased. However, if multiple disjoint shapes were created as a single markup, the entire disjoint shape on which you click will be erased.

If you select the **Annotation Eraser** button from the **Eraser Hover Bar**, which is the last button, and click on the free form shape or highlight that has multiple end-connected segments, all the segments will be erased.

If you want to erase only a part of the free form shape or highlight, you can use the first three buttons on the **Eraser Hover Bar**.

***Tip**: Similar to the text highlight, you can click on the free form pen markup and press the DELETE key to delete all the shapes of that markup.*

REUSING MARKUP TOOLS

If you have to add multiple markups of the same type in Revu, by default you have to manually invoke that tool again after using it. To avoid this, you can use the **Reuse Markup Tools** functionality on the **Status Bar**, shown in Figure 20. As mentioned earlier, the **Status Bar** is displayed below the **Main Workspace** and can be turned on or off using the F8 key. If this functionality is turned on, the markup tool that you selected will automatically be activated again after you finish using it. This process will continue until you press the ESC key.

Figure 20** The **Reuse Markup Tools** button on the **Status Bar

Hands-on Tutorial	*In this tutorial, you will complete the following tasks:* *1. Use the pinned **Tutorial Files** folder to open a PDF file.* *2. Add a text box markup and customize it.* *3. Add another text box markup using the custom tool saved on the **Recent Tools** tool set.* *4. Add a typewriter markup and customize its properties.* *5. Add a callout markup and use the **Format Painter** tool to match its **Appearance** properties with those of the text box markup.* *6. Add a note markup.* *7. Use the **Highlight** tool to highlight some content of the PDF file.*

Section 1: Adding a Text Box Markup

In this section, you will open the **Architecture_Plan_02.pdf** file from the **C02** folder. You will then add a text box markup in this file and customize the properties of that markup.

1. Start Revu 2019.

2. Restore the **Revu Training** profile that you created in the previous chapter.

   This profile displays the **Properties** panel on the right side of the **Main Workspace**.

3. Invoke the **File Access** panel > **Explorer Mode**.

4. From the drop-down list at the top, select the pinned **Tutorial Files** folder; the content of this folder is displayed.

5. Double-click on the **C02** folder; the **Architecture_Plan_02.pdf** file is displayed.

6. Double-click on the **Architecture_Plan_02.pdf** file to open it. **This is a sample PDF file provided by Bluebeam for its users to learn various tools in Revu.**

   The default cursor mode is set to **Pan**. As a result, the current cursor is displayed as the pan cursor. You will first change it to the select cursor because you can pan by holding down the Scroll Wheel button on the mouse.

7. Press the **V** key to change the cursor to the select cursor.

   You will now zoom to the top left of the plan where the text **OPEN OFFICE 200** appears.

8. Hover the cursor near the top left of the plan where the text **OPEN OFFICE 200** appears.

9. Scroll forward to zoom to that area and then pan to bring that area to the center of the **Main Workspace**, refer to Figure 21.

10. From the **Menu Bar**, click **Tools > Markup > Text Box**. Alternatively, you can press the **T** key on the keyboard to invoke the **Text Box** tool; the cursor changes to the text box cursor.

*Tip: Revu is one of the most user-friendly software available for PDF markups. Every time Revu wants you to do something, it is displayed as a prompt sequence on the left side of the **Status Bar**. For example, when you invoke the **Text Box** tool, the left side of the **Status Bar** displays a prompt sequence "Select region to place text box".*

11. Press and hold down the Left Mouse Button and drag two opposite corners to define the text box, as shown in Figure 21.

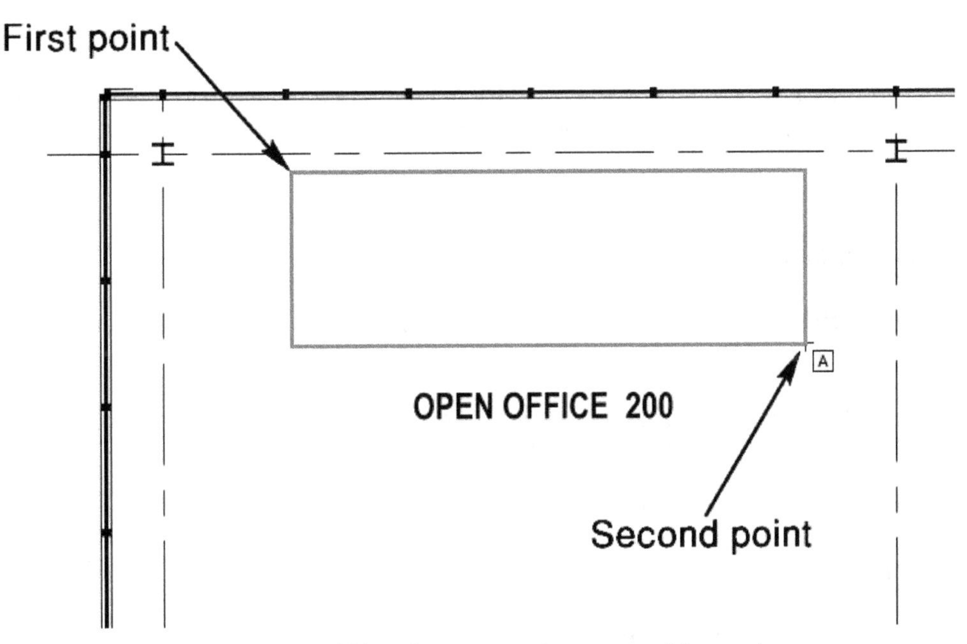

Figure 21 *Specifying the two opposite corners of the text box*

On doing so, the realtime text editor is displayed where you can type the text. But before you start typing the text, you need to change its appearance properties in the **Properties** panel displayed on the right of the **Main Workspace**.

12. In the **Appearance** area of the **Properties** panel displayed on the right of the **Main Workspace**, click the **Color** swatch and select **Blue**; the color of the text box frame changes to Blue.

13. From the **Fill Color** swatch, select **Blue Violet**; the text box is filled with this color.

    With the fill color specified, the content of the PDF file behind the text box is hidden. Therefore, you need to reduce the fill opacity.

14. In the **Fill Opacity** spinner, type **30** as the value and then press ENTER; the fill opacity is changed in the text box.

15. In the **Line Width** spinner, type **0.5** as the value and then press ENTER; the line thickness of the text box frame is reduced.

16. From the **Font** drop-down list, select **Arial**.

    Remember that the fonts in the **Font** drop-down list are displayed in three sections. The first section shows the fonts used in the currently selected markup. The second section shows the list of standard fonts supported by all PDF viewers. The third section shows all other available fonts. Note that if you use the fonts from the third section, they will have to be embedded in the PDF file for viewing in other PDF viewers.

17. Make sure the value in the **Font Size** drop-down list is set to **12**.

    The font color needs to be changed as well. But at this stage, you will write the text with the default font color and change it later while editing the text.

    You are now ready to write the text.

18. Enter the following text in the text box:

    The open office area needs at least two windows on each wall. Please check with the Architect if that is Ok.

19. Once the text is written, click anywhere outside the text box to exit the **Text Box** tool; the text is written and appears as shown in Figure 22.

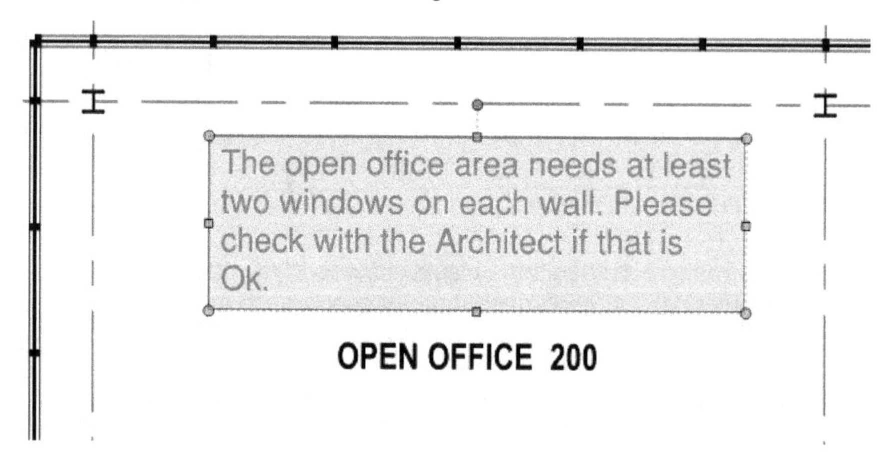

Figure 22 *The text box markup added to the PDF file*

With the text box markup still selected, you will now edit the text font color.

20. From the **Properties Panel** > **Text Color** swatch, select **Blue** to change the text font color to Blue.

    This now completes the text box markup customization.

21. If need be, drag the middle-right control handle of the text box to ensure the text is written in a maximum of four lines.

Depending on the size of the text box that was defined while adding this markup, the text box may be bigger than what is required for the content of the markup. You will now autosize the text box.

22. Right-click on the text box markup; a shortcut menu is displayed.

23. Select **Autosize Text Box** from the shortcut menu; the text box is automatically resized to match the content.

24. Press the ESC key to deselect the text box markup.

Section 2: Using the Recent Tools Tool Set to Add Another Text Box Markup

Whenever you customize any markup tool, the custom version of it is automatically saved on the **Recent Tools** tool set. This tool set is available on the **Tool Chest** panel. In this section, you will use the custom version of the **Text Box** tool to add another text box markup.

1. From the **Panel Access Bar** on the left side of the **Main Workspace**, click the **Tool Chest** button; the **Tool Chest** panel is displayed with various tool sets.

   Notice that the **Recent Tools** tool set shows the custom version of the **Text Box** tool you invoked in the previous section, see Figure 23. You will now use this tool to add another text box markup.

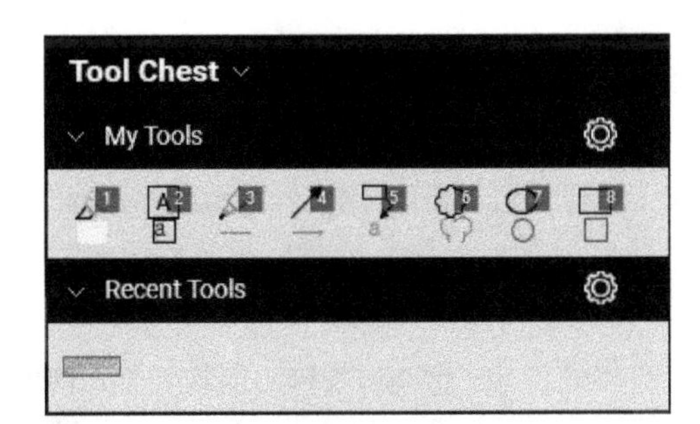

*Figure 23 The **Recent Tools** tool set with the custom version of the **Text Box** tool*

2. Press and hold down the Scroll Wheel button and then pan down to the lower left corner of the plan where the text **OPEN OFFICE 244** appears, refer to Figure 24.

   As discussed earlier in this chapter, the tools in the tool sets are saved in the **Drawing** mode or the **Properties** mode. If the tool is in the **Drawing** mode, it allows you to add the exact same copy of the markup, as if you were copying and pasting it. If the tool is in the **Properties** mode, you are allowed to add the new markup with new content, still maintaining the **Appearance** properties of the custom markup. By default, the tools in the **Recent Tools** tool set are added in the **Drawing** mode. This is evident by the button of the tool displayed as the smaller copy of the markup you added earlier.

3. Hover the cursor over the custom version of the **Text Box** tool in the **Recent Tools** toolset; the name of the markup tool is displayed, along with the content of the text box markup you added in the previous section.

4. Click on this tool; the exact same copy of the previous text box markup is attached to the cursor.

5. Click below the **OPEN OFFICE 244** text to place the text box markup, as shown in Figure 24.

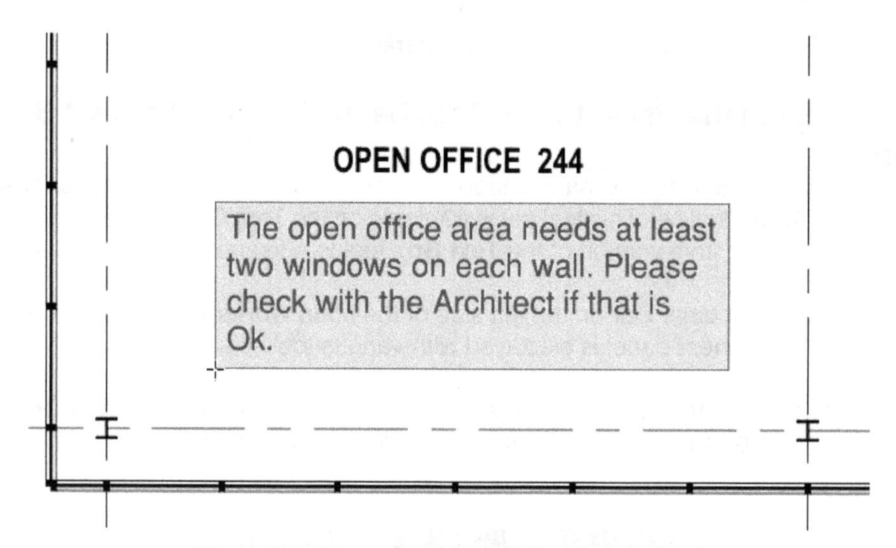

Figure 24 A second text box markup being added

The text box markup is added and is currently selected. You will now deselect this markup.

6. Press the ESC key to deselect the text box markup.

   Next, you will change this tool to the **Properties** mode in the **Recent Tools** tool set so you can add a markup with new content but the same **Appearance** properties. But before you do that, you will navigate to the top right corner of the plan.

7. Using the Scroll Wheel mouse button, invoke the zoom and pan tools to navigate to the top right of the plan where the text **OPEN OFFICE 218** appears, refer to Figure 25.

8. Double-click on the **Text Box** tool in the **Recent Tools** tool set; the icon of the tool changes to the original **Text Box** tool and the icon of the custom appearances is displayed below it. Also, the cursor changes to the text box cursor and you are prompted to select the region to place the text box. This prompt sequence is displayed on the left side of the **Status Bar**.

9. Drag two opposite corner points of the text box above the text **OPEN OFFICE 218**, refer to Figure 25.

10. Enter the following text:

    If the Architect agrees to add windows to Open Offices 200 and 244, do the same thing to this open office as well as Open Office 222.

11. Click anywhere outside the text box to finish adding the markup; the text box is still selected, as shown in Figure 25.

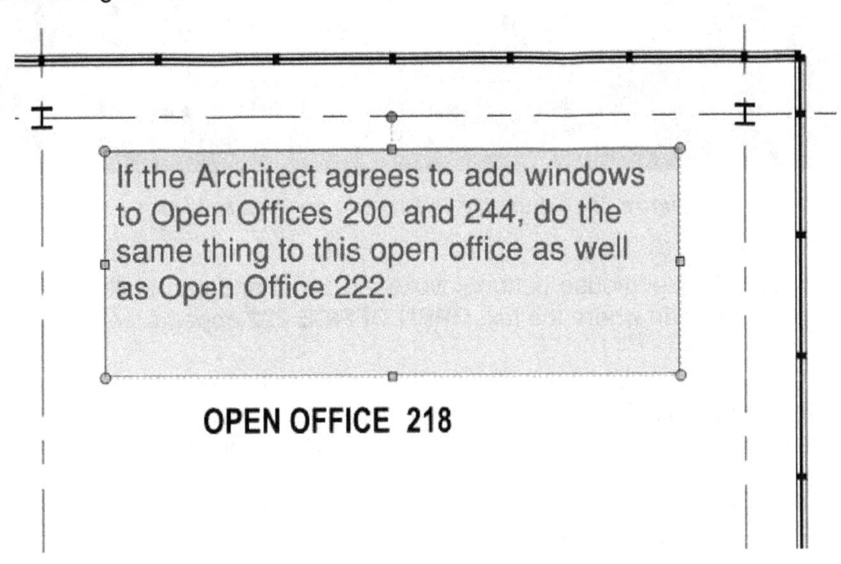

***Figure 25** The text box markup with custom content*

Notice the text box is bigger than that required for the content of the markup. You will now fix this by autosizing the text box.

12. Right-click on the text box markup and select **Autosize Text Box** from the shortcut menu; the text box is resized to the content of the markup.

    You would have noticed that every time you finish the text box markup, you need to resize the text box. As mentioned earlier in this chapter, you can configure the Revu preferences to automatically resize the text box of all markups that allow you to add content in a text box. It is recommended to configure that preference now.

13. From the top left, click **Revu > Preferences**; the **Preferences** dialog box is displayed.

14. From the left pane, click **Tools**; the options related to various markup tools are displayed in the **Markup** tab in the right pane of the dialog box, refer to Figure 26.

15. From the right pane, select **Autosize Text Box and Callout Markups**, as shown in Figure 26.

16. Click **OK** in the dialog box.

    Next, you will add another text box markup near the bottom right corner of the plan.

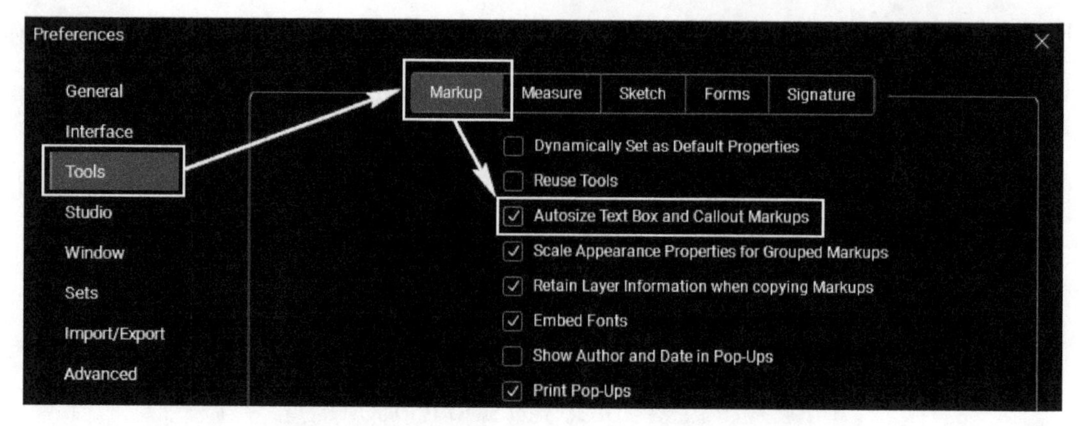

Figure 26 *Setting the preference to autosize text box*

17. Using the Scroll Wheel mouse button, invoke the zoom and pan tools to navigate to the bottom right of the plan where the text **OPEN OFFICE 222** appears, refer to Figure 27.

    Because you added a new text box markup, the tool for that is also displayed in the **Drawing** mode in the **Recent Tools** tool set. However, you will use the tool in the **Properties** mode to add the markup as you need to add different content to this markup.

18. From the **Recent Tools** tool set, click on the **Text Box** tool that is displayed in the **Properties** mode; the cursor changes to the text box cursor.

19. Drag two opposite corners below the text **OPEN OFFICE 222**; the realtime text editor is displayed.

20. Enter the following text:

    Refer to the markup in OPEN OFFICE 218 and apply similar changes.

21. Click anywhere outside the text box; the markup is added and the text box is automatically resized to match the content of the markup, as shown in Figure 27.

22. Press the ESC key to deselect the markup.

    It is better to save this PDF file before proceeding further.

23. From the menu bar at the top of the Revu window, click **File > Save** to save the current PDF file with the markups. Alternatively, you can press CTRL + S keys.

24. Double-click the Scroll Wheel button on the mouse to zoom to the extents of the PDF file.

Section 3: Adding Typewriter Markups

In this section, you will add typewriter markups to get the Architectural designer to review all the room numbers and basins that you will highlight later in this tutorial.

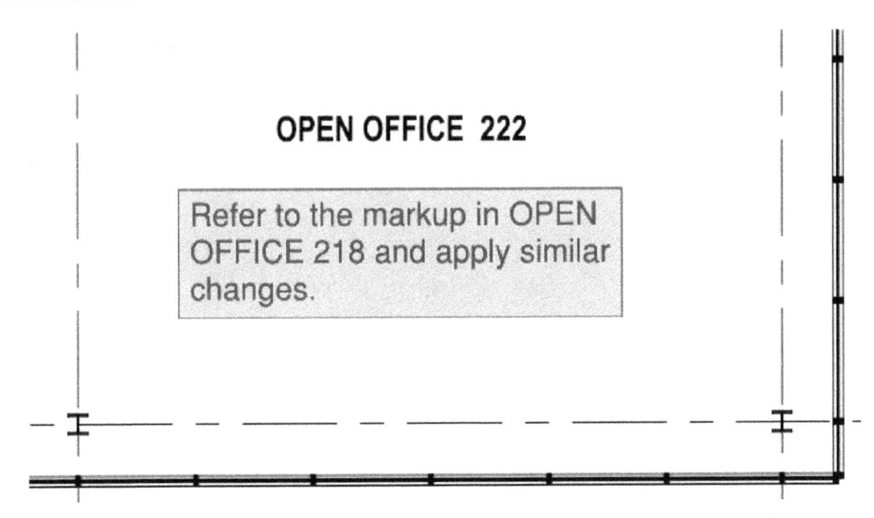

Figure 27 The text box markup added and automatically resized

1. From the **Menu Bar**, click **Tools > Markup > Typewriter**. Alternatively, press the **W** key to invoke the **Typewriter** markup tool; you are prompted to select the location to place the typewritten text.

2. Using the Scroll Wheel mouse button, invoke the zoom and pan tools to navigate to the area near the top left of the border of the plan, refer to Figure 28.

3. Click near the top left corner of the border; the realtime text editor is displayed. Also, the **Properties** panel shows various properties that can be customized.

4. From the **Properties** panel > **Appearance** area, **Font** drop-down list, select **Arial**.

5. From the **Font Size** drop-down list, select **24** to increase the font size.

6. Click the **Text Color** swatch and select **Blue** to change the color of the text.

7. Enter the following text:

   Please review all the tags of all the room numbers that are highlighted.

8. Press the ESC key; the text is written.

9. Press the ESC key again to deselect the markup. Figure 28 shows the typewritten text.

   Notice that the custom version of the **Typewriter** tool is automatically added in the **Drawing** mode to the **Recent Tools** tool set. You will now switch this tool to the **Properties** mode and use it add another typewriter markup.

10. Double-click on the custom version of the **Typewriter** tool in the **Recent Tools** tool set; the tool switches to the **Properties** mode and the cursor changes to the typewriter cursor.

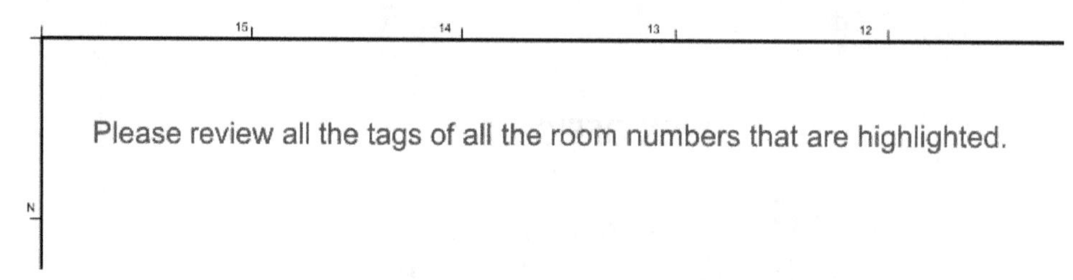

Figure 28 The typewriter markup added to the PDF file

11. Click below the start point of the previous typewriter markup; the realtime text editor is displayed.

12. Enter the following text:

    Change the basins highlighted in the restrooms to the specs in the BIM Execution Plan.

13. Press the ESC key; the typewriter markup is added, as shown in Figure 29.

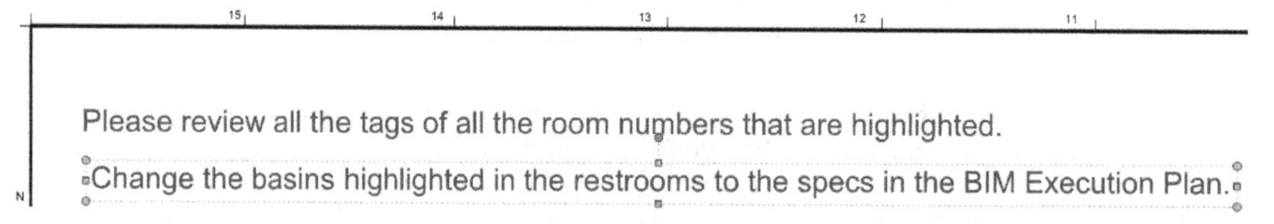

Figure 29 The custom version of typewriter markup added

Chances are that the two markups you added are not aligned properly. You will now align the two markups.

14. With the bottom typewriter markup still selected, hold down the SHIFT key and select the top typewriter markup; the control handles of the top markup are displayed in Green.

    Next, you will right-click to display the shortcut menu to align the markups. It is important to note that the markup that is to be used as the parent for the alignment is the one on which you have to right-click.

15. Right-click on the top typewriter markup and select **Alignment > Align Left** from the shortcut menu, as shown in Figure 30; the start of the bottom markup is aligned to that of the top markup.

16. Press the ESC key to deselect the two markups.

17. Double-click the Scroll Wheel button on the mouse to zoom to the extents of the PDF file.

Section 4: Adding Callout Markups

In this section, you will add a callout markup and then use the **Format Painter** tool to match

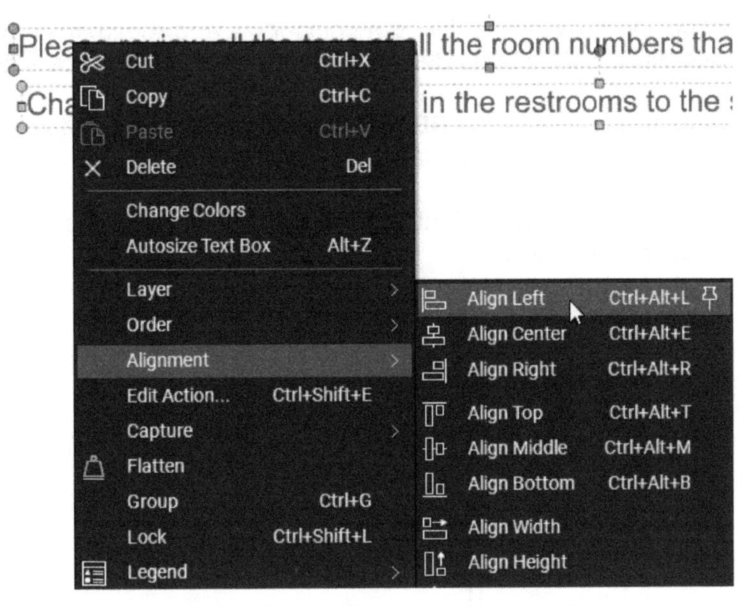

Figure 30 *Aligning the two typewriter markups*

the **Appearance** properties of the callout markup with those of the text box markup. You will then use the custom version of this markup from the **Recent Tools** tool set to add more callout markups.

The callout markup comprises of a leader line with an arrowhead and a text box. To ensure that the leader arrowhead snaps to the content of the PDF file, you need to turn on the **Snap to Content** button on the **Status Bar**.

1. From the **Status Bar**, click on the **Snap to Content** button to turn it on, as shown in Figure 31.

Figure 31 *Turning on* ***Snap to Content***

2. Using the Scroll Wheel mouse button, invoke the zoom and pan tools to navigate to the **MEETING 208** and **MEETING 209** text near the top center of the plan, refer to Figure 32.

3. From the **Menu Bar**, click **Tools > Markup > Callout**. Alternatively, press the **Q** key on the keyboard; the **Callout** tool is invoked and the cursor changes to the callout cursor.

4. Click on the curved wall to define the start point of the arrowhead, refer to Figure 32; the leader line with the arrowhead is added and the preview of the text box is displayed.

5. Move the cursor to the right of the **RECEPTION 255** text to ensure the text box is being placed on the right side of the leader landing, refer to Figure 32.

6. When the text box appears on the right of the leader landing, click to place the text box; the realtime text editor is displayed.

7. Enter the following text:

   This curved wall needs to be changed to glass partition wall

8. Click anywhere outside the text box; the callout markup is added with the default properties, as shown in Figure 32.

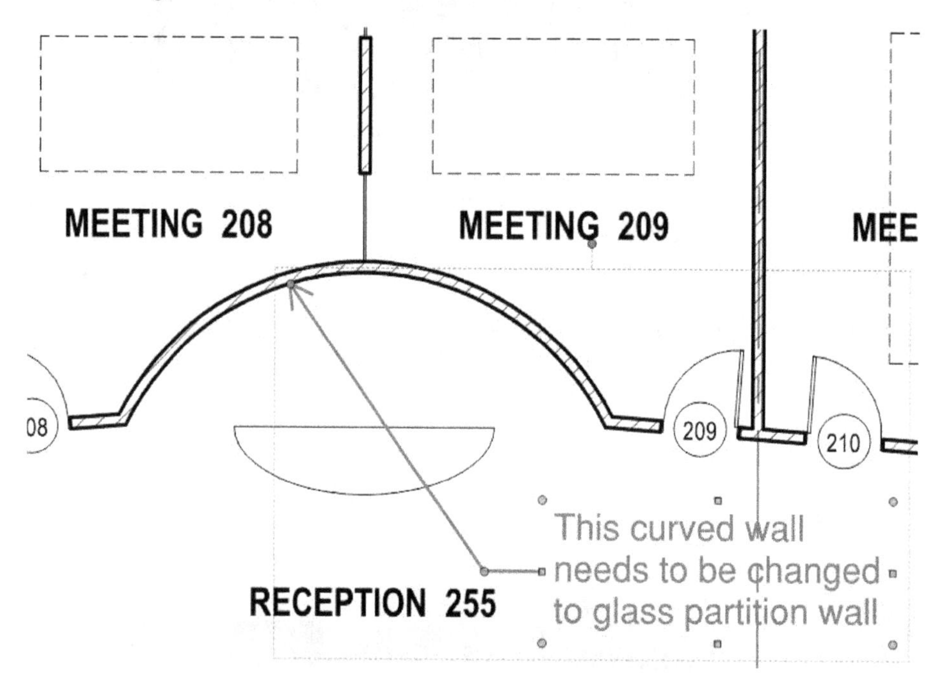

Figure 32 The callout markup with default properties

You will now use the **Match Properties** tool to match the properties of this callout markup with that of one of the existing text box markups.

9. Navigate to the text box markup in the **OPEN OFFICE 218** room on the top right of the plan.

10. Right-click on the text box markup and select **Format Painter** from the shortcut menu.

11. Navigate back to the callout markup you added earlier.

12. Hover the cursor over the text of the callout markup; the cursor changes to the paintbrush cursor.

13. Click on the text of the callout; the **Appearance** properties of the markup are modified to match those of the text box callout.

14. Press the ESC key to clear the current selection.

    Most properties of the callout markup were modified. However, you will still have to manually change the arrowhead style as the text box markup did not have any settings related to the arrowhead.

15. Click the callout markup to select it; the **Properties** panel displays the properties of this markup.

16. From the **Appearance** area > **Line Width** edit box, enter **1** as the value; the line width of the leader line and text box are modified.

17. From the **End** drop-down list, select the third arrowhead type, which creates a closed filled arrowhead.

18. If required, drag the middle right control handle of the text box to fit the text in three lines.

    Chances are that by dragging the text box, its size may have become bigger than what is required to fit the content. You will not resize this text box.

19. Right-click on the text box and select **Autosize Text Box** from the shortcut menu.

20. Press the ESC key to deselect the callout markup. Figure 33 shows this markup after editing its properties.

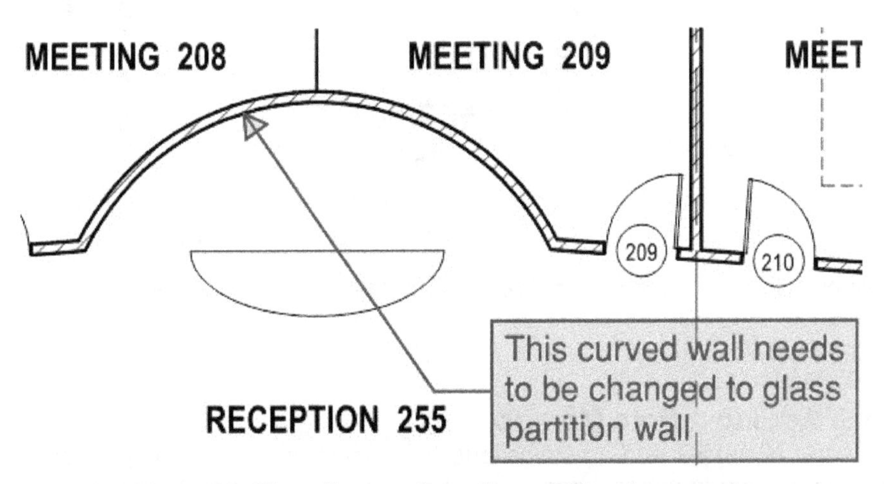

Figure 33 *The callout markup after modifying its properties*

The custom version of this tool is saved in the **Drawing** mode in the **Recent Tools** tool set. You will now change this to the **Properties** mode and use it to add more callout markups.

21. In the **Recent Tools** tool set, double-click on the **Callout** tool to switch it to the **Properties** mode; the cursor changes to the callout cursor.

22. Navigate to the **STAIR 251** text, which is on the left of the **RECEPTION 255** text.

23. Click on the circle of the door tag **251** as the location of the arrowhead; the arrowhead is placed and the preview of the text box is displayed.

24. Move the cursor to the top right of this text to ensure the text box is being placed on the right of the leader landing, refer to Figure 34.

25. Click to place the text box on the right of the landing and then type the following text:

    This needs to be a fire safe door of the specs defined in BEP

26. Click anywhere outside to finish adding the markup.

27. Press the ESC key to deselect the markup. Figure 34 shows the callout markup.

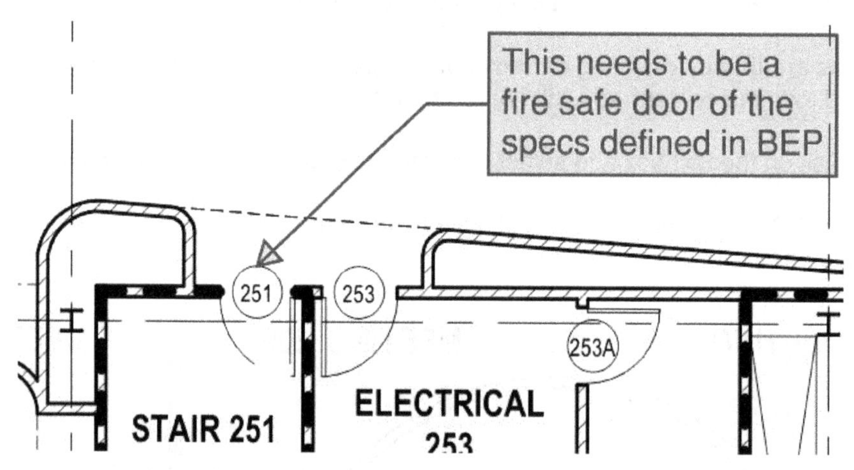

Figure 34 Another callout markup with custom properties

28. Double-click the Scroll Wheel button on the mouse to zoom to the extents of the PDF file.

29. Save the PDF file.

Section 5: Adding a Note Markup

In this section, you will add a note callout. The content in the **Note** pop-up window will be copied and pasted from a text document provided to you in the **C02** folder.

1. Using Windows Explorer, browse to the **Tutorial Files > C02** folder.

2. Open the **Field-Test.txt** file and copy its content.

3. Return to the Revu window.

4. Navigate to the bottom center of the plan where the **BREAK ROOM 233** text appears.

5. From the **Menu Bar**, click **Tools > Markup > Note**. Alternatively, press the **N** key to invoke the **Note** tool; the cursor changes to the note cursor and the **Properties** panel shows various properties of this markup.

6. In the **Properties** panel > **Appearance** area, click the **Color** swatch and select **Red** to change the color of the note icon to Red.

7. From the **Icon** drop-down list, make sure **Note** is selected.

8. Click on the right side of the **BREAK ROOM 233** text to place the note icon, refer to Figure 35; the **Note** pop-up window is displayed.

   You will now paste the text that you copied from the text file into this pop-up window.

9. In the pop-up window, right-click to display the shortcut menu and select **Paste**; the content of the text file is copied in the **Note** pop-up window, as shown in Figure 35.

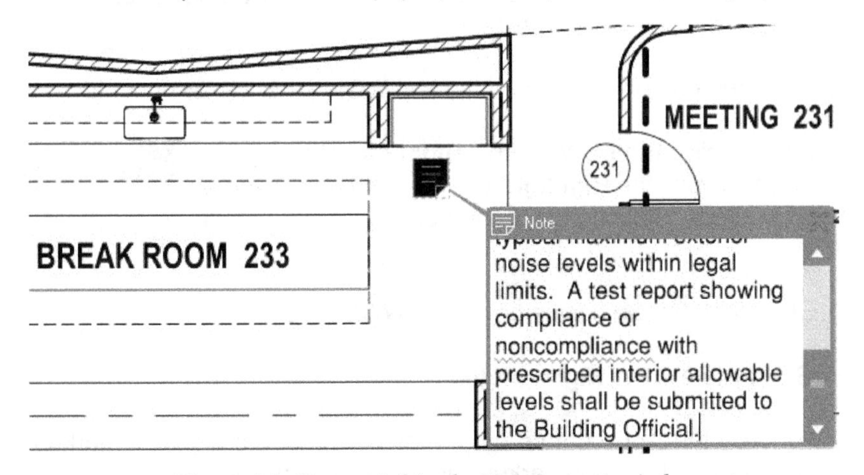

*Figure 35 Text copied in the **Note** pop-up window*

10. Click anywhere outside the pop-up window to finish adding the note markup.

    Notice that the pop-up window is still displayed. You will now close this window.

11. Click on the **X** button on the top right of the **Note** pop-up window to close it.

If you want to read the content of the note, you can hover the cursor over it. However, if there is a lot of content added in the **Note** pop-up window, hovering the cursor over it is not the right way of reading the content. In that case, it is better to open the **Note** pop-up window. You will now do this.

12. Double-click on the **Note** icon that you recently added; the **Note** pop-up window is opened.

13. Once you have reviewed the content of the note, close the pop-up window by clicking on the **X** button on its top right.

14. Press the ESC key to deselect the note markup.

    The custom version of the **Note** tool is added in the **Drawing** mode to the **Recent Tools** tool set.

***Tip:** As mentioned earlier in the chapter, by default the **Recent Tools** tool set only saves twelve recent tools. To change this number, you can click on the cogwheel on the top right of this tool set and select the desired value from the **Maximum Recent** option.*

15. Double-click the Scroll Wheel button on the mouse to zoom to the extents of the PDF file.

16. Save the PDF file.

Section 6: Using the Highlight Markup

In this section, you will highlight the basins in the men's and women's restrooms and then highlight the text of the storage areas.

1. Navigate to the center of the plan where the text **WOMEN'S RR 254** appears.

2. From the **Menu Bar**, click **Tools > Markup > Highlight**. Alternatively, press the **H** key to invoke the **Highlight** tool; depending on where your cursor is located, it will change to the pen highlighter or text highlighter.

3. One by one, highlight the two basins below the **WOMEN'S RR 254** text.

4. Press the ESC key to exit the **Highlight** tool.

   You will now modify the **Appearance** properties of one of the highlight markup and then match the properties of the other markup with the modified one.

5. Click on one of the highlight markups; the **Properties** panel on the right of the **Main Workspace** shows the properties of the markup.

6. In the **Appearance** area, click on the **Color** swatch and select **Blue Violet**; the color of the highlight markup updates.

7. With the highlight markup still selected, right-click on it and select **Format Painter** from the shortcut menu.

8. Click on the other highlight markup to match its properties with those of the modified markup.

9. Press the ESC key to deselect the highlight markups. Figure 36 shows the PDF after highlighting the two basins.

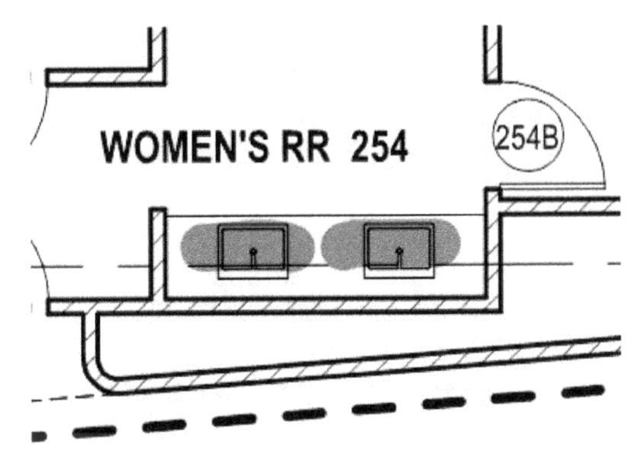

Figure 36 *Basins highlighted using the highlight markup*

Next, you will use the custom version of the **Highlight** tool from the **Recent Tools** tool set to highlight the **STORAGE 253A** and **STORAGE 254A** text.

10. Zoom close to the **STORAGE 253A** and **STORAGE 254A** text.

11. Invoke the custom version of the **Highlight** tool from the **Recent Tools** toolset.

12. Hover the cursor over the **STORAGE 253A** text, the cursor changes to the text highlight cursor.

    While highlighting the text, it is important that you carefully select only the required text. If you are not careful, it might select a lot of unwanted text to highlight as well. If that happens, you can press the CTRL + Z key to undo the highlighting of the text.

13. Carefully select the **STORAGE 253A** text to highlight it.

14. Similarly, select the **STORAGE 254A** text to highlight it.

15. Press the ESC key to deselect the **Highlight** tool. Figure 37 shows the PDF file after highlighting the text.

    Next, you will highlight the basins in the men's restroom, which is located on the right of the women's restroom.

16. Navigate to the right of the women's restroom where the text **MEN'S RR 258** appears.

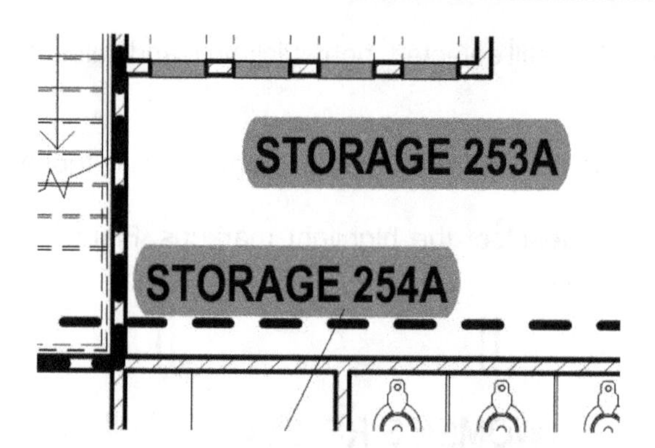

Figure 37 Text highlighted using the highlight markup

17. Using the custom version of the **Highlight** markup on the **Recent Tools** tool set, one by one highlight the two basins below the text **MEN'S RR 258**.

18. Next, highlight the **STORAGE 258A** text, which is located above the men's restroom.

19. Press the ESC key to exit the **Highlight** tool. Figure 38 shows the PDF file after highlighting the basins and the text.

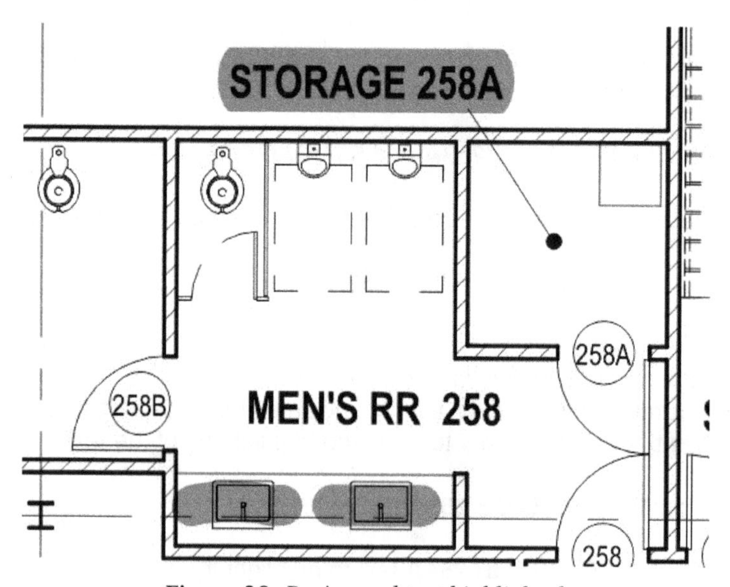

Figure 38 Basins and text highlighted

20. Double-click the Scroll Wheel button to zoom to the extents of the PDF file.

21. Save the PDF file.

22. Close the PDF file.

Skill Evaluation

Evaluate your skills to see how many questions you can answer correctly. The answers to these questions are given at the end of the book.

1. Markups added to a PDF file cannot be modified. (True/False)

2. The text box markup cannot be filled with a color. (True/False)

3. The arrowhead of the **Callout** tool can be changed. (True/False)

4. The **Highlight** tool allows you to highlight text in the PDF file. (True/False)

5. The **Pen** markup can be used to create a free form shape. (True/False)

6. Which tool allows you to add a text with a leader line?

   (A) **Callout** (B) **Text Box**
   (C) **Leader** (D) **None**

7. Which tool allows you to add a paragraph text in a pop-up window that can be hidden?

   (A) **Highlight** (B) **Note**
   (C) **Text** (D) **Window**

8. Which tool is used to match the properties of a source markup with those of the other selected markups?

   (A) **CTRL +C** (B) **Match Properties**
   (C) **Copy** (D) **Format Painter**

9. When you customize a markup, the custom version of that tool is automatically saved in which tool set?

   (A) **My Tools** (B) **Custom Tools**
   (C) **File Tools** (D) **Recent Tools**

10. The custom tools can be saved in the **Recent Tools** tool set in which two modes?

    (A) **Properties** (B) **Model**
    (C) **Appearance** (D) **Drawing**

Class Test Questions

Answer the following questions:

1. Explain briefly the process of changing the icon of the note markup.
2. How will you change a tool from the **Properties** mode to the **Drawing** mode in the **Recent Tools** toolset?
3. Explain briefly the use of a tool in the **Drawing** mode.
4. How will you change the **Appearance** properties of the highlight markup?
5. Explain briefly the process of matching properties of the markups.

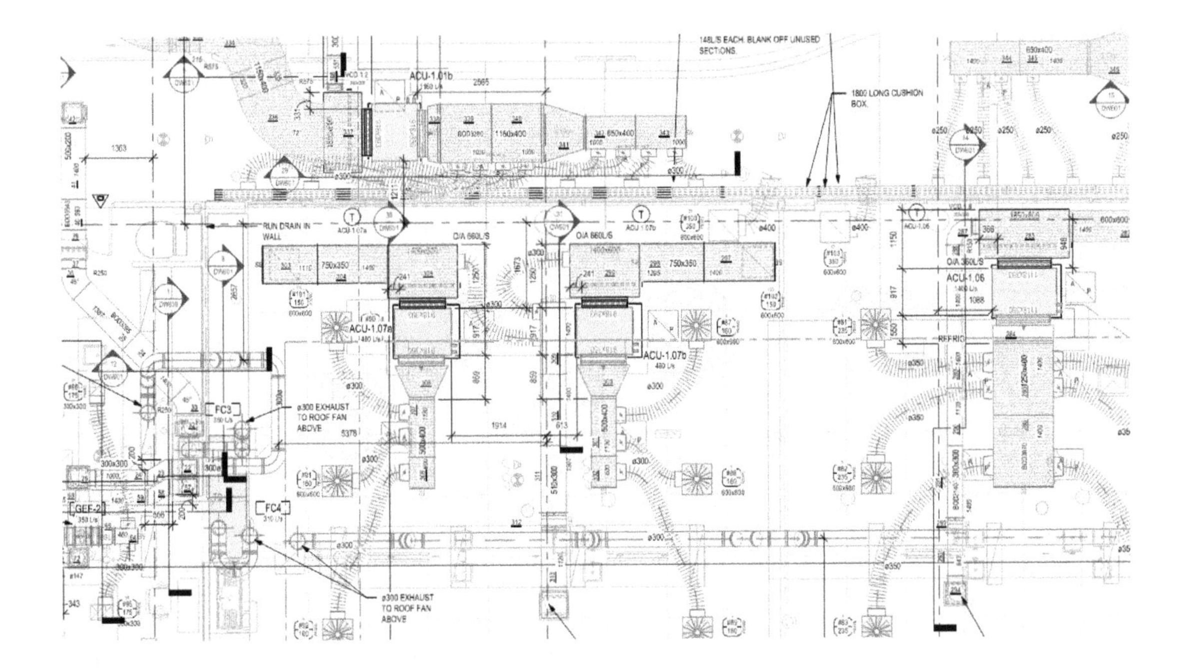

Chapter 3 - Working with the Markup Tools - II

The objectives of this chapter are to:

- √ *Explain how to add line markups*
- √ *Teach you how to create custom line styles*
- √ *Explain how to add arrow markups*
- √ *Explain how to add arc markups*
- √ *Explain how to add polyline markups*
- √ *Explain how to add dimension markups*

WORKING WITH THE LINE MARKUP TOOLS

Revu provides you with five types of line markup tools. They are **Line**, **Arrow**, **Arc**, **Polyline**, and **Dimension**. All these tools are discussed next.

The Line Markup

Menu: Tools > Markup > Line
Keyboard Shortcut: L

This tool allows you to draw a single line segment. When you invoke this tool, the cursor will change to the line markup cursor and you will be prompted to select start and end points of the line. If you hold down the SHIFT key, you can create lines that snap at 45-degrees intervals, including horizontal and vertical lines. You can use the **Properties** panel to modify the **Appearance** properties of the line segment during its creation or after it is created. Revu comes with a number of **Standard** and **Advanced** line styles that are available in the **Style** drop-down list for you to assign to the selected line segment. You can also create custom line styles by clicking on the **Manage** option available at the bottom in the **Style** drop-down list. For example, Figure 1 shows a horizontal line segment with the **Track** line style assigned from the **Advanced** section in the **Style** drop-down list. Figure 2 shows the same line segment with the **Road** line style assigned from the same list.

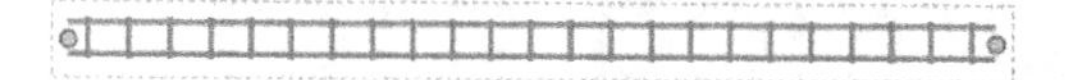

*Figure 1 A line segment with the **Track** line style*

*Figure 2 A line segment with the **Road** line style*

Using the **Start** and **End** lists, you can modify the start and end appearances of the line segment. For example, you can assign arrowheads or slash (architectural ticks) as the start and end appearances of the line segments. Note that if you select the closed filled arrowheads, you can then select the fill color from the **Fill Color** swatch. Figure 3 shows a line segment with a closed filled arrowhead as the start appearance and slash as the end appearance.

Figure 3 A line segment with start and end appearances defined

Creating Custom Line Styles

As mentioned earlier, you can also create custom line styles to be used in Revu. Once created, these line styles can also be shared with other members of your team. The following are the procedures for creating various types of custom line styles.

Procedure for Creating a Custom Line Style with Text

The following is the procedure for creating a custom line style with text.

1. Select an existing line segment; the **Properties** panel will display the properties of the line.

2. Click on the **Style** drop-down list and scroll down to the bottom.

3. Select the **Manage** option; the **Manage Line Style Sets** dialog box will be displayed, as shown in Figure 4.

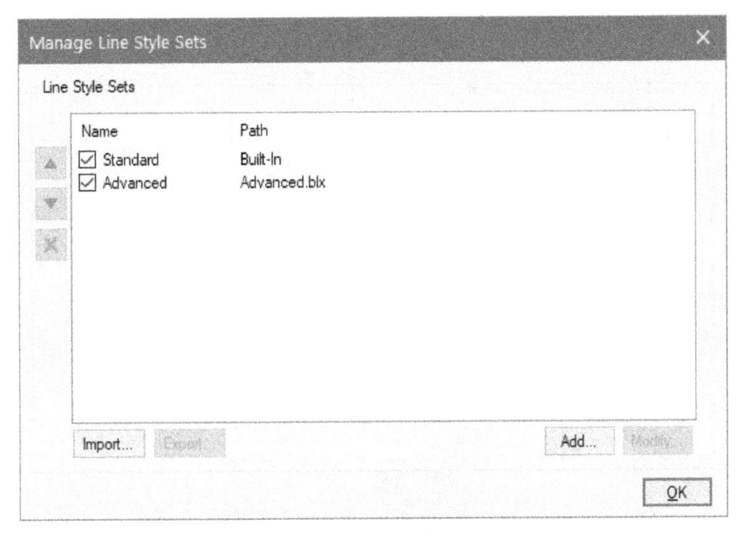

*Figure 4 The **Manage Line Style Sets** dialog box*

Creating a custom line style is a two step process. You first need to create a line style set, which acts as the group. Next, you create the custom line style in that set. So you will first create the line style set.

4. Click the **Add** button in the dialog box; the **Add Line Style Set** dialog box will be displayed.

5. From the **Step 1: Select Type** area in the dialog box, make sure **New** is selected. This ensures that a new set is created.

6. In the **Step 2: Select Location** area > **Title** text box, enter the name of the line style set. Notice that the **Location** display box shows a file with the same name as the name of your line style set and extension .blx. This is the extension of the custom line style set that you can share with other team members by exporting and then importing on their machines. Figure 5 shows this dialog box with a custom line style set name defined.

7. Click **OK** in the dialog box; the **Edit Line Style Set** dialog box will be displayed.

   This is the dialog box in which you add various line styles for the line style set you defined.

8. In the **Edit Line Style Set** dialog box, click the **Add** button; the **Add Line Style** dialog box will be displayed, prompting you to enter the name of the line style.

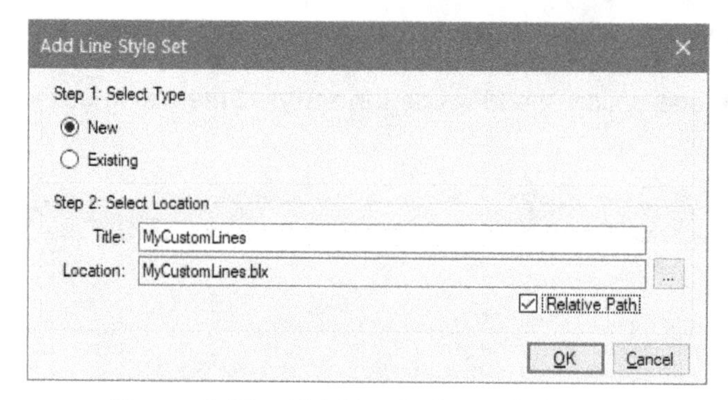

*Figure 5 The **Add Line Style Set** dialog box*

9. Enter the name of the new line style, as shown in Figure 6.

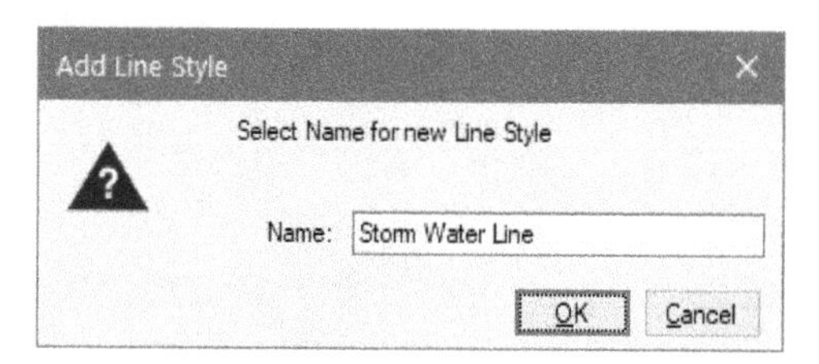

*Figure 6 The **Add Line Style** dialog box*

10. Click **OK** in the **Add Line Style** dialog box; the **Line Style Editor** dialog box will be displayed. This is the dialog box in which you will define the line style.

11. From the **Components** area, click the Green + button, the **Add Component** dialog box will be displayed.

12. From the **Type** drop-down list, select **Dash**, as shown in Figure 7.

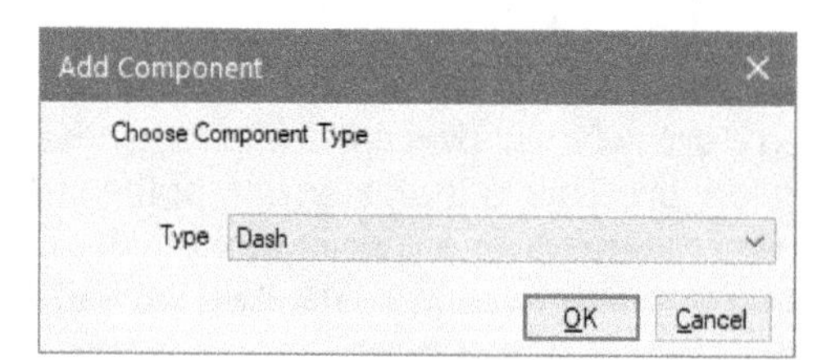

Figure 7 Adding a dash component to the line style

13. Click **OK** in the dialog box; you will be returned to the **Line Style Editor** dialog box.

14. From the **Components** area, click the Green + button, the **Add Component** dialog box will be displayed.

15. From the **Type** drop-down list, select **Space** and then click **OK**; you will be returned to the **Line Style Editor** dialog box. This dialog box now shows a dash and a space component in the **Components** area.

16. With the **Space** component still selected, change the **Width** value to the required value; the preview in the **Properties** area will show a dashed line style.

17. From the **Components** area, click the Green + button, the **Add Component** dialog box will be displayed.

18. From the **Type** drop-down list, select **Text** and click **OK** in the dialog box; you will be returned to the **Line Style Editor** dialog box.

    You will notice that the **Text** area will be added to the right of the **Components** area.

19. In the **Text** field, enter the text you want to display in the line style.

20. From the **Font** drop-down list, select the required font style.

21. From the **Font Size** drop-down list, select the required font size.

22. From the **Font Style** area, select the option to make the text bold or italic, if required.

23. From the **Components** area, click the Green + button, the **Add Component** dialog box will be displayed.

24. From the **Type** drop-down list, select **Space**.

25. Click **OK** in the dialog box to return to the **Line Style Editor** dialog box.

26. In the **Properties** area > **Description** text box, enter the description of this new line style.

27. Select the **Scale with Line Thickness** check box. This ensures that the line style is automatically scaled if you change the width of the line in the **Properties** panel.

    Figure 8 shows the **Line Style Editor** dialog box with the custom line style created.

28. Click **OK** in the **Line Style Editor** dialog box; you will be returned to the **Edit Line Style Set** dialog box. This dialog box will now show the custom line style and its description.

29. Click **OK** in the **Edit Line Style Set** dialog box to return to the **Manage Line Style Sets** dialog box. This dialog box will now show the custom line style set you created.

30. Click **OK** in the **Edit Line Style Set** dialog box; you will be returned back to the PDF file.

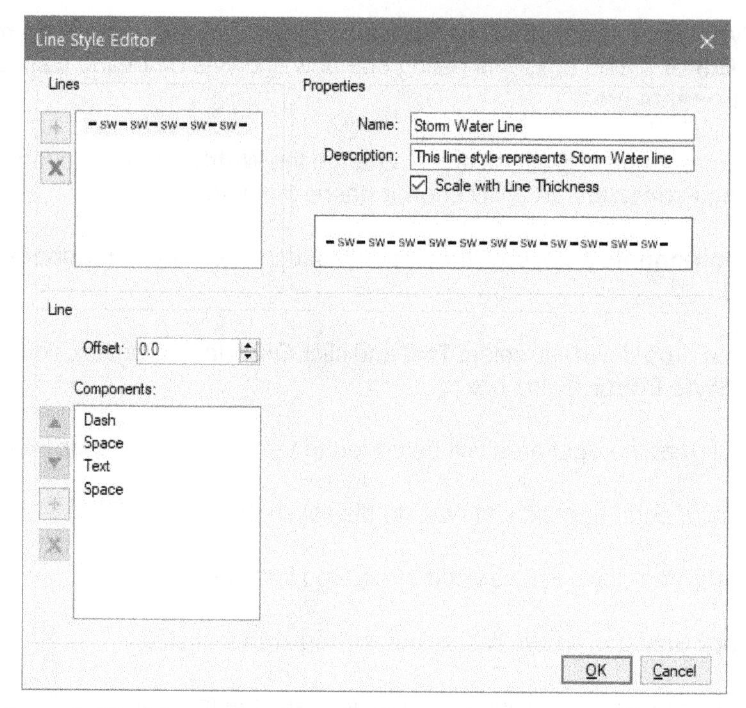

*Figure 8 The **Line Style Editor** dialog box with a custom line style created*

Procedure for Creating a Custom Line Style with a Vector

The following is the procedure for creating a custom line style with a vector.

1. Select an existing line segment; the **Properties** panel will display the properties of the line.

2. Click on the **Style** drop-down list and scroll down to the bottom.

3. Select the **Manage** option; the **Manage Line Style Sets** dialog box will be displayed.

   The line style set you created earlier will be listed in this dialog box. You can add the additional line style to the same set.

4. Double-click on the custom line style set name; the **Edit Line Style Set** dialog box will be displayed.

   This dialog box shows the line styles that already defined in the current set.

5. In the **Edit Line Style Set** dialog box, click the **Add** button; the **Add Line Style** dialog box will be displayed, prompting you to enter the name of the new line style.

6. Enter the name of the new line style.

7. Click **OK** in the **Add Line Style** dialog box; the **Line Style Editor** dialog box will be displayed. This is the dialog box in which you will define the line style.

8. From the **Components** area, click the Green + button, the **Add Component** dialog box will be displayed.

9. From the **Type** drop-down list, select **Dash**.

10. Click **OK** in the dialog box; you will be returned to the **Line Style Editor** dialog box.

11. From the **Components** area, click the Green + button, the **Add Component** dialog box will be displayed.

12. From the **Type** drop-down list, select **Space** and then click **OK**; you will be returned to the **Line Style Editor** dialog box. This dialog box now shows a dash and a space component in the **Components** area.

13. With the **Space** component still selected, change the **Width** value to the required value; the preview in the **Properties** area will show a dashed line style.

14. From the **Components** area, click the Green + button, the **Add Component** dialog box will be displayed.

15. From the **Type** drop-down list, select **Vector** and click **OK** in the dialog box; you will be returned to the **Line Style Editor** dialog box.

    You will notice that the vector creation area will be added to the right of the **Components** area.

16. Using the tools in the vector creation area, create the required vector. Figure 9 shows a vector drawn using the tools in this area.

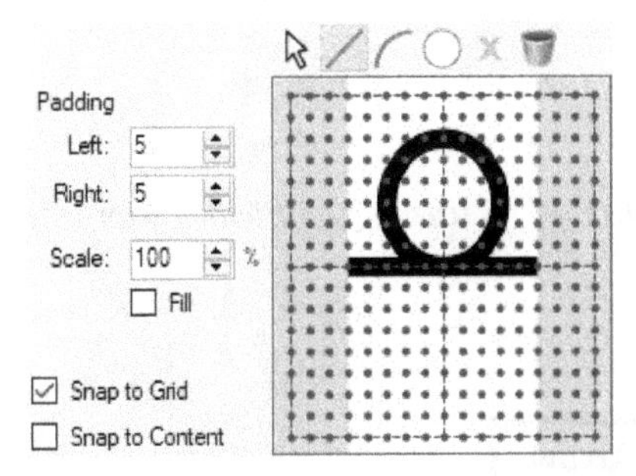

Figure 9 A vector drawn for the custom line style

17. From the **Components** area, click the Green + button, the **Add Component** dialog box will be displayed.

18. From the **Type** drop-down list, select **Space**.

19. Click **OK** in the dialog box to return to the **Line Style Editor** dialog box.

20. In the **Properties** area > **Description** text box, enter the description of this new line style.

21. Select the **Scale with Line Thickness** check box. This ensures that the line style is automatically scaled if you change the width of the line in the **Properties** panel.

22. Click **OK** in the **Line Style Editor** dialog box; you will be returned to the **Edit Line Style Set** dialog box. This dialog box will now show the custom line style and its description.

23. Click **OK** in the **Edit Line Style Set** dialog box to return to the **Manage Line Style Sets** dialog box. This dialog box will now show the custom line style set you created.

24. Click **OK** in the **Manage Line Style Sets** dialog box; you will be returned back to the PDF file.

Procedure for Assigning a Custom Line Style to a Line Segment

The following is the procedure for assigning a custom line style to a line segment.

1. Select the line segment; the **Properties** panel will display the properties of the line.

2. Click on the **Style** drop-down list and scroll down to the bottom; the custom line styles you created are listed under the custom line style set.

3. Select the custom line style; the appearance of the line segment is modified. Figure 10 shows a line segment with a text line style assigned to it that represents storm water and Figure 11 shows the line segment with a vector line style assigned to it.

— SW — SW — SW — SW — SW — SW — SW — SW — SW — SW —

Figure 10 *A line segment with text line style*

Figure 11 *A line segment with vector line style*

The Arrow Markup

Menu: Tools > Markup > Arrow
Keyboard Shortcut: A

This tool allows you to draw an arrow markup. This markup is similar to the line markup, with the difference that the end of this markup will automatically have a closed filled arrow assigned. When you invoke this tool, you will be prompted to specify the start point of the arrow markup. Once you specify the start point, the preview of the arrow markup will be displayed and the arrowhead of the markup will be attached to the cursor. You can click anywhere to locate the arrowhead of the markup. Note that you can hold down the SHIFT key to draw a markup by snapping 45-degrees angle intervals, similar to the line markup. The fill color of the arrowhead can be controlled using the

Fill Color swatch in the **Properties** panel. If required, you can assign a different end style to the arrow markup using the **End** drop-down list and also assign an end style at the start using the **Start** drop-down list. The size of the arrow can be controlled using the drop-down list available on the right of the **Start** and **End** lists. If you have a custom line style created, you can assign that to the arrow markup using the **Style** drop-down list. Figure 12 shows an arrow markup with a closed filled arrow at the end and a slash at the start.

Figure 12 An arrow markup

The Arc Markup

Menu: Tools > Markup > Arc
Keyboard Shortcut: SHIFT+C

Arc Shift+C

This tool allows you to draw an arc markup. By default, the arc markup is drawn by specifying the start point and the endpoint of the arc. This type of arc is generally elliptical in shape. If you hold down the SHIFT key, you can draw a circular arc. Figure 13 shows an elliptical arc drawn using two points and Figure 14 shows a circular arc drawn by holding down the SHIFT key and specifying two points.

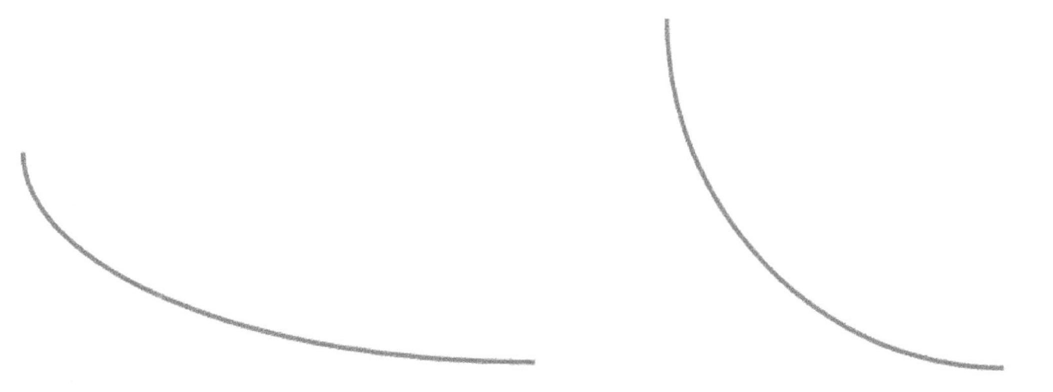

Figure 13 An elliptical arc drawn using two points   *Figure 14 A circular arc drawn using two points*

To draw a more accurate circular arc, you can hold down the ALT key. This allows you to draw an arc by specifying three points: the start point of the arc, a point on the arc, and the endpoint of the arc. Figure 15 shows a three point arc markup being created.

Similar to the other markup tools, you can use the **Appearance** section of the **Properties** panel to change the appearance properties of the arc markup. For example, you can use the **Style** drop-down list to assign a standard, advanced, or custom line style to the arc markup. Using the **Hatch** drop-down list, you can assign a hatch pattern to the section of the arc. The scale of the hatch pattern can be changed using the **Scale** drop-down list. Figure 16 shows the hatch patterns assigned to the arc markups shown in Figures 13 and 14.

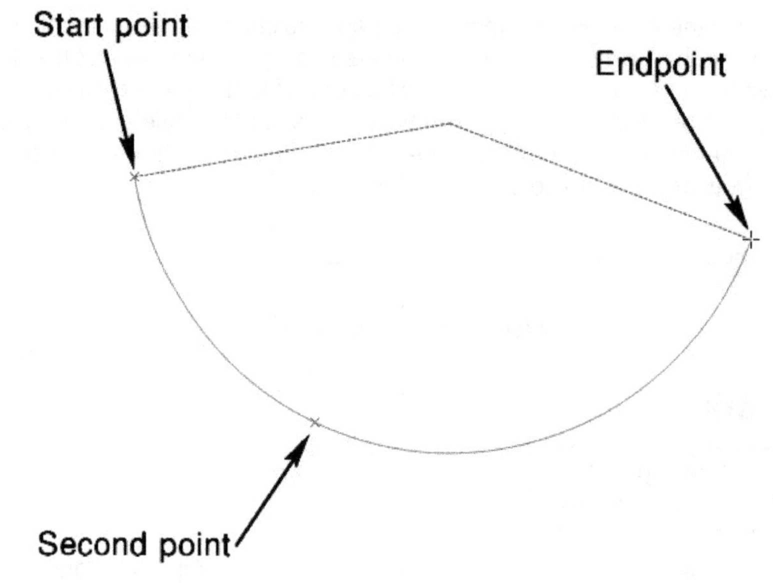

Figure 15 *Drawing a circular arc markup using three points*

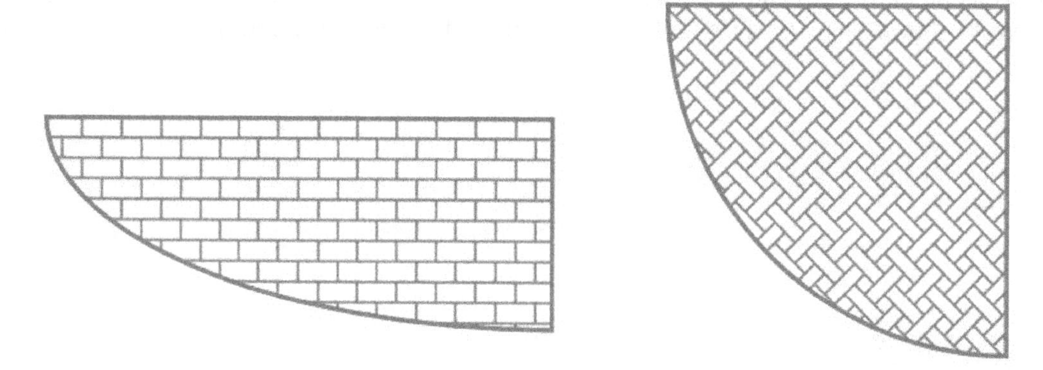

Figure 16 *Hatch patterns assigned to the arcs shown in Figures 13 and 14*

The Polyline Markup

Menu: Tools > Markup > Polyline
Keyboard Shortcut: SHIFT+N

Polyline Shift+N

This tool allows you to draw a markup that comprises of multiple end-connected line segments. When you invoke this tool, you will be prompted to select the vertices of the polyline. You can click at various locations to define the polyline vertices. If you hold down the SHIFT key, you can specify a polyline segment that snaps at 45-degrees intervals. Once you have specified all the vertices, you can press the ENTER key to end the polyline markup and exit the tool. Alternatively, you can click near the start point of the polyline markup to close the shape. Figure 17 shows an open polyline markup and Figure 18 shows a closed polyline markup.

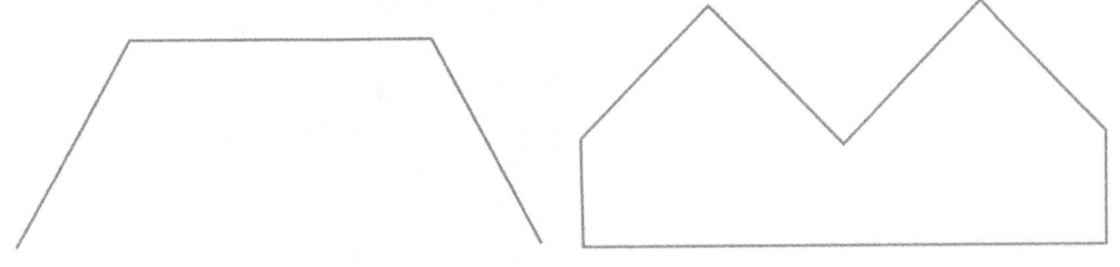

Figure 17 An open polyline markup

Figure 18 A closed polyline markup

The **Polyline** markup tool can also be used to create a rectangle or a square markup. To create a rectangle markup, invoke the **Polyline** tool and then press and hold down the left mouse button and drag to define the two opposite corners. To create a square markup, invoke this tool and then press and hold down the SHIFT key. Next, drag to define the two opposite corners. Figure 19 shows a rectangle markup added using the **Polyline** tool and Figure 20 shows a square markup added using the same tool.

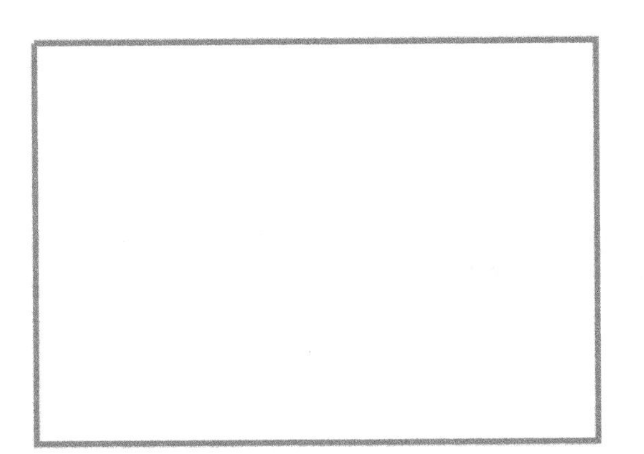

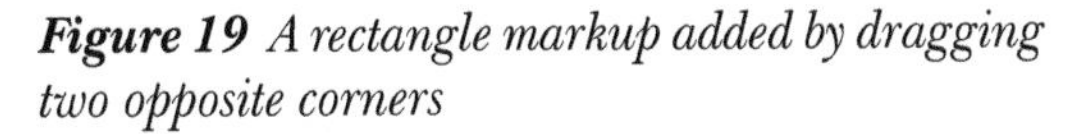

Figure 19 A rectangle markup added by dragging two opposite corners

Figure 20 A square markup added by holding down the SHIFT key and dragging two opposite corners

***Tip**: The closed polyline markups cannot be filled with a color or a hatch pattern. If you want the closed shape to be filled with a color or a hatch pattern, you need to use the **Rectangle** or **Polygon** markup tool. These tools are discussed in the later chapters.*

If a polyline markup consists of line segments that were not drawn horizontal or vertical by holding down the SHIFT key, you can convert them into arc segments. To do that, right-click on a segment that is not horizontal or vertical and select **Convert to Arc** from the shortcut menu, as shown in Figure 21. On doing so, the segment on which you right-clicked will be converted into an arc segment and control handles will be displayed at the two ends of the arc. You can use these control handles to modify the arc.

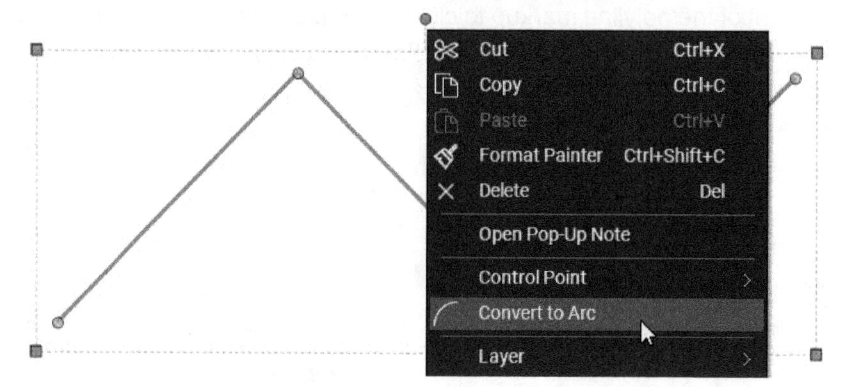

Figure 21 *Converting an inclined polyline segment into arc*

Figure 22 shows a polyline markup with three segments.

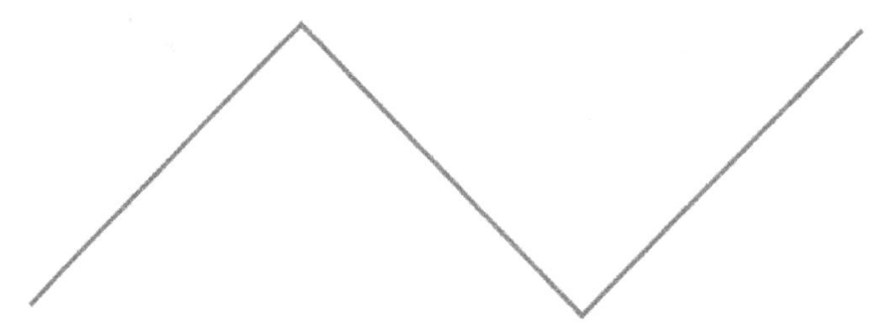

Figure 22 *A polyline markup with three segments*

Figure 23 shows the same markup with the third segment converted into an arc. In this figure, you can also see the control handles at the two endpoints of the arc segment.

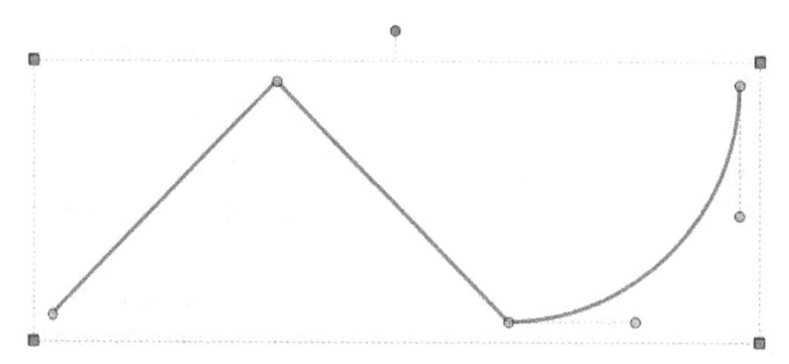

Figure 23 *The third segment of the polyline markup converted into an arc segment*

The Dimension Markup

Menu: Tools > Markup > Dimension
Keyboard Shortcut: SHIFT+L

Dimension Shift+L

This tool allows you to draw a reference dimension markup with custom text. When you invoke this tool, the cursor will change to the dimension cursor and you will be prompted to select the start and end points of the dimension. Once you have specified the two points, you can enter the custom text for the dimension markup. Note that this dimension is not going to be the true measurement between the specified points. Figure 24 shows a couple of dimension markups with custom text.

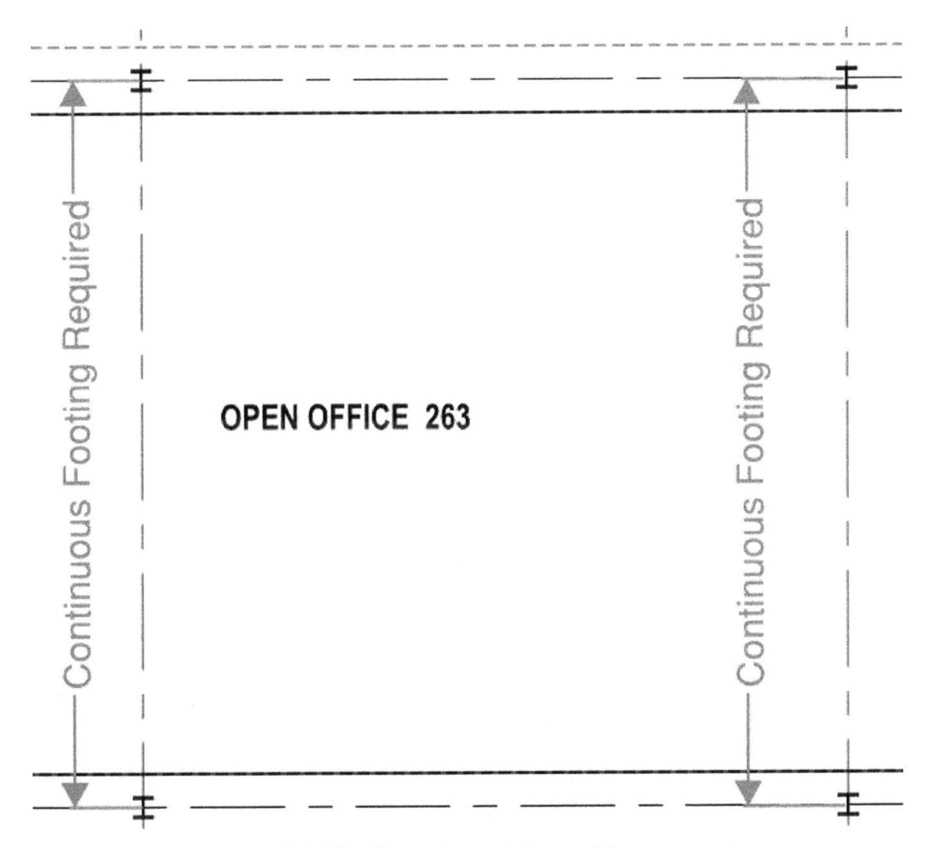

Figure 24 *The dimension markups with custom text*

Similar to the other markups, you can use the **Appearance** area of the **Properties** panel to modify the appearance properties of the dimension markup as well. If you do not want to display the text in the dimension markup, you can clear the **Show caption** check box from the **Properties** panel.

Hands-on Tutorial

In this tutorial, you will complete the following tasks:

1. *Use the pinned **Tutorial Files** folder to open a PDF file.*
2. *Add a line markup and customize it.*
3. *Create a custom line style with text and a custom line style with a vector shape.*
4. *Assign custom line styles to the line markups.*
5. *Add an arrow markup and customize it.*
6. *Add polyline markups and convert one of their segments into arc.*
7. *Add arc markups using three points.*
8. *Add dimension markups and customize them.*

Section 1: Adding a Line Markup

In this section, you will open the **Architecture_Plan_03.pdf** file from the **C03** folder. You will then add a line markup in this file and customize the properties of that markup.

1. Start Revu 2019.

2. Restore the **Revu Training** profile that you created in Chapter 01.

   This profile displays the **Properties** panel on the right side of the main workspace.

3. Invoke the **File Access** panel > **Explorer Mode**.

4. From the drop-down list at the top, select the pinned **Tutorial Files** folder; the content of this folder is displayed.

5. Double-click on the **C03** folder; the **Architecture_Plan_03.pdf** file saved in this folder is displayed.

6. Double-click on the **Architecture_Plan_03.pdf** file to open it.

   The default cursor mode is set to **Pan**. As a result, the current cursor is displayed as the pan cursor. You will first change it to the select cursor because you can pan by holding down the Scroll Wheel button on the mouse.

7. Press the **V** key to change the cursor to the select cursor.

   You will now zoom to the top left of the plan where the text **OPEN OFFICE 200** appears.

8. Navigate to the top left of the plan where the text **OPEN OFFICE 200** appears, refer to Figure 25.

9. From the **Menu Bar**, click **Tools > Markup > Line**. Alternatively, press the **L** key to invoke the **Line** markup tool; the cursor changes to the line markup cursor and you are prompted to select the start point of the line.

10. Click above the text box markup you added in the previous chapter to define the start point of the line markup, refer to Figure 25.

11. Hold down the SHIFT key and specify the endpoint of the line markup on the right of the text box markup; the line markup is added and is selected as shown in Figure 25.

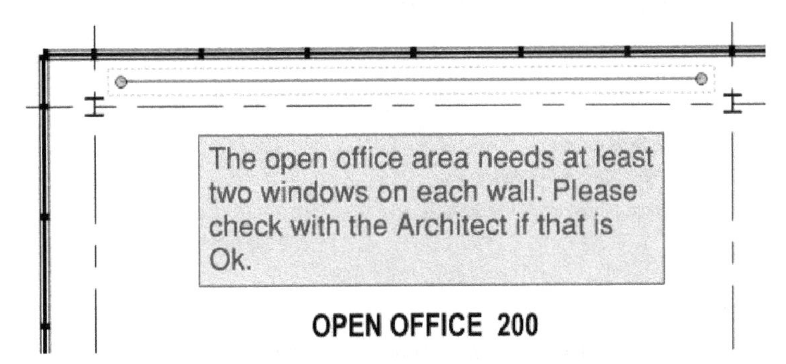

Figure 25 *The line markup added*

With the line markup still selected, the **Properties** panel shows the default properties of this markup. You will now modify its color.

12. In the **Properties** panel, click the **Color** swatch and select the **Blue** color; the color of the line markup changes to Blue.

Section 2: Creating Custom Line Styles

In this section, you will create two custom line styles. The first line style will have text in it and the second line style will have a vector shape.

1. If the line markup is not already selected, click on it to select; the **Properties** panel displays the properties of this markup.

2. Click on the **Style** drop-down list; all the default line styles are displayed.

3. Scroll down to the bottom and click **Manage**; the **Manage Line Style Sets** dialog box is displayed.

   As mentioned earlier in this chapter, creating custom line styles is a two step process. In the first step, you need to create a line style set, which is like a group. In the second step, you need to create the line styles that will be a part of the line style set. You will now create a line style set with the name **Company ABC Line Styles**.

4. In the **Manage Line Style Sets** dialog box, click the **Add** button from the lower right; the **Add Line Style Set** dialog box is displayed.

5. From the **Step 1: Select Type** area, make sure **New** is selected.

6. In the **Step 2: Select Location** area > **Title** edit box, enter **Company ABC Line Styles**; a custom line style set file is created with the name you defined and extension **blx** and is displayed in the **Location** display box show, as shown in Figure 26.

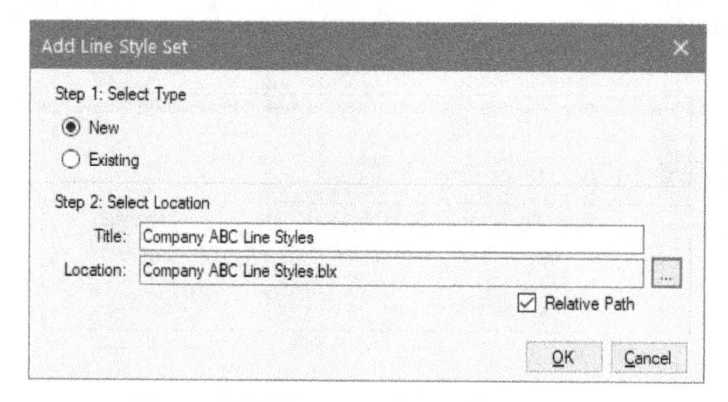

***Figure 26** The custom line style set added*

The **Relative Path** check box is selected by default. This ensures that this line style set file is saved in the same path as that of the custom profile you created in Chapter 01.

7. Click **OK** in the dialog box; the **Edit Line Style Set** dialog box is displayed.

   This is the dialog box in which you will now create a new line style.

8. In the **Edit Line Style Set** dialog box, click **Add** from the lower right corner; the **Add Line Style** dialog box is displayed.

9. Enter **Wall Socket** as the name of the line style and click **OK** in the **Add Line Style** dialog box; the **Line Style Editor** dialog box is displayed to define the new line style.

   The line style is defined by adding various components using the **+** button available on the left of the **Components** area in the **Line Style Editor** dialog box. You will start by adding a dash component, then space, followed by the text, and then another space.

10. Click the **+** button on the left of the **Components** area; the **Add Component** dialog box is displayed.

11. From the **Type** drop-down list in the **Add Component** dialog box, select **Dash**, as shown in Figure 27.

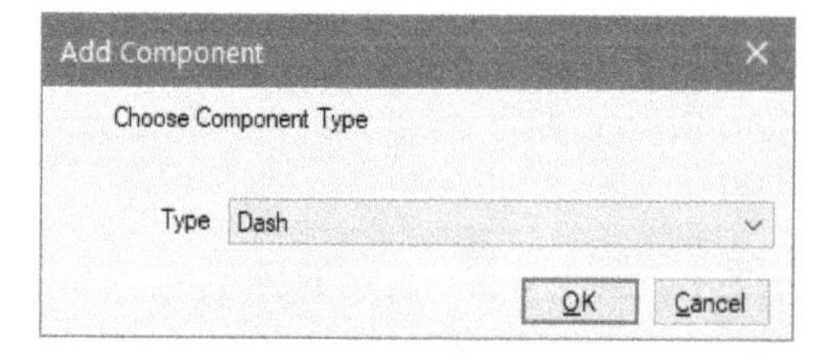

***Figure 27** Selecting the **Dash** component type*

12. Click **OK** in the dialog box; you are returned to the **Line Style Editor** dialog box and the **Dash** component is displayed in the **Components** area.

    Next, you will change the width of this dash component.

13. Change the value in the **Width** spinner to **5**. Do not press ENTER.

*Tip: Pressing the ENTER key after entering the value in the **Width** spinner will close the **Line Style Editor** dialog box. In that case, you can double-click on the **Wall Socket** line style in the **Edit Line Style Set** dialog box to return to the **Line Style Editor** dialog box.*

    Next, you will add a space component.

14. Click the + button on the left of the **Components** area; the **Add Component** dialog box is displayed.

15. From the **Type** drop-down list in the **Add Component** dialog box, select **Space**.

16. Click **OK** in the dialog box; you are returned to the **Line Style Editor** dialog box and the **Space** component is displayed below the **Dash** component in the **Components** area. Also, the preview of the line style changes to the dashed line style.

    You will leave the default width of this component and now add the text component.

17. Click the + button on the left of the **Components** area; the **Add Component** dialog box is displayed.

18. From the **Type** drop-down list in the **Add Component** dialog box, select **Text**.

19. Click **OK** in the dialog box; you are returned to the **Line Style Editor** dialog box and the **Text** component is displayed below the **Space** component in the **Components** area. Also, the **Text** area is displayed on the right of the **Components** area.

    You will now change the text in the **Text** box to **WALL SOCKETS**.

20. In the **Text** box, type **WALL SOCKETS**; the preview of the line shows this text.

21. In the **Font Size** drop-down list, change the size of the text to **6**.

    Next, you will add another space component.

22. Click the + button on the left of the **Components** area; the **Add Component** dialog box is displayed.

23. From the **Type** drop-down list in the **Add Component** dialog box, select **Space**.

24. Click **OK** in the dialog box; you are returned to the **Line Style Editor** dialog box and the **Space** component is displayed below the **Text** component in the **Components** area.

Next, you will add the description of the line style.

25. In the **Description** edit box, enter the following text:

    This line style is used to specify the areas on the wall where the wall sockets are to be installed.

26. Click the **Scale with Line Thickness** check box. This ensures the line style is automatically scaled if you change the thickness of the line markup. The **Line Style Editor** dialog box, after adding this information, is shown in Figure 28.

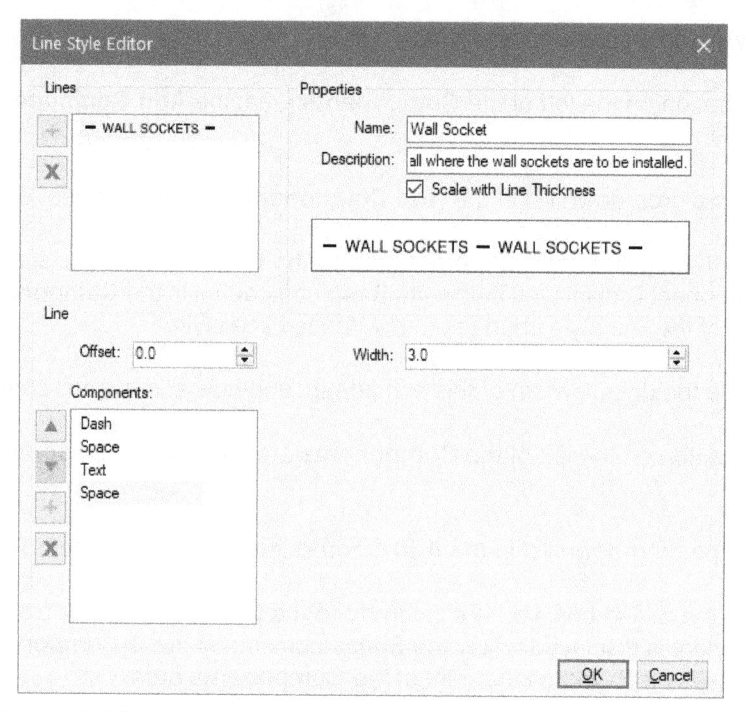

*Figure 28 The **Line Style Editor** dialog box with the text line style defined*

This completes the creation of the line style with text.

27. Click **OK** in the **Line Style Editor** dialog box; the line style is created and you are returned to the **Edit Line Style Set** dialog box. Note the new line style and its description listed in this dialog box, as shown in Figure 29.

    Next, you will create a line style with a vector to represent windows.

28. In the **Edit Line Style Set** dialog box, click **Add**; the **Add Line Style** dialog box is displayed.

29. Enter **Window Line** as the name of the line style in the **Add Line Style** dialog box.

30. Click **OK** in the dialog box; the **Line Style Editor** dialog box is displayed.

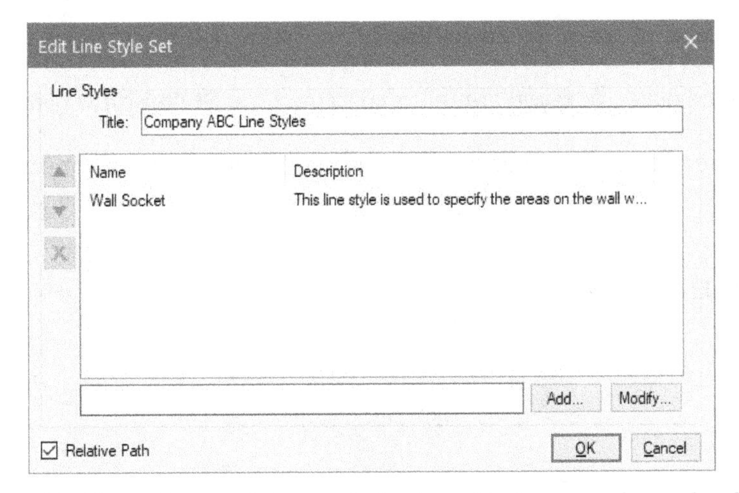

Figure 29 The ***Edit Line Style Set*** *dialog box with the custom line style added*

You will start with adding a dash, followed by a space, then you will draw the vector representing the window, and then end it with another space.

31. Click the + button on the left of the **Components** area; the **Add Component** dialog box is displayed.

32. From the **Type** drop-down list, select **Dash**.

33. Click **OK** in the dialog box; you are returned to the **Line Style Editor** dialog box and the **Dash** component is displayed in the **Components** area.

    Next, you will change the width of this dash component.

34. Change the value in the **Width** spinner to **5**. Do not press ENTER.

    Next, you will add a space component.

35. Click the + button on the left of the **Components** area; the **Add Component** dialog box is displayed.

36. From the **Type** drop-down list, select **Space**.

37. Click **OK** in the dialog box; you are returned to the **Line Style Editor** dialog box and the **Space** component is displayed below the **Dash** component in the **Components** area. Also, the preview of the line style changes to the dashed line style.

    You will leave the default width of this component and now add the vector component.

38. Click the + button on the left of the **Components** area; the **Add Component** dialog box is displayed.

39. From the **Type** drop-down list, select **Vector**.

40. Click **OK** in the dialog box; you are returned to the **Line Style Editor** dialog box.

    Notice the area to draw the vector displayed on the right of the **Components** area. The **Line** tool in this area is active by default, which allows you to draw lines to define the vector shape.

41. Press and hold down the SHIFT key and draw a shape similar to the one shown in Figure 30.

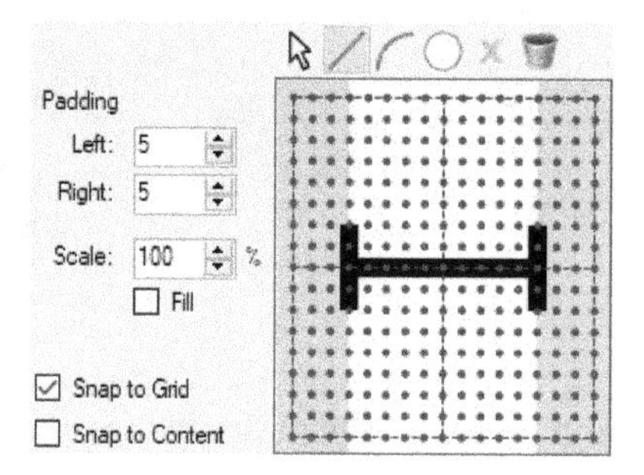

Figure 30 *Drawing the vector shape for the line style*

42. Change the value in the **Scale** spinner to 150.

    The padding values ensure that there is space on both sides of the vector. You will leave the padding values to the default value of **5**. Next, you will add a space component on the right of the vector.

43. Click the **+** button on the left of the **Components** area; the **Add Component** dialog box is displayed.

44. From the **Type** drop-down list, select **Space**.

45. Click **OK** in the dialog box; you are returned to the **Line Style Editor** dialog box. The preview of the line style now shows the vector line style you created.

46. In the **Description** edit box, enter the following text:

    This line style is used to specify the areas on the wall where windows are to be placed.

47. Click the **Scale with Line Thickness** check box. This ensures the line style is automatically scaled if you change the thickness of the line markup. The **Line Style Editor** dialog box, after adding this information, is shown in Figure 31.

48. Click **OK** in the **Line Style Editor** dialog box; you are returned to the **Edit Line Style Set** dialog box where the two line styles are listed, as shown in Figure 32.

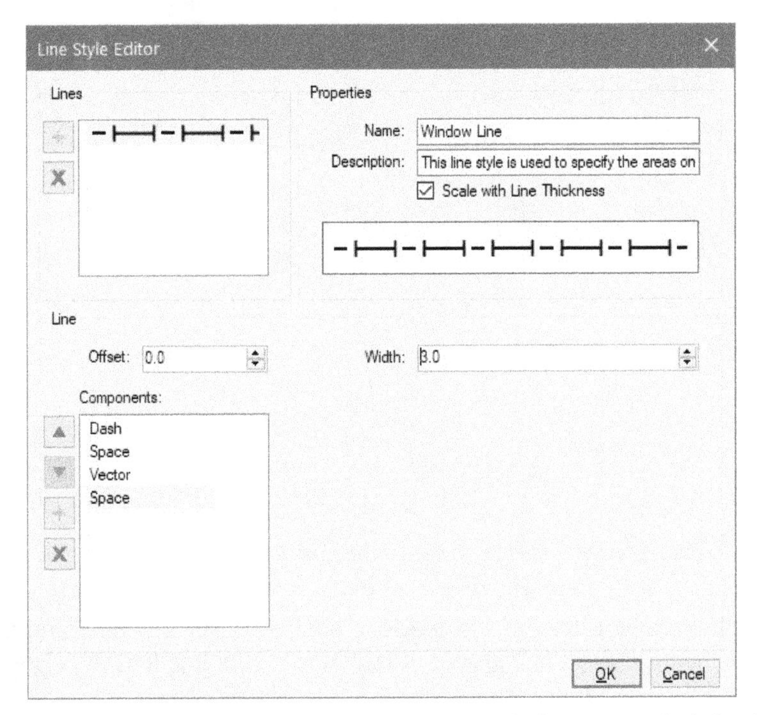

*Figure 31 The **Line Style Editor** dialog box with the vector style defined*

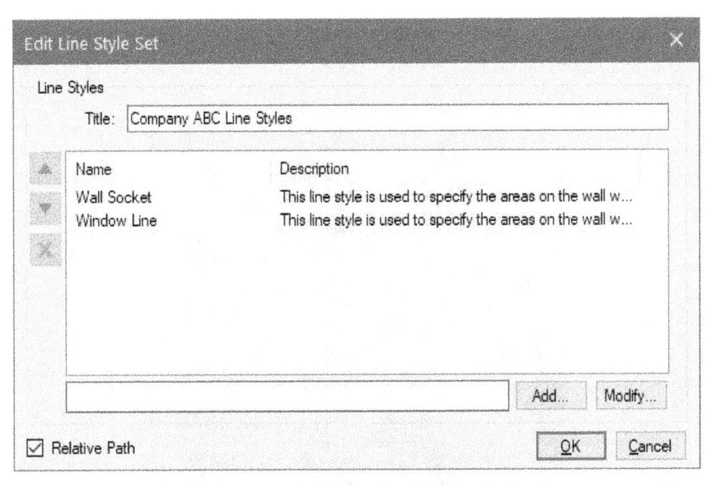

*Figure 32 The **Edit Line Style Set** dialog box with the two custom line styles*

49. Click **OK** in the **Line Style Editor** dialog box; you are returned to the **Manage Line Style Sets** dialog box that now shows the custom line style set that you created, as shown in Figure 33.

50. Click **OK** in the **Manage Line Style Sets** dialog box; you are returned to the PDF file.

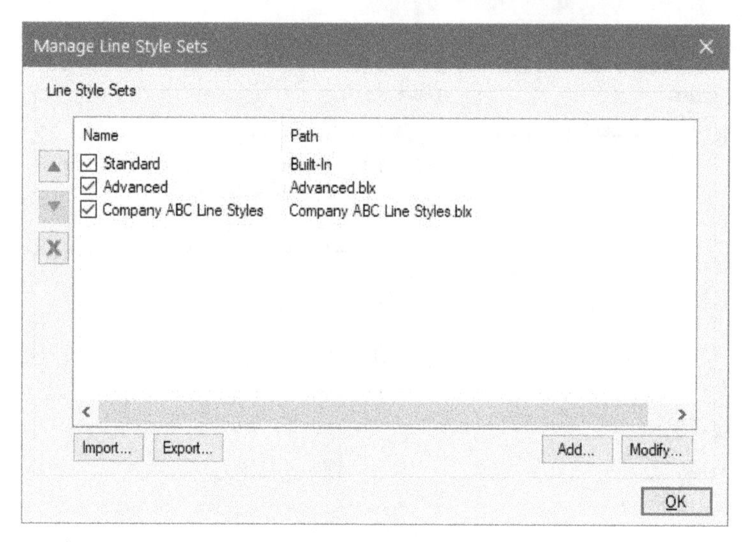

Figure 33 The ***Manage Line Style Sets*** *dialog box with the custom line style set added*

Notice that the line you drew in the previous section is still selected and the **Properties** panel shows the properties of this line. You will now assign the text line style you created to this selected line.

51. In the **Properties** panel, click the **Style** drop-down list.

52. Scroll down to the bottom of this drop-down list; the two custom line styles you created are listed under the **Company ABC Line Styles** section, as shown in Figure 34.

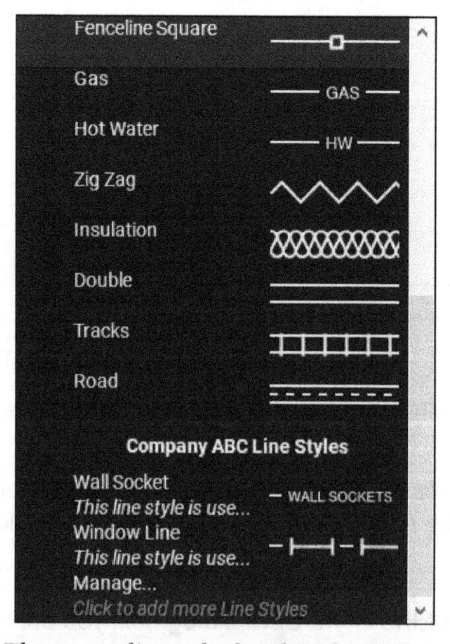

Figure 34 The custom line styles listed in the ***Style*** *drop-down list*

53. Click the **Wall Socket** line style; the style of the line you drew is changed and it now shows the text inside the line, as shown in Figure 35.

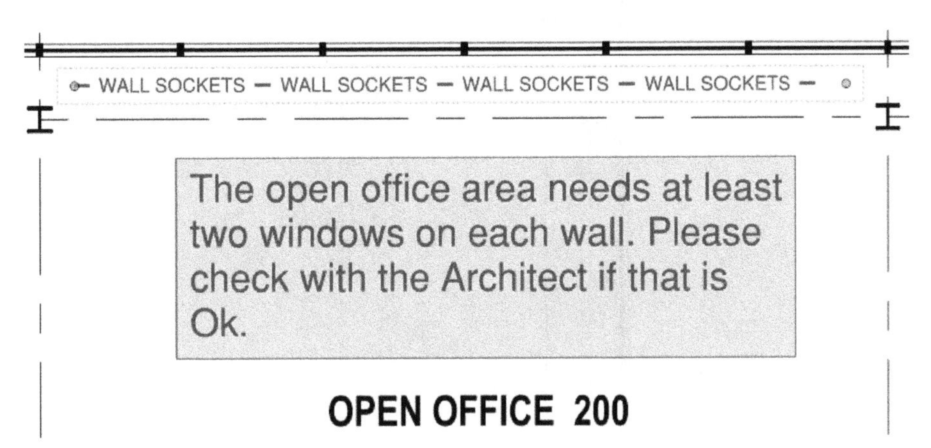

Figure 35 *The line style of the selected line changed to the custom line style*

Tip: *If need be, you can hold down the SHIFT key and drag the right control handle of the line to resize so it appears similar to the one shown in the figure above.*

Next, you will draw a vertical line on the left of the **OPEN OFFICE 200** text and then change its line style to the vector style you created earlier.

54. Press the **L** key to invoke the **Line** markup tool; the cursor changes to the line markup cursor and you are prompted to select the start point of the line.

    Note that because you used the default **Line** markup tool, the line that you will draw will have the default **Appearance** properties.

55. Specify the start point of the line by clicking on the left of the **OPEN OFFICE 200** text, refer to Figure 36.

56. Hold down the SHIFT key and draw a vertical line on the left of the **OPEN OFFICE 200** text, refer to Figure 36.

    With the line still selected, the **Properties** panel displays its default properties. You will now change the color and style of this line.

57. Click on the **Color** swatch and select **Dark Green** to change the color of the line.

58. Click on the **Style** drop-down list and scroll down to the bottom.

59. Select the **Window Line** as the style of this line; the appearance of the line changes in the PDF file, as shown in Figure 36.

60. Press the ESC key to deselect the line.

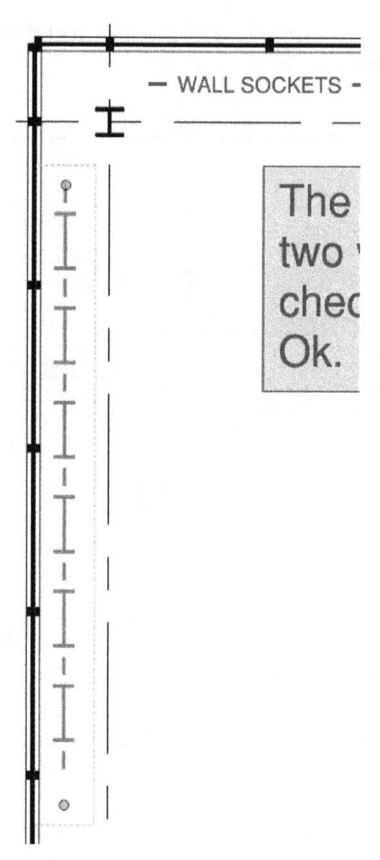

Figure 36 The style of the vertical line changes to the custom vector style

As mentioned in the previous chapter, every time you customize a tool in Revu, the custom version of that tool is added to the **Recent Tools** tool set. In your case, the **Line** tools with both custom lines are listed in the **Recent Tools** tool set. You will now invoke these tools to draw lines in the other three open office areas in the plan.

61. Navigate to the bottom left corner of the plan where the text **OPEN OFFICE 244** appears.

62. From the panel bar on the left of the main workspace, invoke the **Tool Chest** panel; various tool sets are displayed in this panel.

    The custom tools in the **Recent Tools** tool set are currently in the **Properties** mode. This is evident by the icon of these tools that are similar to the original **Line** markup icon. You first need to convert these tools into the **Drawing** mode so you can add the exact same copy of the line markups that you added earlier.

63. Double-click on the two **Line** tools on the **Recent Tools** tool set; both markups are changed to the **Drawing** mode.

64. Click on the horizontal line markup tool in the **Recent Tools** tool set; the preview of the exact same copy of the previous wall socket line is attached to the cursor.

65. Place the line markup below the text box markup you added in the previous chapter, refer to Figure 37.

66. Similarly, click on the vertical line markup tool in the **Recent Tools** tool set and place the line markup on the left of the text box markup, as shown in Figure 37.

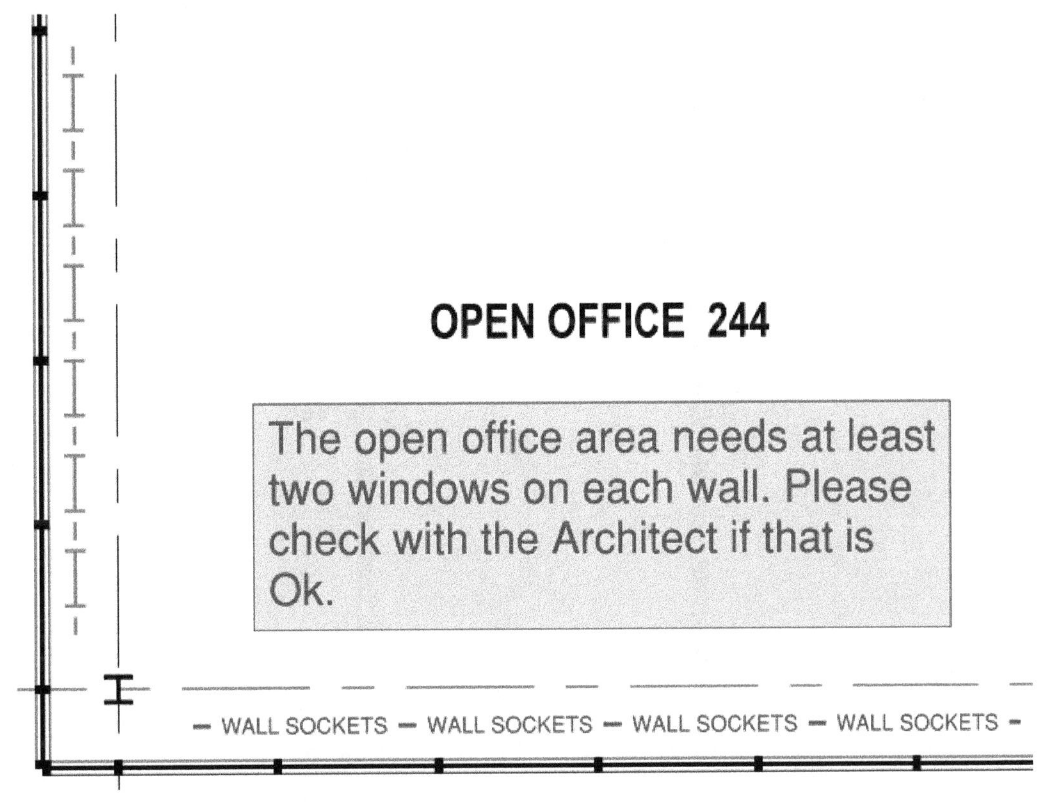

Figure 37 The two custom line markups added to the PDF

67. Similarly, add these two custom line markups to the other two open office spaces on the right side of the plan.

68. Save the PDF file.

Section 3: Adding an Arrow Markup

In this section, you will create an arrow markup and then customize its properties. You will then write a typewriter text on top of the arrow markup.

1. Navigate to the center of the plan where the **LOBBY 256** text appears.

2. From the **Menu Bar**, click **Tools > Markup > Arrow**. Alternatively, press the **A** key to invoke the **Arrow** markup tool; the cursor changes to the arrow markup cursor and you are prompted to specify the start point of the arrow.

3. Click below the **LOBBY** text to specify the start point of the arrow, refer to Figure 38; the arrow markup starts from the specified point and the arrowhead is attached to the cursor.

4. Press and hold down the SHIFT key and move the cursor down; a vertical arrow markup is being added.

5. Click above the centerline to place the arrow markup, as shown in Figure 38.

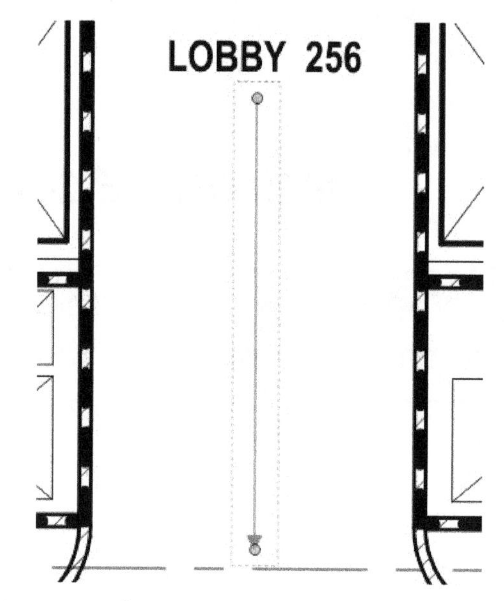

Figure 38 *The arrow markup with default properties*

With the arrow markup still selected, you will now modify the **Appearance** properties of this markup.

6. From the **Properties** panel, click on the **Color** swatch and select **Blue**; the arrow line changes to Blue.

7. Click on the **Fill Color** swatch and select **Blue**; the fill color of the arrowhead changes to Blue.

8. Change the value in the **Line Width** spinner to **2**; the arrow markup is modified and the arrowhead is scaled, as shown in Figure 39.

Next, you will add a typewriter markup and then modify its **Appearance** properties.

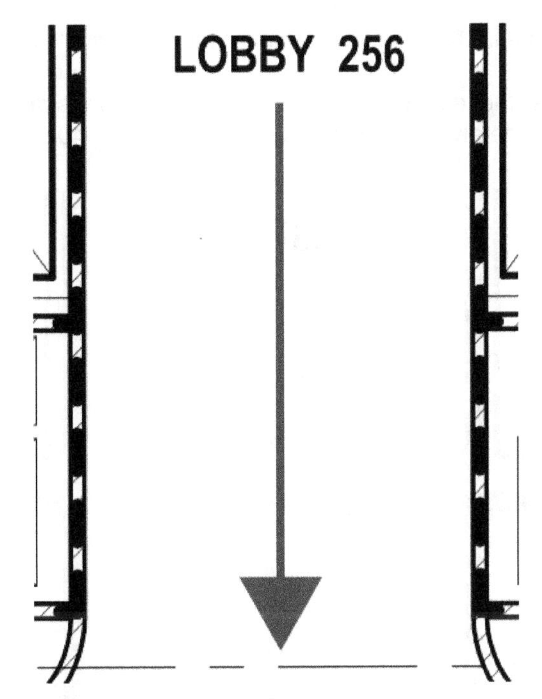

Figure 39 *The* ***Appearance*** *properties of the arrow markup modified*

9. Press the **W** key to invoke the **Typewriter** markup tool; the cursor changes to the typewriter cursor and you are prompted to select the location of the typewritten text.

   You can specify the location of the text anywhere on the left of the arrow. Once you write the text and modify its properties, you will rotate it and move it above the arrow markup.

10. Click on the left of the arrow markup to specify the start point of the text; the realtime text editor is displayed to enter text.

11. Enter the following text:

    SLOPE DIRECTION

12. Press the ESC key to exit the **Typewriter** markup tool.

    You will now change the color of the typewriter markup to Blue.

13. With the typewriter markup still selected, click on the **Text Color** swatch in the **Properties** panel and select **Blue**; the text color changes to Blue.

    Next, you will change the rotation of the markup to **270** in the **Layout** area of the **Properties** panel.

14. Scroll down in the **Properties** panel and expand the **Layout** area, if it is not already expanded.

15. Enter **270** as the value in the **Rotation** spinner; the typewriter markup is rotated.

16. Hover the cursor over the text and when the four-sided arrow is displayed, drag the text and place it on the left of the arrow markup, as shown in Figure 40.

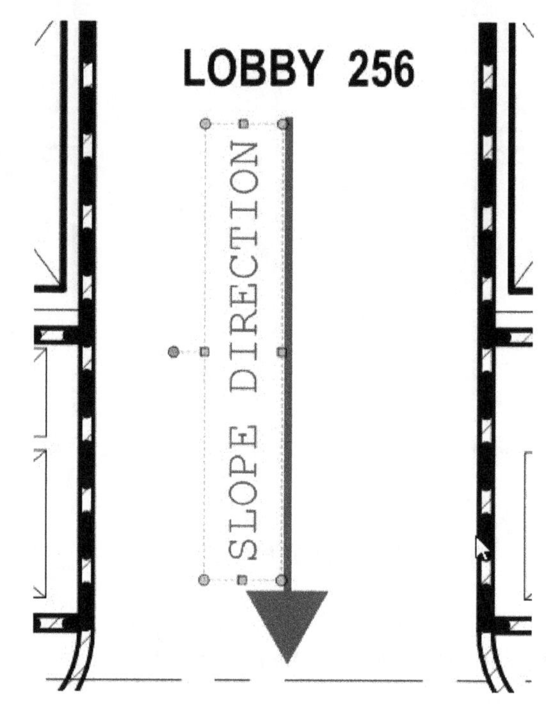

Figure 40 *The typewriter text moved above the arrow markup*

17. Press the ESC key to deselect the typewriter markup.

18. Save the PDF file.

Section 4: Adding Polyline Markups

In this section, you will add a polyline markup and then change its properties. You will then use the custom version of the polyline markup from the **Recent Tools** tool set to add another markup.

1. Navigate to **OPEN OFFICE 248** towards the middle center of the plan, refer to Figure 41.

2. From the **Tools** menu, click **Markup** > **Polyline** to invoke the **Polyline** markup tool; the cursor changes to the polyline markup cursor and you are prompted to select vertices of the polyline.

3. Click on the point labeled as **1** in Figure 41 to specify the start point of the polyline markup.

4. Hold down the shift key and click on the point labeled as **2** in Figure 41 to specify the second point of the polyline.

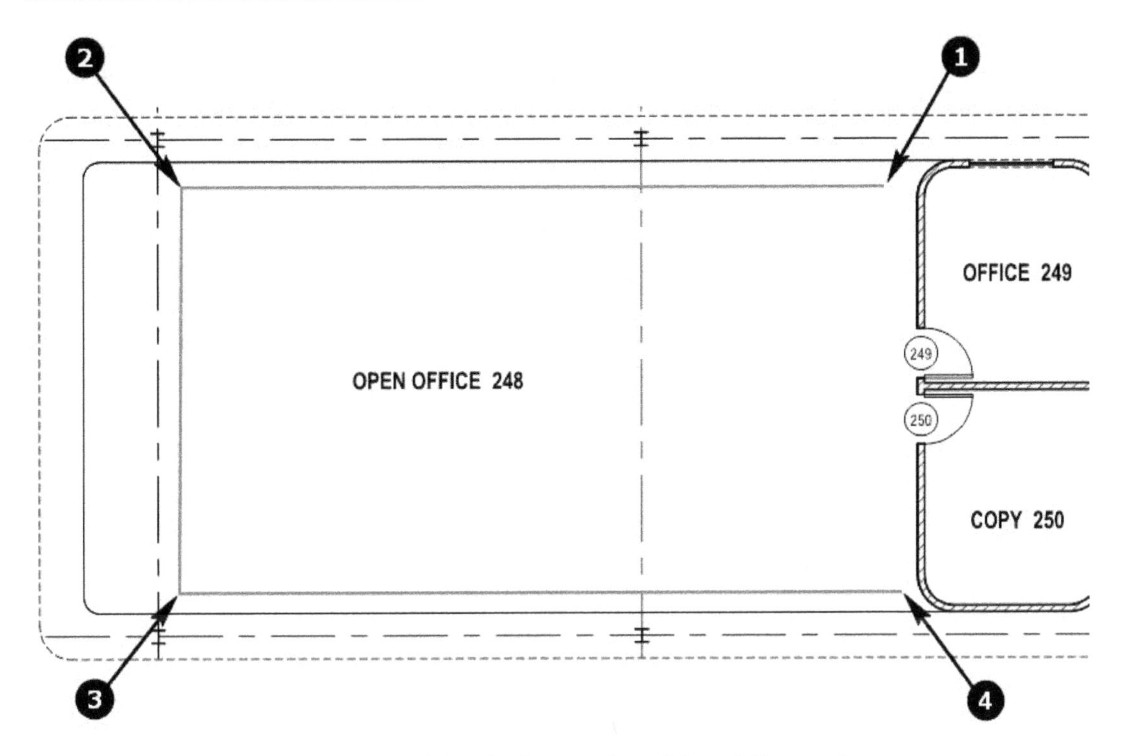

Figure 41 Specifying the four vertices of the polyline markup

To define the next point, you will release the SHIFT key as the next segment should not be a straight line. This is because you will later convert this segment into an arc segment.

5. Release the SHIFT key and then click on the point labeled as **3** in Figure 41 to define the next point of the polyline. Note that if you do not release the SHIFT key, you will not be able to convert this segment into an arc segment later on.

6. Hold down the shift key and click on the point labeled as **4** in Figure 41 to specify the last point of the polyline.

7. Press the ENTER key to create the polyline markup and exit the tool; the polyline markup is still selected.

   You will now modify the **Appearance** properties of this polyline.

8. In the **Properties** panel, click on the **Color** swatch and select **Blue** to change the color of the polyline to Blue.

9. In the **Line Width** spinner, enter **3** as the value.

   Next, you will change the left segment into an arc segment.

10. Right-click on the left segment of the polyline; a shortcut menu is displayed.

11. From the shortcut menu, click **Convert to Arc**; the left segment is converted into the arc segment and the control handle point is displayed at the center of this segment. Note that if the **Convert to Arc** option is grayed out, that means the segment is a straight line segment You need to drag one of the endpoints to make sure it is not a straight segment.

12. Press and drag this control handle to the left to change the shape of the arc segment, as shown in Figure 42. Once the shape is similar to the one shown in this figure, release the control handle. The arc segment, after the modification, is shown in Figure 43.

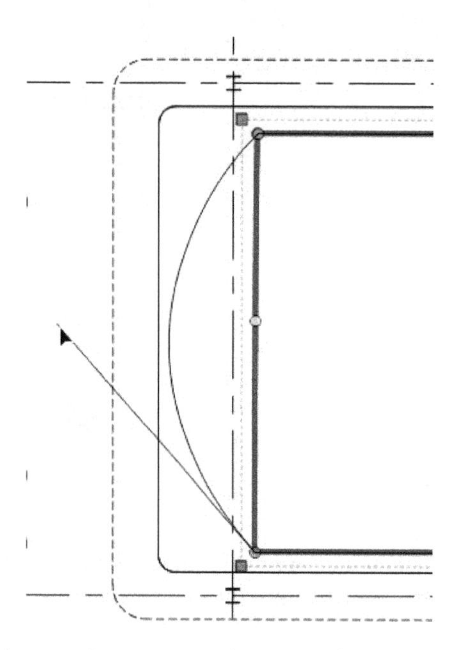

Figure 42 *Dragging the control handle of the arc segment*

Figure 43 *The arc segment after the modification*

13. Press the ESC key to deselect the polyline markup.

14. Use the custom version of the **Polyline** markup to draw a similar markup around the **OPEN OFFICE 263** text, which is on the other side of the plan.

15. Convert the right segment of the polyline into an arc segment, similar to the one shown in Figure 44.

16. Save the PDF file.

Section 5: Adding Arc Markups using Three Points

In this section, you will add the arc markups using three points. Remember that by default, the **Arc** markup tool adds the markup using two points. For this tool to add a three point arc, you need to hold down the ALT key before you specify the first point. Also, to draw this arc, you need to snap to the endpoints of the walls in the plan. Therefore, you need to ensure the **Snap to Content** option is turned on from the status bar.

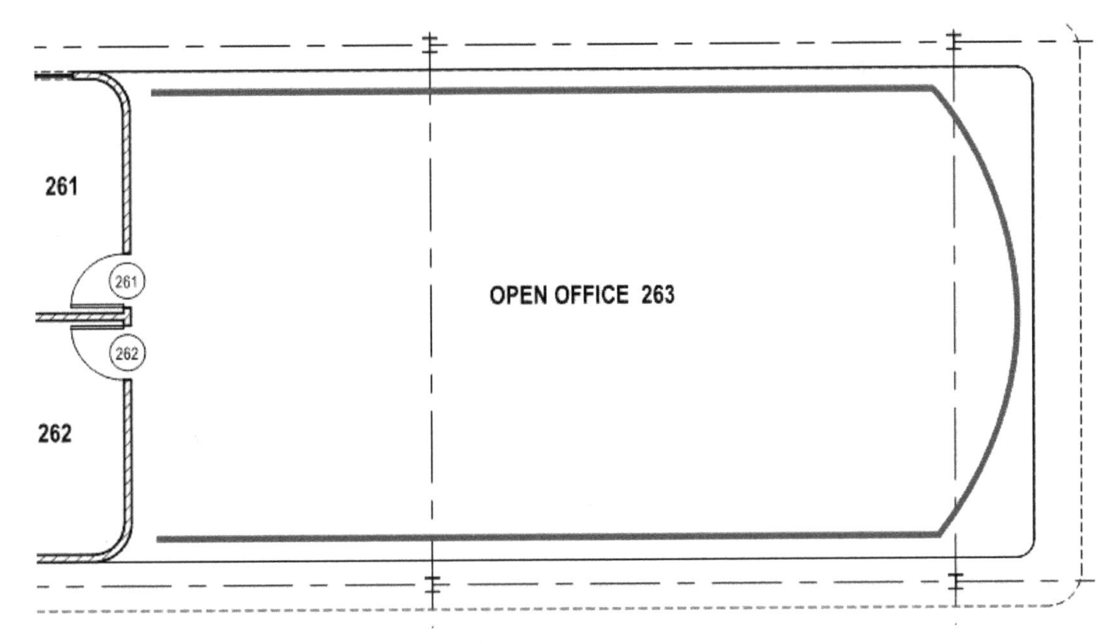

Figure 44 The polyline markup added

1. From the status bar available below the main workspace, make sure the **Snap to Content** option is turned on, as shown in Figure 45.

Figure 45 The ***Snap to Content*** *option on the status bar*

2. Navigate to the middle left of the plan, refer to Figure 46.

3. From the menu bar, click **Tools** > **Markup** > **Arc**; the **Arc** markup tool is invoked and you are prompted to select the region to place the arc.

   As mentioned earlier, you need to hold down the ALT key for this tool to draw the arc using three points.

4. Hold down the ALT key and snap to the point labeled as **1** in Figure 46 as the start point of the arc.

5. Release the ALT key and click on the point labeled as **2** in Figure 46 as the second point of the arc.

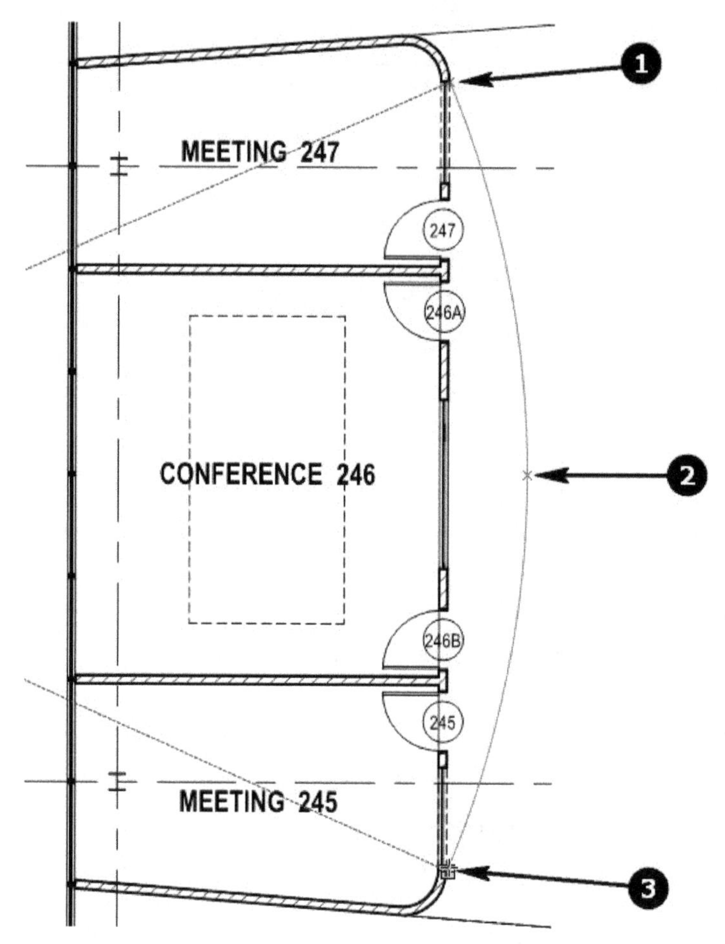

Figure 46 Specifying the three points of the arc

6. Snap to the point labeled as **3** in Figure 46 to define the third point of the arc; the arc markup is drawn and the **Properties** panel displays the properties of this arc.

   Next, you will change the **Appearance** properties of this arc markup.

7. In the **Properties** panel > **Appearance** area, click the **Color** swatch and select **Blue**; the arc color is changed to Blue.

8. In the **Line Width** spinner, enter **2** as the value.

9. Click on the **Style** drop-down list and scroll to the bottom to the **Company ABC Line Styles** section.

   This section displays the line styles you created in the previous chapter. You will now assign the **Window Line** style to the arc.

10. Select the **Window Line** style; the line style of the arc is changed to this line style.

11. Press the ESC key to deselect the arc markup. The plan, after adding and customizing the arc markup, is shown in Figure 47.

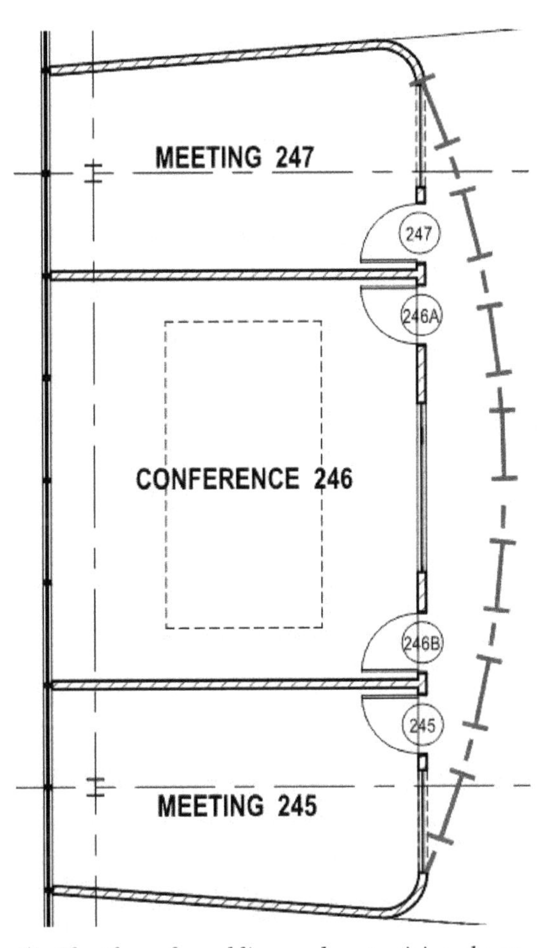

Figure 47 The plan after adding and customizing the arc markup

12. Use the custom version of the **Arc** tool from the **Recent Tools** tool set to add another arc to the other side of the plan around the **CONFERENCE 265** room.

***Tip**: As mentioned in the previous chapter, you can click on the cogwheel on the top right of the **Recent Tools** tool set and select **Properties Mode** to ensure all the custom tools are automatically added to this tool set in the **Properties Mode**.*

13. Double-click the Scroll Wheel mouse button to zoom to the extents of the plan.

14. Save the PDF file.

Section 6: Adding a Dimension Markup

In this section, you will add a dimension markup and then modify its properties.

1. Navigate to the bottom center of the plan, refer to Figure 48.

2. From the menu bar, click **Tools** > **Markups** > **Dimension**; the cursor changes to the dimension cursor and you are prompted to select the start point of the markup.

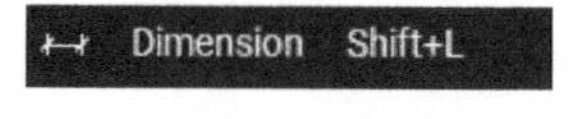

3. Snap to the point labeled as **1** in Figure 48 to define the start point of the dimension markup.

4. Snap to the point labeled as **2** in Figure 48 to define the endpoint of the dimension markup; the dimension line is added and the text editor is displayed on top of it to add content to this markup.

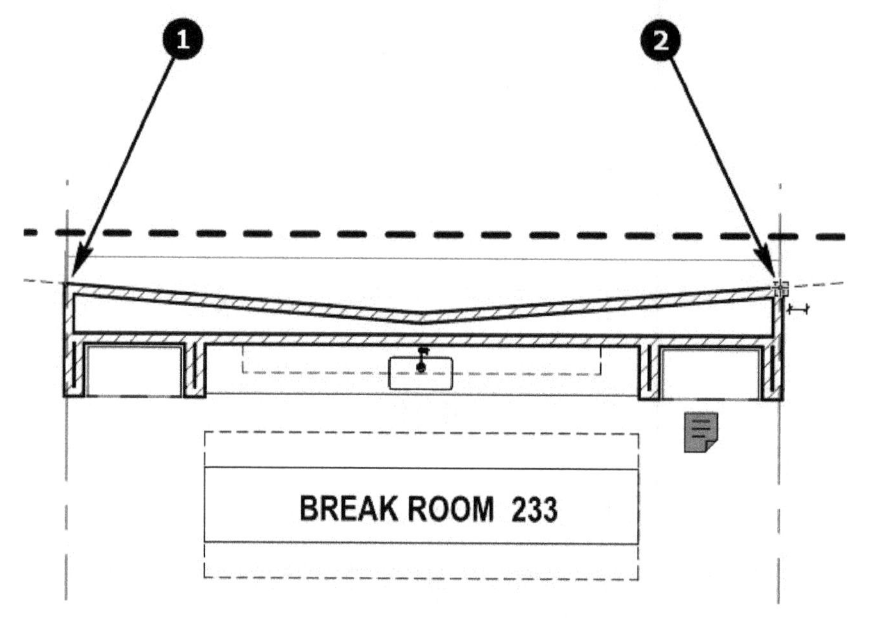

***Figure 48** Specifying the start and end points of the dimension markup*

5. Enter the following text:

   FIRE RATED WALL

6. Click somewhere outside the dimension markup to end the markup tool; the text is added to the dimension markup.

   The dimension markup is still selected. You will now modify the **Appearance** properties of this markup.

7. In the **Properties** panel > **Appearance** area, click the **Color** swatch and select **Blue** to change the color of the dimension line to Blue.

8. Click the **Fill Color** swatch and select **Blue Violet** to change the fill color of the arrowheads.

9. Select the **Highlight** check box on the right of the **Fill Color** swatch. This ensures that any content of the PDF file hidden behind the arrowheads is visible.

10. Click the **Text Color** swatch and select **Blue** to change the text color to Blue.

    The dimension markup is sitting too close to the wall. You will drag it up so it sits above the dashed line of the PDF file. Note that to ensure the endpoints of the dimension do not move, you need to drag the Yellow circle at the center of the dimension text.

11. Hover the cursor over the Yellow circle on the dimension text; the cursor changes to a small arrow cursor, which is the move text cursor.

12. Press and drag the dimension text above the dashed line of the PDF file, refer to Figure 49.

13. Press the ESC key. The plan, after adding and customizing the dimension markup, is shown in Figure 49.

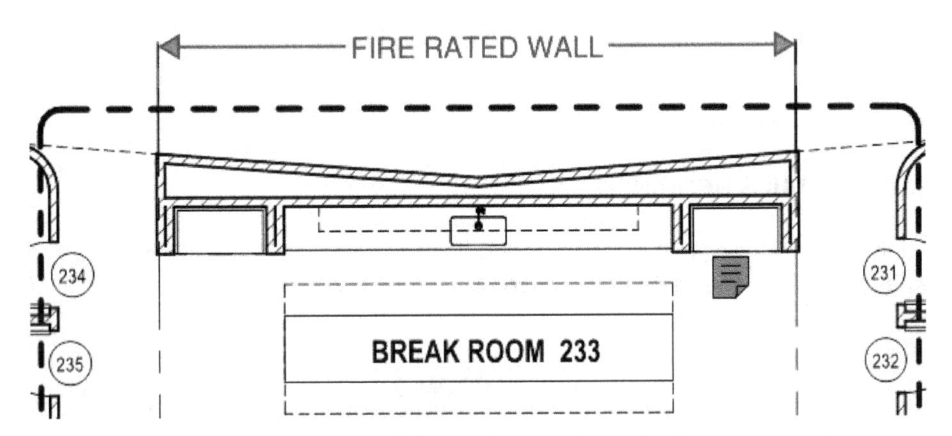

Figure 49 *The plan after adding and customizing the dimension markup*

Next, you will place another dimension markup between the columns displayed below the **BREAK ROOM 233** text. This markup will be added using the custom **Dimension** tool in the **Recent Tools** tool set.

14. Pan down to the sectioned columns displayed below the **BREAK ROOM 233** text, refer to Figure 50.

    The direction of defining the start and end points of the dimension markup is saved with the custom settings. In the dimension markup that you will add now, the text needs to be placed below the selection points. Therefore, you will define the dimension markup location points from right to left.

15. From the **Recent Tools** tool set, invoke the custom **Dimension** tool; the cursor changes to the dimension cursor.

16. Snap to the point labeled as **1** in Figure 50 to define the start point of the dimension markup.

17. Snap to the point labeled as **2** in Figure 50 to define the endpoint of the dimension markup; the dimension line is added and the text editor is displayed on top of it to add content to this markup.

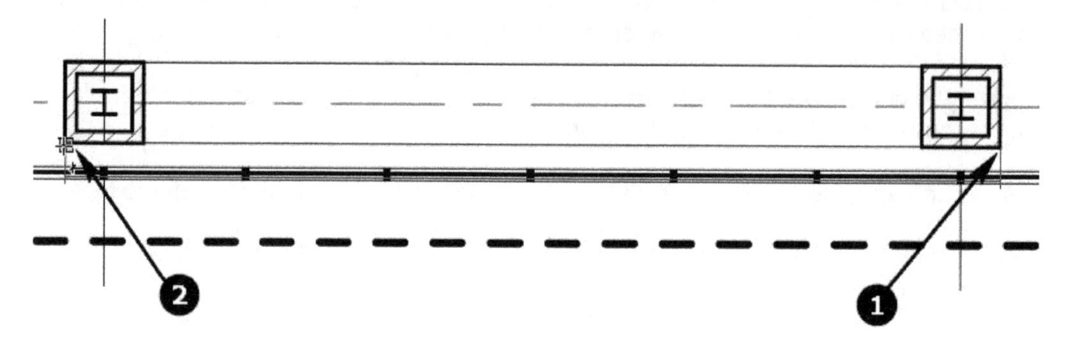

Figure 50 *Specifying the start and end points of the dimension markup*

18. Enter the following text:

COLUMN DETAILS REQUIRED

19. Press the ESC key; the text is added to the dimension markup and the markup is no more selected. The plan, after adding this markup, is shown in Figure 51.

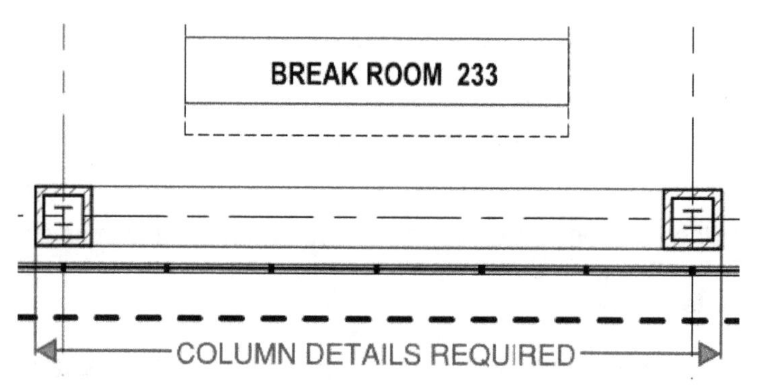

Figure 51 *The custom dimension markup added to the plan*

20. Double-click the Scroll Wheel mouse button to zoom to the extents of the PDF file.

21. Save the PDF file and close it.

Skill Evaluation

Evaluate your skills to see how many questions you can answer correctly. The answers to these questions are given at the end of the book.

1. The line markups added to a PDF file can be assigned custom line styles. (True/False)

2. The line style of the arrow markup cannot be modified. (True/False)

3. The arrowheads of the dimension markups cannot be changed. (True/False)

4. A horizontal polyline segment added by holding down the SHIFT key cannot be converted into an arc segment. (True/False)

5. You can only add arc markups using two points. (True/False)

6. Which tool allows you to add a dimension markup with custom text?

   (A) **Text** (B) **Line**
   (C) **Leader** (D) **Dimension**

7. Which tool allows you to add multiple end-connected lines?

   (A) **Polyline** (B) **Line**
   (C) **Multiline** (D) **Dimension**

8. To create the arc markup using three points, which key needs to be held down before specifying the first point?

   (A) **CTRL** (B) **SHIFT**
   (C) **TAB** (D) **ALT**

9. Which drop-down list in the **Properties** panel is used to assign a custom line style to a line or a polyline markup?

   (A) **Linetype** (B) **Color**
   (C) **Width** (D) **Style**

10. Which tool is used to add arrow markups?

    (A) **Leader** (B) **Arrow**
    (C) **Text** (D) **Multiline**

Class Test Questions

Answer the following questions:

1. Explain briefly the process of adding a custom line style with text.
2. How will you place a dimension markup?
3. Explain briefly the process of converting a polyline segment into an arc segment.
4. Explain briefly the process of drawing an arc markup using three points.
5. Explain briefly the process of changing the line style of a polyline markup to a custom line style.

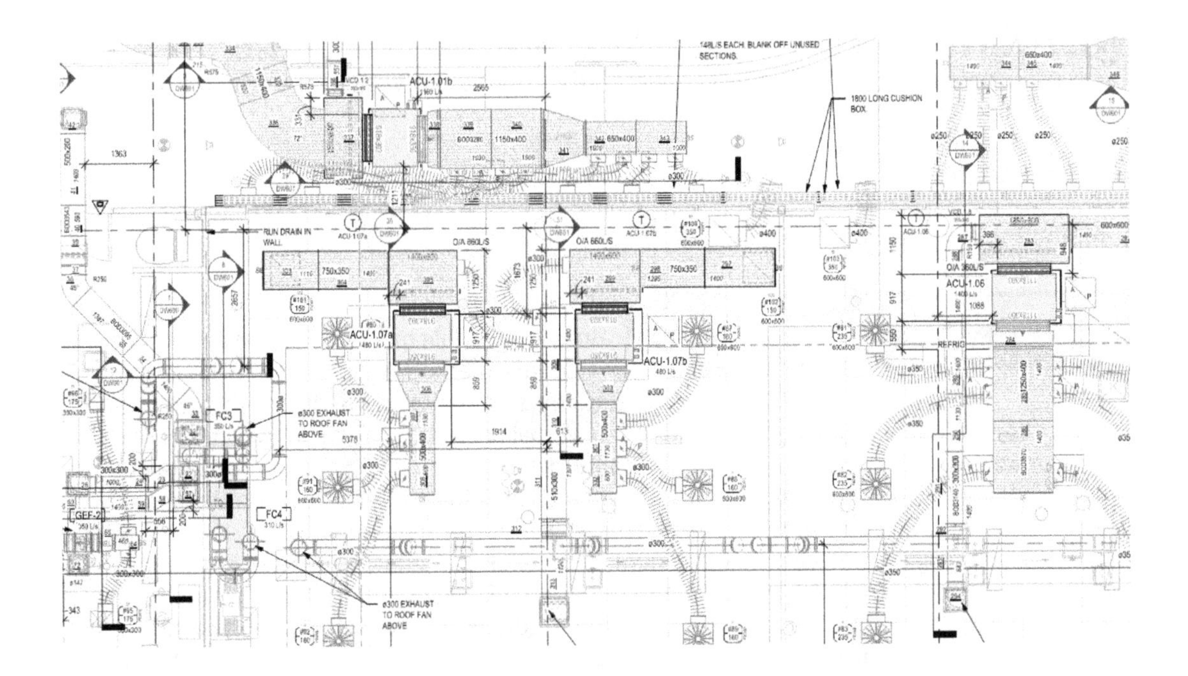

Chapter 4 - Working with the Markup Tools - III

The objectives of this chapter are to:

- √ *Explain various shape markups*
- √ *Explain how to add rectangle markups*
- √ *Explain how to create custom hatch patterns or import them from AutoCAD*
- √ *Explain how to add ellipse and circle markups*
- √ *Explain how to add polygon markups*
- √ *Explain how to add cloud markups*
- √ *Explain how to add cloud+ markups*

WORKING WITH THE SHAPE MARKUP TOOLS

Revu provides six shape markup tools: **Rectangle**, **Ellipse**, **Polygon**, **Cloud**, and **Cloud+**. All these tools are discussed next.

The Rectangle Markup

Menu: Tools > Markup > Rectangle
Keyboard Shortcut: R

This tool allows you to draw a rectangle markup. You can also hold down the shift key and draw a square markup using this tool. The rectangles or squares drawn using this tool can also be filled with a color and a hatch pattern. When you invoke this tool, the cursor will change to the rectangle cursor and you will be prompted to select the region to place the rectangle. You can click two opposite corners on the PDF file to draw a rectangle or hold down the SHIFT key after specifying the first corner to draw a square. The **Appearance** area of the **Properties** panel allows you to appearance properties of the markup. Revu provides a number of hatch patterns that you can fill the rectangle with. You can also import hatch patterns from an AutoCAD hatch pattern file or create your own custom hatch patterns. This can be done using the **Hatch** drop-down list in the **Properties** panel. Figure 1 shows a rectangle markup with the fill color and hatch pattern defined. Note that in this case, the **Highlight** check box on the right of the **Fill Color** swatch was selected to ensure the text of the PDF content was not hidden behind the fill color and hatch.

Figure 1 *A rectangle markup with fill color and hatch pattern*

Working with Custom Hatch Patterns

As mentioned earlier, Revu allows you to import hatch patterns from an AutoCAD hatch pattern file or create your own custom hatch patterns. The following are the procedures for this.

Importing Hatch Patterns from an AutoCAD Hatch Pattern File

The AutoCAD hatch pattern files are saved in the **PAT** format. It is important to note that if the **PAT** file contains multiple hatch pattern definitions, all the hatch patterns will be imported into Revu. The following is the procedure for importing hatch patterns from the **PAT** file into Revu.

1. Using Windows Explorer, copy the **PAT** file into a directory that you can access from Revu.

2. In Revu, click on the shape feature that needs to be filled with the custom hatch pattern.

3. In the **Properties** panel, click on the **Hatch** drop-down list.

4. Scroll down and select **Manage**; the **Manage Hatch Pattern Sets** dialog box will be displayed.

   Similar to the line styles, hatch patterns also require sets to be created, which are like groups. In those sets, you import hatch patterns or create custom hatch patterns.

5. In the **Manage Hatch Pattern Sets** dialog box, click **Add**; the **Add Hatch Pattern Set** dialog box will be displayed.

6. To create a new hatch pattern set, click **New** from the **Step 1: Select Type** area.

7. Enter the name of the hatch pattern set in the **Title** edit box in the **Step 2: Location** area; the location display box will show a new .bhx file being created with the same name as the hatch pattern set you defined, as shown in Figure 2.

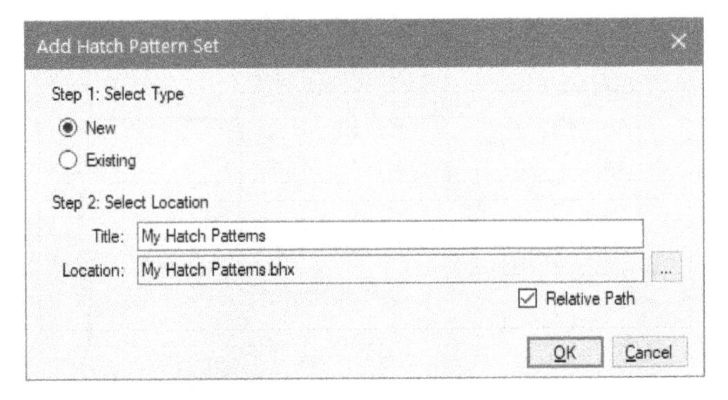

Figure 2 *Creating a new hatch pattern set*

   Selecting the **Relative Path** check box ensures that the custom hatch pattern set file is saved in the same folder as the current profile.

8. Click **OK** in the dialog box; the **Hatch Pattern Editor** dialog box will be displayed.

   By default, this dialog box is blank. You can create a hatch pattern from scratch in this dialog box or import them from the AutoCAD hatch pattern file.

9. From the top right, click the **Import AutoCAD Pattern** button; the **Open** dialog box will be displayed.

10. Browse to the **PAT** file and then double-click on it; all the hatch patterns that were defined in that file will be imported and displayed in the **Patterns** area of the **Hatch Pattern Editor** dialog box.

11. Click the name of one of the imported hatch patterns, the **New Pattern** area of the **Hatch Pattern Editor** dialog box will show the preview of that pattern, as shown in Figure 3.

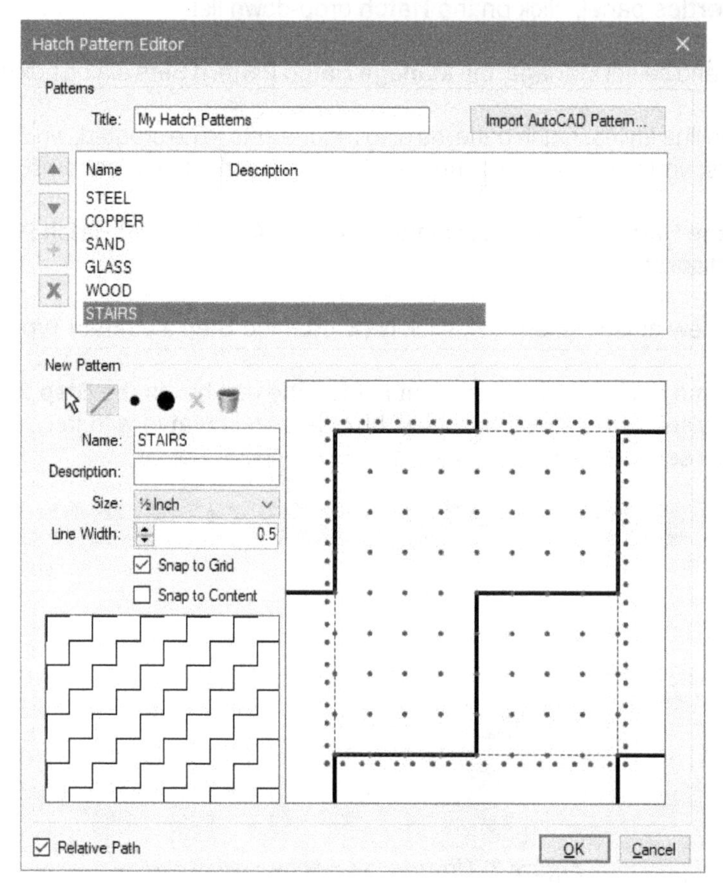

Figure 3 The preview of one of the imported hatch patterns

12. Click **OK** in the **Hatch Pattern Editor** dialog box to return to the **Manage Hatch Pattern Sets** dialog box.

13. Click **OK** in the **Manage Hatch Pattern Sets** dialog box to return to the PDF file.

14. With the shape markup still selected, click on the **Hatch** drop-down list; all the imported hatch patterns are listed under the name of your custom hatch pattern set at the bottom of the **Hatch** drop-down list.

15. Select one of the imported hatch patterns; the shape is filled with the custom hatch pattern.

16. If required, change the color of the hatch pattern using the **Color** swatch on the right of the **Hatch** drop-down list.

17. If required, change the scale in the **Scale** spinner. Figure 4 shows a rectangle mark up filled with a custom hatch pattern.

Figure 4 A rectangle markup filled with a custom hatch pattern

Creating Custom Hatch Patterns in Revu

The following is the procedure for creating a custom hatch pattern in Revu.

1. In Revu, click on the shape feature that needs to be filled with the custom hatch pattern.

2. In the **Properties** panel, click on the **Hatch** drop-down list.

3. Scroll down and select **Manage**; the **Manage Hatch Pattern Sets** dialog box will be displayed.

   If the hatch pattern set is already created, you can double-click on it to add more hatch patterns to that set.

4. In the **Manage Hatch Pattern Sets** dialog box, double-click on the name of the hatch pattern set to which you want to add more patterns; the **Hatch Pattern Editor** dialog box will be displayed.

   All the existing hatch patterns will be listed in this dialog box. To create a new hatch pattern, you need to click on the + button on the left of the **Name** field in the top half of this dialog box.

5. Click on the + button on the left of the **Name** field; the **Add Hatch Pattern** dialog box will be displayed.

6. Enter the name of the new hatch pattern in the **Name** edit box, as shown in Figure 5.

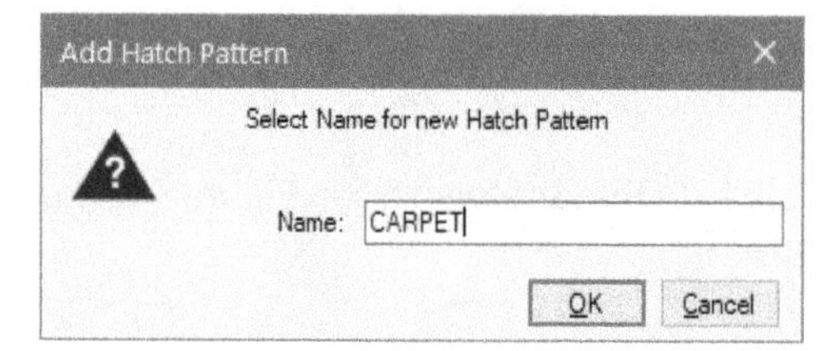

***Figure 5** Adding a new custom hatch pattern*

7. Click **OK** in the dialog box; you will be returned to the **Hatch Pattern Editor** dialog box.

   The name of the custom hatch pattern will be selected in the upper section of the dialog box and the tools required to create a hatch pattern will be available in the lower section.

8. Using the tools available in the lower section of the dialog box, draw the hatch pattern. Figure 6 shows a hatch pattern with four lines and a dot.

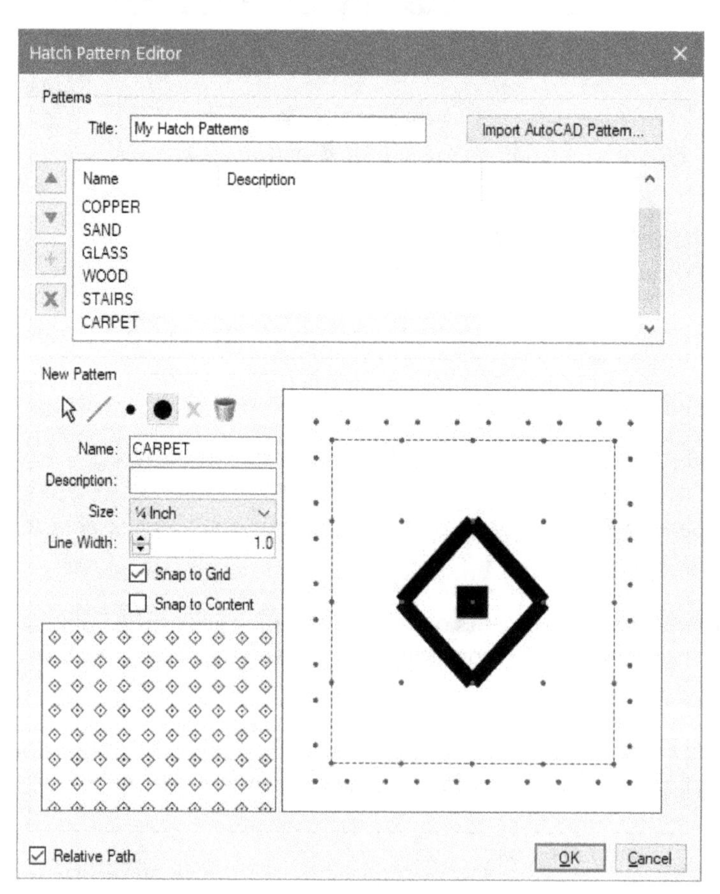

***Figure 6** A custom hatch pattern created*

9. Click **OK** in the **Hatch Pattern Editor** dialog box to return to the **Manage Hatch Pattern Sets** dialog box.

10. Click **OK** in the **Manage Hatch Pattern** dialog box to return to the PDF file.

11. With the shape markup still selected, click on the **Hatch** drop-down list; all the imported and custom hatch patterns are listed under the name of your custom hatch pattern set at the bottom of the **Hatch** drop-down list.

12. Select the custom hatch pattern; the shape is filled with it.

13. If required, change the color of the hatch pattern using the **Color** swatch on the right of the **Hatch** drop-down list.

14. If required, change the scale in the **Scale** spinner. Figure 7 shows a rectangle mark up filled with a custom hatch pattern.

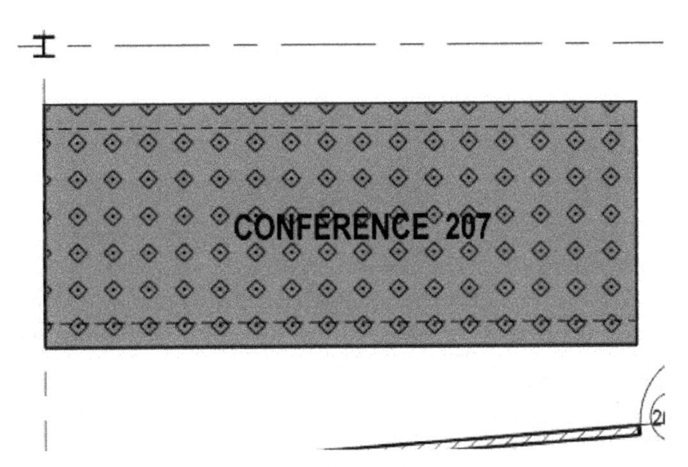

Figure 7 *A rectangle markup filled with the custom hatch pattern*

The Ellipse Markup

Menu: Tools > Markup > Ellipse
Keyboard Shortcut: E

Ellipse E

This tool allows you to draw an ellipse or a circle. By default, you can drag two opposite corners to draw an ellipse. However, if you hold down the shift key and drag two opposite corners, this tool draws a circle. Alternatively, if you hold down the ALT key, this tool allows you to draw a center point circle. The ellipses or circles drawn using this tool can also be filled with a color and a hatch pattern. When you invoke this tool, the cursor will change to the ellipse cursor and you will be prompted to select the region to place the ellipse. You can click two opposite corners on the PDF file to draw an ellipse or hold down the SHIFT key after specifying the first corner to draw a circle. Figure 8 shows an ellipse with custom properties and hatch fill and Figure 9 shows a center point circle being drawn by holding the ALT key before specifying the first point.

Figure 8 A ellipse markup filled with the color and hatch pattern

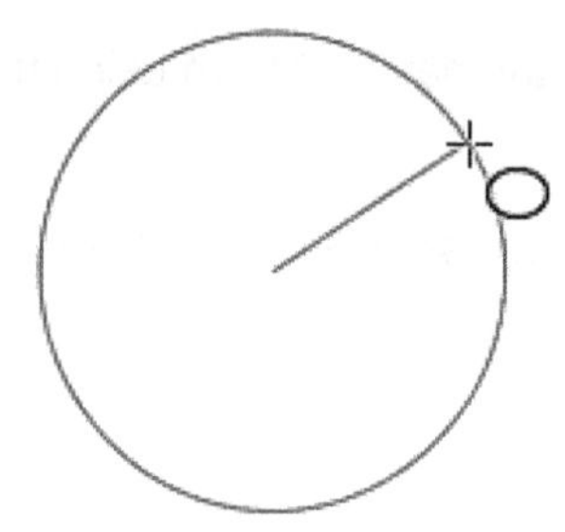

Figure 9 A center point circle markup being drawn by holding down the ALT key

> **What I do**
>
> *Every time I customize a markup with fill color, I select the **Highlight** check box that is available on the right of the **Fill Color** swatch. This ensures that the content of the PDF file is not hidden behind the color fill and is highlighted for the users to see.*

The Polygon Markup

Menu: Tools > Markup > Polygon
Keyboard Shortcut: SHIFT+P

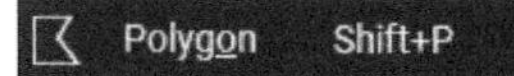

This tool allows you to draw a closed multi-sided shape that can be filled with color and a hatch pattern. When you invoke this tool, the cursor will change to the polygon cursor and you will be prompted to select the vertices to define the polygon. You can click on the PDF file to define various segments of the polygon. Holding down the SHIFT key will allow you to draw segments that snap at 45-degrees intervals. It is important to note that if you have specified a vertex at the wrong location, you can press BACKSPACE to remove that point. Once you have specified all the vertices of the polygon segments, press ENTER to close the shape and exit the tool.

If the polygon segments were not drawn horizontal or vertical by holding down the SHIFT key, you can convert them into arc segments. To do this, right-click on the segment and select **Convert to Arc** from the shortcut menu. Figure 10 shows a polygon markup filled with color and a hatch pattern. In this markup, the top segment is converted into an arc segment by right-clicking on it and selecting **Convert to Arc** from the shortcut menu. You can also see the control handles of the polygon markup in this figure.

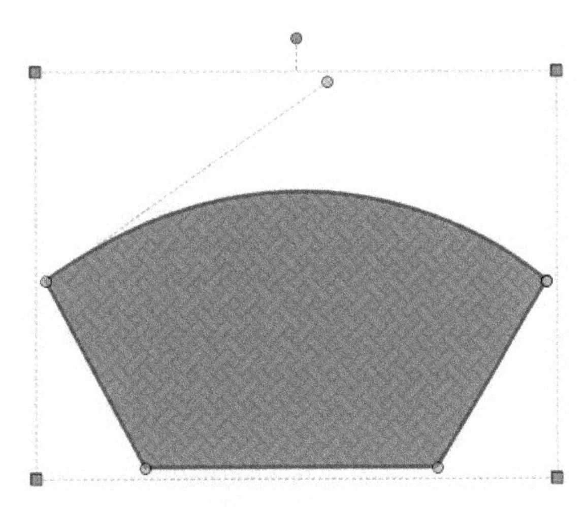

Figure 10 *A polygon markup with top segment converted into arc*

As shown in Figure 10, the polygon markups have two sets of control handles. These are discussed next.

Yellow Control Handles: These are available at the endpoints of the segments and are used to resize that particular segment by dragging.

Blue Control Handles: These are available at the corners of the bounding box of the polygon markup and are used to uniformly scale the polygon markup.

Note that instead of clicking to define the vertices of the polygon, you can press and hold down the Left Mouse Button and drag two opposite corners to draw a rectangle. You can also hold down the SHIFT key and then drag to draw a square. Similar to the other shape markups, you can also fill the polygon with color and a hatch pattern. Figure 11 shows a rectangle created using the **Polygon** markup tool by dragging two opposite corner points and Figure 12 shows a square created using the **Polygon** tool by holding down the SHIFT key and dragging two opposite corner points.

Figure 11 *A rectangle created using the* ***Polygon*** *markup tool*

Figure 12 *A square created using the* ***Polygon*** *markup tool*

The Cloud Markup

Menu: Tools > Markup > Cloud
Keyboard Shortcut: C

This tool allows you to draw a closed multi-sided revision cloud by specifying the vertices of the cloud segments. When you invoke this tool, the cursor will change to the cloud cursor and you will be prompted to select the vertices to define the cloud region. Similar to the polygon tool, you can click on the PDF file to define the vertices of the cloud segments and then press ENTER to close the revision cloud and exit the tool. You can press and drag the Left Mouse Button to draw a rectangular revision cloud or hold down the SHIFT key and then drag to define a square revision cloud.

***Tip**: Similar to the polygon markup, you can also press the BACKSPACE while drawing the cloud markup to remove the last vertex that you defined.*

Just like other markups, you can modify the properties of the cloud markup also using the **Properties** panel. Figure 13 shows a multi-sided cloud markup filled with color. Figure 14 shows a square cloud markup drawn by holding down the SHIFT key and dragging to define the two opposite corner points. This markup is also filled with color.

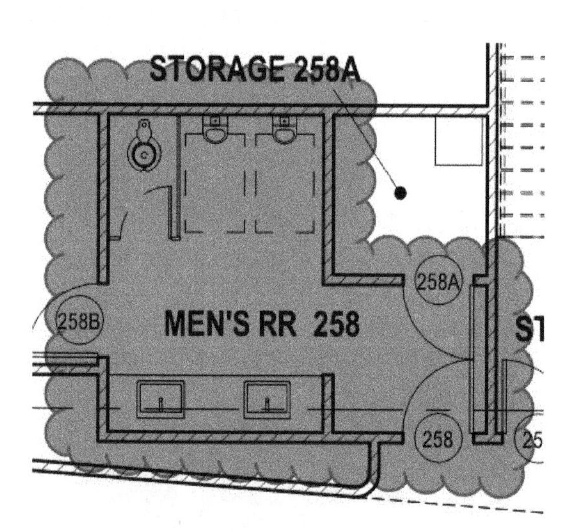

***Figure 13** A multi-sided cloud markup*

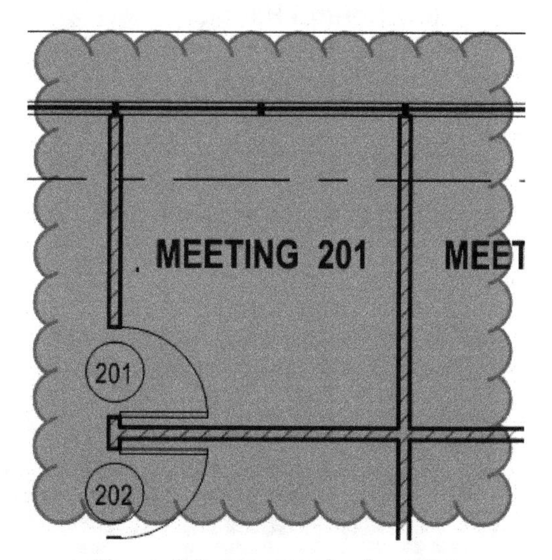

***Figure 14** A square cloud markup*

Selecting the **Invert** check box on the right of the **Size** spinner in the **Properties** window allows you to invert the cloud markup. Figure 15 shows an inverted cloud.

Similar to the polygon markup, the cloud markup also has two sets of control handles, as shown in Figure 16. The Yellow control handles are used to stretch the segment of the cloud and the Blue control handles are used to uniformly scale the cloud.

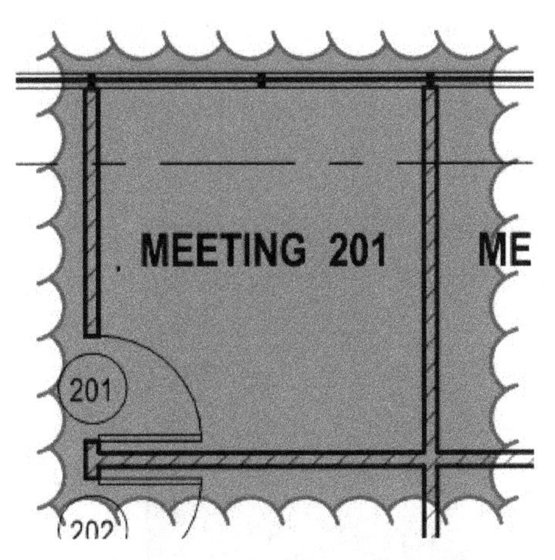

Figure 15 An inverted cloud

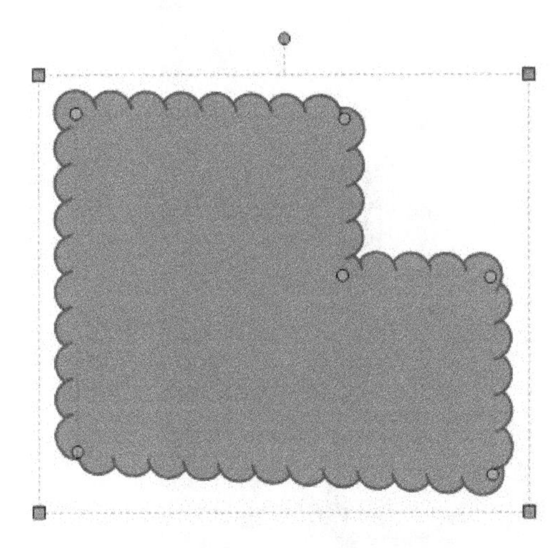

Figure 16 The two sets of control handles of the cloud

The Cloud+ Markup

Menu: Tools > Markup > Cloud+
Keyboard Shortcut: K

This tool is used to add a revision cloud with a callout attached to it using a leader line. When you invoke this tool, you will be prompted to select the vertices to define the cloud region. You can click on the PDF file to define the segments and then enter to close the cloud. On doing so, a callout will be displayed attached to the cloud using a leader line and you will be prompted to select a region to place the callout. Once you place the callout, you will be able to write the content of the callout in the realtime text editor. Similar to the cloud markup, you can also drag two opposite corners to define the shape of the cloud and then place the callout.

To modify the properties of the cloud+ markup, the **Appearance** area of the **Properties** panel is divided into two sections: the **Cloud Properties** section to modify the cloud properties and the **Callout Properties** section to modify the properties of the leader line, text, and text box, as shown in Figure 17.

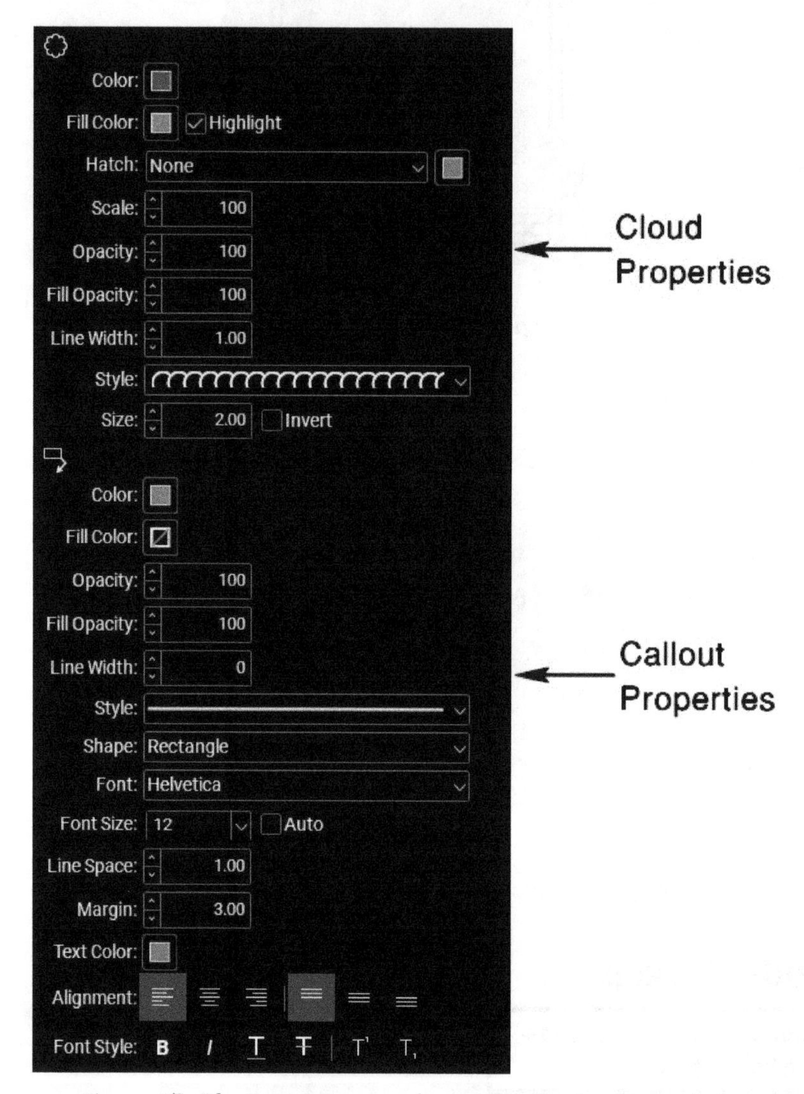

Figure 17 *The* ***Properties*** *panel to modify the cloud+ markup*

Figure 18 shows a cloud+ markup with the modified cloud and callout properties.

***Tip**: If you already have a cloud markup with all the required properties, you can use the **Format Painter** tool to copy those properties and assign them to the cloud+ markup. However, you will still have to modify the callout properties manually.*

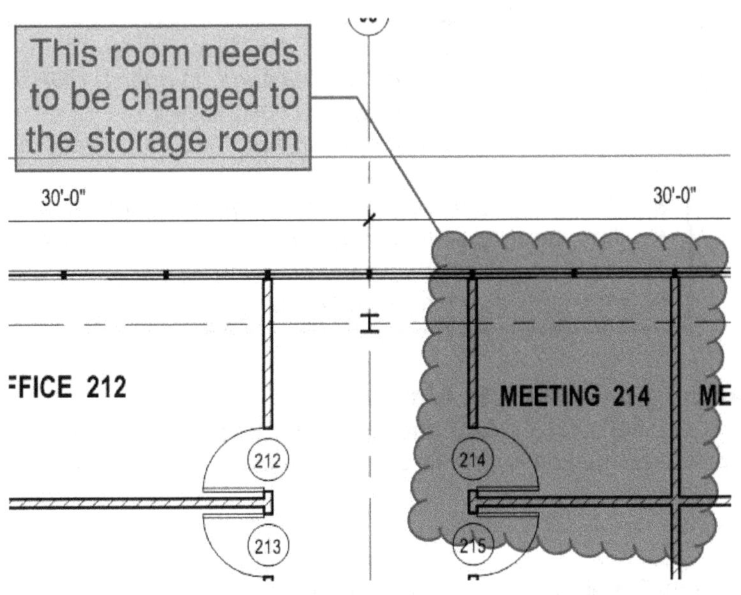

Figure 18 A cloud+ markup with modified properties

Hands-on Tutorial

In this tutorial, you will complete the following tasks:

1. *Use the pinned **Tutorial Files** folder to open a PDF file.*
2. *Add a rectangle markup and customize it.*
3. *Import hatch pattern from an AutoCAD hatch pattern file.*
4. *Create a new custom hatch pattern.*
5. *Assign custom hatch patterns to the rectangle markups.*
6. *Use the **Ellipse** tool to add circle markups and customize them.*
7. *Add polygon markups and customize them.*
8. *Add cloud+ markups and customize them.*

Section 1: Adding a Rectangle Markup

In this section, you will open the **Architecture_Plan_04.pdf** file from the **C04** folder. You will then add a rectangle markup in this file and customize the properties of that markup.

1. Start Revu 2019.

2. Restore the **Revu Training** profile that you created in Chapter 01.

3. Invoke the **File Access** panel > **Explorer Mode**.

4. From the drop-down list at the top, select the pinned **Tutorial Files** folder; the content of this folder is displayed.

5. Double-click on the **C04** folder; the **Architecture_Plan_04.pdf** file saved in this folder is displayed.

6. Double-click on the **Architecture_Plan_04.pdf** file to open it.

   The default cursor mode is set to **Pan**. As a result, the current cursor is displayed as the pan cursor. You will first change it to the select cursor because you can pan by holding down the Scroll Wheel button on the mouse.

7. Press the **V** key to change the cursor to the select cursor.

   You will now zoom to the top center of the plan where the text **CONFERENCE 207** appears and draw a rectangle markup.

8. Navigate to the top of the plan where the text **CONFERENCE 207** appears, refer to Figure 19.

   The rectangle markup needs to be placed on top of the rectangle around the **CONFERENCE 207** text. Therefore, you need to ensure the **Snap to Content** option is turned on from the **Status Bar**.

9. From the **Status Bar**, ensure the **Snap to Content** option is turned on.

10. From the **Menu Bar**, click **Tools > Markup > Rectangle** or press the **R** key to invoke the **Rectangle** markup tool; the cursor changes to the rectangle cursor and you are prompted to select the region to place the rectangle.

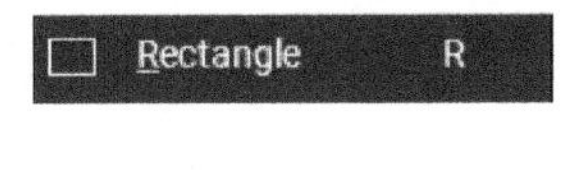

11. Click on the point labeled as **1** in Figure 19 as the first corner of the rectangle.

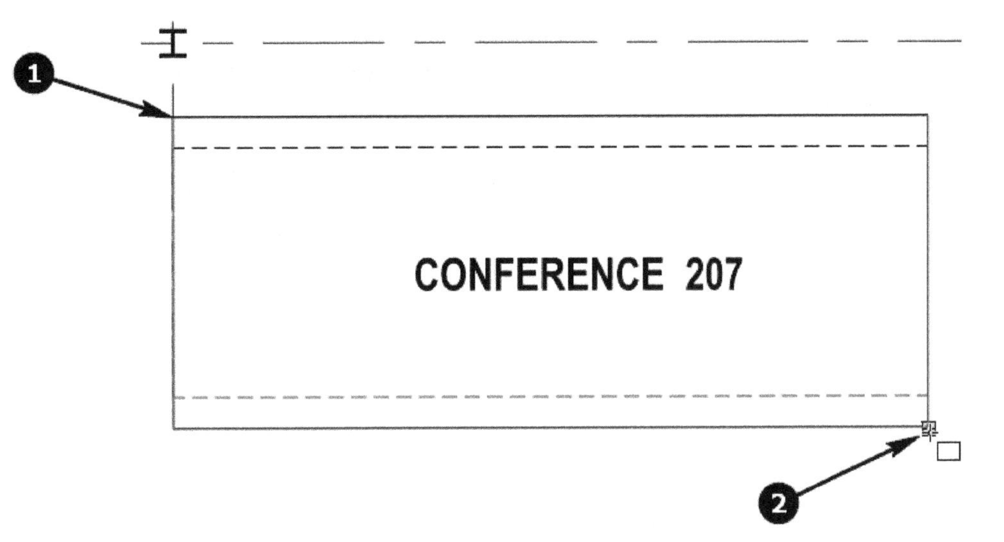

Figure 19 *Drawing a rectangle markup*

12. Click on the point labeled as **2** in Figure 19 as the second corner of the rectangle; the rectangle markup is added and is selected.

    Next, you will change the color of the rectangle and also assign a fill color to it.

13. With the rectangle markup still selected, click the **Color** swatch in the **Properties** window and select **Blue**; the rectangle color changes to Blue.

14. Click the **Fill Color** swatch and select **Blue Violet** to assign the fill color to the rectangle.

    With the fill color assigned, the content of the PDF behind the rectangle is hidden. You will now select the **Highlight** check box on the right of the **Fill Color** swatch to show the content of the PDF that is hidden because of the fill color.

15. Select the **Highlight** check box on the right of the **Fill Color** swatch; the content of the PDF is displayed even though the rectangle is filled with color.

Section 2: Importing Hatch Patterns from AutoCAD

In this section, you will create a new hatch pattern set and then import some of the AutoCAD hatch patterns from a **PAT** file available in the **C04** folder.

1. With the rectangle markup still selected in the PDF file, click on the **Hatch** drop-down list; all the available hatch patterns are available in this drop-down list.

2. Scroll down and select **Manage**; the **Manage Hatch Pattern Sets** dialog box is displayed.

3. From the lower right in this dialog box, click **Add**; the **Add Hatch Pattern Set** dialog box is displayed.

4. From the **Step 1: Select Type** area, make sure **New** is selected, as shown in Figure 20.

5. In the **Step 2: Select Location** area > **Title** edit box, enter the name of the set as **Company ABC Patterns**; the **Location** field displays the **Company ABC Patterns.bhx** file being created, as shown in Figure 20.

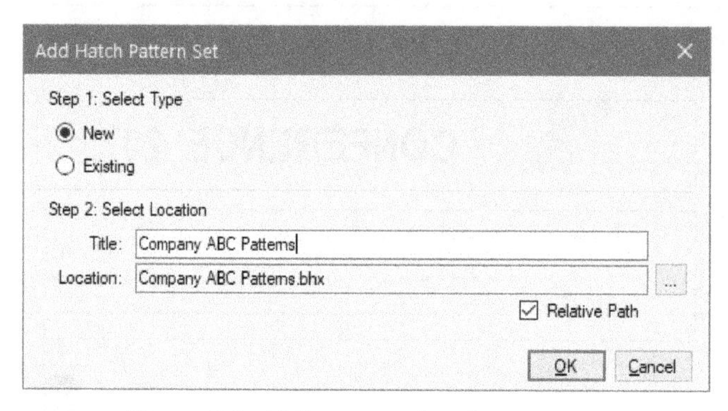

Figure 20 *Adding a new hatch pattern set*

The **Relative Path** check box in this dialog box is selected to ensure the hatch pattern .bhx file is being saved in the same location as the current profile.

6. Click **OK** in the **Add Hatch Pattern Set** dialog box; the **Hatch Pattern Editor** dialog box is displayed.

   The **Import AutoCAD Pattern** button on the top right of this dialog box is used to import the hatch patterns from the **PAT** file containing the AutoCAD hatch pattern definitions. For this tutorial, you are provided a hatch pattern file in the **C04** folder.

7. From the top right in the **Hatch Pattern Editor** dialog box, click **Import AutoCAD Pattern**; the **Open** dialog box is displayed, as shown in Figure 21.

8. Browse to the **Tutorial Files > C04** folder; the **Custom-Hatches.pat** file is displayed in this folder.

9. Double-click on the **Custom-Hatches.pat** file; six hatch patterns are imported into Revu and are listed in the upper half of the **Hatch Pattern Editor** dialog box.

10. Click on the **Stairs** hatch pattern in the upper half of the dialog box; the lower half shows the preview of this hatch pattern, as shown in Figure 22.

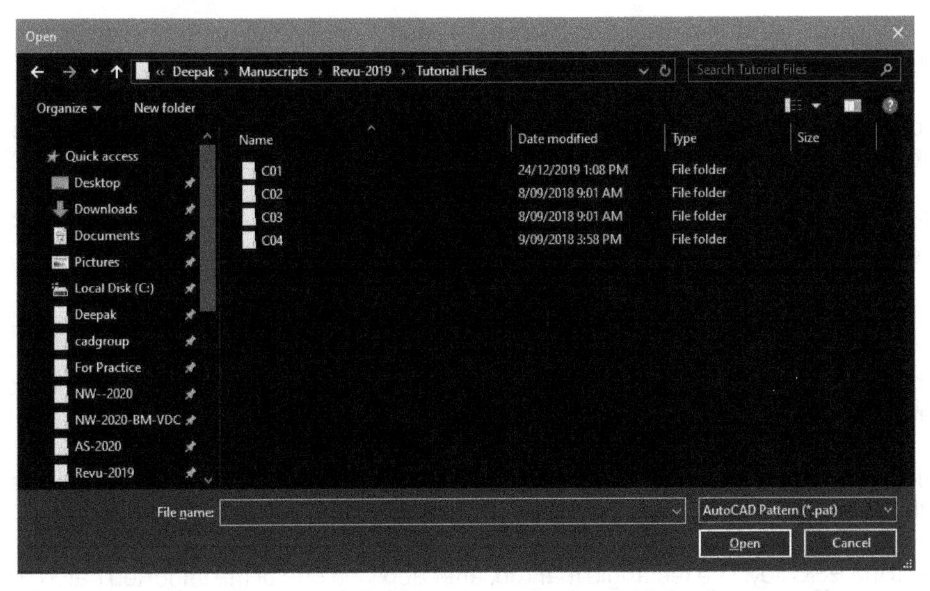

*Figure 21 The **Open** dialog box to open the AutoCAD hatch pattern file*

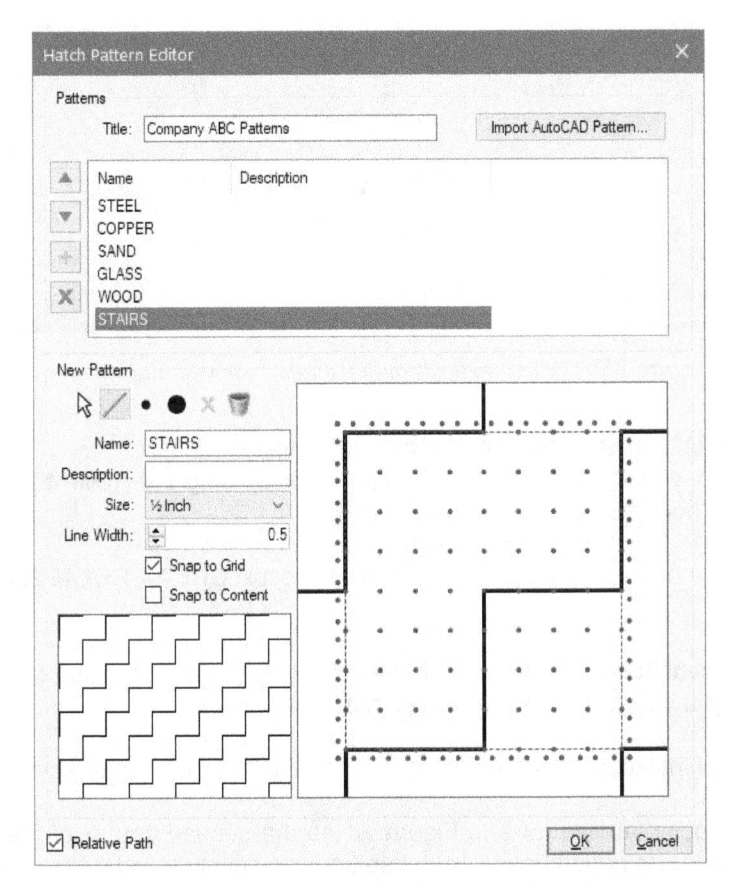

Figure 22 The preview of one of the imported hatch patterns

Next, you will close this dialog box and then assign one of the imported hatch patterns to the rectangle markup you added earlier.

11. Click **OK** in the **Hatch Pattern Editor** dialog box; you are returned to the **Manage Hatch Pattern Sets** dialog box where the **Company ABC Patterns** set is displayed.

12. Click **OK** in the **Manage Hatch Pattern Sets** dialog box; you are returned to the PDF file where the rectangle markup you added earlier is highlighted.

13. From the **Properties** panel, click on the **Hatch** drop-down list.

14. Scroll down in this drop-down list and you will notice the imported hatch patterns listed under the **Company ABC Patterns** category.

15. Select the **WOOD** pattern from the drop-down list; the selected hatch pattern is assigned to the rectangle markup.

16. Press the ESC key. The rectangle markup, after applying one of the imported hatch patterns, is shown in Figure 23.

Figure 23 *The rectangle markup with the imported hatch pattern*

Section 3: Creating a Custom Hatch Pattern

In this section, you will add a new rectangle markup using the **Recent Tools** toolset and then create a new hatch pattern to assign to it.

1. Navigate to the bottom center of the plan where the text **BREAK ROOM 233** appears, refer to Figure 24.

2. From the **Recent Tools** tool set, invoke the custom **Rectangle** markup tool that got saved there when you created the last rectangle markup.

3. Click on the point labeled as **1** in Figure 24 as the first corner of the rectangle.

4. Click on the point labeled as **2** in Figure 24 as the second corner of the rectangle; the rectangle markup is added based on the settings you customized earlier and is selected.

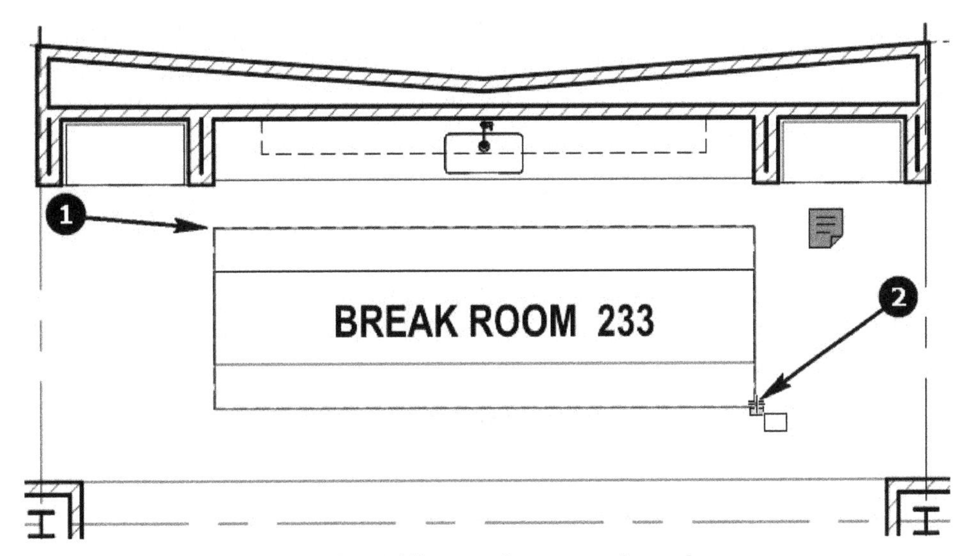

Figure 24 Adding another rectangle markup

Notice that this markup has the same hatch pattern assigned as the previous rectangle markup. You will now create a custom hatch pattern to assign to this markup.

5. With the rectangle markup still selected, click on the **Hatch** drop-down list in the **Properties** panel.

6. From the bottom in the **Hatch** drop-down list, select **Manage**; the **Manage Hatch Pattern Sets** dialog box is displayed.

   The custom hatch pattern you will create will also be added to the **Company ABC Patterns** set.

7. Double-click on the **Company ABC Patterns** set; the **Hatch Pattern Editor** dialog box is displayed.

8. From the **Patterns** area in the dialog box, click the Green + button; the **Add Hatch Pattern** dialog box will be displayed.

9. Enter **CARPET** as the name of the hatch pattern in the dialog box, as shown in Figure 25.

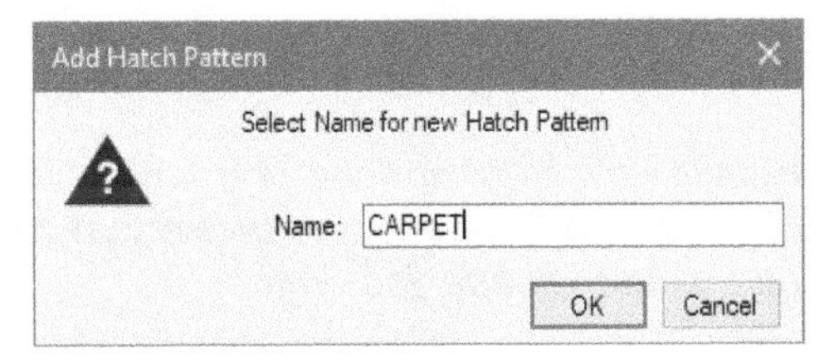

Figure 25 Adding a new hatch pattern

10. Click **OK** in the dialog box; you are returned to the **Hatch Pattern Editor** dialog box.

    The **New Pattern** area in the lower half of the dialog box provides you with the tools required to create a custom hatch pattern.

11. With the **Line** tool active in the **New Pattern** area, draw four lines to form a diamond shape, refer to Figure 26.

12. Invoke the **Large Dot** tool and place a dot at the center of the diamond, as shown in Figure 26. Notice that the lower left area of the **Hatch Pattern Editor** dialog box shows how the hatch pattern will look like.

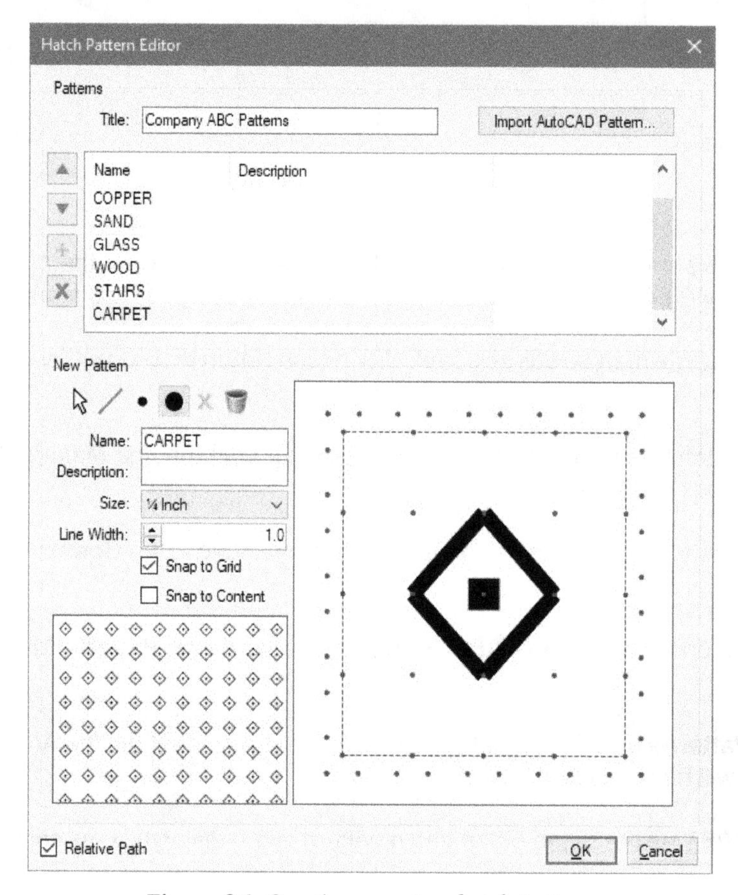

***Figure 26** Creating a custom hatch pattern*

***Tip**: If you make a mistake while drawing lines, you can invoke the **Select** tool and select the segment and then click the **Delete** tool to delete that segment. Clicking the **Clear** tool will clear the entire pattern shape you drew.*

This completes the custom hatch pattern. You will now close the **Hatch Pattern Editor** dialog box.

13. Click **OK** in the **Hatch Pattern Editor** dialog box; you are returned to the **Manage Hatch Pattern Sets** dialog box.

14. Click **OK** in the **Manage Hatch Pattern Sets** dialog box; you are returned to the PDF file.

15. With the rectangle markup still selected, click on the **Hatch** drop-down list.

16. From the bottom of this drop-down list, select **CARPET**; the hatch pattern is assigned to the rectangle markup.

    Notice that the pattern scale is too big for the rectangle. You will now change the scale of the pattern.

17. In the **Properties** panel > **Appearance** area > **Scale** field, enter **30** as the value; the hatch pattern scale is modified.

18. Press the ESC key to deselect the rectangle markup. This markup, after applying the custom hatch pattern, is shown in Figure 27.

Figure 27 *The rectangle markup with the custom hatch pattern*

19. Zoom to the extents of the PDF file by double-clicking the Scroll Wheel button on the mouse.

20. Save the PDF file.

Section 4: Adding Circle Markups

In this section, you will use the **Ellipse** markup tool to add a circle markup. You will then match the properties of the circle markup with those of the last rectangle markup. You will then add another copy of the same circle using the **Recent Tools** tool set.

1. Navigate to the left of the plan where the text **OPEN OFFICE 248** appears, refer to Figure 28.

2. From the **Menu Bar**, click **Tools > Markup > Ellipse** or press the **E** key to invoke the **Ellipse** markup tool; you are prompted to select the region to place the ellipse markup.

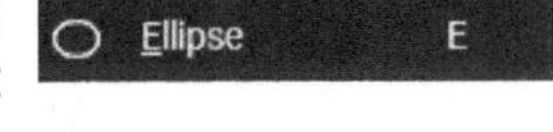

    You will hold down the ALT key to draw a center point circle markup.

3. Hold down the ALT key and specify the center point of the circle, labeled as **1** in Figure 28, and then a point on the circle, labeled as **2** in Figure 28.

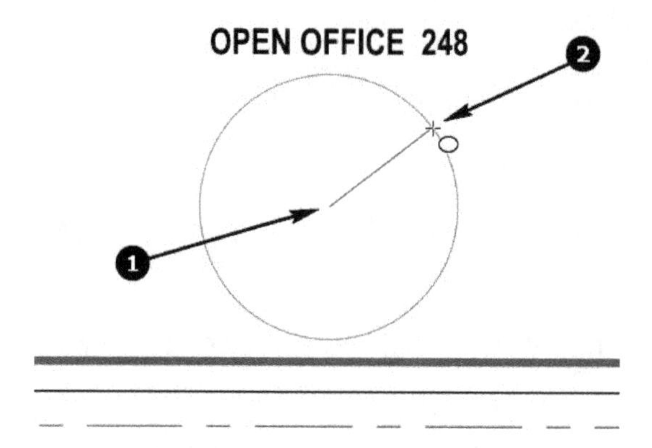

*Figure 28 Adding a circle markup using the **Ellipse** tool*

You will now use the last rectangle markup as the source object and assign its properties to the circle markup.

4. Navigate to the last rectangle markup that you added.

5. Right-click on it and select **Format Painter** from the shortcut menu.

6. Navigate back to the circle markup that you added earlier in this section.

7. Hover the cursor over the circle; the cursor changes to the paintbrush cursor.

8. Once the paintbrush cursor is displayed, click on the circle; the appearance properties of the circle are changed to those of the rectangle markup.

9. Press the ESC key to exit out of the **Format Painter** tool.

   The custom version of the **Ellipse** tool is added to the **Recent Tools** tool set. You will now change this tool to the drawing mode and place the exact same copy of the circle markup you added earlier.

10. In the **Recent Tools** tool set, double-click on the custom **Ellipse** tool to ensure it changes to the drawing mode. In this mode, the icon will show a circle similar to the one you placed earlier.

11. Place another copy of the circle markup above the **OPEN OFFICE 248** text.

    Chances are the second circle you placed may not be vertically aligned with the first circle. You will now align the two circles.

12. Hold down the SHIFT key and select the two circle markups.

13. Right-click on the bottom circle and from the shortcut menu, select **Alignment > Align Center**, as shown in Figure 29.

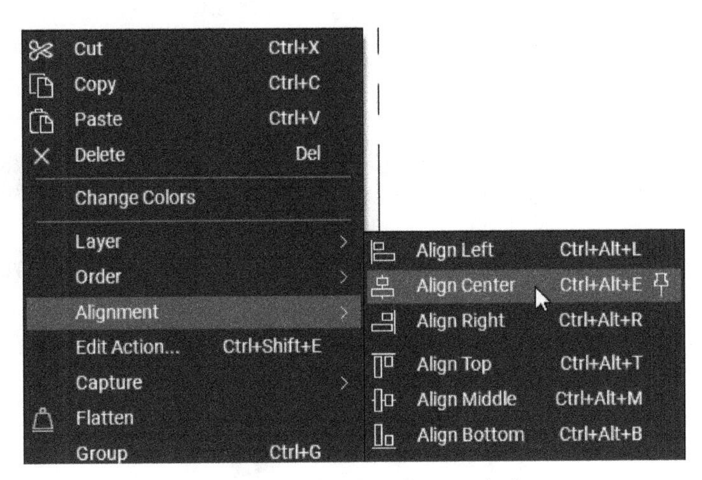

Figure 29 *Aligning the two circles*

14. Press the ESC key to deselect the two circles. Figure 30 shows the two circles, after aligning.

Figure 30 *The two circles aligned*

Tip: *You can also hold down the CTRL + SHIFT key and drag a markup to create its copies in the orthogonal directions.*

Section 5: Adding Polygon Markups

In this section, you will add a polygon markup and then customize it. You will then use the custom version of this tool to add more polygon markups

1. Navigate to the top center of the plan where the text **MEETING 208** appears, refer to Figure 31.

2. From the **Menu Bar**, click **Tools > Markup > Polygon**; the **Polygon** tool is invoked and you are prompted to select vertices for polygon.

   Polygon Shift+P

   You will now click the internal points of meeting room 208 to create a polygon to define the floor spacing of this room

3. Click the five vertices labeled as **1** to **5** in Figure 31 to define the polygon.

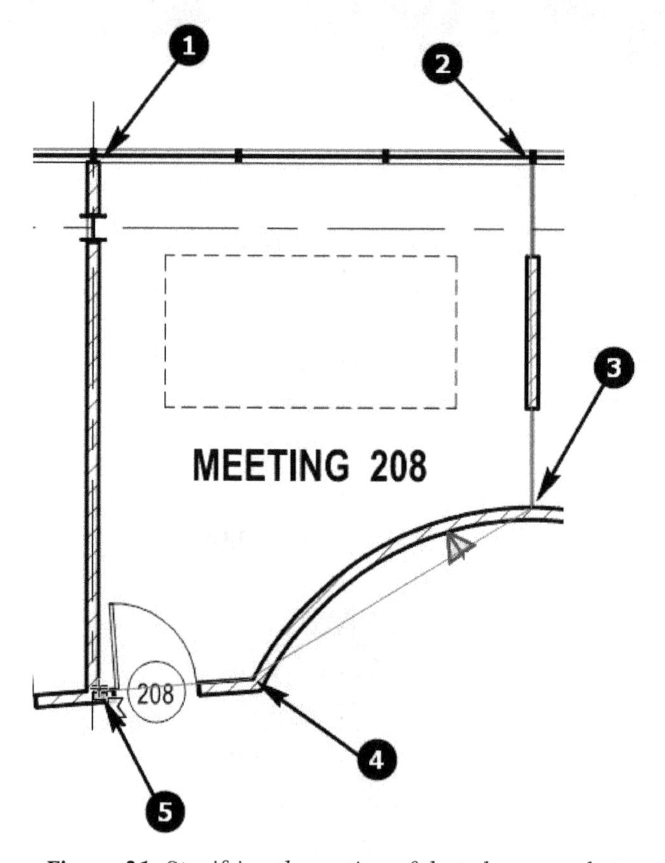

Figure 31 *Specifying the vertices of the polygon markup*

***Tip**: If you select an incorrect point as the polygon vertex by mistake, you can press the BACKSPACE key to clear that vertex.*

4. Press ENTER to close the polygon and exit the tool; the polygon is created and is selected.

Notice that a straight segment is drawn between vertices 3 and 4. You will now convert this into a curved segment and then use its handlebars to ensure it matches the shape of the curved wall.

5. Right-click on the segment between vertices 3 and 4 and select **Convert to Arc** from the shortcut menu; the segment is converted into an arc segment and the handlebars are displayed at the two ends of this segment.

   Before you start dragging these handlebars, it is better to turn off the **Snap to Content** option. This ensures the handlebars are not snapping to the content of the PDF file so you can freely modify the shape of the curved segment.

6. From the **Status Bar**, turn off the **Snap to Content** option.

7. Drag the handlebars at the two ends of the curved segment to match the shape of this segment with that of the underlying curved wall, as shown in Figure 32.

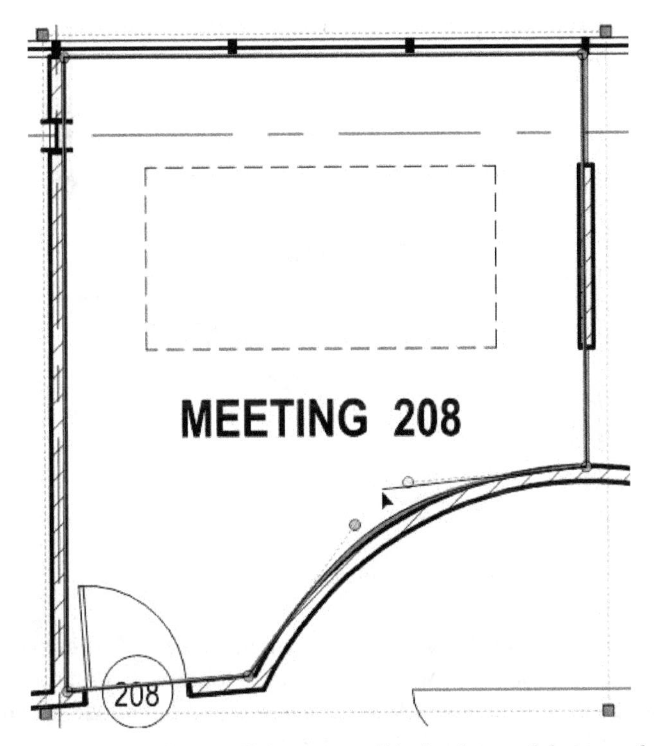

Figure 32 Dragging the handlebars to modify the shape of the curved segment

You will now modify the appearance properties of this polygon markup using the **Properties** panel.

8. With the polygon markup still selected, click the **Color** swatch in the **Properties** panel and select **Blue**; the color of the polygon segments is changed to Blue.

9. Click the **Fill Color** swatch and select **Blue Violet**; the polygon is filled with this color.

10. Select the **Highlight** check box on the right of the **Fill Color** swatch; the content of the PDF that was hidden behind the fill color is displayed.

11. In the **Fill Opacity** spinner, enter **50** as the value.

    This completes the process of customizing the polygon markup.

12. Press the ESC key to deselect the polygon markup. Figure 33 shows the customized polygon markup.

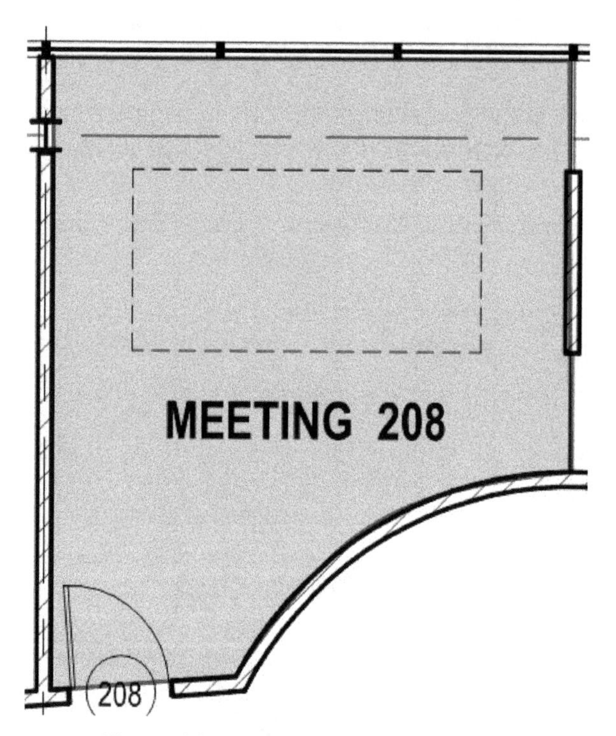

Figure 33 *Customized polygon markup*

13. Use the custom version of the **Polygon** tool from the **Recent Tools** tool set to add a similar markup to the meeting room 209, which is adjacent to the meeting room 208. You will have to turn on the option to snap to content while specifying the vertices and then turn it off while dragging the handlebars of the curved segments.

> ***What I do***
>
> *Using the **Revu > Preferences > Tools > Markup** tab, you can change the drag behavior for the shapes to **Drag Curves**. This allows you to drag a vertex while creating the polygon markup to add curved segments. However, I prefer leaving this option to **Drag Rectangle** and changing the segments to arc segments after adding the markup.*

Section 6: Adding Cloud+ Markups

In this section, you will add a cloud+ markup and then customize it. You will then use the custom version of this tool to add more cloud+ markups

1. Navigate to the area where the text **STAIR 259** appears, which is on the right of the **MEN'S RR 258** text, refer to Figure 34.

2. From the **Menu Bar**, click **Tools > Markup > Cloud+** or press the **K** key on the keyboard; the **Cloud+** tool is invoked and you are prompted to select vertices for cloud region.

   The cloud+ markup has two components: the cloud and the callout. You will first drag two opposite corner points to around the stairs to draw the cloud and then place the callout on the right of the cloud.

3. Press and hold down the Left Mouse Button and drag the two opposite corners similar to the ones labeled as **1** and **2** in Figure 34 to define the cloud.

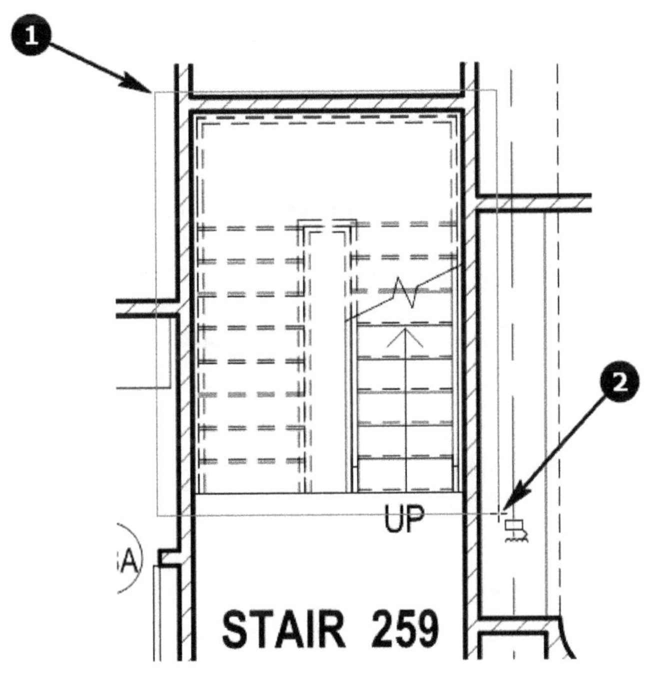

Figure 34 Specifying the two opposite corners of the cloud

4. Once the cloud is created, click on the right of it to place the callout, refer to Figure 35. Once you place the callout, the realtime text editor is displayed.

5. Enter the following text:

   Reverse the direction of the stairs so they start in front of the door

6. Click anywhere outside the callout box; the cloud+ markup is added, as shown in Figure 35.

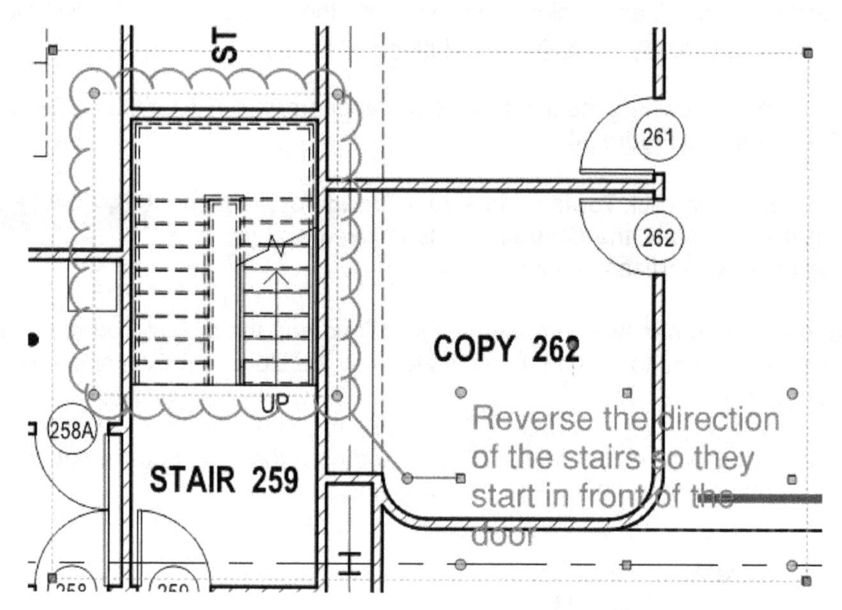

Figure 35 A cloud+ markup added to the PDF

You will now customize the properties of this markup. As mentioned earlier in this chapter, the **Appearance** area of the **Properties** panel of the cloud+ markup is divided into two sections: the **Cloud Properties** section to modify the cloud properties and the **Callout Properties** section to modify the properties of the leader line, text, and text box.

7. In the **Properties** panel > **Appearance** > **Cloud Properties** section, modify the following properties:

   Color: **Blue**
   Fill Color: **Blue Violet**
   Highlight: **Checked**
   Fill Opacity: **50**
   Size: **1**

8. In the **Callout Properties** section, modify the following properties:

   Color: **Blue**
   Fill Color: **Blue Violet**
   Fill Opacity: **50**
   Line Width: **0.5**
   Text Color: **Blue**

9. Drag the text box to ensure the text is in three lines.

10. Autosize the text box. The modified cloud+ markup is shown in Figure 36.

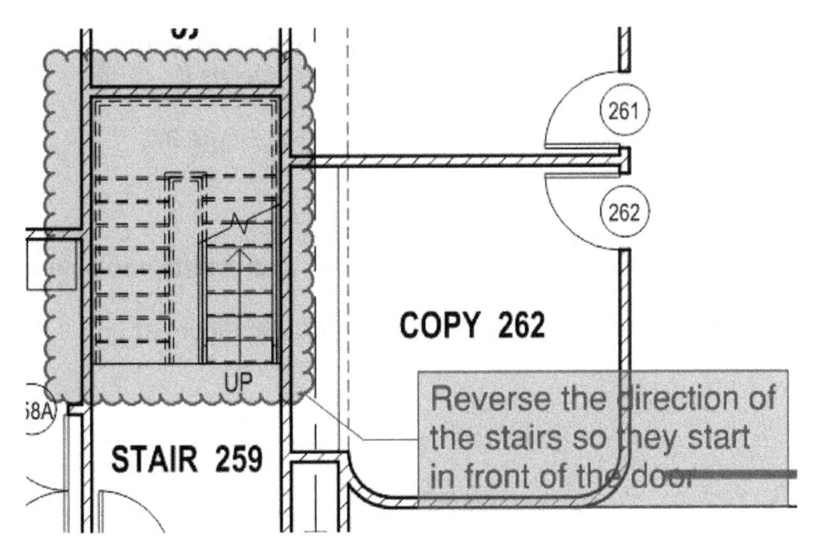

Figure 36 The cloud+ markup after modifying the appearance

You will now use the custom version of this cloud+ tool to markup another area of the file.

11. Pan down to the **MEETING 231** and **OFFICE 229** rooms, which is below the area you are currently in, refer to Figure 37.
12. Invoke the custom version of the **Cloud+** tool from the **Recent Tools** tool set.
13. Specify the four vertices labeled **1** to **4** in Figure 37 and then press enter to draw the cloud.

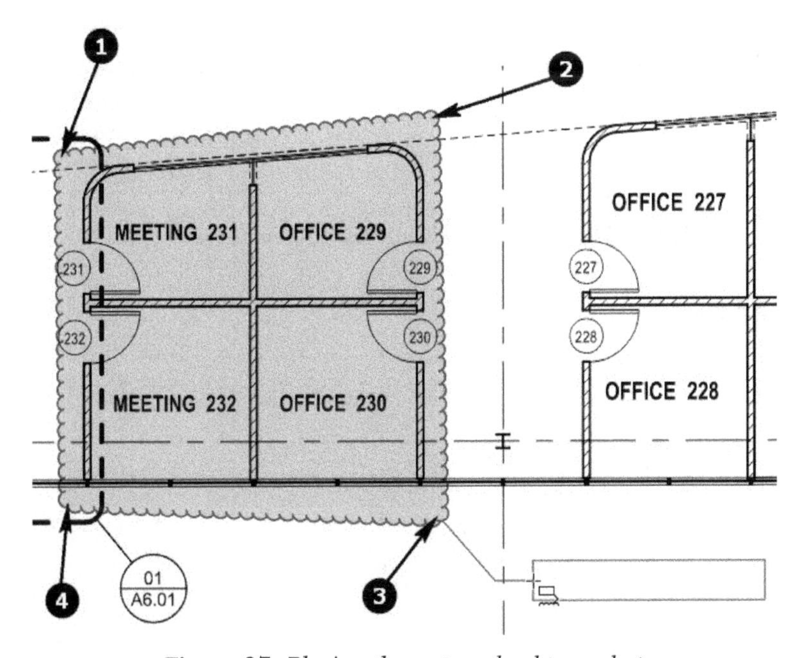

Figure 37 Placing the custom cloud+ markup

***Tip**: If you make a mistake in specifying the vertex to place the cloud, you can press the BACKSPACE key to deselect that vertex and redefine it.*

14. Place the callout on the lower right of the cloud, refer to Figure 37.

15. Enter the following text:

    Change this area to staff recreational area with a kitchenette and a ping-pong table

16. Resize the text box to fit the text in four lines, as shown in Figure 38.

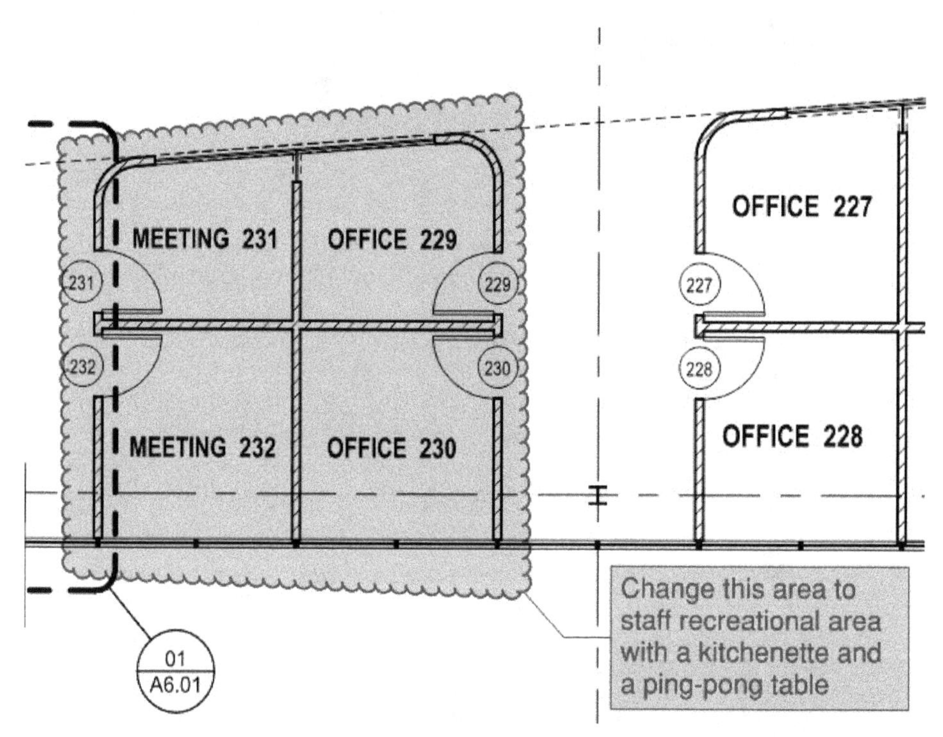

***Figure 38** The custom cloud+ markup added to the file*

17. Save the PDF file.

18. Close Revu.

Skill Evaluation

Evaluate your skills to see how many questions you can answer correctly. The answers to these questions are given at the end of the book.

1. Holding down the SHIFT key while drawing a rectangle callout does not do anything. (True/False)
2. Revu allows you to import hatch patterns from AutoCAD. (True/False)
3. To create custom hatch patterns in Revu, you need to be a programming expert. (True/False)
4. There is no difference between the **Cloud** and **Cloud+** markup tools. (True/False)
5. There is no difference between the **Polyline** and **Polygon** markup tools. (True/False)
6. Which key allows you to draw a center point circle using the **Ellipse** markup tool?

   (A) **ALT** (B) **SHIFT**
   (C) **CTRL** (D) **None**

7. Which tool allows you to add a cloud markup with a callout?

   (A) **Cloud** (B) **Leader**
   (C) **Markup** (D) **Cloud+**

8. Which of the following options are correct in Revu?

   (A) **You can import AutoCAD hatches** (B) **You cannot import AutoCAD hatches**
   (C) **You cannot create custom hatches** (D) **You can create your custom hatches**

9. Which drop-down list in the **Properties** panel is used to assign a hatch pattern to a polygon, cloud or a cloud+ markup?

   (A) **Linetype** (B) **Hatch**
   (C) **Fill** (D) **Style**

10. What is the shortcut key to add the cloud+ markup?

    (A) **B** (B) **C**
    (C) **K** (D) **M**

Class Test Questions

Answer the following questions:

1. Explain briefly the process of importing an AutoCAD hatch pattern.
2. How will you create a cloud+ markup?
3. Explain briefly the process to create a custom hatch pattern.
4. Explain briefly the process to draw a center point circle markup.
5. Explain briefly the process of changing the cloud into a reverse cloud.

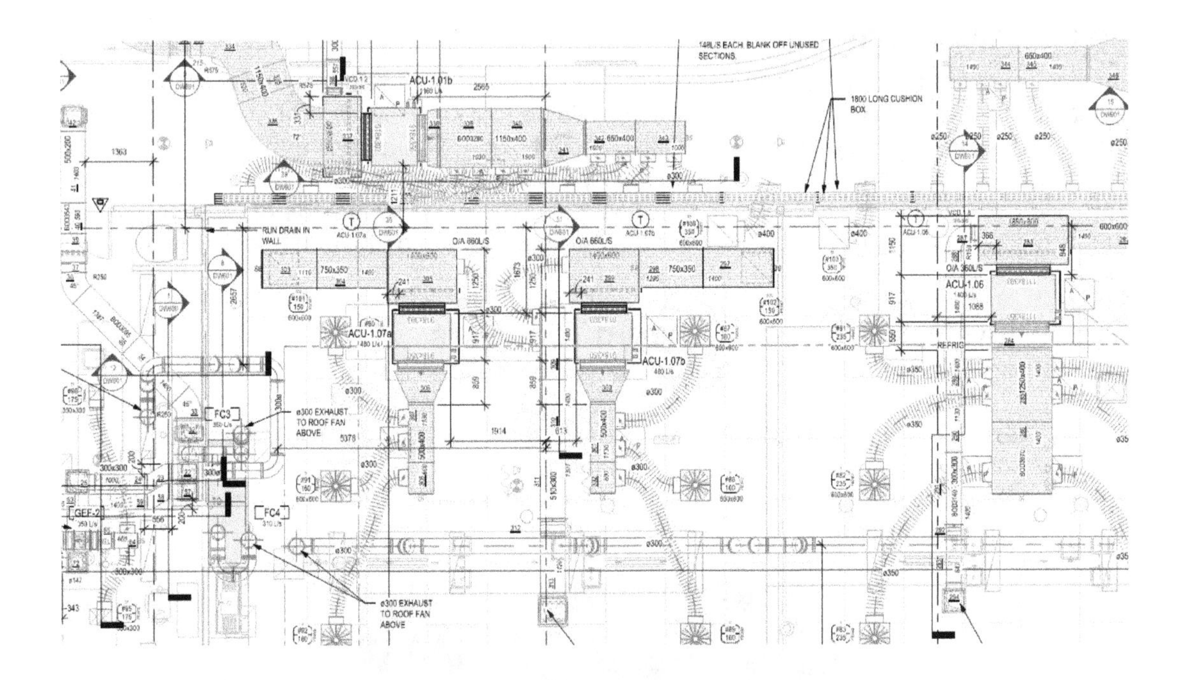

Chapter 5 - Creating and Managing Custom Tool Sets

The objectives of this chapter are to:

- √ *Explain the process of creating custom tool sets*
- √ *Show how to add tools to the custom tool sets*
- √ *Explain the process of editing custom tool sets*
- √ *Explain how to export and import custom tool sets*
- √ *Explain how to create a custom tool set by importing punchkeys for creating Punch Lists*

CUSTOM TOOL SETS

As discussed in the earlier chapters, when you customize a markup tool, the custom version of that tool is automatically saved in the **Recent Tools** tool set. However, when you close Revu and restart it, the **Recent Tools** tool set is wiped out clean. This means that the tools that you customized are no more available. To avoid this, it is important to create your own custom tool sets and add the customized tools to your tool sets. The other advantage of creating custom tool sets is that you can distribute it among your team to standardize the design review and markup process.

Procedure for Creating Custom Tool Sets

The following is the procedure for creating custom tool sets.

1. From the **Panel Bar**, click on the **Tool Chest** button to display the **Tool Chest** panel.

2. From the top of the tool sets, click **Tool Chest > Manage Tool Sets**, as shown in Figure 1; the **Manage Tool Sets** dialog box will be displayed.

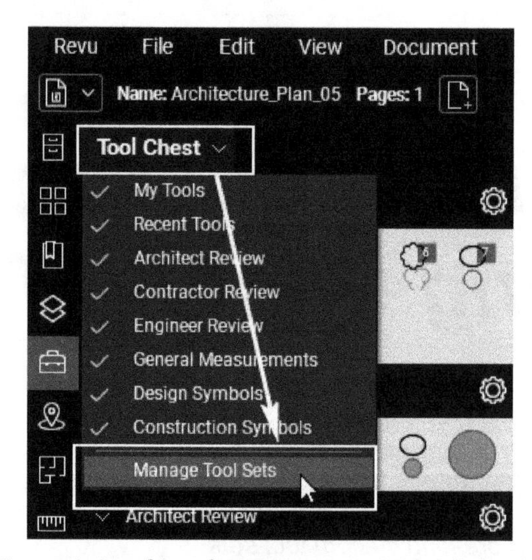

Figure 1 *Invoking the* ***Manage Tool Sets*** *dialog box*

The **Manage Tool Sets** dialog box displays all the tool sets that are currently available. The tool sets with a check box on the left of their names are the ones that are available in the current profile. Note that there are a number of other tool sets that are not displayed in the current profile.

3. From the bottom right in the **Manage Tool Sets** dialog box, click **Add**; the **Add Tool Set** dialog box will be displayed.

   By default, **New** is selected in the **Type** area. This ensures that a new tool set is going to be created.

4. In the **Title** field, type **Project Manager Review.** This will be the name of the new tool set.

5. In the **Options** area, select the **Show in All Profiles** check box, as shown in Figure 2. This ensures that this tool set will be displayed in all profiles.

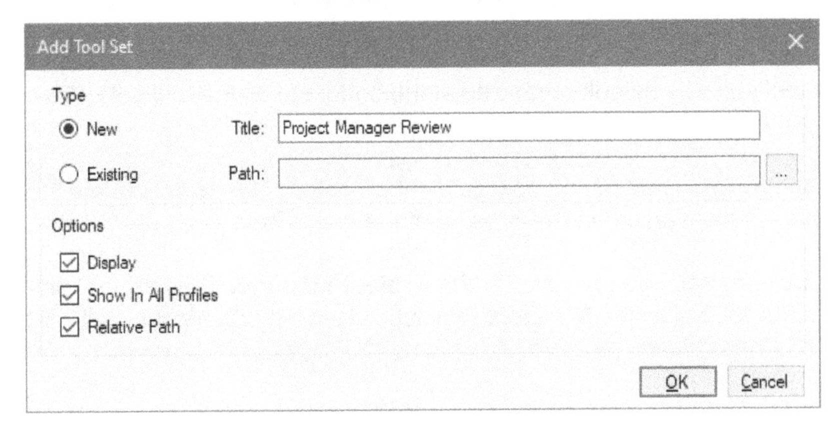

Figure 2 Creating a new tool set

6. Click **OK** in the **Add Tool Set** dialog box; the **Save As** dialog box is displayed that prompts you to save the tool set as a **Project Manager Review.btx** format file, as shown in Figure 3.

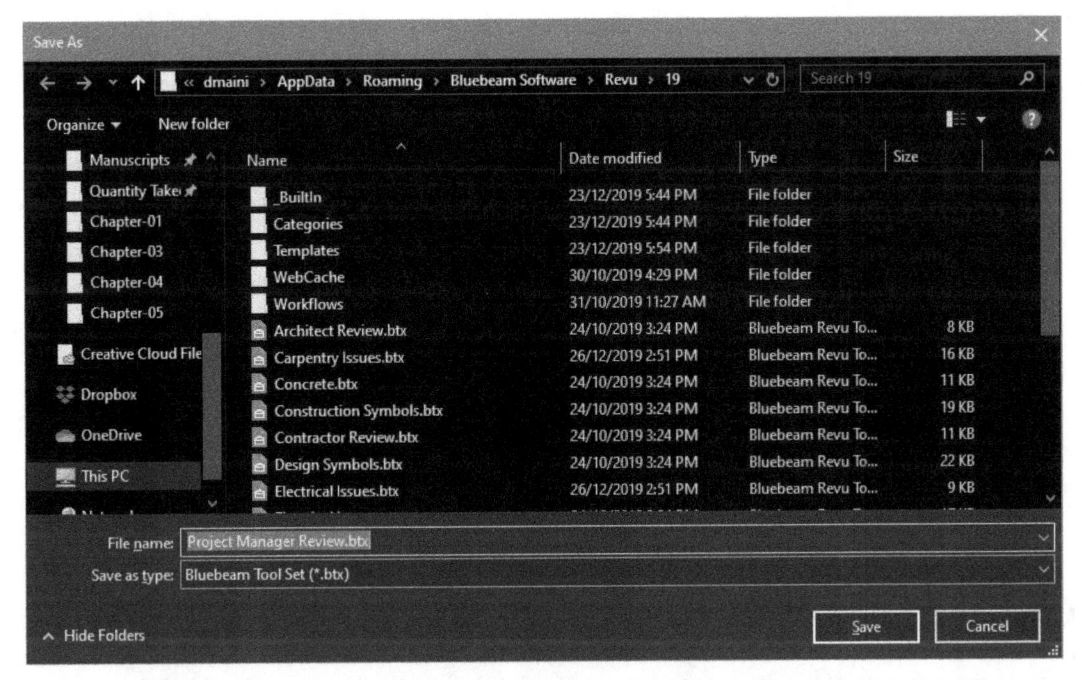

Figure 3 Saving the BTX file of the new tool set

The reason a new **.btx** format file is created is so that you can distribute this file to the rest of your team and standardize the process of reviewing and marking up the PDF files in your organization. This file is by default saved in the roaming profile folder of the current user.

7. Click **Save** in the **Save As** dialog box; the new profile is created and you are returned to the **Manage Tool Sets** dialog box.

   The new tool set is by default organized at the bottom of all the tool sets. You can reorder it and move it above the rest of the tool sets.

8. In the **Manage Tool Sets** dialog box, select the **Project Manager Review** tool set; the **Up** arrow button is highlighted on the left side of the dialog box.

9. Use the **Up** arrow button to move up the **Project Manager Review** tool set and place it second in the list, below the **My Tools** tool set. This is the order in which it will be displayed in the **Tool Chest** panel.

***Note:** The **My Tools** tool set has some special properties that allow you to assign shortcut keys to the tools. You will learn more about this tool set later in this chapter.*

10. Similarly, create additional tool sets and organize them in the order in which you want them to appear in the **Tool Chest** panel.

11. Click **OK** in the **Manage Tool Sets** dialog box; the new tool set is created and is displayed in the **Tool Chest** panel below the **My Tools** tool set, as shown in Figure 4.

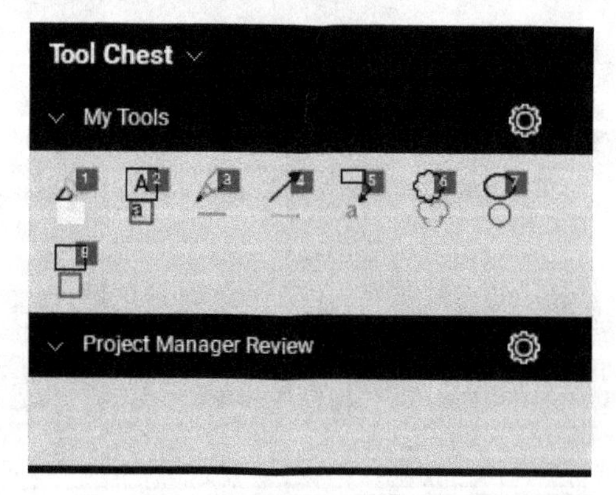

***Figure 4** The new tool set displayed in the **Tool Chest** panel*

CHANGING THE SUBJECT OF THE MARKUPS

It is important to note that before you add your custom markups to your tool sets, you should change the subjects of the markups to more sensible subjects. For example, the default subject of a cloud+ markup is **Cloud+**. So before adding this markup to your tool set, you should change it to a more sensible subject, such as **Architecture Cloud+**, or **Engineer Cloud+**, and so on.

This ensures that you can later on filter these markups based on the disciplines that added the markups.

Procedure for Changing the Subject of a Single Markup

The following is the procedure for changing the subject of a single markup that you customized.

1. Click on the custom markup that you added to select it; the **Properties** panel shows the properties of the selected markup.

2. Scroll to the top where it shows the **Author** and **Subject** fields.

   It is important to leave the **Author** field as is so it is automatically populated based on the user name of the person who added the markup.

*Tip: You can review and modify the user details in the **Preferences** dialog box > **General** tab > **Options** area.*

3. In the **Subject** field, type the required subject for the markup.

4. Press ENTER.

5. Click anywhere outside on the PDF file to deselect the markup.

6. Click on the same markup again and notice the subject is changed.

   Figure 5 shows the subject of the custom cloud+ markup changed to **Project Manager's Cloud+**.

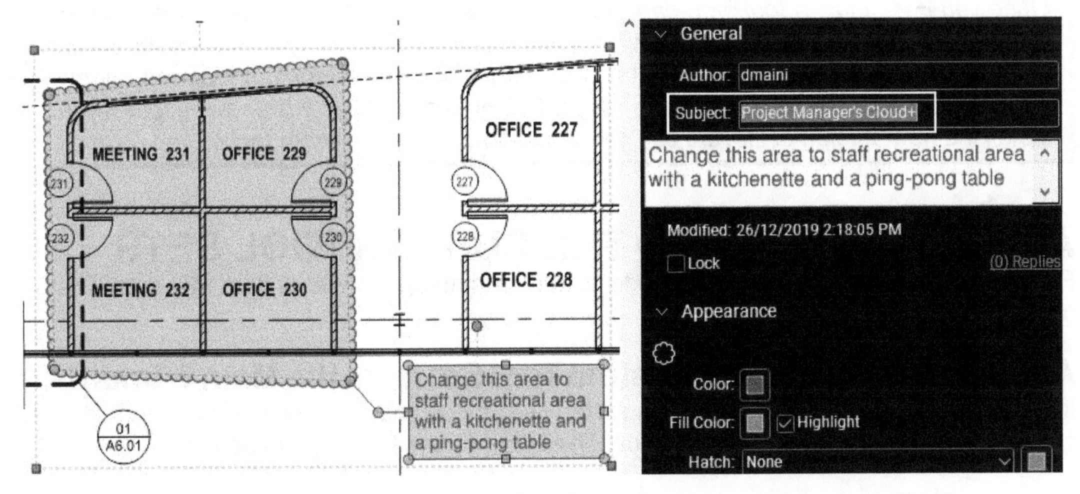

*Figure 5 The subject of a cloud+ markup changed to **Project Manager's Cloud+***

7. Similarly, click on another markup to change the subject.

Procedure for Changing the Subject of Multiple Markups of the Same Type

The following is the procedure for changing the subject of multiple markups of the same type.

1. Hold down the SHIFT key and select multiple markups of the same type; the **Properties** panel is displayed.

   Note that when you select multiple markups, only the properties that can be modified together are highlighted. The properties that cannot be modified together are greyed out.

2. Scroll to the top where it shows the **Subject** field.
3. In the **Subject** field, type the required subject for the selected markups.
4. Press ENTER; the subject is disappeared.

   The reason the subject is disappeared is that you have selected multiple markups.

5. Click anywhere outside on the PDF file to deselect the markups.
6. Click on one of the markups you selected earlier and notice the subject is changed.
7. Similarly, one by one select the other markups and notice the subject of all the selected markups is changed.

> ***What I do***
>
> *Once you customize a tool, you can then right-click on it and select the* ***Set as Default*** *option from the shortcut menu. This will make the custom version of the tool as the default tool. However, I am not a big fan of this as I prefer adding the custom tools on my own custom tool sets that I can share with the rest of my teams. Also, this ensures that when I install the new version of Revu, I can still import and use my custom tools.*

ADDING MARKUPS TO THE CUSTOM TOOL SETS

Revu provides you a number of methods of adding markups to the custom tool sets. The two main methods are discussed next.

Adding a Markup to the Custom Tool Set from the Main Workspace

If you have customized a markup and updated its subject, you can then add that markup to the custom tool set from the main workspace. The following are the steps to do this:

1. Right-click on the markup in the main workspace; a shortcut menu is displayed.
2. In the shortcut menu, click **Add to Tool Chest > <Name of your Tool Set>**. Figure 6 shows a cloud+ markup being added to the **Project Manager Review** tool set.

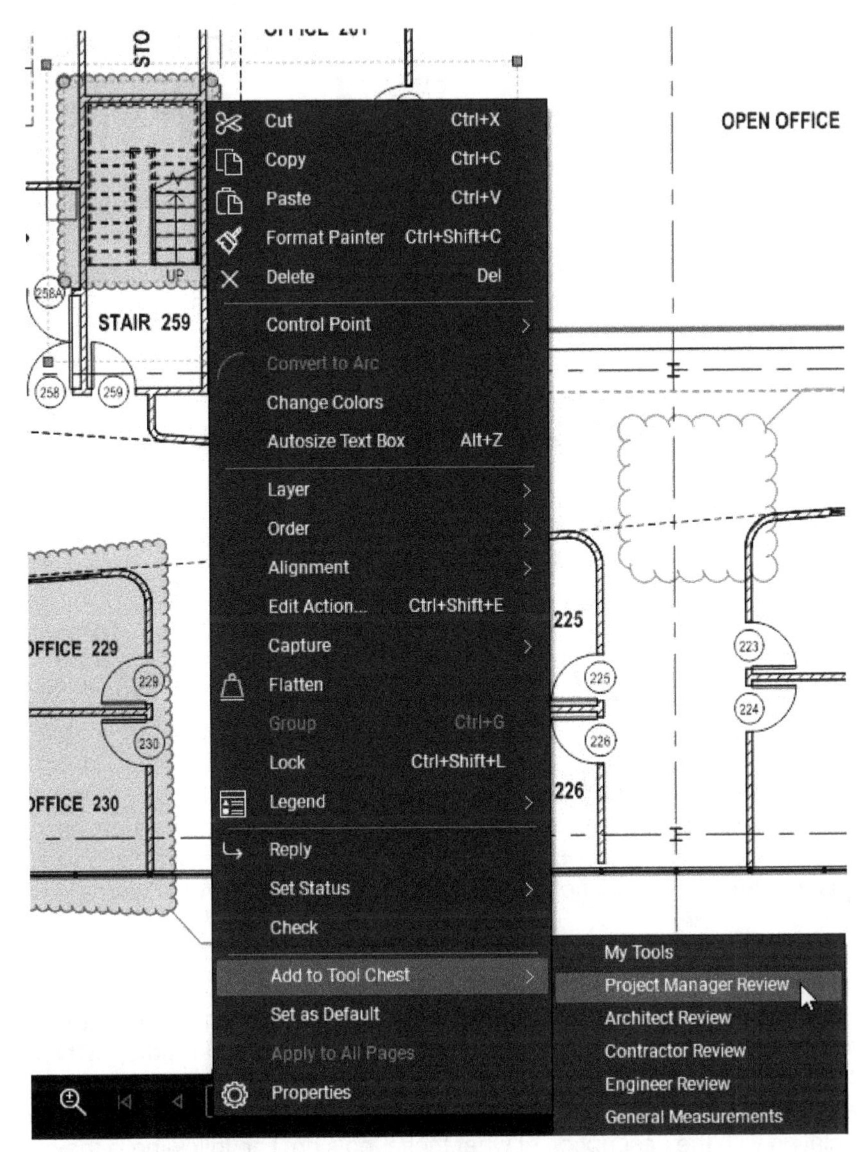

***Figure 6** Adding a custom markup to a custom tool set*

On doing this, the custom markup is added to the custom tool set. By default, the markup is added in the **Drawing** mode. As discussed in earlier chapters, you can double-click on the markup to change it to the **Properties** mode. Figure 7 shows the custom cloud+ markup added to the **Project Manager Review** tool set and changed to the **Properties** mode.

***Figure 7** The cloud+ markup added to the **Project Manager Review** tool set*

Adding a Markup to the Custom Tool Set from the Recent Tools Tool Set

If you have customized a markup and it is available on the **Recent Tool** tool set, then you can add it to your custom tool set. The following are the steps to do this:

1. Press and hold down the left mouse button on the custom tool in the **Recent Tools** tool set.

2. Drag and drop the custom tool on to your custom tool set. Figure 8 shows a text box markup being dragged and dropped from the **Recent Tools** tool set on to the **Project Manager Review** tool set.

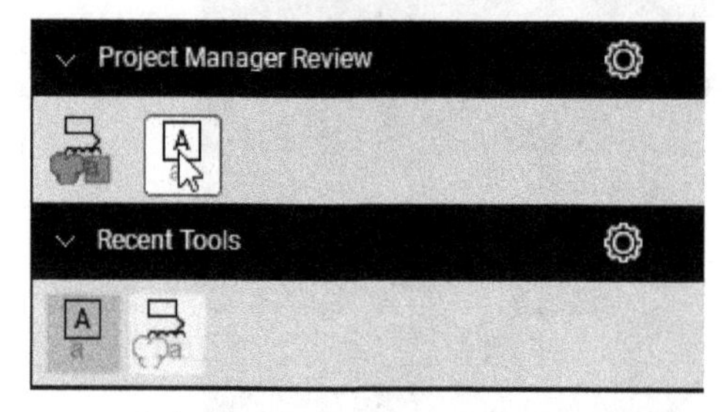

Figure 8 Dragging and dropping a custom tool from the ***Recent Tools*** *tool set*

MANAGING CUSTOM TOOL SETS

Once you have created your custom tool sets, you can change their view mode, export tool sets, set the scales for the tools in the tool set, change the size of their symbols, and so on. Some of these tools to manage the tool sets are discussed next and the remaining are discussed in later chapters.

Changing the View Mode of a Tool Set

By default, the tools are displayed in a tool set in the **Symbol** view. In this view, the tools appear as the icon of the original tool (**Properties** mode) or as the original markup **(Drawing** mode). However, if you are not familiar with various tool icons, it gets hard to figure out which icon in the tool set does what. In this case, these tools should be displayed in the **Detail** view mode in which they are displayed with the description of what these tools do. The following is the procedure for changing the view mode of a tool set:

1. Add your custom tools to your tool set. Figure 9 shows multiple tools added to the **Project Manager Review** tool set. These tools are currently displayed in the **Symbol** view.

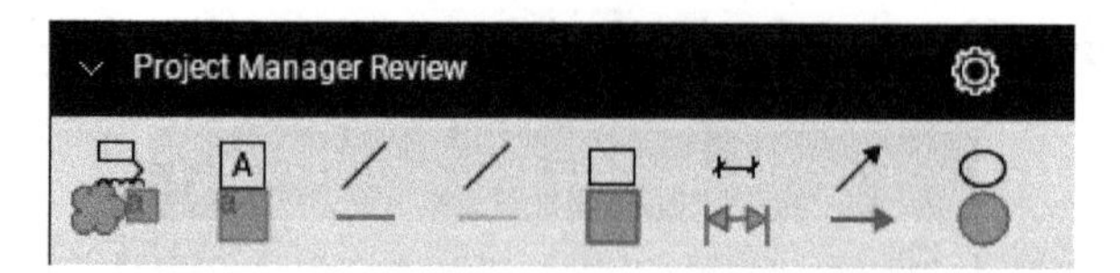

Figure 9 The tools in the custom tool set displayed in the ***Symbol*** *view*

2. To change the view of the tools in the tool set, click on the cogwheel on the top right of the tool set and select **Detail** from the shortcut menu, as shown in Figure 10.

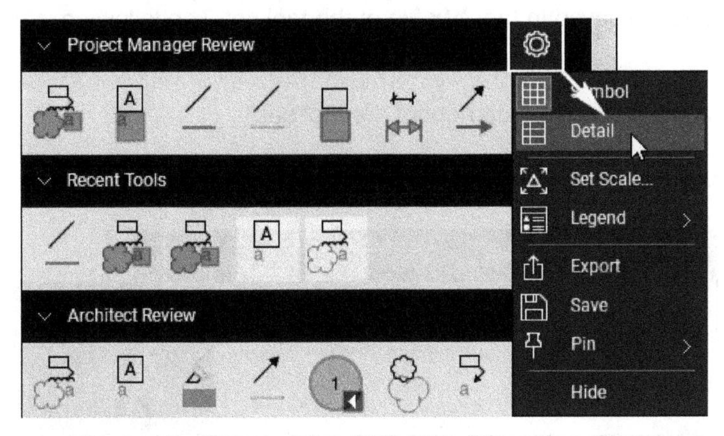

Figure 10 Changing the view mode of the tools to ***Detail***

On doing so, the custom tools are displayed with details, as shown in Figure 11.

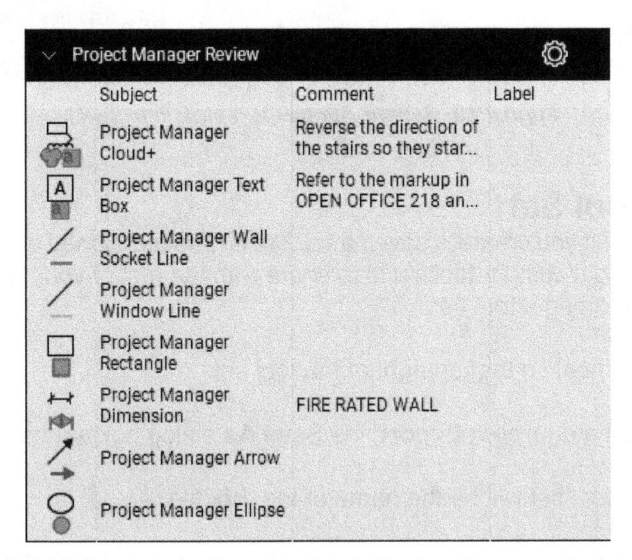

Figure 11 The tools in the tool set displayed in ***Detail*** *view*

Tip*: In the* ***Detail*** *view, you can double-click on the subject or comment of the tool to edit those values. However, it is noticed that changing the subject using this method does not assign that subject to the new markup added using this tool.*

Saving Changes to a Tool Set

Once you make changes to a tool set or add new tools to it, you need to save the tool set to ensure the changes are saved in the **.btx** file of the tool set. The following is the procedure for doing this:

1. Click on the cogwheel on the top right of the tool set.

2. From the shortcut menu, click **Save**, as shown in Figure 12.

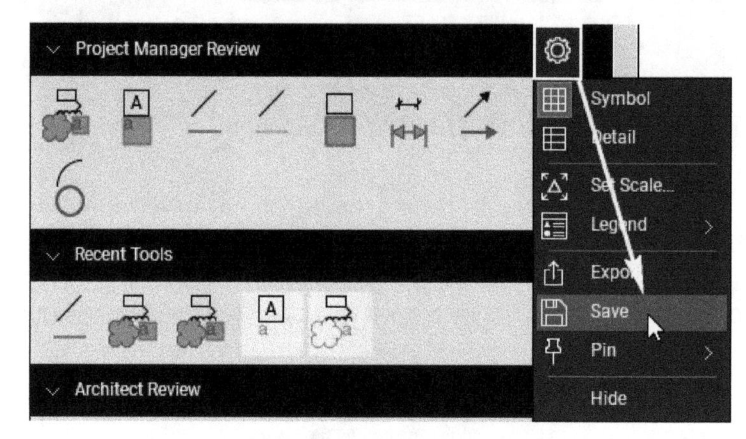

Figure 12 *Saving changes to a custom tool set*

Exporting a Tool Set

The custom tool set that you create is saved as a **.btx** file. As mentioned earlier, one of the main advantages of creating a custom tool set is to share with the rest of your team. The following is the procedure for exporting a tool set:

1. Click on the cogwheel on the top right of the tool set.

2. From the shortcut menu, click **Export**; the **Save As** dialog box will be displayed.

   The name of the tool set will be the name of the **.btx** file.

3. Browse to the location where you want to save the tool set and click **Save** in the dialog box.

Importing a Tool Set

To standardize the design review and markup process in your organization, you can import the tool sets exporter by your team member. The following is the procedure for doing this:

1. From the top of the tool sets, click **Tool Chest > Manage Tool Sets**, as shown in Figure 13; the **Manage Tool Sets** dialog box is displayed.

2. From the bottom left in the dialog box, click **Import**; the **Open** dialog box is displayed.

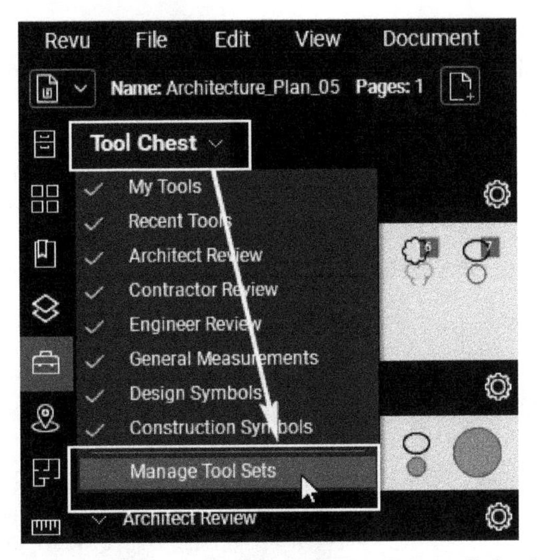

Figure 13 Invoking the ***Manage Tool Sets*** *dialog box*

3. Browse to the folder where the **.btx** file of the tool set is saved.

4. Double-click on the **.btx** file; the tool set is added at the bottom of the list in the **Manage Tool Sets** dialog box, as shown in Figure 14.

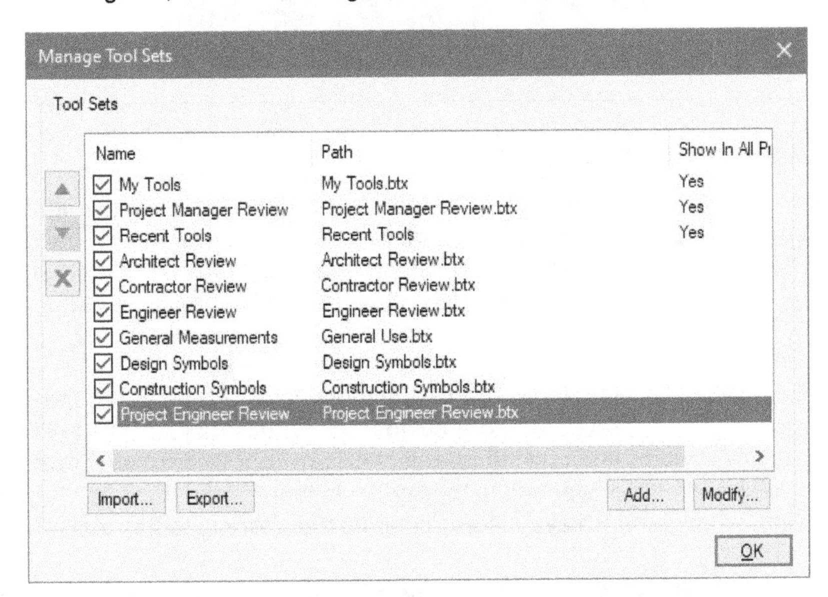

Figure 14 The imported tool set listed in the dialog box

5. Select the tool set and then use the **Up** or **Down** keys to move this tool set up or down in the list.

6. Click **OK** in the **Manage Tool Sets** dialog box.

Deleting a Tool from a Tool Set

If you added a tool to a tool set by mistake, you can simply delete it from the tool set. The following is the procedure for doing this:

1. In the custom tool set, right-click on the tool that was added by mistake; a shortcut menu is displayed.

2. From the shortcut menu, click **Delete**, as shown in Figure 15; the selected tool will be deleted from the tool set.

Figure 15 *Deleting a tool from a tool set*

THE MY TOOLS TOOL SET

The **My Tools** tool set is a permanent part of Revu and is available in all profiles. This tool set has a special property that automatically assigns a numeric shortcut key to the tools added to this tool set. The shortcut key is displayed on the top right of the tool in this tool set. As you drag and drop the tools before or after another tool in this toolset, their shortcut keys are automatically updated.

By default, this tool set has eight tools added and those tools are assigned the numeric keys from 1 to 8 as their shortcut keys. To add a custom tool to this tool set, right-click on it in the main workspace and from the shortcut menu, select **Add to Tool Chest > My Tools**, as shown in Figure 16. On doing so, the tool is added to the **My Tools** tool set and the next available numeric key or a combination of numeric keys are assigned as the shortcut to this tool.

Figure 17 shows the **My Tools** tool set with six additional tools added, which are assigned numeric keys 9 to 14 as their shortcut keys. You can press these keys on the keyboard to invoke their respective markup tools.

***Tip**: The slide bar available at the bottom of the **Tool Chest** panel is used to increase or decrease the size of the tools in all the tool sets.*

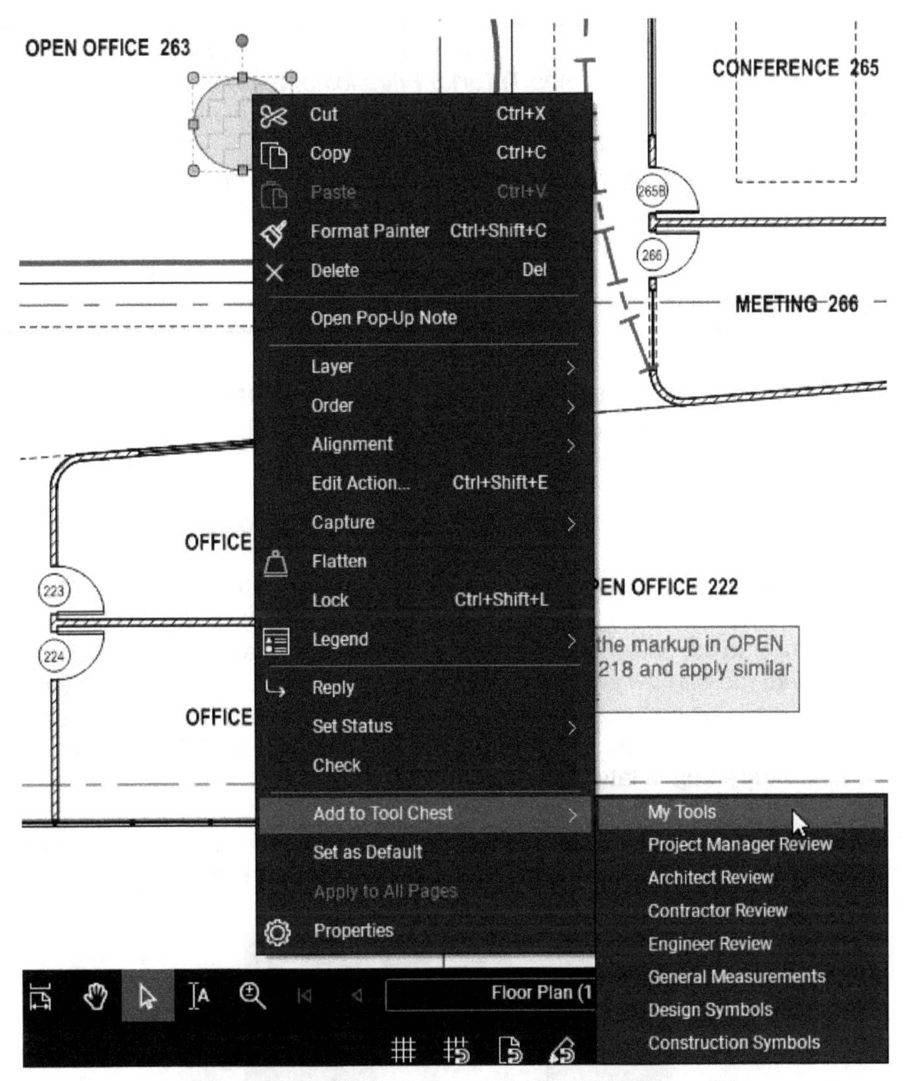

Figure 16 *Adding a custom tool to the* ***My Tools*** *tool set*

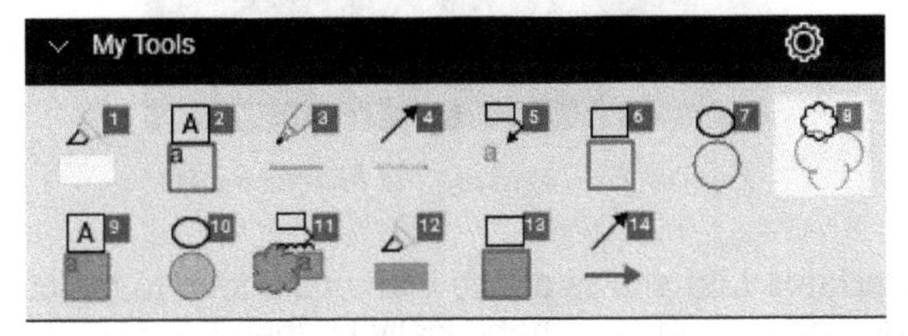

Figure 17 *The* ***My Tools*** *tool set with additional tools added*

Hands-on Tutorial

In this tutorial, you will complete the following tasks:

1. *Use the pinned* ***Tutorial Files*** *folder to open a PDF file.*
2. *Change the subject of various markups in the PDF file.*
3. *Add two custom tool sets.*
4. *Add the custom tools from the PDF file to one of the custom tool sets.*
5. *Change the view mode of the custom tool set.*
6. *Import a tool set.*
7. *Create a Punchlist tool set by importing a CSV file.*
8. *Add tools to the* ***My Tools*** *tool set.*

Section 1: Changing the Subject of the Custom Markups

In this section, you will open the **Architecture_Plan_05.pdf** file from the **C05** folder. You will then change the subject of the custom tools in this PDF file.

1. Start Revu 2019 and restore the **Revu Training** profile that you created in Chapter 1.

2. From the **File Access > Explorer** tab, open the **C05 > Architecture_Plan_05.pdf** file.

   This file is similar to the one you worked on in the previous chapter and has various custom markups added. You will now edit the subject of these markups. It is easier to use the **Markup List** to select markups and change their subject. It is available below the **Main Workspace** by default and displays all the markups added to the current PDF file. You will learn more about the **Markups List** in later chapters.

3. From the bottom left of the Revu window, click the **Markups** button, as shown in Figure 18; the **Markups List** is displayed below the main workspace.

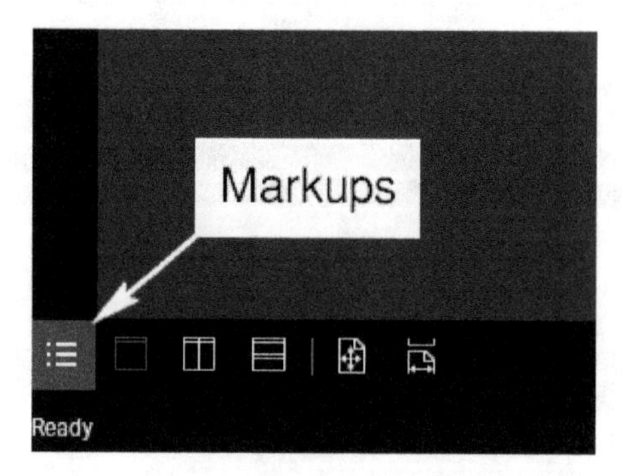

Figure 18 Invoking the ***Markup List***

Notice that the **Markups List** shows all the markups added to the current PDF file. These markups are sorted and grouped alphabetically by their subjects, as shown in Figure 19. You can use the divider line to resize the **Markups List**. You will now select the markups and change their subjects. In this case, you will add **Project Manager's** as the prefix to the existing subject of the markups.

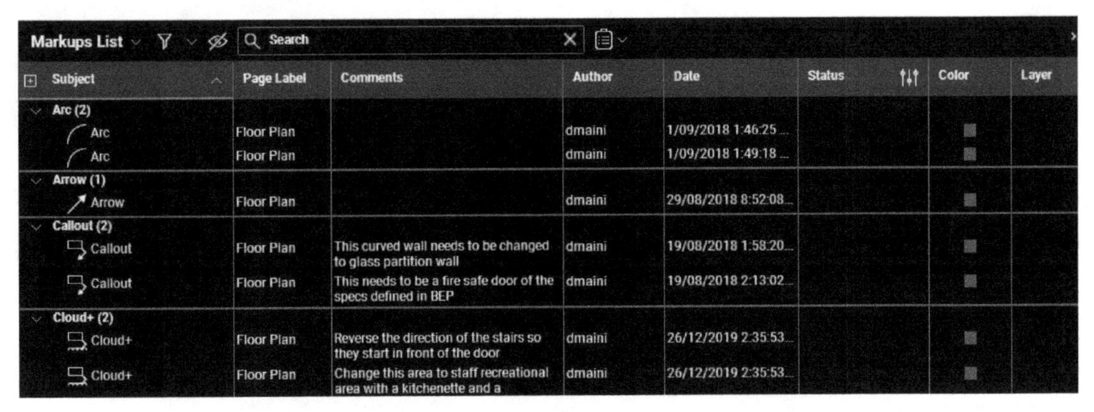

*Figure 19 The **Markups List** showing the markups sorted and grouped by their subjects*

4. Hold down the SHIFT key and select the two **Arc** markups from the **Markups List**; the two markups are highlighted in the PDF file and also the **Properties** panel shows their subject as **Arc**.

5. In the **Properties** panel > **Subject** field, add **Project Manager's** as the prefix to the existing subject. Make sure you add a space after the prefix so that the subject now reads **Project Manager's Arc**, as shown in Figure 20.

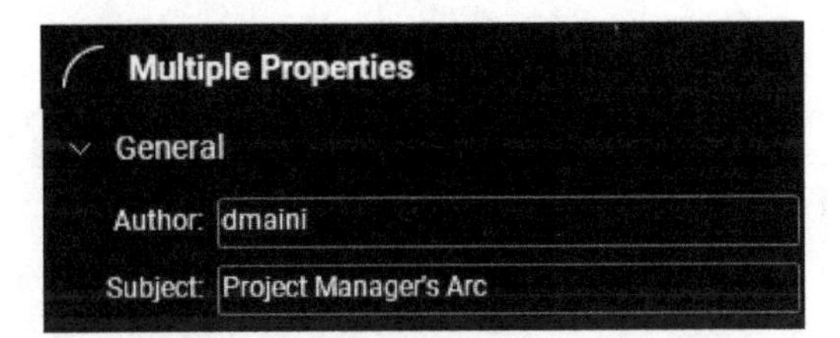

Figure 20 The subject of the two arcs changed

Once the subject is changed, the two arcs will move down in the **Markups List** as the markups in this list are sorted alphabetically by their subjects.

6. Scroll to the top of the **Markups List** and select the **Arrow** markup.

7. In the **Properties** panel > **Subject** field, add **Project Manager's** as the prefix to the existing subject. Make sure you add a space after the prefix so that the subject now reads **Project Manager's Arrow**, as shown in Figure 21.

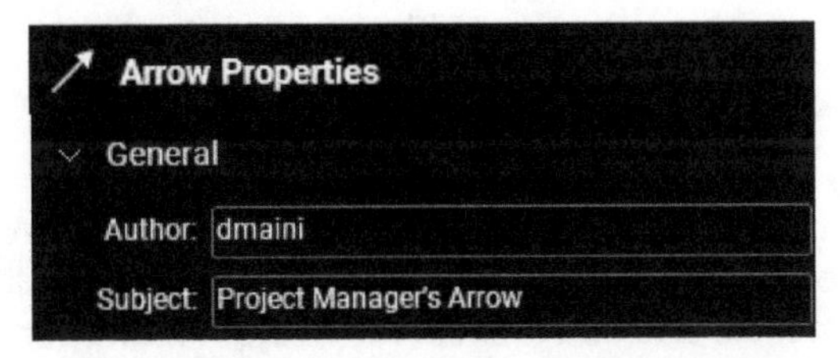

Figure 21 The subject of the arrow markup changed

8. Similarly, change the subject of all the rest of the markups, except the **Line** markups. Make sure you scroll down in the **Markups List** and update the **Rectangle**, **Text Box**, and **Typewriter** markups as they are displayed below the various **Project Manager's** markups.

   The reason you did not change the **Line** markups is that there are two different types of lines. These lines will be assigned different subjects.

9. Hold down the CTRL key and select the four Blue color Wall Socket lines in the **Markups List**.

10. In the **Properties** panel > **Subject** field, enter **Project Manager's Wall Socket** as the prefix to the current subject. The subject field should read **Project Manager's Wall Socket Line** as the subject.

11. In the **Markups List** select the remaining lines which are Green color and add **Project Manager's Window** as the prefix to the current subject. The subject field should read **Project Manager's Window Line** as the subject.

    The **Markups List**, after changing the subject of all the **Line** markups, is shown in Figure 22.

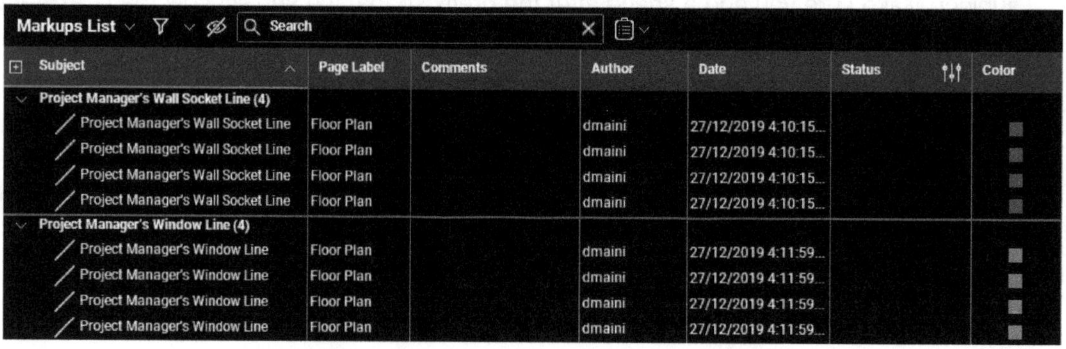

Markups List | Search

Subject	Page Label	Comments	Author	Date	Status	Color
Project Manager's Wall Socket Line (4)						
Project Manager's Wall Socket Line	Floor Plan		dmaini	27/12/2019 4:10:15...		
Project Manager's Wall Socket Line	Floor Plan		dmaini	27/12/2019 4:10:15...		
Project Manager's Wall Socket Line	Floor Plan		dmaini	27/12/2019 4:10:15...		
Project Manager's Wall Socket Line	Floor Plan		dmaini	27/12/2019 4:10:15...		
Project Manager's Window Line (4)						
Project Manager's Window Line	Floor Plan		dmaini	27/12/2019 4:11:59...		
Project Manager's Window Line	Floor Plan		dmaini	27/12/2019 4:11:59...		
Project Manager's Window Line	Floor Plan		dmaini	27/12/2019 4:11:59...		
Project Manager's Window Line	Floor Plan		dmaini	27/12/2019 4:11:59...		

*Figure 22 The **Markup List** after changing the subject of the **Line** markups*

12. Click the **Markups** button again to hide the **Markups List**, refer to Figure 18.

13. Zoom to the extents of the PDF file by double-clicking the Wheel Mouse Button.

14. Click on some of the markups in the main workspace and notice their subjects that show **Project Manager's** as a prefix to the original subject of that markup.

15. Save the PDF file.

Section 2: Creating Custom Tool Sets

In this section, you will create two custom tool sets with the name **Project Manager Markups** and **Project Manager Measurements**. Later in this book, you will learn how to customize various measurement tools and add them to the **Project Manager Measurements** tool set.

1. If the **Tool Chest** is not already turned on, click on the **Tool Chest** button from the **Panel Bar** to turn it on.

2. From the top of the tool sets, click **Tool Chest > Manage Tool Sets**, as shown in Figure 23; the **Manage Tool Sets** dialog box is displayed.

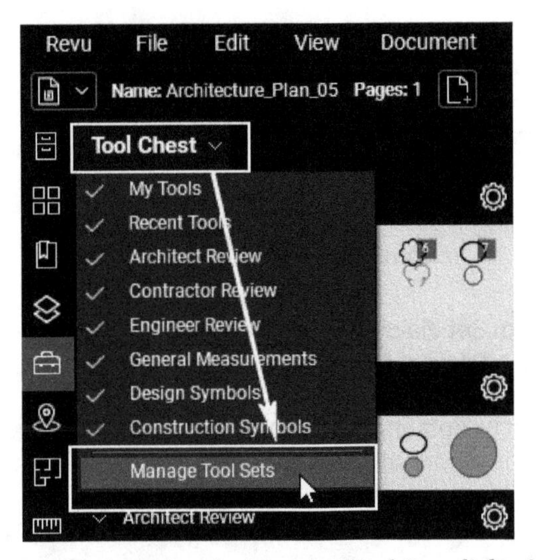

*Figure 23 Invoking the **Manage Tool Sets** dialog box*

The **Manage Tool Sets** dialog box displays all the tool sets that are currently available. The tool sets with a check box on the left of their names are the ones that are available in the current profile. Note that there are a number of other tool sets that are not displayed in the current profile and are not listed in the dialog box.

3. From the bottom right in the **Manage Tool Sets** dialog box, click **Add**; the **Add Tool Set** dialog box is displayed.

   By default, **New** is selected in the **Type** area. This ensures that a new tool set is going to be created.

4. In the **Title** field, type **Project Manager Markups**, as shown in Figure 24. This will be the name of the new tool set.

5. In the **Options** area, select the **Show in All Profiles** check box, as shown in Figure 24. This ensures that this tool set will be displayed in all profiles.

***Tip:** It is important to remember that if a tool set has the option to **Show in All Profiles** turned on, then you cannot delete this tool set. You will first have to double-click on the tool set and then in the **Modify Tool Set** dialog box, clear the **Show in All Profiles** check box to be able to delete the tool set.*

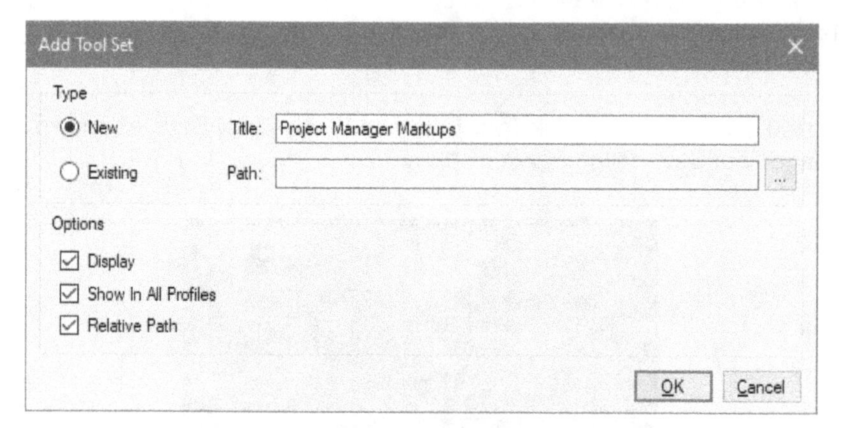

Figure 24 Creating a new tool set

6. Click **OK** in the **Add Tool Set** dialog box; the **Save As** dialog box is displayed that prompts you to save the tool set as a **Project Manager Markups.btx** format file, as shown in Figure 25.

   Notice in Figure 25 that by default, this file is being saved in the roaming profile of the current user. This is the location where all the default tool sets are also saved.

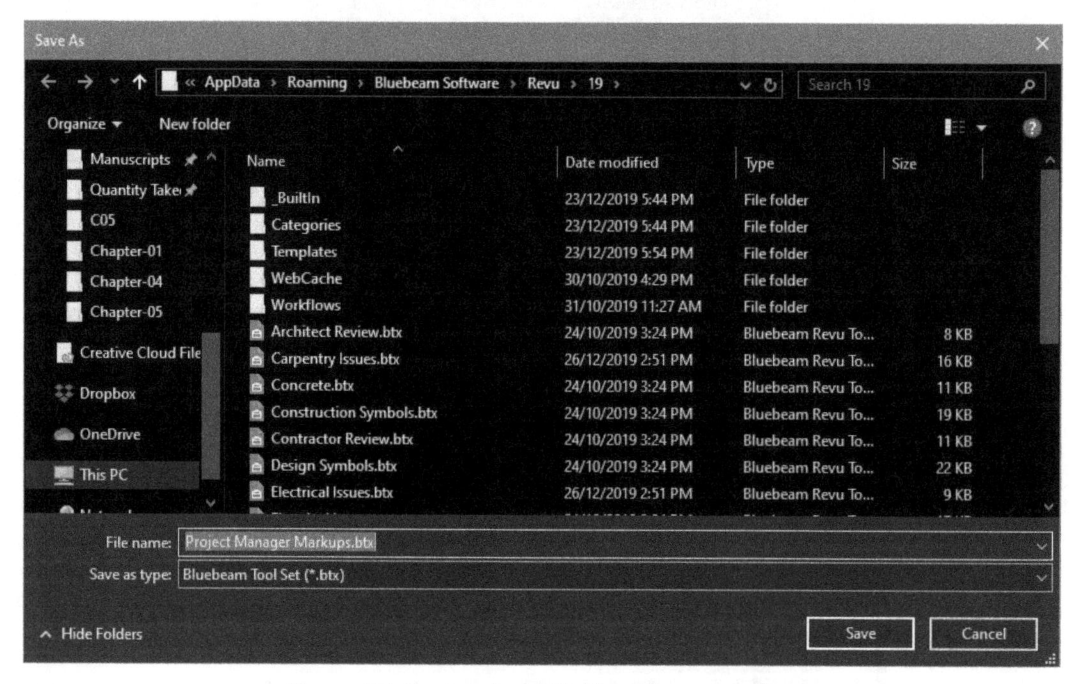

Figure 25 Saving the BTX file of the new tool set

The reason a new **.btx** format file is created is so that you can distribute this file to the rest of your team and standardize the process of reviewing and marking up the PDF files in your organization. This file is by default saved in the roaming profile folder of the current user.

7. Click **Save** in the **Save As** dialog box; the new tool set is created and you are returned to the **Manage Tool Sets** dialog box.

   The new tool set is by default organized at the bottom of all the tool sets. You can reorder it and move it above the rest of the tool sets.

8. In the **Manage Tool Sets** dialog box, select the **Project Manager Markups** tool set; the **Up** arrow button is highlighted on the left side of the dialog box, labeled as **1** in Figure 26.

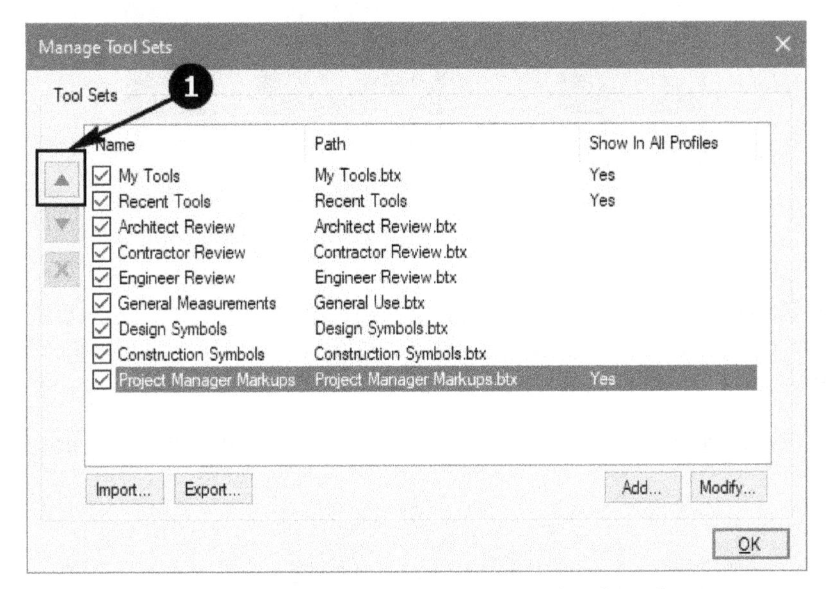

*Figure 26 The **Up** arrow button to reorder the tool set*

9. Use the **Up** arrow button to move up the **Project Manager Markups** tool set and place it second in the list, below the **My Tools** tool set. This is the order in which it will be displayed in the **Tool Chest** panel.

***Tip:** As you reorder the tool sets in the **Manage Tool Sets** dialog box, you can see them moving up and down in the **Tool Chest** panel as well.*

10. Similarly, create another tool set with the name **Project Manager Measurements** and organize it below the **Project Manager Markups** tool set.

11. Click **OK** in the **Manage Tool Sets** dialog box; the new tool sets are displayed in the **Tool Chest** panel below the **My Tools** tool set, as shown in Figure 27.

    At this stage, these two tool sets are blank. You will now add the customized markups from the PDF file to the **Project Manager Markups** tool set.

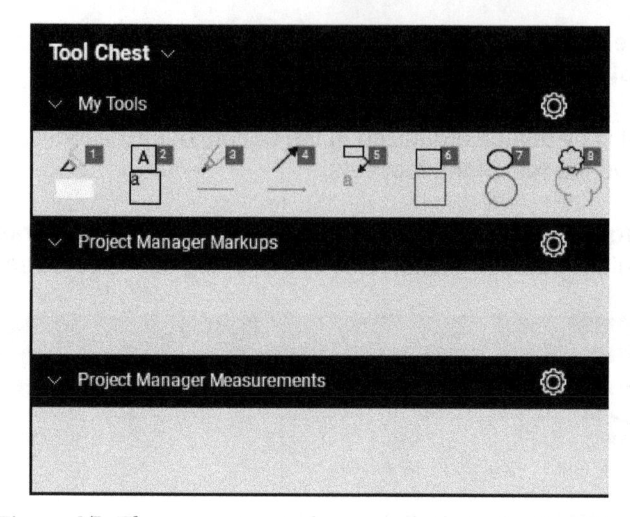

Figure 27 The two custom tool sets available in the ***Tool Chest***

Section 3: Adding Customized Markups to the Custom Tool Set

In this section, you will add the customized markups in the PDF file to the **Project Manager Markups** tool set.

1. Navigate to the center of the PDF file so you can see most of the markups, as shown in Figure 28.

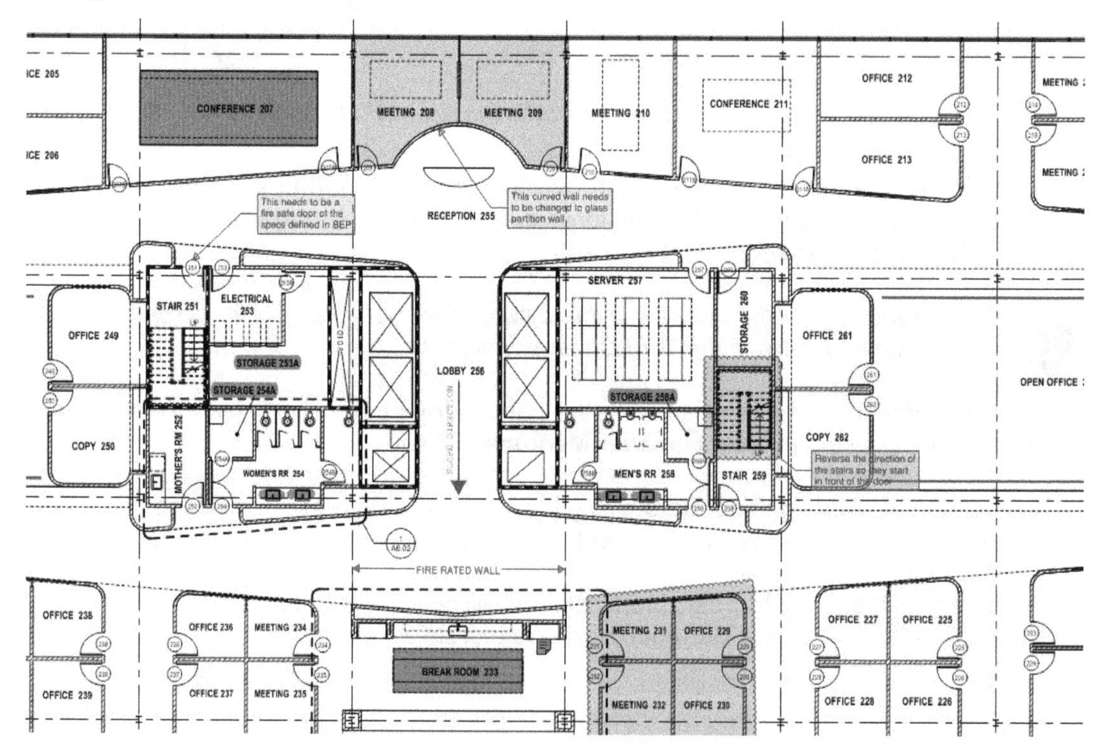

Figure 28 Navigating to the center of the PDF file

The first customized markup that you will add to the **Project Manager Markups** tool set is the polygon markup placed in the **MEETING 209** room at the top center of the PDF file.

2. Select the polygon markup created in the **MEETING 209** room; the subject in the **Properties** panel is displayed as **Project Manager's Polygon**.

3. Right-click on this markup and from the shortcut menu, select **Add to Tool Chest > Project Manager Markups**, as shown in Figure 29.

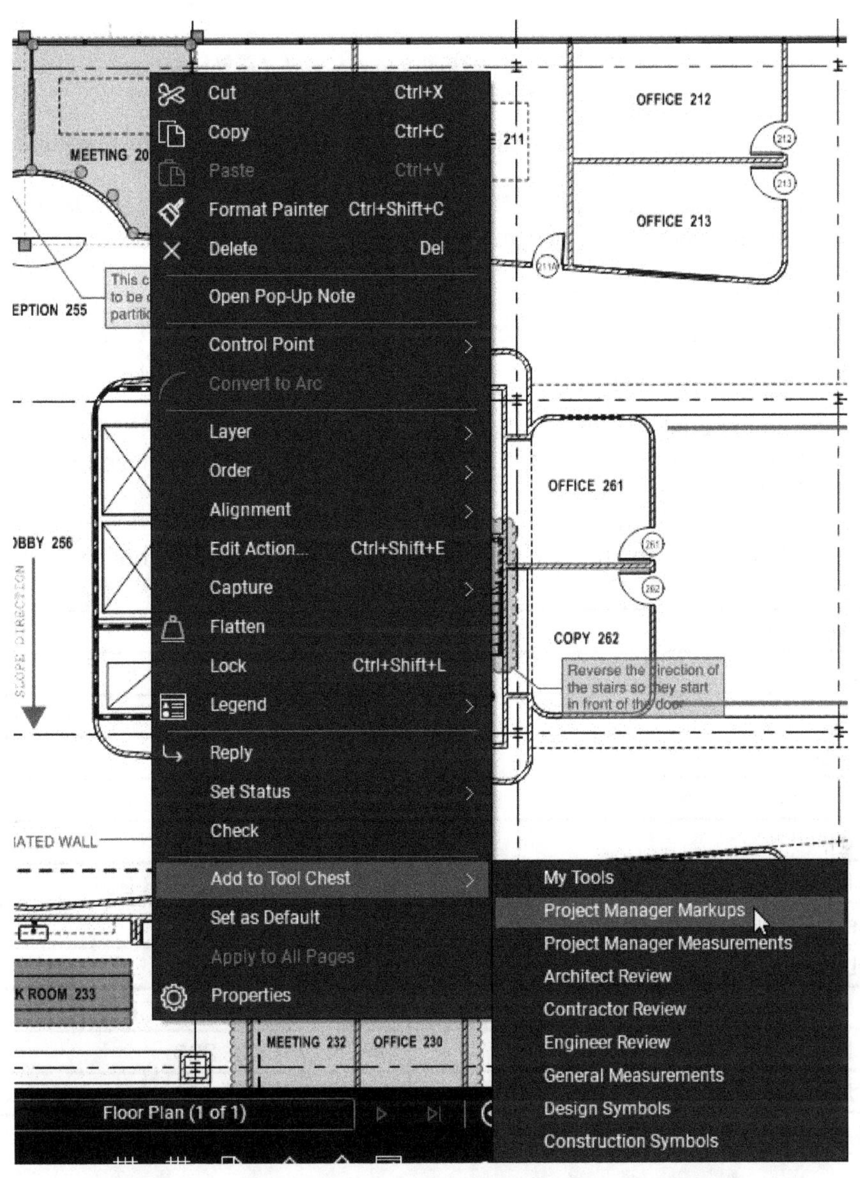

Figure 29 Adding the polygon markup to the custom tool set

On doing so, the customized polygon tool is added to the **Project Manager Markups** tool set, as shown in Figure 30. As evident in this figure, the polygon tool is added in the **Drawing** mode. In this case, you will add all the customized markup tools to the tool set first and then change all of them to the **Properties** mode.

Figure 30 *The customized polygon tool added to the tool set*

Next, you will add the customized callout tool to the **Project Manager Markups** tool set.

4. On the callout markup below **MEETING 209** room, right-click and select **Add to Tool Chest > Project Manager Markups**, refer to Figure 29; the customized callout tool is added to the tool set.

   Next, you will add the customized rectangle markup added to **CONFERENCE 207** room to the custom tool set.

5. Right-click on the customized rectangle markup added to **CONFERENCE 207** room and select **Add to Tool Chest > Project Manager Markups**; the customized rectangle tool is added to the tool set.

6. Similarly, add the following tools to the **Project Manager Markups** tool set:

Markup Tool	**Room No**
Typewriter	LOBBY 256
Arrow	LOBBY 256
Dimension	Below LOBBY 256
Note	BREAK ROOM 233
Cloud+	STAIR 259
Highlight	Sink below MEN'S RR 258
Ellipse	OPEN OFFICE 248
PolyLine	OPEN OFFICE 248
Arc	Right of CONFERENCE 246
Text Box	OPEN OFFICE 200
Window Line	OPEN OFFICE 200
Wall Socket Line	OPEN OFFICE 200

Note*: Remember that the highlight markup added to a text cannot be added to a custom tool set. Only the highlight added to non-text areas of the PDF can be added to the custom tool set.*

Figure 31 shows the **Project Manager Markups** tool set after adding all the customized markup tools.

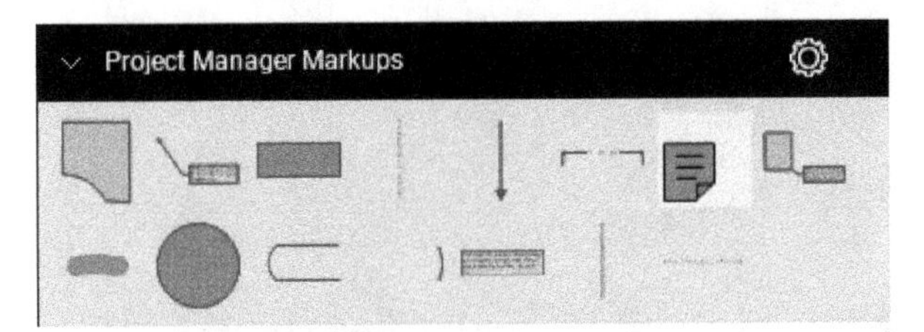

*Figure 31 The **Project Manager Markups** tool set after adding the customized tools*

Section 4: Managing Custom Tool Set

In this section, you will change all the customized markup tools added to the **Project Manager Markups** tool set to the **Properties** mode. You will then use the **Size** slider bar at the bottom of the tool sets to change the size of the tool buttons in the tool sets. Finally, you will use the cogwheel available on the top right of the **Project Manager Markups** tool set to manage this custom tool set.

1. One by one, double-click on all the customized tools in the **Project Manager Markups** tool set to change them to the **Properties** mode. After double-clicking on the last tool, press the ESC key to exit out of that tool. Figure 32 shows the **Project Manager Markups** toolset after changing all the tools to the **Properties** mode.

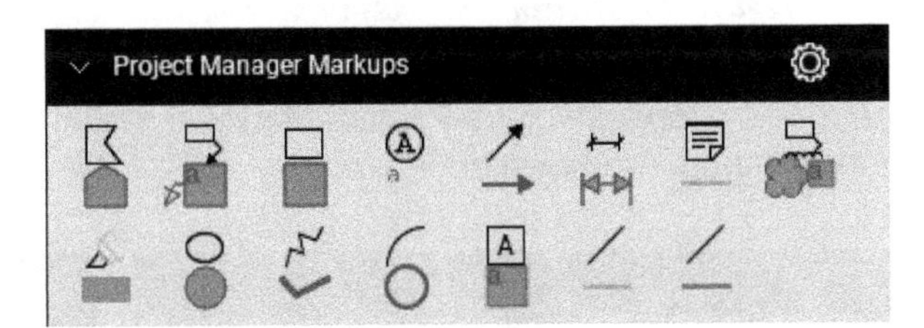

*Figure 32 The **Project Manager Markups** tool set after changing all the tools to the **Properties** mode*

You will now use the **Size** slider bar at the bottom of the tool sets to change the size of the tools in the tool sets.

2. From the bottom of the tool sets, use the **Size** slider, labeled as **1** in Figure 33, to change the size of the tool buttons in the tool sets.

   Notice that dragging the **Size** slider to the right increases the size of the tool buttons in the tool sets and dragging it to the left reduces the size of the tool buttons.

3. Adjust the **Size** slider to size the tool buttons that you prefer. I prefer leaving it to around a quarter of the way.

Next, you will change the view of the **Project Manager Markups** tool set to **Details** view.

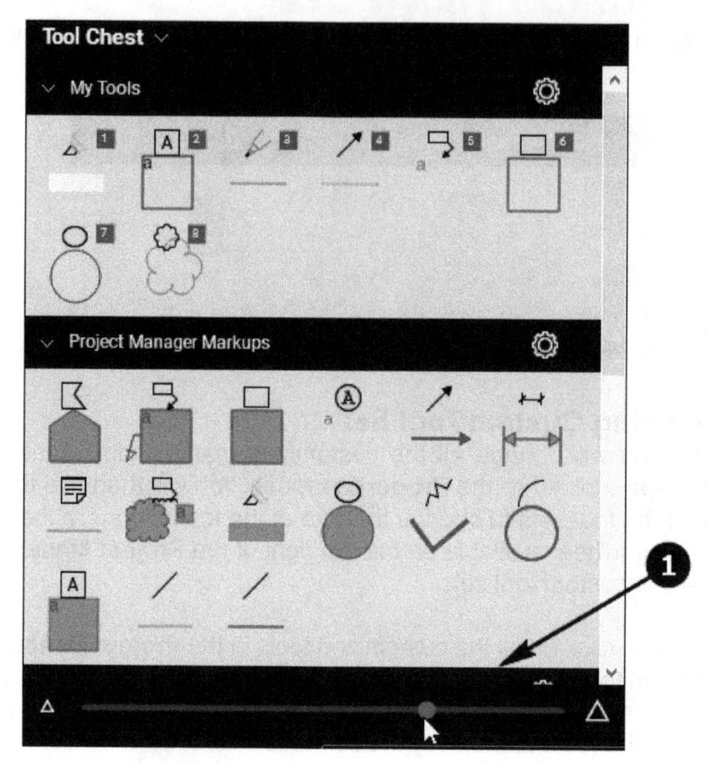

Figure 33 Resizing the tools in the tool sets

4. Click on the cogwheel on the top right of the **Project Manager Markups** tool set and select **Detail** from the shortcut menu, as shown in Figure 34.

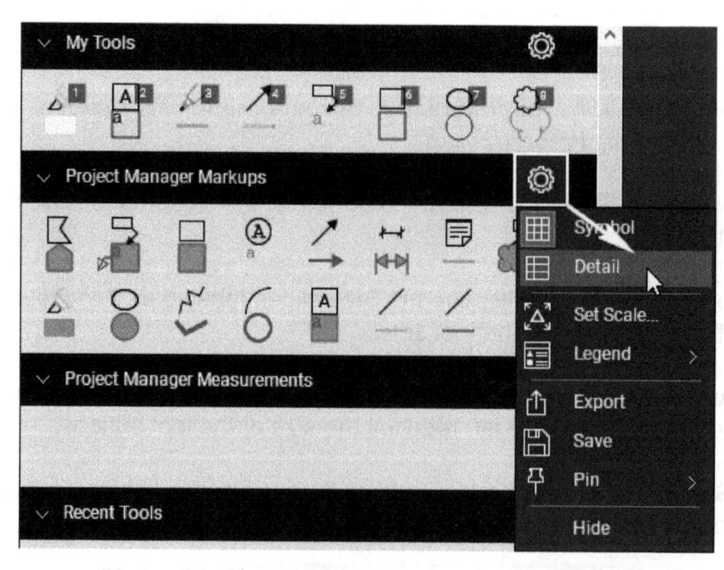

*Figure 34 Changing the tool set to the **Detail** view*

On doing so, the tools in the **Project Manager Markups** tool set are displayed with details, as shown in Figure 35. Notice that the subject of all the tools have **Project Manager's** as a prefix.

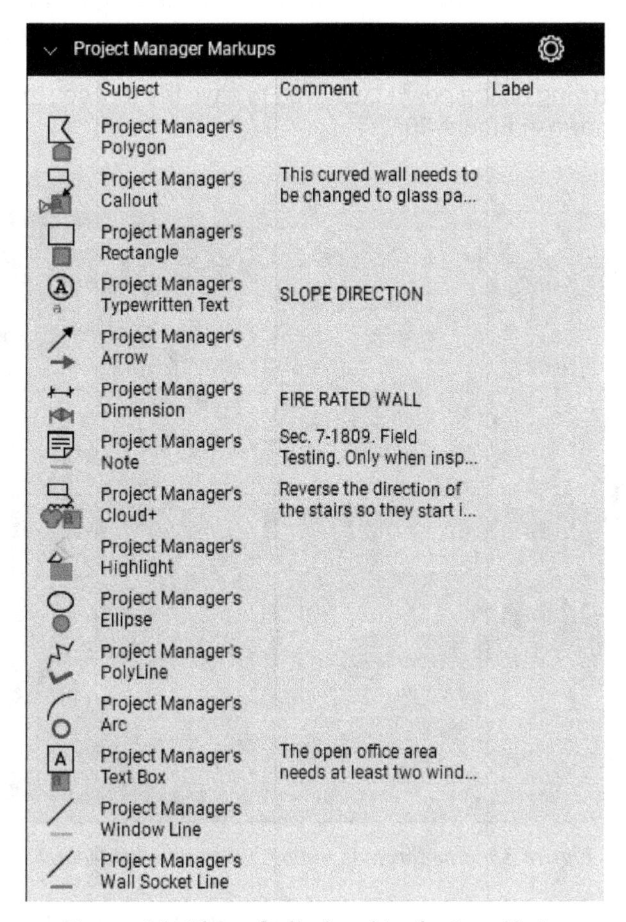

***Figure 35** The tools displayed in the **Detail** view*

5. Use the **Size** slider at the bottom of the tool sets to resize the tools to your preferred size.

   Next, you will save this tool set.

6. Click on the cogwheel on the top right of the **Project Manager Markups** tool set and select **Save** from the shortcut menu; the changes made to the toolset are now saved.

   Next, you will export this tool set so you can share it with the rest of your team. But before doing that, you will restore the **Symbol** view.

7. Click on the cogwheel on the top right of the **Project Manager Markups** tool set and select **Symbol** from the shortcut menu; the tools in the tool set are changed back to the symbols.

8. Click on the cogwheel on the top right of the **Project Manager Markups** tool set and select **Export** from the shortcut menu.

   In case the tool set is not saved after you made changes to it, a **Warning** dialog box may be displayed informing you that you must save your tool set before exporting.

9. Click **OK** if the **Warning** dialog box is displayed to save the tool set; the **Save As** dialog box is displayed, as shown in Figure 36.

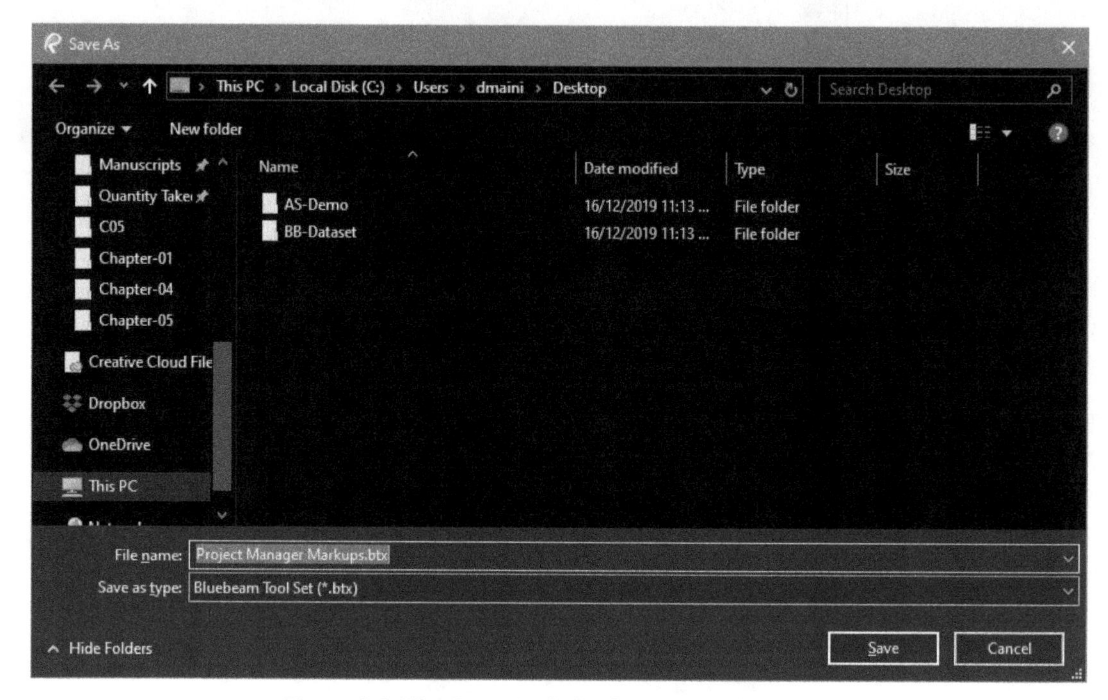

*Figure 36 The **Save As** dialog box to save the tool set*

Notice that the file is being saved with the name of the tool set and extension **.btx**. You will save this tool set in the **Tutorial Files > C05** folder.

10. Browse to the **Tutorial Files > C05** folder.

    Notice that this folder already has a tool set saved in the BTX format. You will import this tool set later in this tutorial.

11. Click the **Save** button in the **Save As** dialog box; the tool set is saved with the name **Project Manager Markups.btx** in the **C05** folder and you are returned to the Revu window.

Section 5: Importing a Tool Set

In this section, you will import tool set from the **Tutorial Files > C05** folder. You will then reorder this tool set to place it below the **Project Manager Measurements** tool set.

1. From the top of the tool sets, click **Tool Chest > Manage Tool Sets**; the **Manage Tool Sets** dialog box is displayed.

2. From the bottom left in this dialog box, click **Import**; the **Open** dialog box is displayed, as shown in Figure 37.

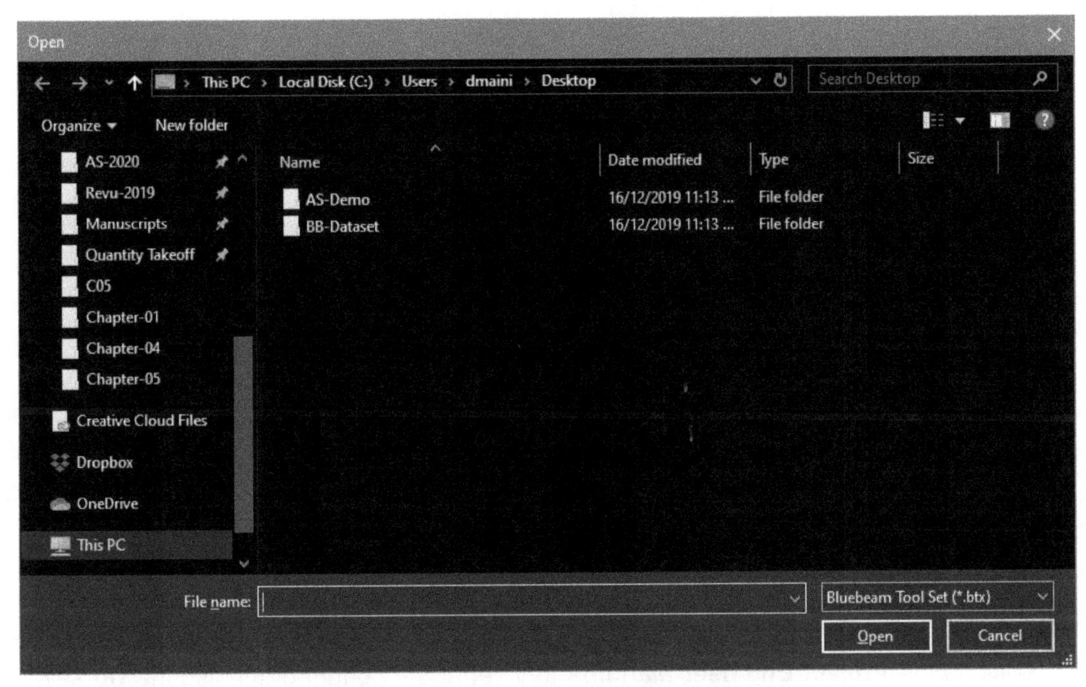

*Figure 37 The **Open** dialog box to import a tool set*

The BTX file of the tool set that you need to import is saved in the **Tutorial Files > C05** folder. You will now browse to this folder.

3. Browse to the **Tutorial Files > C05** folder.

   Notice the two BTX files in this folder. The **Project Engineer Markups.btx** file is the one you need to import. The other file is that of the tool set you saved in the previous section.

4. Double-click on the **Project Engineer Markups.btx** file; this tool set is added at the bottom of the **Manage Tool Sets** dialog box.

   Before you change the order of the tool sets, you will double-click on the name of the **Project Engineer Markups** tool set and turn on the option to display it in all profiles.

5. In the **Manage Tool Sets** dialog box, double-click on the **Project Engineer Markups** tool set; the **Modify Tool Set** dialog box is displayed.

6. In the dialog box, select the **Show in All Profiles** check box, enclosed in a rectangle in Figure 38.

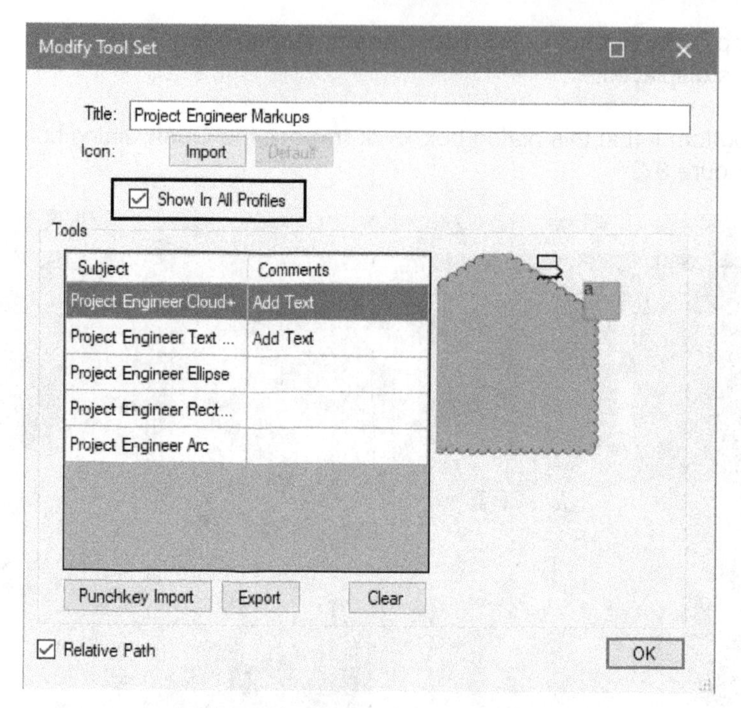

*Figure 38 The **Modify Tool Set** dialog box*

7. Click **OK** in the **Modify Tool Set** dialog box to return to the **Manage Tool Sets** dialog box.

8. Click on the **Project Engineer Markups** tool set at the bottom of the list; the **Up** key is highlighted in the dialog box.

9. Using the **Up** key in the **Manage Tool Sets** dialog box, move the tool set up and place it below the **Project Manager Measurements** tool set.

   As mentioned earlier, while reordering the tool sets in the **Manage Tool Sets** dialog box, you can see those changes live in the **Tool Chest** panel in the Revu window.

10. Click **OK** in the **Manage Tool Sets** dialog box. Figure 39 shows the **Tool Chest** panel with the imported **Project Engineer Markups** tool set.

Section 6: Creating a Punchkey (Defects Symbols) Tool Set Based on a CSV File

Revu allows you to create a new tool set with punchkey symbols, also known as defects symbols, by importing a CSV file that has the content for those symbols. For this tutorial, you are provided with a CSV file to be imported. However, before you import that file, you will open the CSV file to have a look at its content.

1. Using Windows Explorer, browse to the **Tutorial Files > C05** folder.

2. Double-click on the **Punchkeys.csv** file to open it. Figure 40 shows this CSV file.

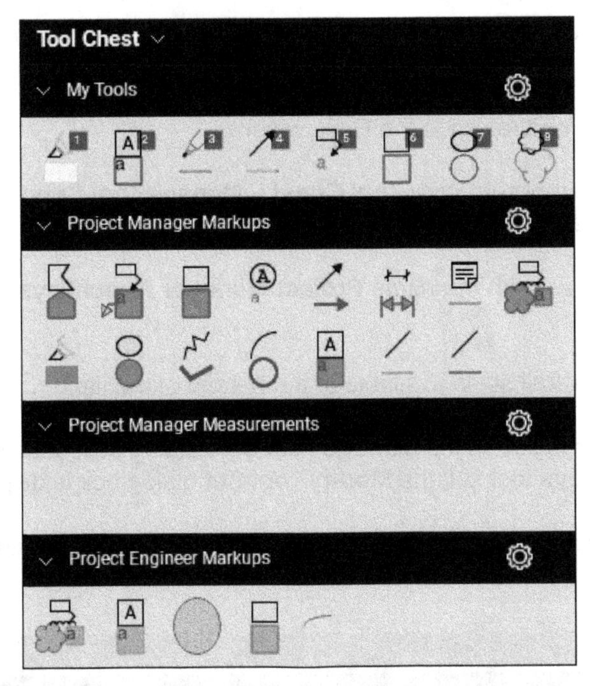

*Figure 39 The imported **Project Engineer Markups** tool set*

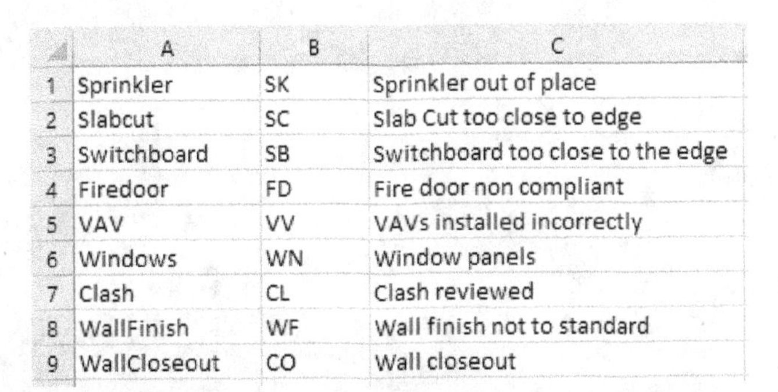

	A	B	C
1	Sprinkler	SK	Sprinkler out of place
2	Slabcut	SC	Slab Cut too close to edge
3	Switchboard	SB	Switchboard too close to the edge
4	Firedoor	FD	Fire door non compliant
5	VAV	VV	VAVs installed incorrectly
6	Windows	WN	Window panels
7	Clash	CL	Clash reviewed
8	WallFinish	WF	Wall finish not to standard
9	WallCloseout	CO	Wall closeout

Figure 40 The CSV file showing the content of punchkeys

Description of the CSV File Content

The following is the description of the content of the CSV file:

Column A

This column has the content that will be used as the subject of the punchkeys or defect symbols.

Column B

This column shows the Characters that will be included in the punchkeys or defect symbols.

Column C

This column will be used as the comments of the punchkeys or defect symbols.

Next, you need to close the CSV file because if it is left open, you will not be able to import its content in Revu.

3. Close the CSV file and return to the Revu window.

4. From the top of the tool sets, click **Tool Chest > Manage Tool Sets**; the **Manage Tool Sets** dialog box is displayed.

5. Create a new tool set with the name **Project Manager Punchkeys** and save the BTX file of this tool set.

   By default, the new tool set is available at the bottom of the list.

6. From the bottom of the list in the **Manage Tool Sets** dialog box, double-click on the **Project Manager Punchkeys** tool set; the **Modify Tool Set** dialog box is displayed.

7. From the bottom left in the **Modify Tool Set** dialog box, click **Punchkey Import**; the **Open** dialog box is displayed, as shown in Figure 41.

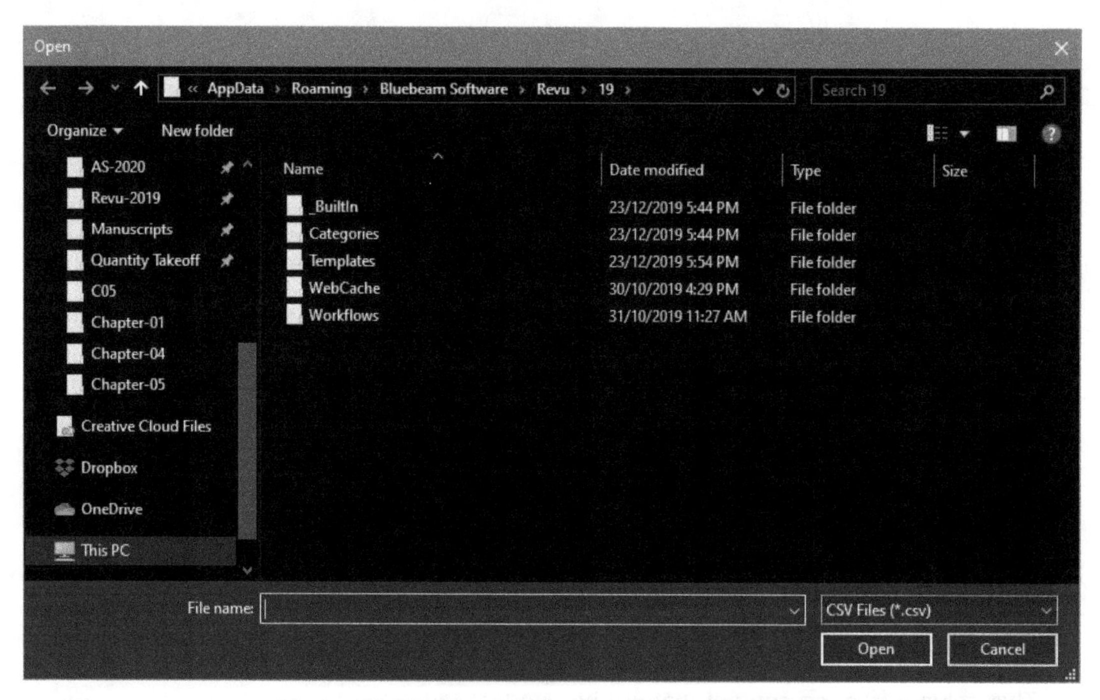

*Figure 41 The **Open** dialog box to import a CSV file*

8. Browse to the **Tutorial Files > C05** folder; the **Punchkeys.csv** file is listed in the folder.

9. Double-click on the **Punchkeys.csv** file; you are returned to the **Modify Tool Sets** dialog box. Notice that the subject and comments of the punchkey symbols are populated in the **Tools** area, as shown in Figure 42.

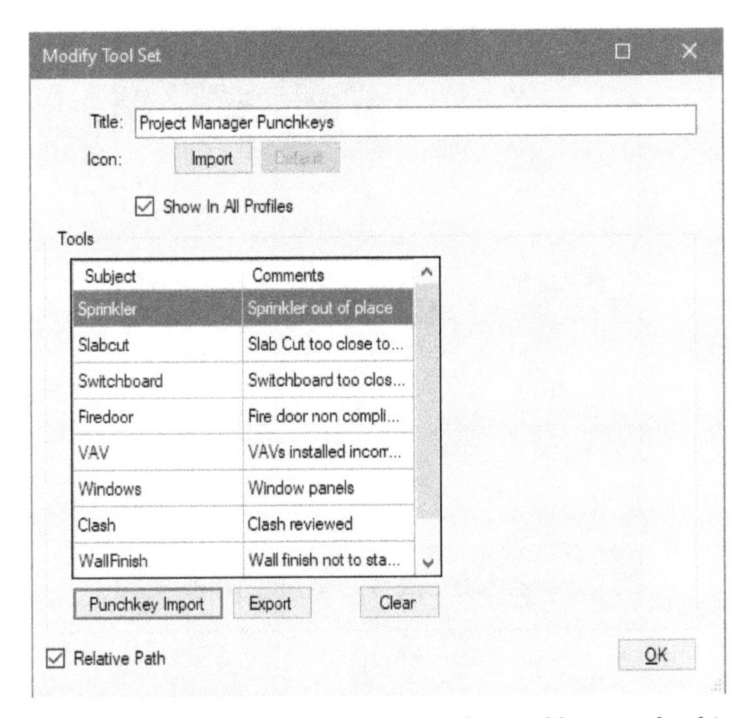

Figure 42 *The subject and comments of the punchkeys populated in the **Modify Tool Set** dialog box*

10. Make sure the **Show in All Profiles** check box is selected in the **Modify Tool Set** dialog box, as shown in Figure 42.

11. Click **OK** in the **Modify Tool Set** dialog box; you are returned to the **Manage Tool Sets** dialog box.

    You will now reorder the new tool set and place it below the **Project Manager Markups** tool set.

12. Select the **Project Manager Punchkeys** tool set from the bottom of the list in the **Manage Tool Sets** dialog box.

13. Use the **Up** key to move the new tool set up and place it below the **Project Manager Markups** tool set.

14. Click **OK** in the **Manage Tool Sets** dialog box. Figure 43 shows the **Tool Chest** panel with the new **Project Manager Punchkeys** tool set.

15. Click on the cogwheel on the top right of the **Project Manager Punchkeys** tool set and select **Details** from the shortcut menu.

    Notice the subject and comments imported from the CSV file are listed on the right of the punchkeys in the tool set, as shown in Figure 44.

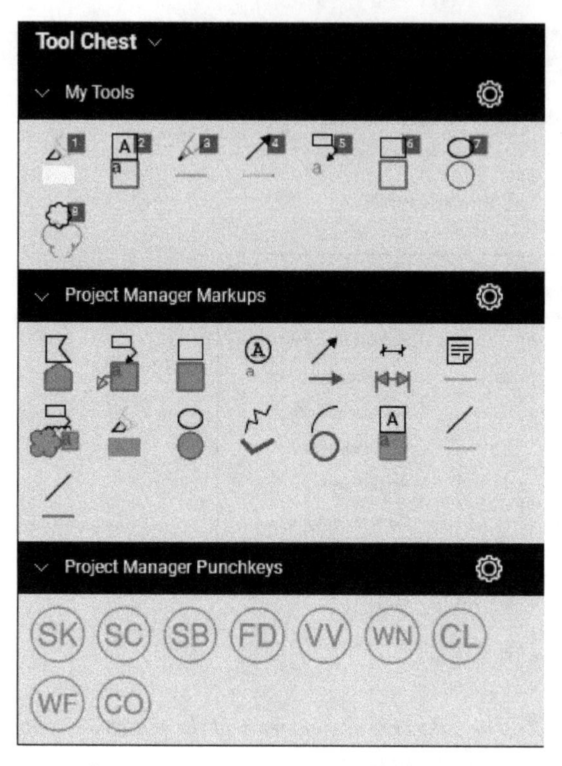

Figure 43 *The* ***Project Manager Punchkeys*** *tool set created by importing a CSV file*

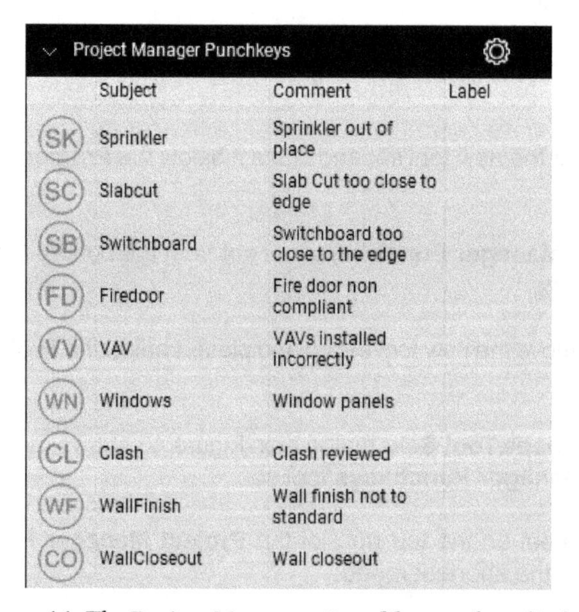

Figure 44 *The* ***Project Manager Punchkeys*** *tool set displayed in the* ***Detail*** *view*

16. Click on the cogwheel on the top right of the **Project Manager Punchkeys** tool set and select **Symbol** from the shortcut menu to change the view of this tool set back to **Symbol**.

17. Try placing some punchkeys using the **Project Manager Punchkeys** tool set on the PDF file. Notice that the punchkeys appear in a Red circle.

> ***What I do***
>
> *Depending on the situation, sometimes I need to change the colors of the punchkeys. In that case, I place the punchkey on the PDF file and then in the **Properties** panel, use the **Change Color** button to change the color from Red to the preferred color. Once changed, I add the punchkey back to the tool set and delete the Red color punchkey from the tool set.*

18. Delete the punchkeys placed on the PDF file.

19. Click on the cogwheel on the top right of the **Project Manager Punchkeys** tool set and select **Save** from the shortcut menu to save the tool set.

20. Save the PDF file.

Section 7: Adding Customized Markup Tool to the My Tools Tool Set

As mentioned earlier, the **My Tools** tool set has a special property where the tools added to it are automatically assigned a numeric shortcut key. This tool set already has eight default tools that are assigned numeric shortcut keys from 1 to 8. You will now add your customized markup tools to this tool set and notice that they are automatically assigned numeric shortcut keys.

1. Scroll to the top in the **Tool Chest** panel and notice that the **My Tools** tool set has eight tools in it. These tools are assigned numeric shortcut keys from 1 to 8, as shown in Figure 45.

***Figure 45** The default **My Tools** tool set*

2. Press the **2** key on the keyboard; the **Text Box** markup tool is invoked and you are prompted to select a region to place the text box.

   This is because the **My Tools** tool set has the numeric key 2 assigned as the shortcut key to invoke the default **Text Box** tool.

3. Press the **5** key on the keyboard; the **Callout** markup tool is invoked because the **My Tools** tool set has the numeric key 5 assigned as the shortcut key to invoke the default **Callout** tool.

You will now add the **Project Manager's Cloud+** tool from the **Project Manager Markup** tool set to the **My Tools** tool set.

4. Drag and drop the **Project Manager's Cloud+** tool from the **Project Manager Markup** tool set to the **My Tools** tool set, as shown in Figure 46.

***Figure 46** Dragging and dropping the customized markup tool from the **Project Manager Markups** tool set to the **My Tools** tool set*

Once the tool is added to the **My Tools** tool set, you will notice number 9 displayed on the top right of the **Project Manager's Cloud+** tool in this tool set. This indicates that the numeric key 9 has been assigned as the shortcut key to invoke this tool.

***Tip:** You can also right-click on a customized markup in the main workspace and select **Add to Tool Chest > My Tools** from the shortcut menu to add it to the **My Tools** tool set.*

5. Similarly, drag and drop the **Project Manager's Text Box** tool from the **Project Manager Markup** tool set to the **My Tools** tool set.

   This tool is automatically assigned the numeric 10 key as the shortcut key. Figure 47 shows the **My Tools** tool set after adding these two customized tools.

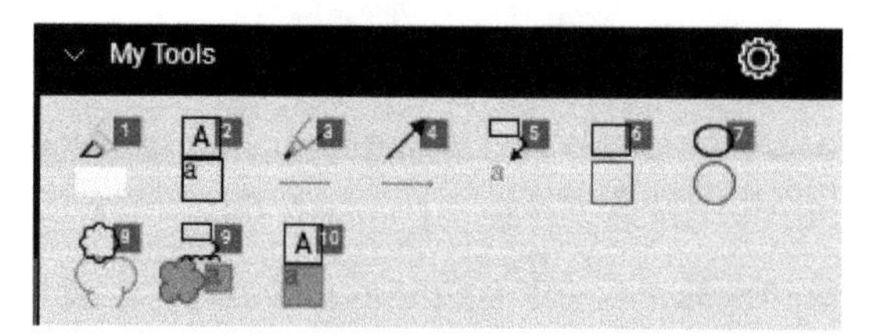

***Figure 47** The **My Tools** tool set after adding additional tools*

Next, you will test the shortcut keys assigned to your customized tools.

6. Press the **9** key on the keyboard; the **Project Manager's Cloud+** tool is invoked.

7. Add the cloud+ markup using this custom tool and notice it has all the properties of the **Project Manager's Cloud+** tool.

8. Similarly, press the **1** and **0** key within a short duration of each other; the **Project Manager's Text Box** tool is invoked.

9. Place the text box and notice it has the same properties as that of the **Project Manager's Text Box** tool.

   This shows that you can add the customized version of your markup tools to the **My Tools** tool set so they are assigned numeric shortcut keys for easy access.

10. Delete the markups added in steps 7 and 9 above.

11. Similarly, add a few more customized markup tools from the **Project Manager Markups** tool set to the **My Tools** tool set.

12. Click on the cogwheel on the top right of the **My Tools** tool set and select **Detail** from the shortcut menu; this tool set is displayed in with the details of various tools, as shown in Figure 48.

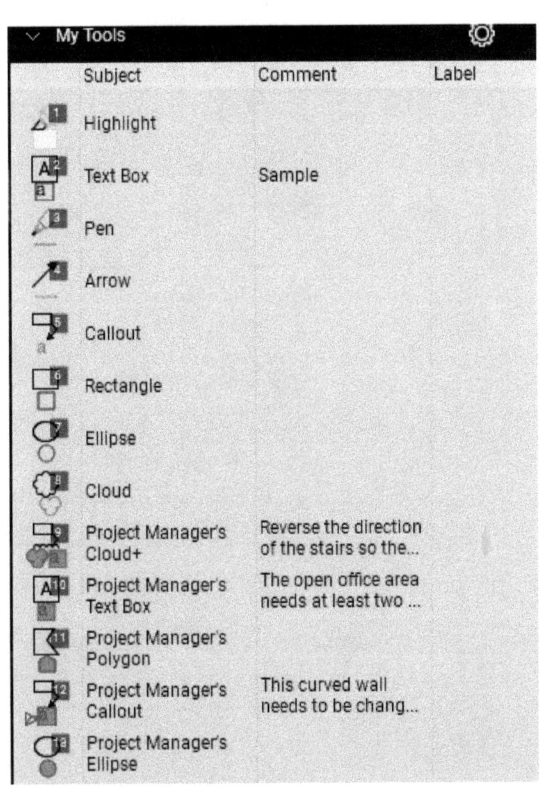

*Figure 48 The **My Tools** tool set displayed in the **Detail** view*

13. Click on the cogwheel on the top right of the **My Tools** tool set and change the **My Tools** tool set back to the **Symbol** view.

14. Click on the cogwheel on the top right of the **My Tools** tool set and save this tool set.

15. Zoom to the extents of the PDF file by double-clicking on the Wheel Mouse Button.

16. Save the PDF file and close Revu.

Skill Evaluation

Evaluate your skills to see how many questions you can answer correctly. The answers to these questions are given at the end of the book.

1. Revu does not allow you to add tools to a custom tool set. (True/False)

2. Revu allows you to import or export custom tool sets. (True/False)

3. The **My Tools** tool set automatically assigns shortcut keys to the tools that are added to this tool set. (True/False)

4. A tool set can be displayed in the **Detail** or **Symbol** view. (True/False)

5. A tool once added to a tool set cannot be deleted. (True/False)

6. Which format is the tool set file saved in?

   (A) TTX (B) PTX
   (C) BTX (D) RTX

7. Which format file can be imported to create a punchkeys tool set?

   (A) CSV (B) DWG
   (C) RVT (D) DOC

8. Which tool set automatically assigns numeric keys as the shortcut keys to the tools added to in?

   (A) **My Tools** (B) **Recent Tools**
   (C) **Both** (D) **None**

9. What is the default view mode of the tool sets?

   (A) **Details** (B) **Symbol**
   (C) **Small Icons** (D) **Large Icons**

10. Which two columns from a CSV file are mapped during punchkey import?

    (A) **Subject** (B) **Shape**
    (C) **Color** (D) **Comments**

Class Test Questions

Answer the following questions:

1. Explain briefly the process of creating a tool set using punchkey import.
2. How will you change a tool set from the **Detail** to **Symbol** view?
3. Explain briefly the process of adding a customized tool to a tool set.
4. Explain briefly how to export a custom tool set.
5. Why should you change the subject or a tool before adding it to your tool set?

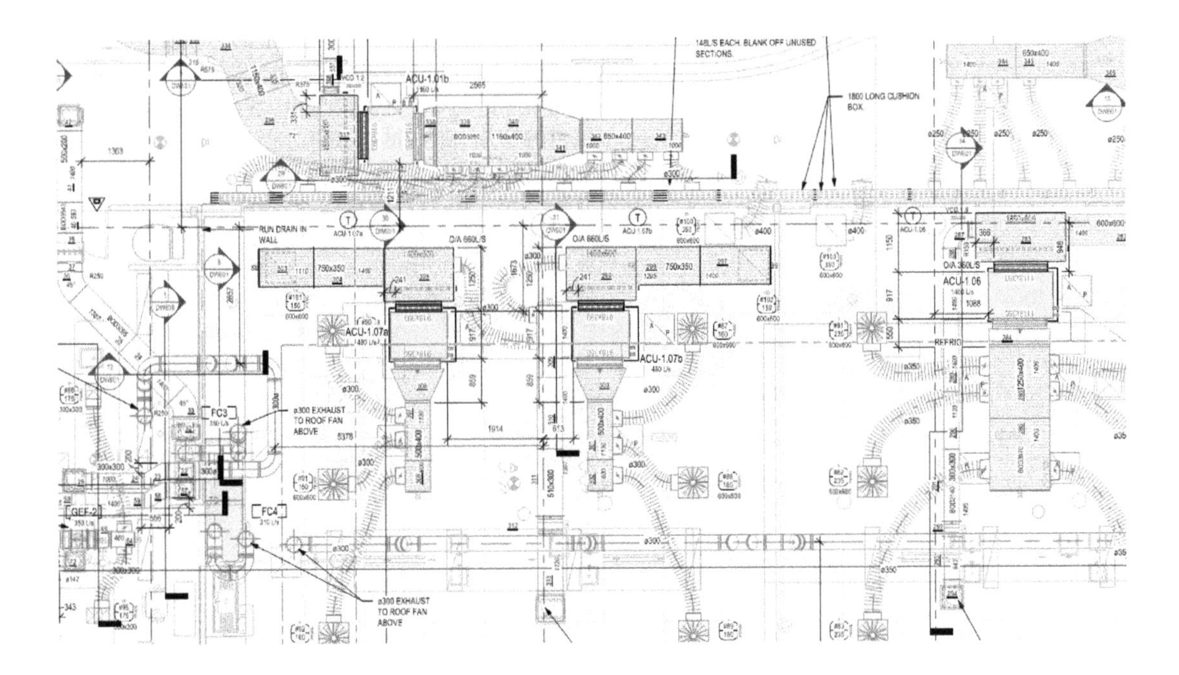

Chapter 6 - Working with the Measurement Tools - I

The objectives of this chapter are to:

- √ *Explain the process of specifying the scale for measurement tools*
- √ *Explain the process of calibrating a sheet*
- √ *Explain the process of adding a viewport with a different scale on the sheet*
- √ *Teaching you how to disable the line weights*
- √ *Explain the process of how to add and customize the length measurement*
- √ *Explain the process of how to add and customize the polylength measurement*
- √ *Explain the process of how to add and customize the perimeter measurement*

QUANTITY TAKEOFF (QTO)

Quantity takeoff (QTO) is a process of measuring the quantities of items in a set of construction documents. It is an important aspect of the pre-construction phase where the estimators use the construction documents provided to them to measure the quantities of items, such as the total carpet area, the total tiled area, the number of lighting fixtures, and so on. These measurements then form the basis of their estimates for bidding for the jobs.

Revu has some excellent set of tools that allow you to perform measurements on PDF files. These tools can be customized to perform some specific types of measurements mentioned above. Combined with custom columns, these customized tools can also provide you with realtime costing of the quantities that you takeoff from the PDF files. You can then add a legend that provides you a sum of all the quantities and costs.

The following flowchart outlines the process of quantity takeoff in Revu. Some of the steps of this flowchart are discussed in this chapter and the remaining are discussed in the next chapter.

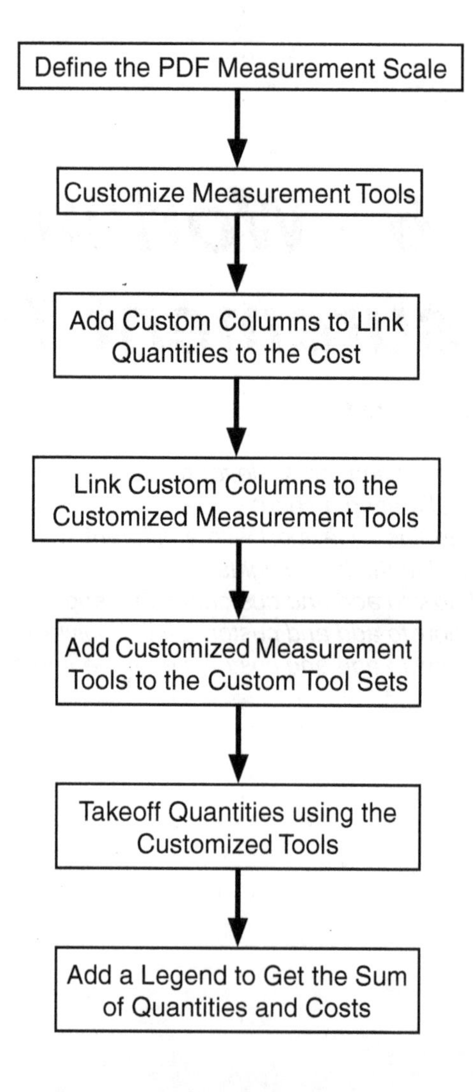

THE IMPORTANCE OF THE VECTOR PDF FILES WHILE PERFORMING QUANTITY TAKEOFF

As mentioned in the earlier chapters, the vector PDFs are the ones generated directly from CAD software and the raster PDFs are the ones that are printed on a sheet of paper and then scanned as PDFs. The Vector PDFs represent the content of the PDF in the vector form and the raster PDFs represent the content as an image.

The vectors in the PDFs allow you to snap to the content of PDFs, which results in accurate quantity takeoffs. However, if you work with the raster PDFs, you will not be allowed to snap to the content of the PDFs. As a result, the quantities that you takeoff will not be accurate. Hence, it is extremely important that wherever possible, you should work with the Vector PDFs to perform quantity takeoffs.

SETTING SCALES FOR THE PDF FILES

Before you start taking off quantities from PDFs, it is important to specify the measurement scale. This allows Revu to use that scale and provide you with the measurement values once you start taking off quantities. By default, when you open the PDF file, the right side of the **Navigation Bar** shows **Scale Not Set**, labeled as **1** in Figure 1. The same information is also available on the **Status Bar**, labeled as **2** in Figure 1.

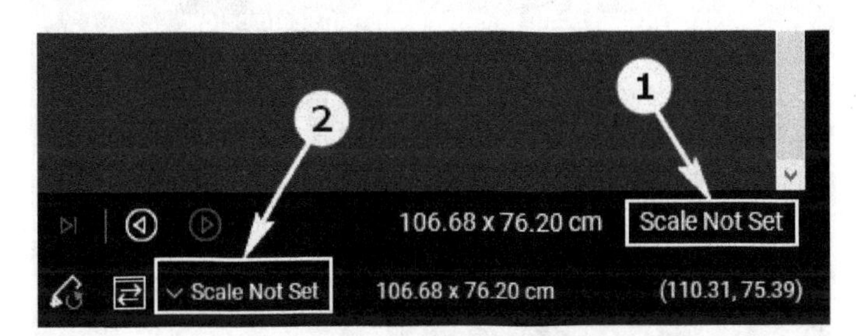

***Figure 1** Reviewing the information about the PDF scale not set*

If you start performing any measurement on this PDF, you will be first prompted to set the scale for the PDF file. In Revu 2019, there are two methods of calibrating PDFs: defining a preset or custom scale and calibrating using two points. Both these methods are discussed next.

Defining a Preset or Custom Scale for the PDF

This method was introduced in Revu 2019 and is used when you are a hundred percent sure of the PDF scale. In this case, you can select from a list of standard scales or you can enter your own custom scale. The procedure for doing this is discussed next.

1. On the **Navigation Bar** or the **Status Bar**, click on the text that says **Scale Not Set**, labeled as **1** and **2** respectively in Figure 1 above. On doing so, the **Set Scale** dialog box will be displayed, as shown in Figure 2.

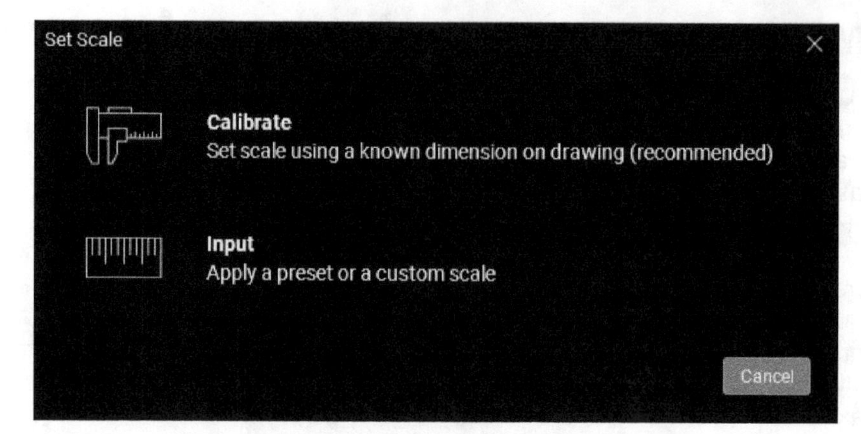

*Figure 2 The **Set Scale** dialog box to select the method to specify sheet scale*

2. From the **Set Scale** dialog box, click **Input**; another **Set Scale** dialog box will be displayed that allows you to select a preset scale or enter a custom scale, as shown in Figure 3.

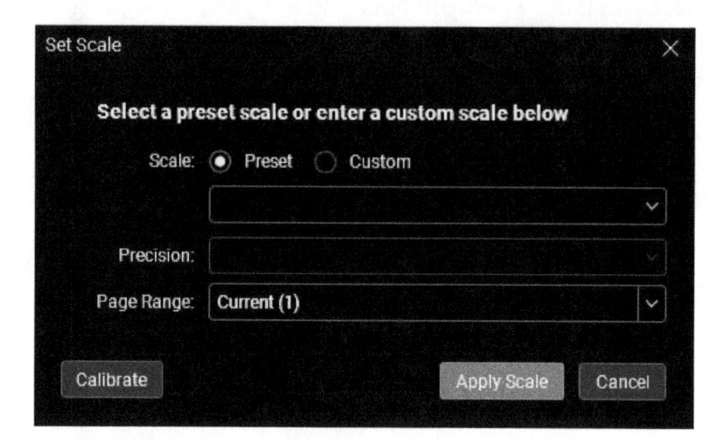

*Figure 3 The **Set Scale** dialog box to select a preset or a custom scale*

By default, the **Preset** option is selected in this dialog box. As a result, you can select a preset scale from the drop-down list in this dialog box.

3. Click on the drop-down list available below the **Preset** radio button and select the required preset scale, as shown in Figure 4.
4. Alternatively, select the **Custom** radio button and enter the custom scale.
5. Select the required precision from the **Precision** drop-down list.
6. From the **Page Range** drop-down list, select the option to apply the scale to the current sheet, all sheets in the PDF, or to a page range.

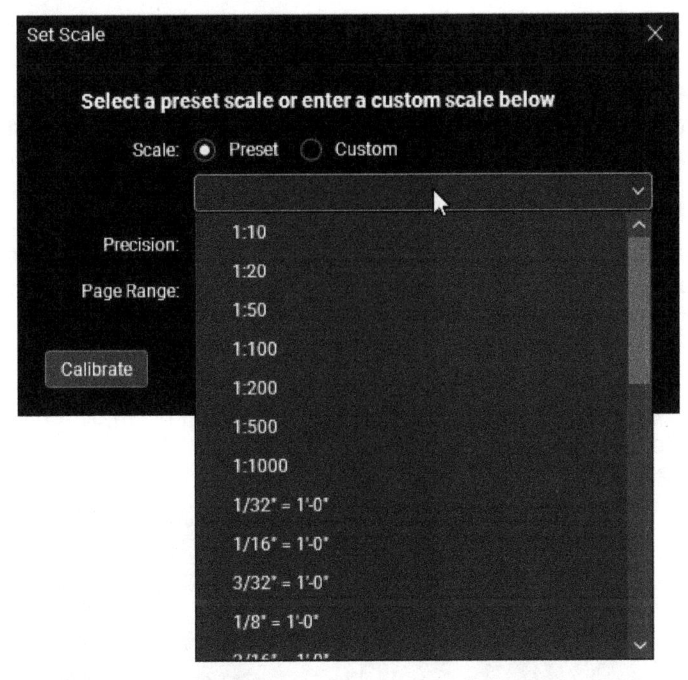

Figure 4 Selecting a preset scale from the drop-down list

7. Click the **Apply Scale** button in the **Set Scale** dialog box; the scale is set and the right side of the **Navigation Bar** and the **Status Bar** will show the sheet scale, as enclosed in rectangles in Figure 5.

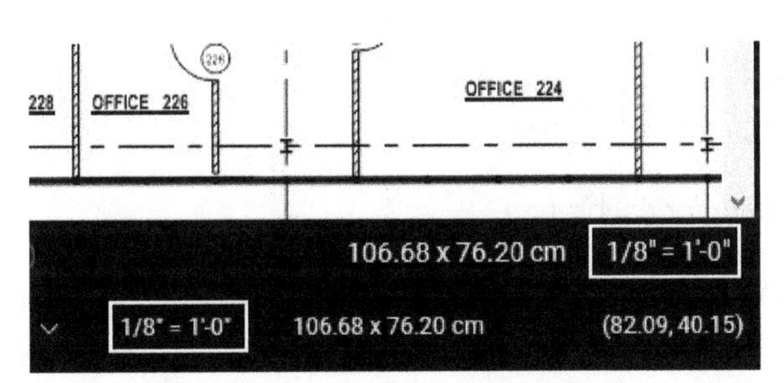

Figure 5 The sheet scale displayed on the ***Navigation Bar*** *and the* ***Status Bar***

> ***What I do***
>
> *Unless you are hundred percent sure that all the sheets of a PDF file has the same scale, I would not recommend applying the same scale to all the sheets. In that case, you are better off going to each sheet of the PDF and then specifying its scale. Although this means more work for you, but this guarantees accurate takeoffs.*

Calibrating a PDF Using Two Points

You can never be a hundred percent sure that the PDFs you are provided to takeoff quantities were printed to the right scale. As a result, it is hard for you to specify a standard or custom scale. So unless you are a hundred percent sure of the sheet scale, it is recommended to calibrate the sheet using two points of a known dimension value and then specify that as the sheet scale. The procedure for doing this is discussed next.

1. On the **Navigation Bar** or the **Status Bar**, click on the text that says **Scale Not Set**, labeled as **1** and **2** respectively in Figure 1 above. On doing so, the **Set Scale** dialog box will be displayed, as shown in Figure 2.

2. From the **Set Scale** dialog box, select **Calibrate**; the **Calibrate** dialog box will be displayed that will inform you that you need to select two points of a known dimension to calibrate the measurement tool, as shown in Figure 6.

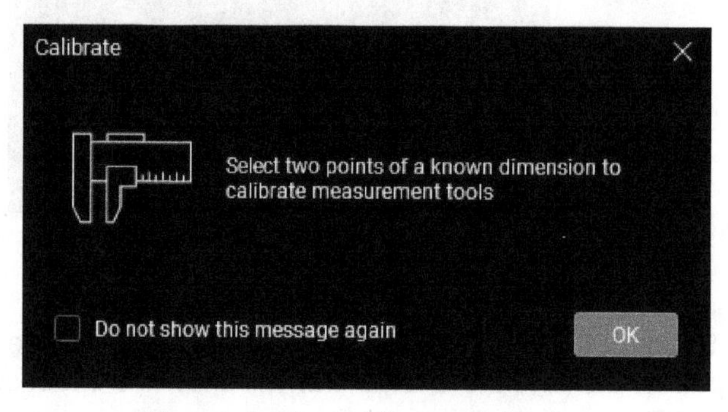

***Figure 6** The **Calibrate** dialog box*

3. Click **OK** in the **Calibrate** dialog box; you will be prompted to select the first point of calibration and drag to select the second point.

4. Snap to the first point for calibration.

5. Snap to the second point for calibration; the **Calibration** dialog box is redisplayed with the **Custom** radio button selected.

   The left side of the scale shows the current measured value on the PDF and the right side is where you have to enter the known dimension value.

6. On the right side of the custom scale, enter the known distance value between the two points and then select the units of measurement. Figure 7 shows the **Calibrate** dialog box where the distance between the two calibration points is specified as **36 Inches**.

7. Select the required precision from the **Precision** drop-down list.

8. From the **Page Range** drop-down list, select the option to apply the scale to the current sheet, all sheets in the PDF, or to a page range.

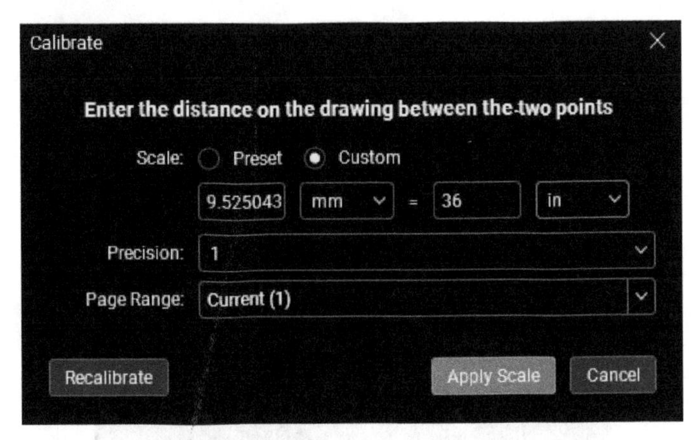

*Figure 7 The **Calibration** dialog box with the calibrated distance specified*

9. Click the **Apply Scale** button in the **Calibrate** dialog box; the scale is set and the right side of the **Navigation Bar** and the **Status Bar** will show the sheet scale, as enclosed in rectangles in Figure 8.

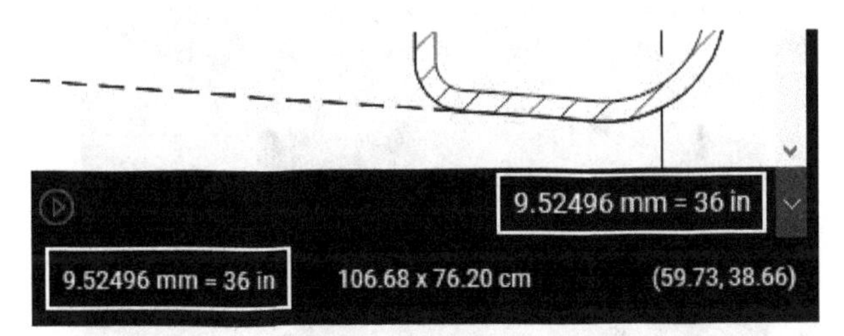

*Figure 8 The sheet scale displayed on the **Navigation Bar** and the **Status Bar***

Applying a Separate Vertical Scale

Sometimes while working with Civil or Infrastructure files, you may need to apply a separate vertical scale for the PDF. The following is the procedure for doing this.

1. On the **Panel Bar**, click the **Measurements** button to display the **Measurements** panel.

2. On the **Measurements** panel, select the **Separate Vertical Scale** check box, enclosed in a rectangle in Figure 9.

   On doing so, the **Measurements** panel will show the **X Scale** and **Y Scale** areas, as shown in Figure 10. The **X Scale** area will show the scale you specified or calibrated. The **Y Scale** area is where you have to specify or calibrate the vertical scale.

3. In the **Y Scale** area, use the **Calibrate**, **Preset**, or **Custom** option to specify the vertical scale.

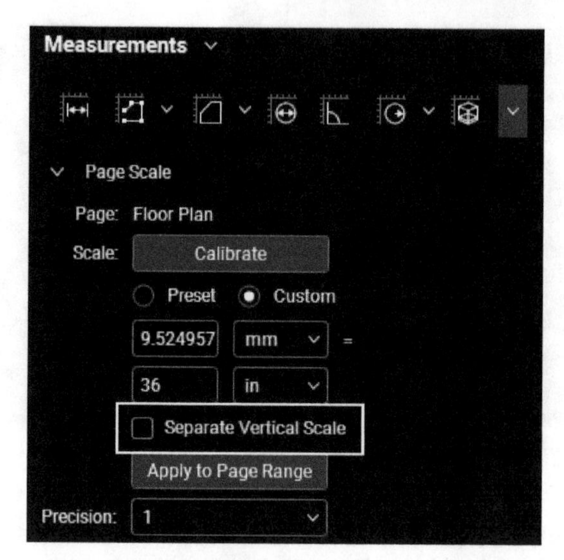

Figure 9 The ***Separate Vertical Scale*** *check box to specify a separate vertical scale*

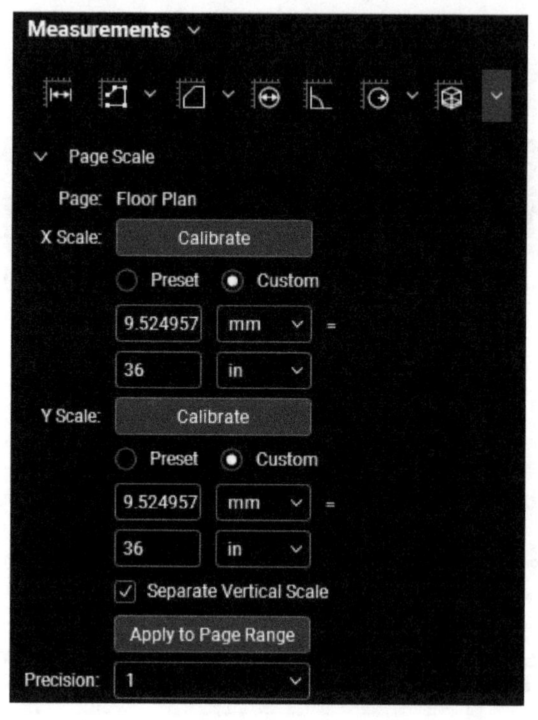

Figure 10 *Specify a separate vertical scale for measurements*

Creating Viewports on the Sheet with Different Scales

A number of PDFs display smaller intricate areas of the drawing in magnified views outside the main view of the sheet. These magnified views normally have scales different from the main view. As a result, the scale calibrated for the rest of the sheet does not work in those magnified views. In Revu, you can create viewports around those magnified views and specify a different scale to those viewports. The following is the procedure for doing this.

1. Scroll down in the **Measurements** panel and expand the **Viewports** area, if it is not already expanded.

2. Click the **Add Viewport (+)** button below the **Viewports** area; the **Add Viewport** dialog box will be displayed informing you to select a region to define a viewport, as shown in Figure 11.

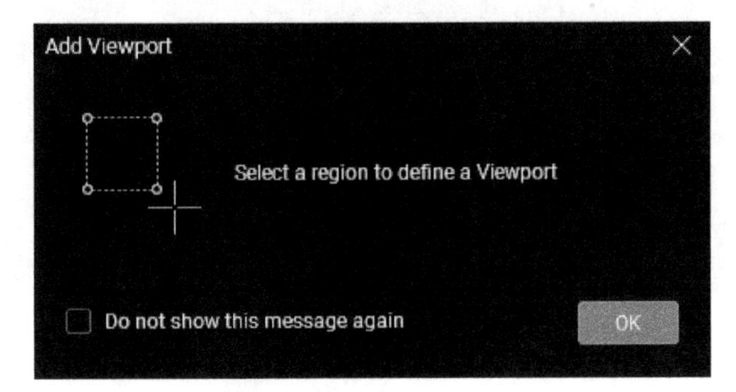

Figure 11 *The **Add Viewport** dialog box*

3. Click **OK** in the **Add Viewport** dialog box; you will be returned to the PDF file and will be prompted to select a region to place a viewport.

4. Drag two opposite corners around the magnified view to create the viewport, as shown in Figure 12.

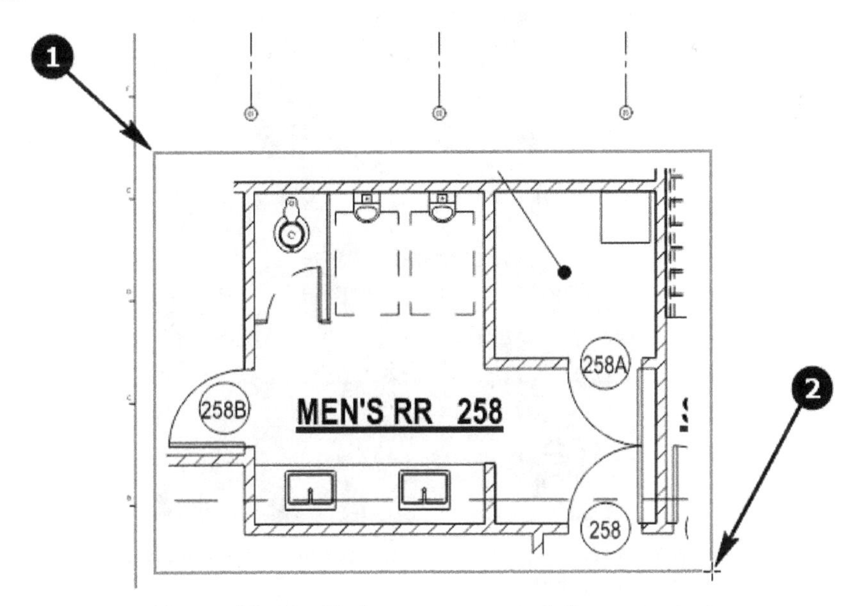

Figure 12 *Specify the two corners to define a viewport*

On specifying the second corner point, the **Add Viewport** dialog box will be redisplayed.

5. In the **Add Viewport** dialog box, change the name to a relevant name, as shown in Figure 13.

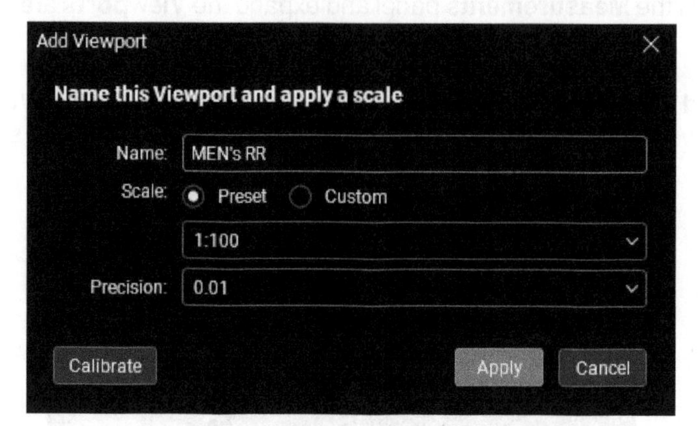

Figure 13 Specifying the name for the viewport

6. Using the preset, custom, or calibrate method, specify the scale for the new viewport.

7. Click the **Apply** button in the **Add Viewport** dialog box; the new viewport will be created and highlighted in Blue on the PDF file. Also, the name of the new viewport will be listed in the **Viewports** area of the **Measurements** panel, along with the options to redefine the scale for this viewport, as shown in Figure 14.

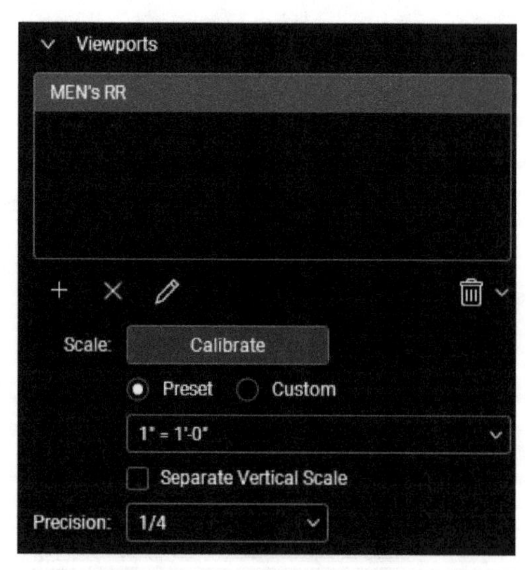

*Figure 14 The new viewport displayed in the **Viewports** area of the **Measurements** panel*

***Tip**: Using the buttons available below the **Viewports** list, you can add a new viewport, delete the selected viewport, edit the selected viewport, or delete all viewports from the page of PDF.*

Disabling the Line Weights Before Performing Measurements

Menu: View > Disable Line Weights

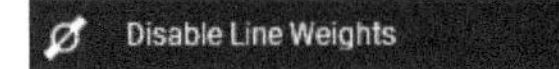

Most drawings have elements that are drawn with a line weight value defined, which makes them appear thicker. The line weights are important to distinguish these elements from the rest of the elements in the drawing but make it hard to take accurate measurements in PDFs. As a result, it is recommended to disable line weights before performing measurements. To disable line weights, from the **Menu Bar**, click **View > Disable Line Weights**. Figure 15 shows a section of a PDF file with the line weights enabled and Figure 16 shows the same section with line weights disabled.

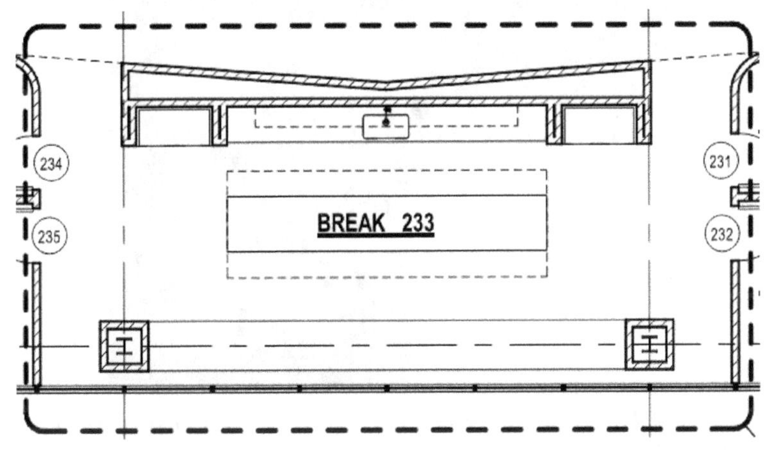

***Figure 15** A section of a PDF with line weights enabled*

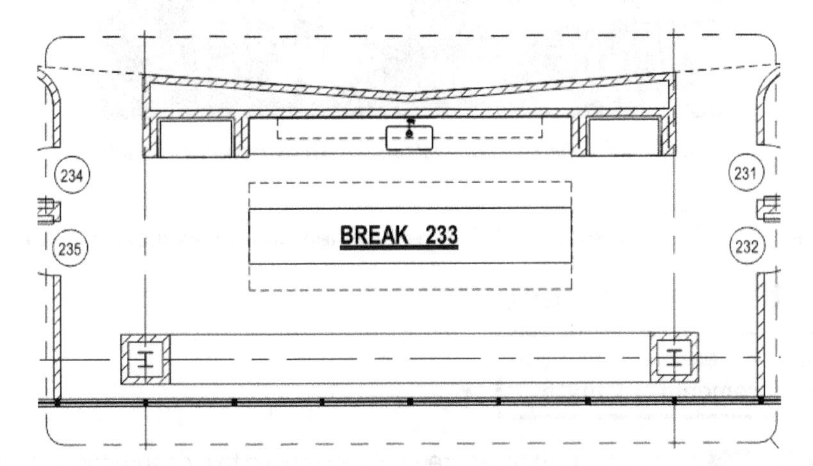

***Figure 16** The same section of the PDF with line weights disabled*

THE MEASUREMENT TOOLS

Revu provides you with a number of measurement tools that are available on the **Tools > Measure** menu, as shown in Figure 17, and on the toolbar at the top of the **Measure** panel, as shown in Figure 18.

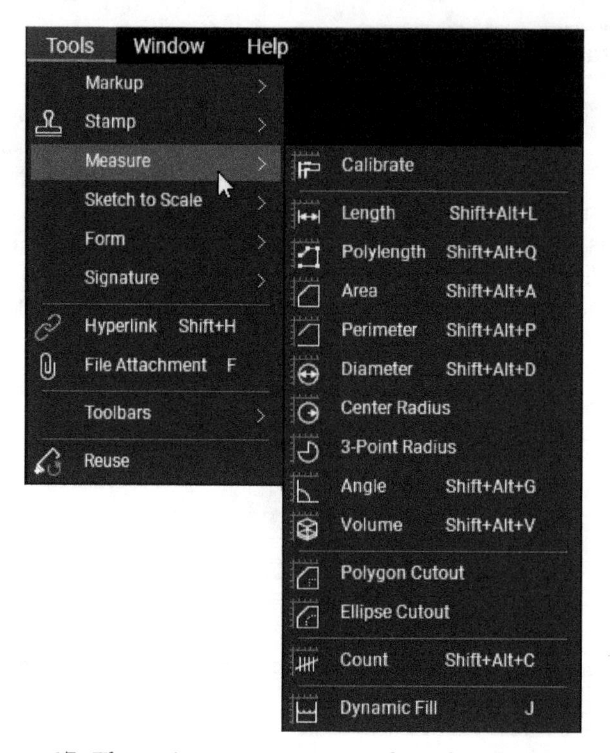

*Figure 17 The various measurement tools in the **Measure** menu*

*Figure 18 The measure tools in the **Measurements** panel*

Some of these tools are discussed next. The rest of them will be discussed in later chapters.

The Length Measurement

Menu: Tools > Measure > Length
Panel: Measurements > Length

The **Length** measurement tool is used to measure the distance between two specified points. When you invoke this tool, you will be prompted to select the start point and drag to the endpoint. Holding down the

SHIFT key while specifying the second point will let you place a linear measurement or let you snap 45-degree angles for measurements. Once you specify the second point of the length measurement, the distance between those two points is displayed as a measurement on the PDF file. Note that if the measurement scale for the PDF file has not been set yet, you will be prompted to calibrate the sheet or set the scale for the sheet. Depending on whether you specify two points in a linear direction or in an aligned direction, the same tool can be used to place a linear measurement or an aligned measurement. Figure 19 shows a linear measurement between two points and Figure 20 shows an aligned measurement between two points.

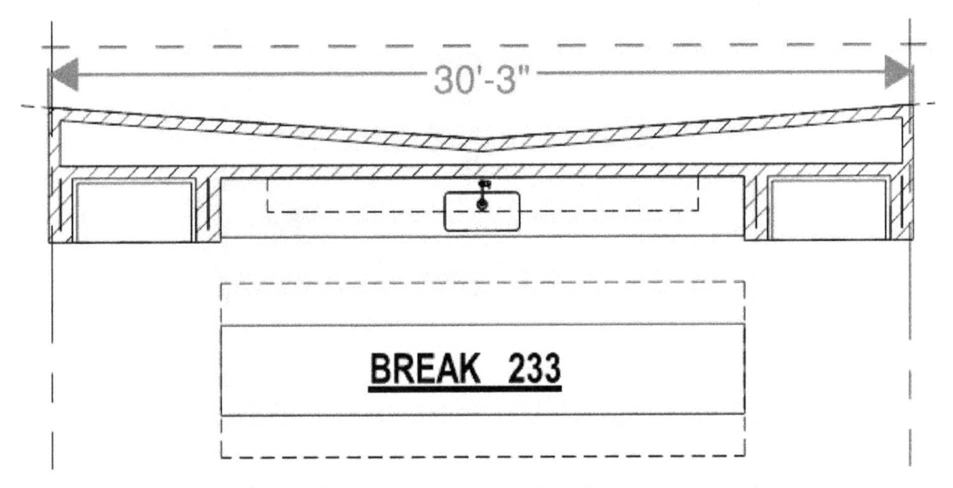

Figure 19 *A linear measurement placed between two points*

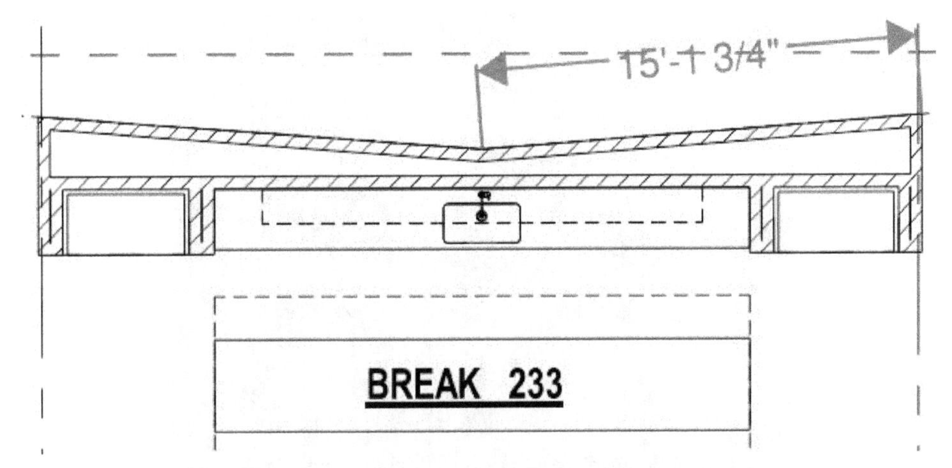

Figure 20 *An aligned measurement placed between two points*

Adding Depth and Slope to the Length Measurement

The **Length** measurement is an extremely versatile tool and can also be used to measure wall areas by specify a depth value to the length measurement or measure distances over an inclined plane by specifying the slope value. To do this, scroll down in the **Measurements** panel and expand the **Length Measurement Properties** area, if not already expanded. In this area, you can specify the depth value, slope in degrees, pitch, or grade, and define the units for the area measurement (for wall areas when you specify the depth). It is important to note that if you specify the slope value, the **Depth** edit box will be disabled. This is because you cannot define both slope and depth values for the length measurement. Figure 21 shows the **Length Measurement Properties** area with a depth value of **10'** specified for the selected measurement. Note that as soon as you enter the slope value, it is displayed on the dimension in the PDF file.

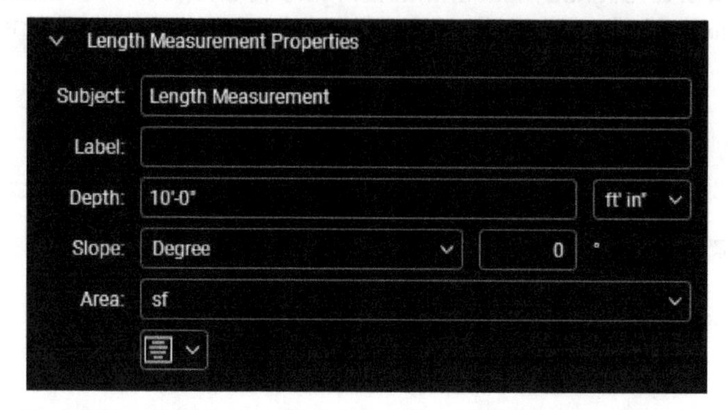

Figure 21 The depth value defined for the length measurement

However, the depth and wall area values are not automatically displayed on the dimension. To display these values, you need to turn their visibility on from the **Show Caption** flyout at the bottom of the **Length Measurement Properties** area, as shown in Figure 22. Note that clearing the **Show Caption** option in the flyout below will turn off the visibility of all captions of the dimension.

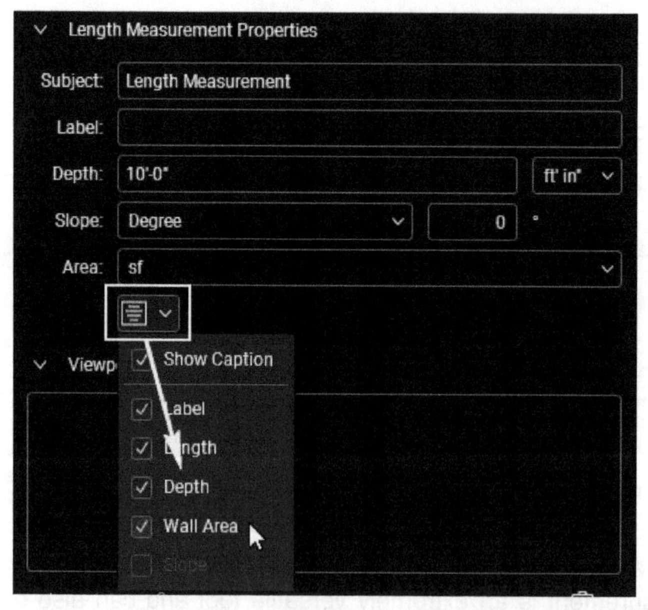

Figure 22 Selecting the ***Depth*** *and* ***Wall Area*** *captions*

Figure 23 shows the length measurement showing the length, depth, and wall area values on the PDF.

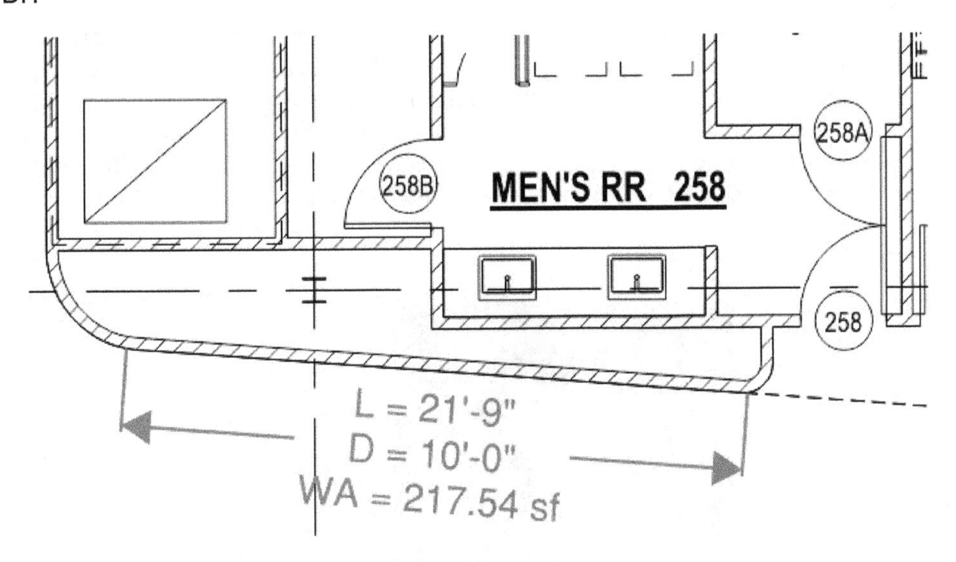

Figure 23 The length measurement showing additional captions

Customizing Appearance of the Length Measurement

Similar to the markup tools, the measurements can also be customized by changing their appearances in the **Properties** panel or the **Properties** toolbar displayed below the **Menu Bar**. Some of the additional appearance properties that can be changed for the measurements include the start and end styles and their sizes. Figure 24 shows the length dimension with the start and end styles changed to architectural ticks and their sizes set to 50%. Also, the text height in this dimension is changed from 12 to 10.

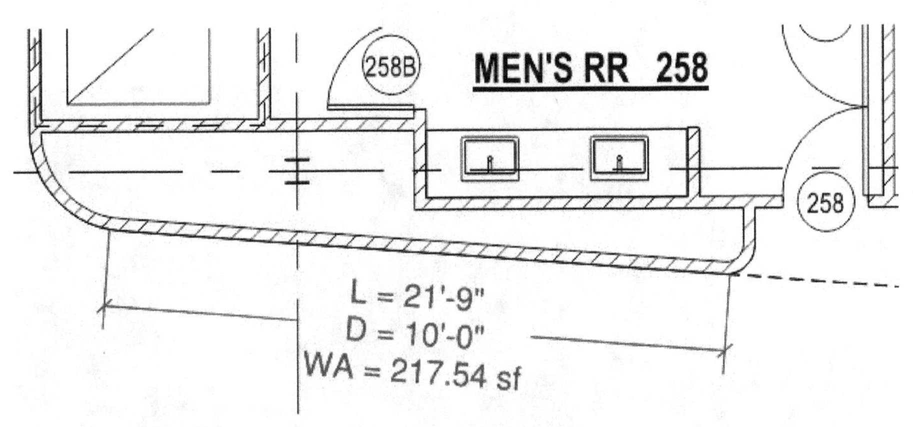

Figure 24 The length measurement with customized appearance

Adding the Customized Version of the Measurement Tools to the Custom Tool Sets

Similar to the markup tools, the customized version of the measurement tools are also listed in the **Recent Tools** tool set. However, as discussed in the earlier chapters, it is recommended to add your customized measurement tools to your own tool sets so that you do not lose them when you close Revu. To do this, right-click on the customized measurement and select **Add to Tool Chest > Your Custom Tool Set**. Figure 25 shows the customized length measurement being added to the **Project Manager Measurements** tool set.

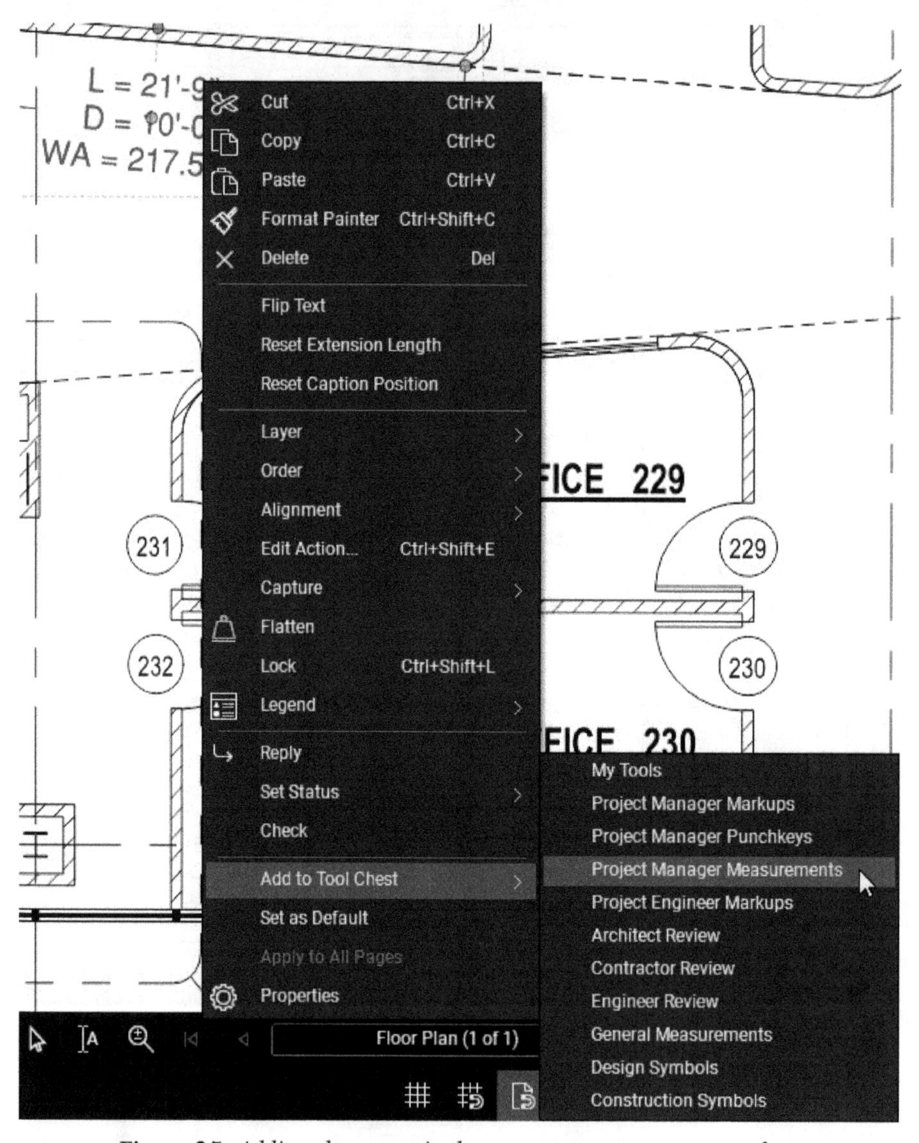

Figure 25 Adding the customized measurement to a custom tool set

The Polylength Measurement

Menu: Tools > Measure > Polylength
Panel: Measurements > Polylength

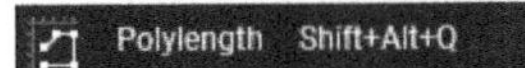

The **Polylength** measurement tool is similar to the **Length** measurement tool, with the difference that this tool allows you to place multiple end-connected measurements. This tool also displays the cumulative length of all the measurement segments. When you invoke this tool, you will be prompted to select vertices of the perimeter to measure. You can one by one specify the vertices of the multiple end-connected measurements. Once you have specified the last point, you need to press ENTER to finish the process and place the measurements. Alternatively, you can press the **C** key on the keyboard or double-click on the last point to finish the process. Figure 26 shows a polylength measurement.

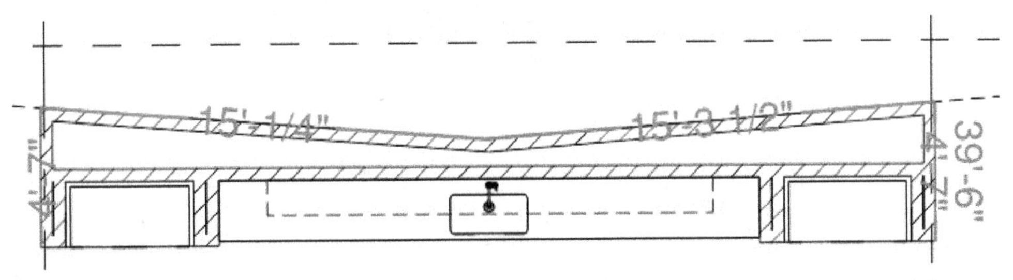

Figure 26 A polylength measurement showing the lengths of individual segments and the sum of all segments

Moving the Measurement Captions

Notice that in Figure 26, the measurement captions overlap the content of the PDF file. To avoid this, you can move the measurement captions. However, you need to make sure you hold the SHIFT key down before moving the values or the entire measurement will move. Figure 27 shows the same polylength measurement as in Figure 26, but the captions moved out by holding the SHIFT key. Also, the total polylength value in Figure 27 is moved to the center and also rotated using the handle directly above this value.

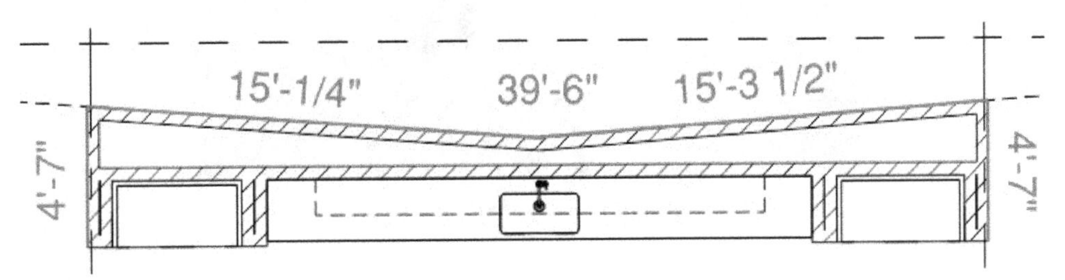

Figure 27 A polylength measurement showing the lengths of individual segments and the sum of all segments

Hiding Segment Values

In some cases, you might just want to show the cumulative length of the polylength measurement and not the individual segment values. To do this, you can turn off the **Show Segment Values** button at the bottom of the **Polylength Measurement Properties** button in the **Measurements** panel, as shown in Figure 28.

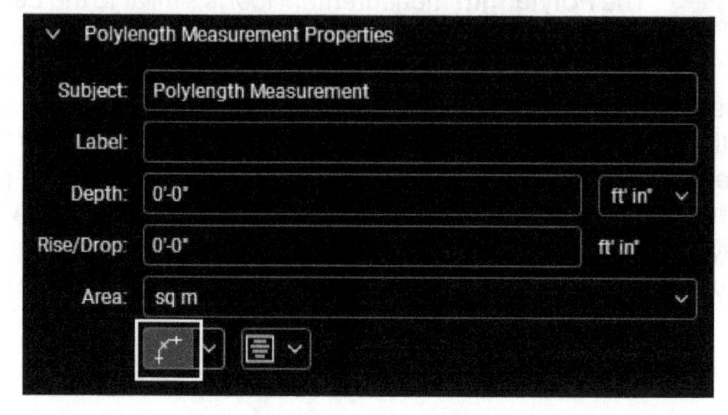

Figure 28 *The button to turn off the polylength segment values*

Changing the Polylength Caption Alignment

By default, the polylength measurement captions are aligned to the segment on which they are placed, as shown in Figures 26 and 27. If required, these captions can be placed horizontally by changing the alignment of these captions. To do this, click on the flyout on the right of the **Show Segment Values** button and select **Align Horizontally**, as shown in Figure 29.

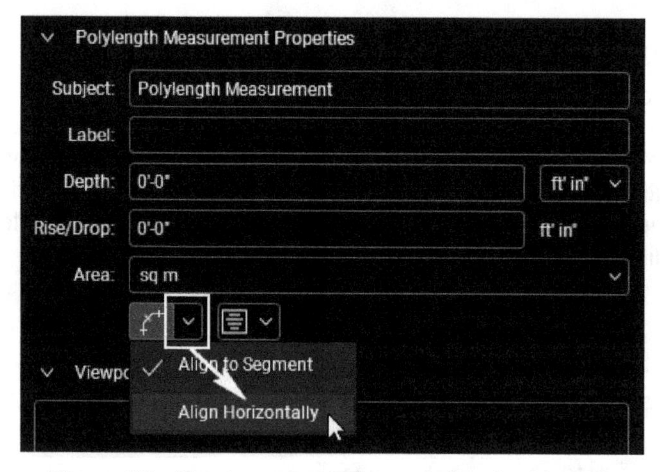

Figure 29 *Aligning measurement captions horizontally*

Adding Rise/Drop Value to the Polylength Measurement

The **Polylength** tool allows you to assign a rise or drop value to the overall length of the measurement. This is done in the **Rise/Drop** field in the **Polylength Measurement Properties** area of the **Measurements** panel. Note that as soon as you specify the rise or drop value, the individual segment values are disabled and the total polylength measurement and the rise/drop values are displayed, as shown in Figure 30.

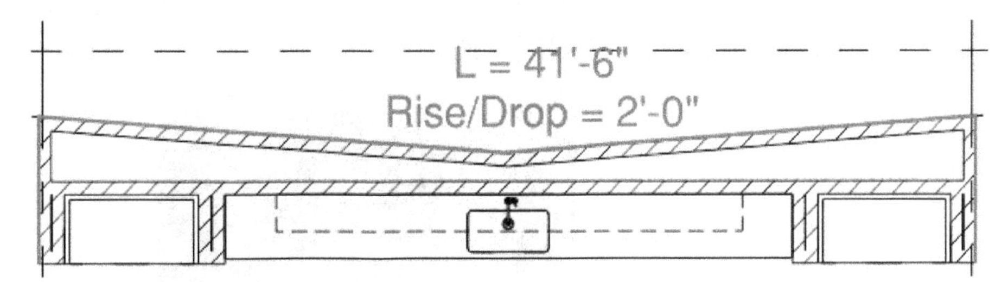

Figure 30 The total length and rise/drop values displayed

> **What I do**
>
> *I prefer customizing the polylength measurement and use it to take off quantities of items such as ductwork or pipework rather than the length measurement. This is because the polylength measurement allows me to add rise and drops, which is essential for items such as ductwork and pipework.*

Converting a Straight Polylength Segment into an Arc Segment

The polylength segments that are not drawn horizontal or vertical can be converted into arc segments. To do this, select the polylength segment and right-click to display the shortcut menu. From this shortcut menu, click **Convert to Arc**, as shown in Figure 31.

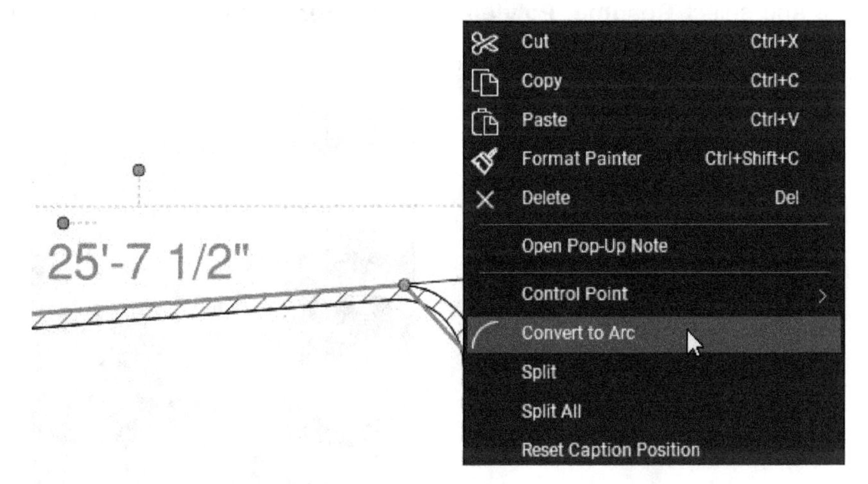

Figure 31 Converting a polylength segment into an arc segment

Once you have converted a segment into an arc segment, you can use the control handles at the two ends to change the shape of the arc segment. This is similar to changing the shape of the arc markup that was discussed in the earlier chapters.

Splitting Polylength Segments into Individual Segments

Sometimes you need to specify separate rise/drop values for the polylength segments. In that case, you can split one of the multiple end-connected polylength segments or split all of them by right-clicking and selecting the required option, as shown in Figure 32.

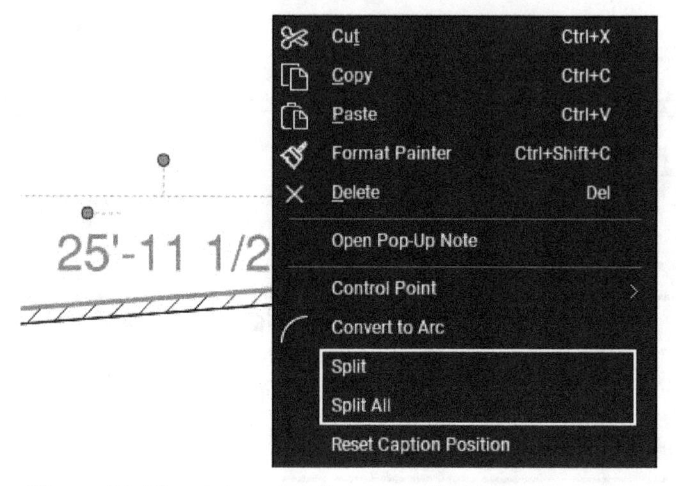

Figure 32 The options to split one or all polylength segments

Resuming the Polylength Measurement from the Last Point

If you have mistakenly ended a polylength measurement, you can resume it from the last point you specified. To do this, right-click on the Yellow control point at the end of the polylength measurement and select **Resume 'Polylength Measurement'** from the shortcut menu, as shown in Figure 33.

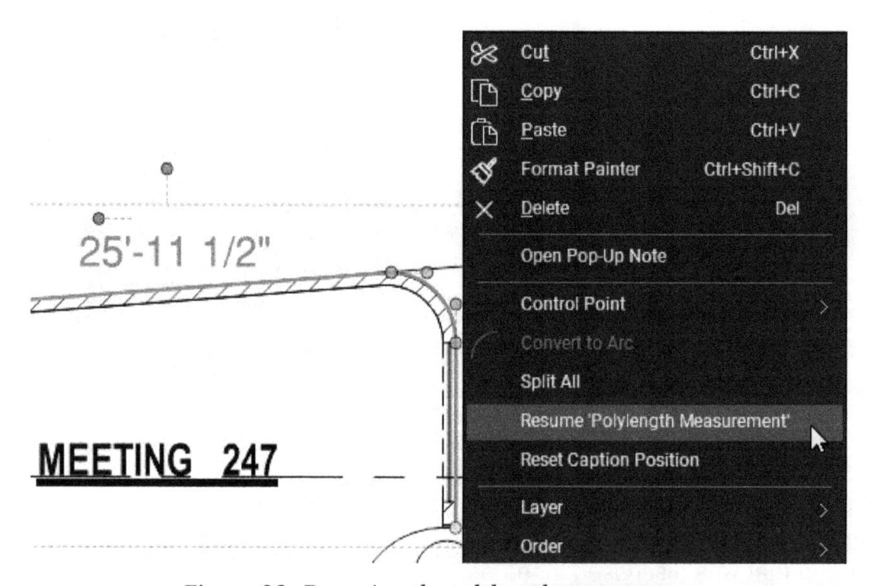

Figure 33 Resuming the polylength measurement

Note: *Similar to the polyline markup, you can also add additional control points or remove an existing control point from the polylength measurement.*

The Perimeter Measurement

Menu: Tools > Measure > Perimeter
Panel: Measurements > Polylength flyout > Perimeter

The **Perimeter** measurement tool is used to measure perimeters by specifying multiple vertices. This tool works similar to the **Polylength** measurement tool, with the following main differences:

- **Closed Loops**: The perimeter measurements can be closed loops. To close the perimeter loop, you can press the **C** key on the keyboard or click close to the start point of the perimeter measurement. It is important to note that although the perimeter measurements can be created as closed loops, they cannot be filled with a color or a hatch pattern. The **Fill Color** option that appears in the **Properties** panel when you select a perimeter measurement is used to specify the fill colors of the arrowheads, if selected at the start and end of the measurement.

- **Drag Two Opposite Corners**: The **Perimeter** tool allows you to drag two opposite corners to create a perimeter measurement of a rectangular region. Alternatively, you can hold down the SHIFT key to create a perimeter measurement of a square region.

- **No Individual Segment Values**: By default, the **Perimeter** tool does not show the length of the individual segments. However, you can turn on the **Show Segment Values** button in the **Perimeter Measurement Properties** area of the **Measurements** panel to show the individual segment values.

- **No Rise/Drop Values**: The perimeter measurements cannot be assigned the rise and drop values as these fields are not available for the perimeter measurements. However, you can define the depth values, which makes it a handy tool to take off wall area measurements of a room.

Figure 34 shows the perimeter measurement of a region in which some segments are converted into arc segments.

What I do

I prefer customizing the perimeter measurement to take off quantities such as skirting of a room. This is because by default, it shows the cumulative length of skirting and if required, I can show the individual length of segments as well.

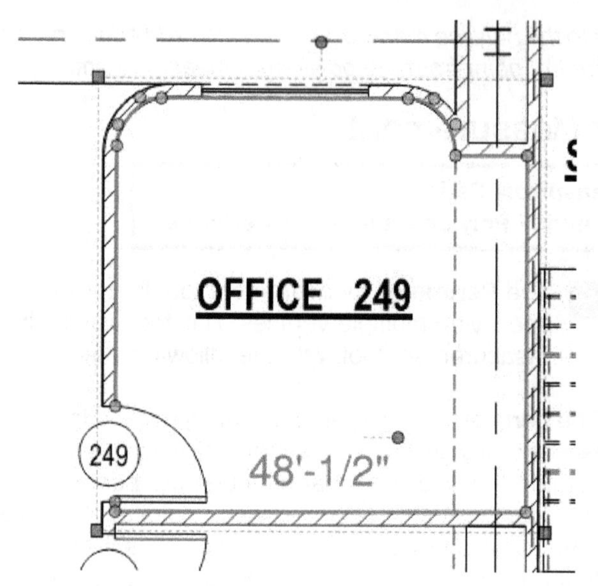

Figure 34 Perimeter measurement of a region

Hands-on Tutorial

In this tutorial, you will complete the following tasks:
1. *Create a new profile for Quantity takeoff.*
2. *Calibrate a PDF sheet.*
3. *Create a viewport and calibrate it to a different scale.*
4. *Add a length measurement and customize it.*
5. *Add a perimeter measurement and customize it.*
6. *Add a polylength measurement and customize it.*

Section 1: Opening a PDF File and Create a New Profile

In this section, you will open the **Architecture_Plan_06.pdf** file from the **C06** folder. You will then create a new profile with the **Measurements** panel attached to the right side of the **Main Workspace**.

1. Start Revu 2019 and open the **Architecture_Plan_06.pdf** file from the **C06** folder.

   You will now attach the **Measurements** panel to the right side of the **Main Workspace**. This way you can have the **Tool Chest** panel with your custom tool set displayed on the left of the **Main Workspace** and the **Measurements** panel displayed on the right of the **Main Workspace**.

2. On the **Panel Access Bar**, right-click on the **Measurements** button and select **Attach > Right** from the shortcut menu, as shown in Figure 35. On doing so, the **Measurements** panel is displayed on the right side of the **Main Workspace**, along with the **Properties** panel.

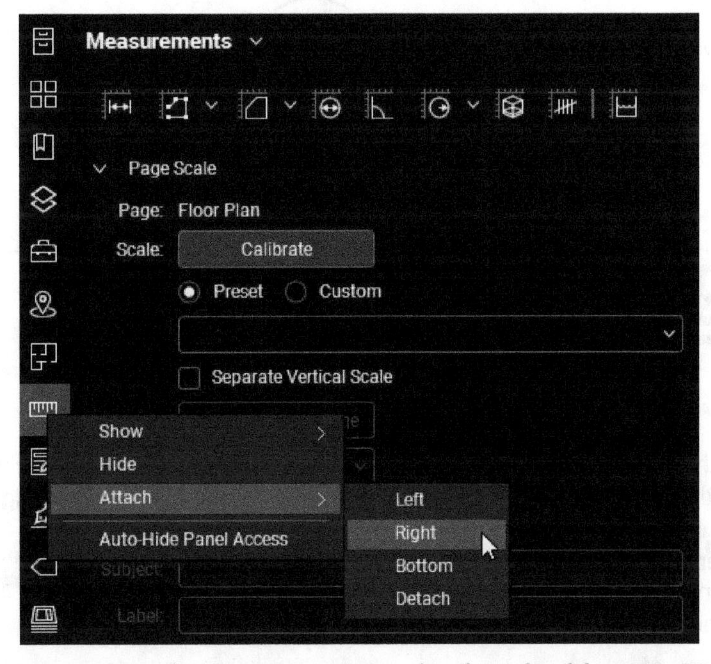

*Figure 35 Attaching the **Measurements** panel to the right of the **Main Worspace***

You will now save this as a profile.

3. On the top left, click **Revu > Profiles > Manage Profiles** and save this as a new profile with the name **Training QTO**. Refer to the tutorial in Chapter 1 of this book to revisit the details of how to create a new profile.

Section 2: Calibrating the PDF File

Notice that this PDF file has a main view and a smaller view on the lower left. In this section, you will calibrate the sheet. Because this sheet does not have any available dimensions, you will use two endpoints of the door to calibrate it. The reason you will use a door is that the doors have standard sizes. In this case, it is 36" if you use Imperial units or 920 mm if you use Metric units. However, before you calibrate the sheet, you will disable the line weights.

1. Navigate close to the door with the tag **245** on the middle left of the main view, as shown in Figure 36.

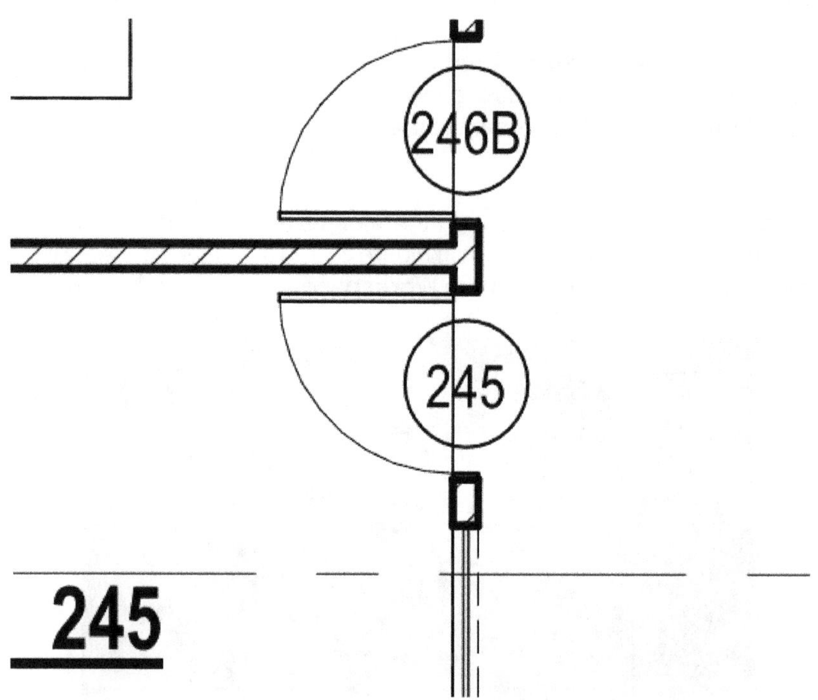

*Figure 36 The door with tag **245** to be used for calibration*

Notice that the walls around the door are displayed with line weights. This will make it hard to calibrate the sheet properly. Therefore, you will first disable the line weights.

2. From the **Menu Bar**, click **View > Disable Line Weights**, which is the last options in the menu; the line weights are disabled from the sheet.

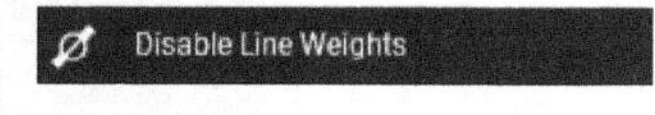

Notice that the right side of the **Navigation Bar** and the **Status Bar** shows **Scale Not Set**. You will now click on one of these messages to start the process of setting the sheet scale.

3. On the right side of the **Navigation Bar**, click **Scale Not Set**, labeled as **1** in Figure 37. Note that if you are using a version older than Revu 2019, you can select the option to calibrate the sheet by going to the **Menu Bar > Measure > Calibrate**.

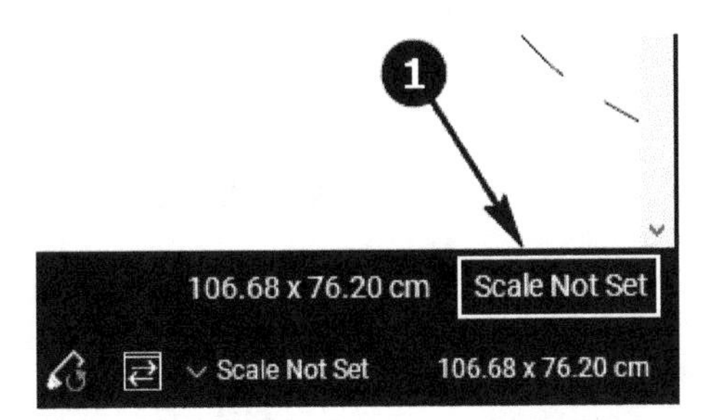

Figure 37 Starting the process of calibrating the sheet

On doing so, the **Set Scale** dialog box is displayed, as shown in Figure 38.

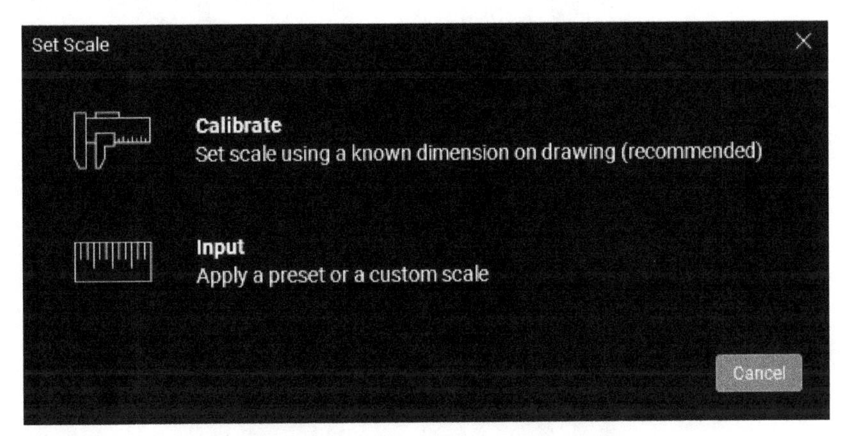

*Figure 38 The **Set Scale** dialog box*

4. From the **Set Scale** dialog box, select **Calibrate**; the **Calibrate** dialog box is displayed prompting you to select two points of the known dimension to calibrate the measurement tools, as shown in Figure 39.

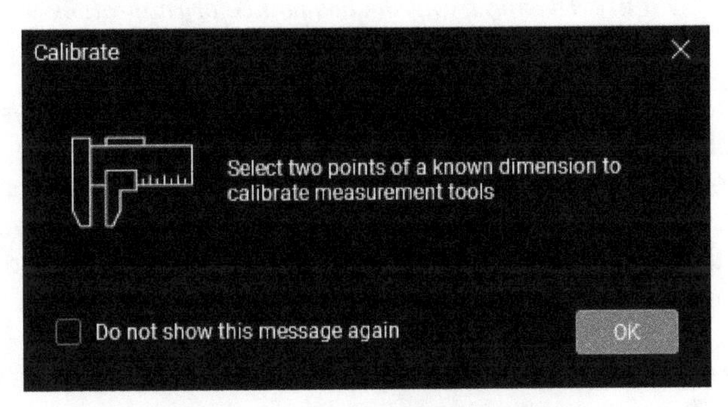

*Figure 39 The **Calibrate** dialog box*

5. Click **OK** in the **Calibrate** dialog box; you are returned to the **Main Workspace** and are prompted to select the start point and drag to specify the endpoint.

   You will now snap to the two endpoints of the door opening and specify the calibration distance. Note that you will be able to snap to the vertices even though the **Snap to Content** option may or may not be turned on from the **Status Bar**.

6. Snap to the point labeled as **1** in Figure 40 as the first point of calibration.

   To select the second point, you will hold down the SHIFT key so the calibration line is drawn vertically down.

7. Hold down the SHIFT key and snap to the point labeled as **2** in Figure 40 as the second point of calibration.

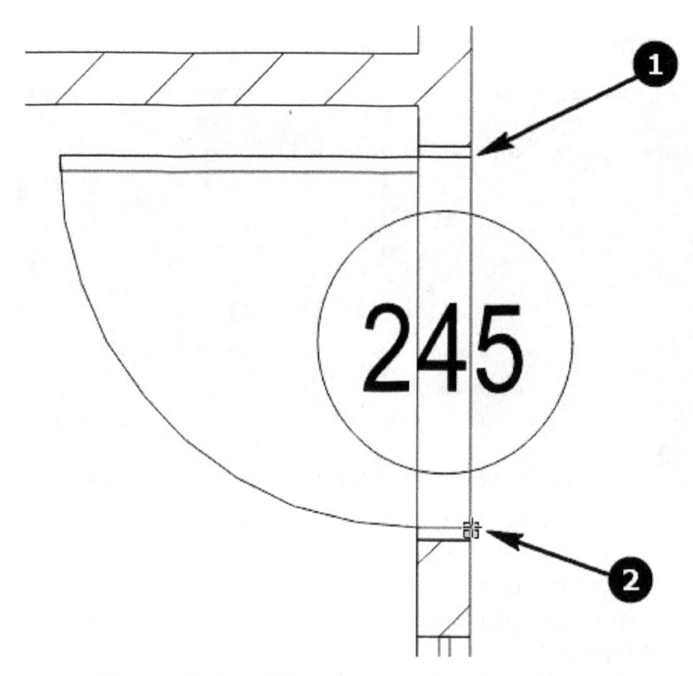

Figure 40 *Specifying the two points for calibration*

On specifying the second point, the **Calibrate** dialog box is displayed with the **Custom** radio button automatically selected. Below this radio button are two sets of values and units separated by the = symbol. The value on the left of the symbol represents the distance between the two points you measured. The value on the right is the actual value you need to specify. Note that depending on whether you use Imperial or Metric units, you need to enter the value based on those units.

8. On the right side of the = symbol, enter **36** if you use the Imperial units of **920** if you use the Metric units.

9. From the units drop-down list on the right of the value, select **in** for **Imperial** units and **mm** for **Metric** units. Figure 41 shows the **Calibrate** dialog box with **36 in** selected for the **Imperial** units.

10. From the **Precision** drop-down list, select **1**; as shown in Figure 41. This is because you do not need any decimal places for the measurement values.

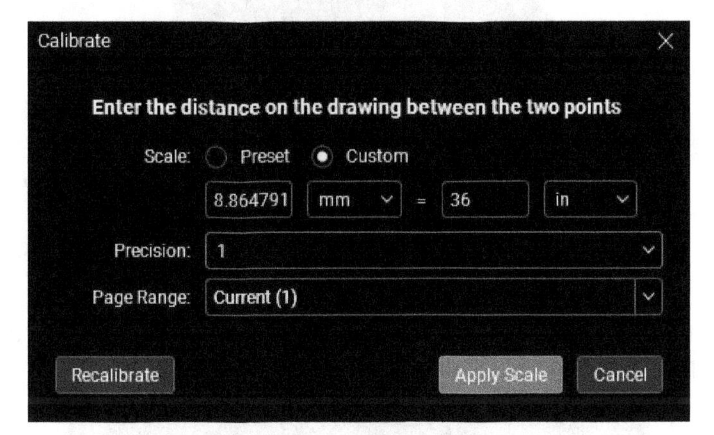

Figure 41 *Specifying the calibration value and precision*

Because this PDF file only has one sheet, you do not need to change anything in the **Page Range** drop-down list.

11. Click **Apply Scale** in the **Calibrate** dialog box; the right side of the **Navigation Bar** and the **Status Bar** now show this scale value, as shown in Figure 42.

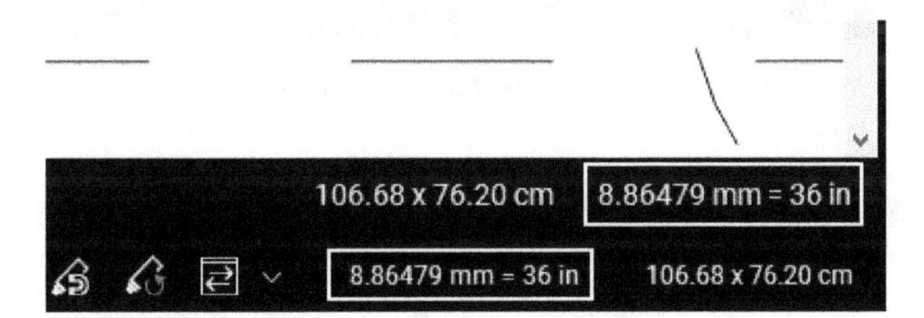

Figure 42 *The sheet scale displayed on the* ***Navigation Bar*** *and the* ***Status Bar***

12. Zoom to the extents of the sheet and then save the file.

Section 3: Creating a Viewport and Calibrating its Scale

In this section, you will create a viewport around the detail view on the lower left of the sheet. You will then use the door with the tag **258B** to calibrate the scale of this viewport.

1. Navigate to the detail view on the lower left of the sheet.

2. On the right side of the **Main Workspace**, click on the **Measurements** button to display the **Measurements** panel, if not already displayed.

3. Click the **Add Viewport** button available below the **Viewports** area in the **Measurements** panel, as shown in Figure 43.

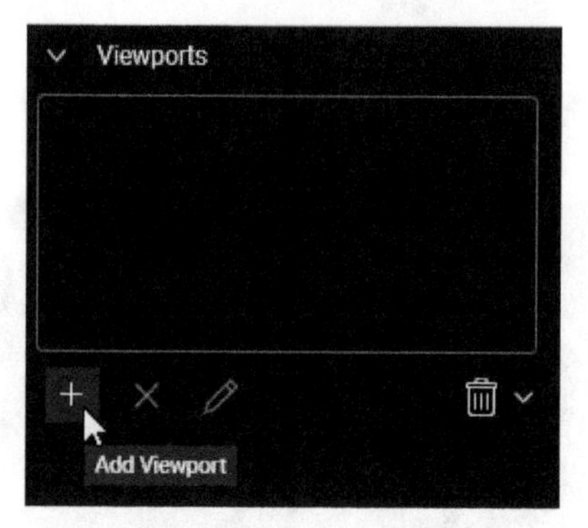

Figure 43 Adding a new viewport

On doing so, the **Add Viewport** dialog box is displayed prompting you to select a region to define a viewport, as shown in Figure 44.

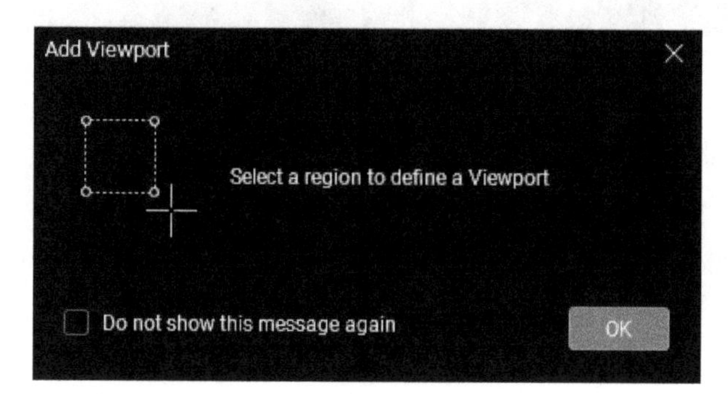

*Figure 44 The **Add Viewport** dialog box*

4. Click **OK** in the **Add Viewport** dialog box; you are returned to the **Main Workspace** and are prompted to select a region to place viewport.

5. Specify the two opposite corners labeled as **1** and **2** in Figure 45 to define the viewport.

   On doing so, the **Add Viewport** dialog box is redisplayed, but this time with a number of options to set the scale for the viewport. You will use the **Calibrate** option in this dialog box and use the two endpoints of the opening of the door with the tag **258B** in the viewport to set the scale.

6. From the lower left of the **Add Viewport** dialog box, click **Calibrate**; the **Calibrate** dialog box is displayed prompting you to select two points of a known dimension to calibrate the measurement tools.

7. Click **OK** in the **Calibrate** dialog box; you are returned to the **Main Workspace**.

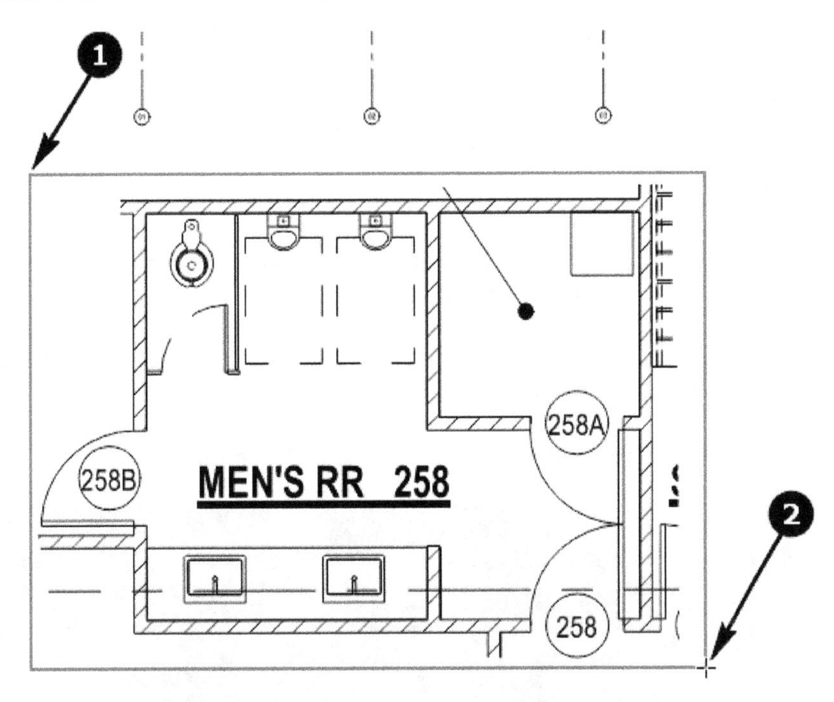

Figure 45 Specifying two points to define a viewport

8. Zoom close to the door with the tag **258B** in the viewport and snap to the point labeled as **1** in Figure 46 as the first point of calibration.

9. Hold down the SHIFT key and snap to the point labeled as **2** in Figure 46 as the second point of calibration.

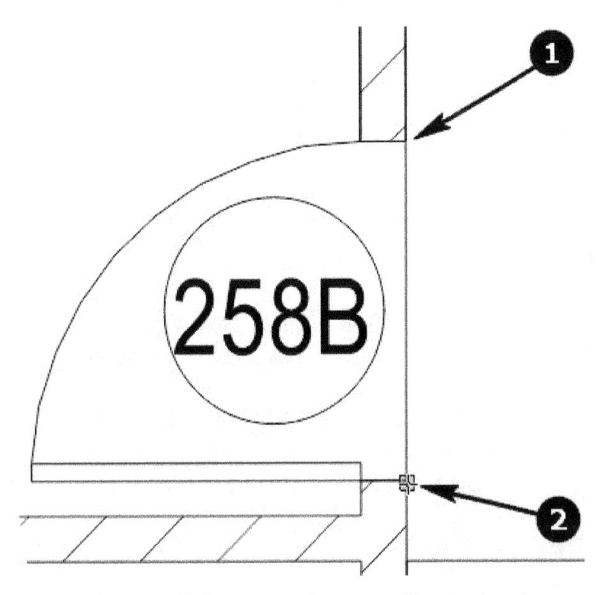

Figure 46 Specifying two points to calibrate the viewport

On doing so, the **Add Viewport** dialog box is redisplayed.

10. In the **Name** field of the **Add Viewport** dialog box, type **MEN'S RR** as the name of the viewport.

11. Select the **Custom** radio button, if not already selected.

12. In the field on the right of **=**, type **36** for the Imperial units or **920** for the Metric units. Figure 47 shows the dialog box with the **Imperial** units.

13. From the drop-down list on the right of the value you typed, select **in** for the Imperial units and **mm** for the Metric units, as shown in Figure 47.

14. From the **Precision** drop-down list, select **1**, as shown in Figure 47.

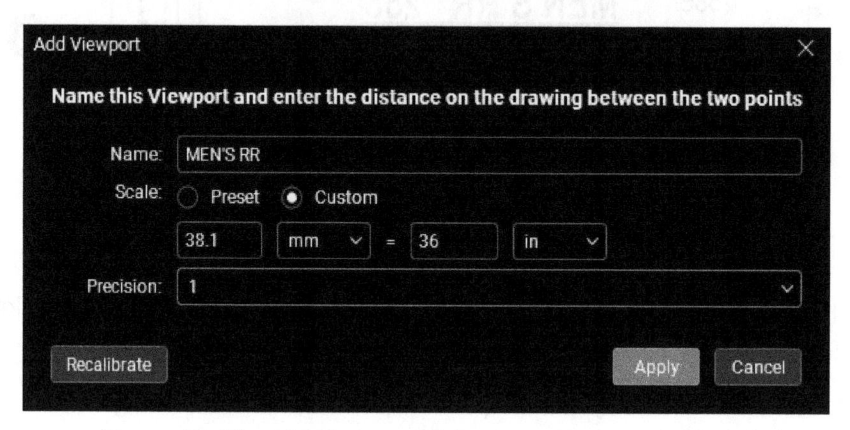

***Figure 47** The **Add Viewport** dialog box with the scale defined*

15. Click **Apply** in the **Add Viewport** dialog box; the viewport is created and calibrated. It is highlighted in Blue on the **Main Workspace** and is also listed in the **Viewports** area of the **Measurements** panel.

16. Press the ESC key to deselect the viewport.

17. Zoom to the extents of the PDF and then save it.

***Note**: The **Navigation Bar** and the **Status Bar** shows the scale of the sheet and not that of the viewport, even though the viewport might be selected and highlighted.*

Section 4: Adding and Customizing the Length Measurement

In this section, you will navigate to the **BREAK 233** room and then add a length measurement to the lines in that room. You will then customize the length measurement.

1. Navigate to the **BREAK 233** room located at the bottom center of the plan, as shown in Figure 48.

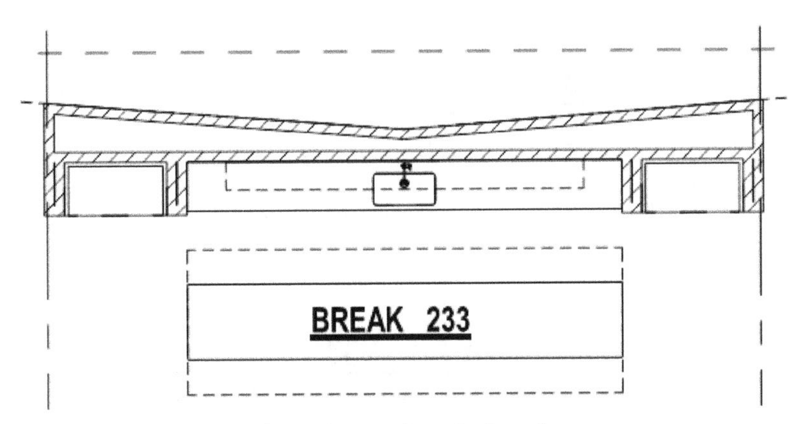

Figure 48 The region to place the length measurement

IMPORTANT NOTE: PLEASE NOTE THAT THE DIMENSION VALUES ON YOUR MACHINE MAY BE SLIGHTLY DIFFERENT FROM THE ONES SHOWN IN THIS TUTORIAL.

2. From the **Menu Bar**, click **Tools > Measure > Length** to invoke this tool; you are prompted to select the start point and drag to the endpoint.

3. Zoom close to the vertex labeled as **1** in Figure 49 and snap to it to select it as the first point of the length measurement. Make sure you do not select the centerline.

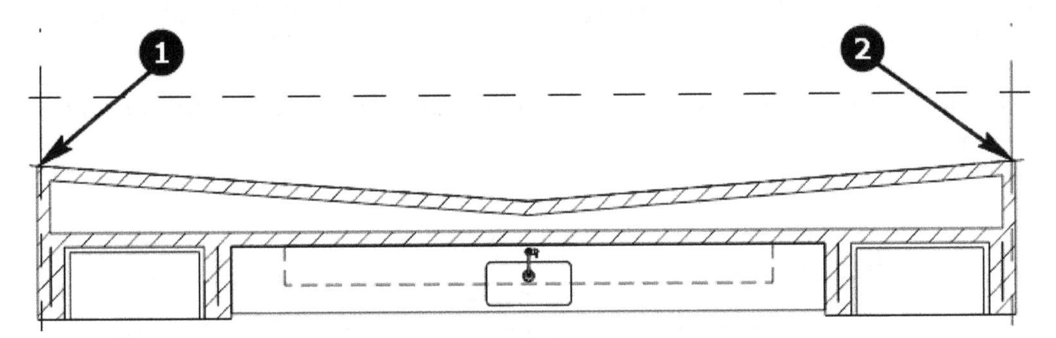

Figure 49 The two vertices to place the length measurement

4. Next, zoom close to the vertex labeled as **2** in Figure 49.

5. Now, hold down the SHIFT key and snap to the point labeled as **2** in Figure 49 to select it as the second point of the length measurement. Make sure you do not select the centerline.

   As soon as you snap to the second point, the length measurement is placed between the two points. If you are using the Imperial units, you will see the length dimension placed. Figure 50 shows the length measurement with Imperial units.

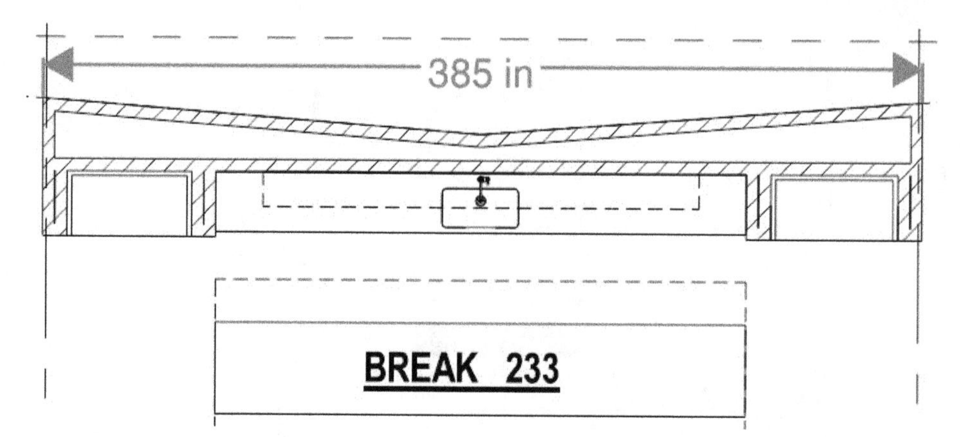

Figure 50 The length measurement placed between the two points

While working with the Metric units, it is acceptable to display the linear dimensions in mm. However, for the Imperial units, it is preferred to display the linear dimensions in feet and inches. Therefore, you will update the calibration for the Imperial units now.

6. For the Imperial units only, select the length measurement, if it is not already selected.

7. In the area below the **Calibrate** button in the **Measurements** panel, change the **36** value to **3**.

8. From the units drop-down list on the right of the value, select **ft' in"**; the length measurement in the **Main Workspace** changes to **32'-1"**, as shown in Figure 51.

   IMPORTANT NOTE: your value may be a little different from the one shown in this figure.

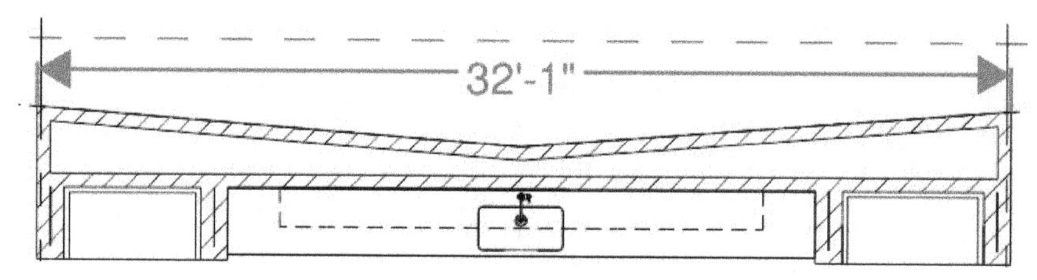

Figure 51 The Imperial length measurement updated to the new scale

9. With the length measurement still selected, press and hold down the Left Mouse Button on the Yellow handle on top of the measurement value and drag it above the centerline, refer to Figure 52.

   You will now change the subject of this length measurement and customize its appearance properties.

10. From the right of the Main Workspace, activate the **Properties** panel.

11. In the **Subject** field, enter **Project Manager's Length** as the value.

12. In the **Appearance** area, make the following changes:

Color:	**Blue**
Highlight:	**Checked**
Line Width:	**0.5**
Start:	**Architectural Tick (last option)**
End:	**Architectural Tick (last option)**
Font Size:	**10**
Text Color:	**Blue**

Figure 52 shows the customized length measurement in Imperial units and Figure 53 shows this measurement in Metric units.

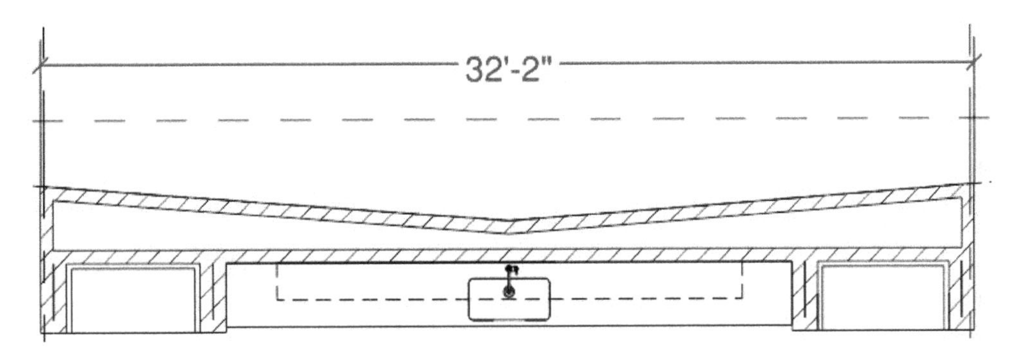

Figure 52 *The customized length measurement in the Imperial units*

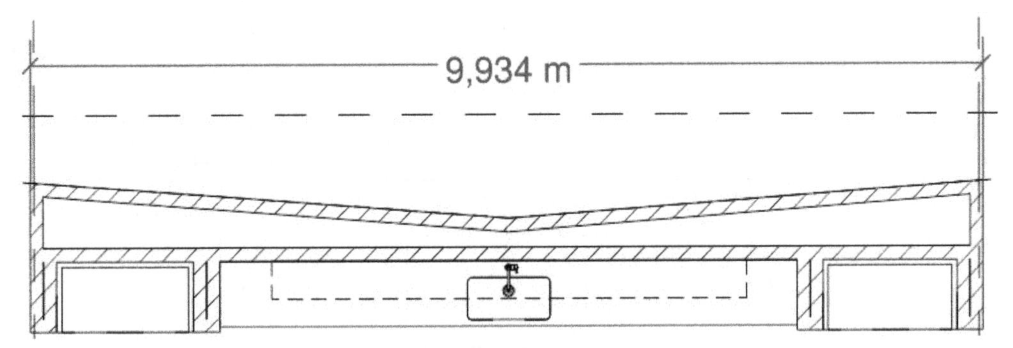

Figure 53 *The customized length measurement in the Metric units*

Notice that the customized version of this length measurement is now added to the **Recent Tools** tool set. At this stage, you will not add this customized tool to your custom tool set. In later chapters, you will create custom columns and then link those columns to the customized tools and then add those customized tools to the tool sets.

Section 5: Adding and Customizing the Perimeter Measurement

In this section, you will navigate to the **MEETING 247** room located at the middle left of the plan and then add a perimeter measurement to that room. You will then customize this perimeter measurement.

1. Navigate to the **MEETING 247** room located at the middle left of the plan, refer to Figure 54.

2. From the **Menu Bar**, click **Tools > Measure > Perimeter**; this tool is invoked and you are prompted to select vertices of the perimeter to measure.

   You will now select the points on the inside walls of this room to place the perimeter measurement.

3. One by one, snap to the points labeled **1** to **7** in Figure 54 to specify the perimeter vertices.

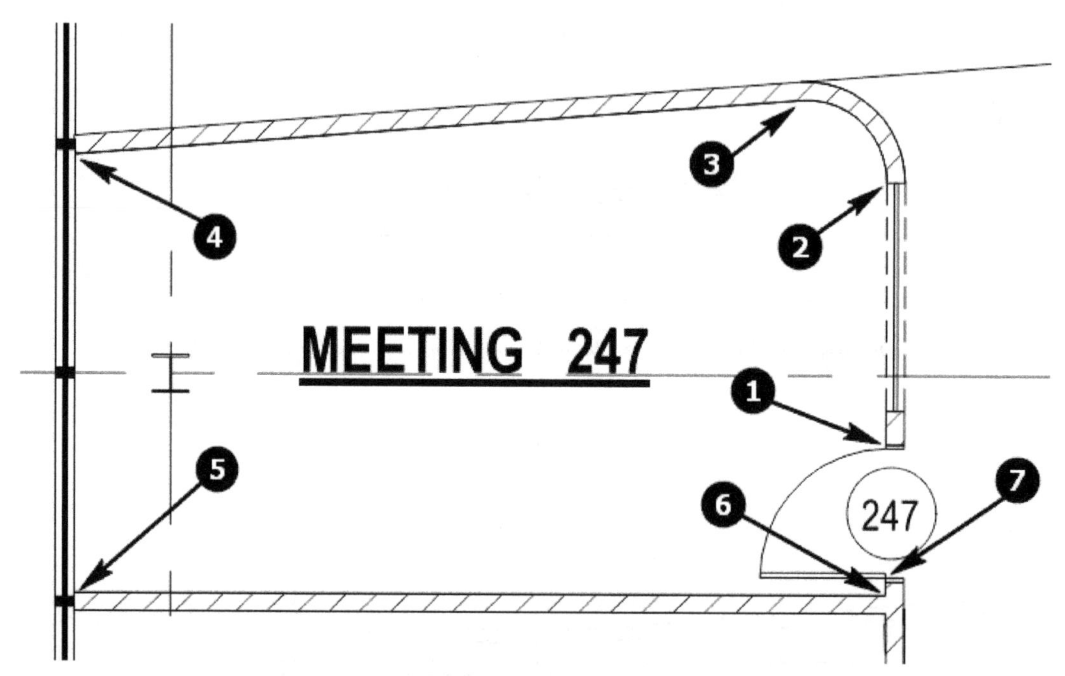

Figure 54 *The vertices for the perimeter measurement*

4. After specifying the 7th point, press the ENTER key, the perimeter measurement is placed with the default appearance and subject.

5. Press the ESC key to exit the **Perimeter** tool.

   Notice that between vertices **2** and **3**, it is a straight line segment. You will now convert this segment into an arc segment.

6. Right-click on the segment between vertices **2** and **3** to display the shortcut menu.

7. From the shortcut menu, select **Convert to Arc**, as shown in Figure 55.

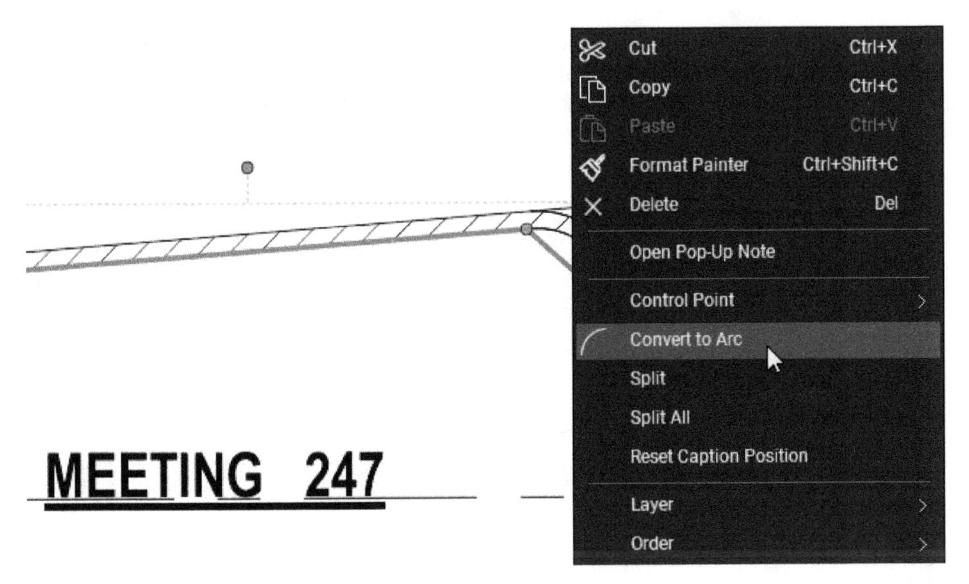

Figure 55 *Converting a straight segment into an arc segment*

On doing so, the segment is converted into an arc segment.

8. If required, turn off the **Snap to Content** option from the **Status Bar** and then use the handles at the two ends of the arc segment to match its shape with that of the curved section of the wall.

   You will now change the subject of this perimeter measurement and customize its appearance properties.

9. From the right of the **Main Workspace**, activate the **Properties** panel.

   This perimeter measurement will be used to calculate the skirting of the room. So you will change the subject of this measurement to **Skirting**.

10. In the **Subject** field, enter **Skirting** as the value.

11. In the **Appearance** area, make the following changes:

    Color: **Blue**
    Highlight: **Not Checked**
    Line Width: **0.5**
    Font Size: **10**
    Text Color: **Blue**

    Next, you will drag the perimeter value text and place it along the first segment.

12. Hold down the SHIFT key and drag the perimeter value text to place it along the first segment.

13. Press the ESC key to deselect the measurement. Figure 56 shows the perimeter measurement in the Imperial units and Figure 57 shows it in the Metric units. Note that the values in your case may be a little different, depending on the arc segment shape in your measurement.

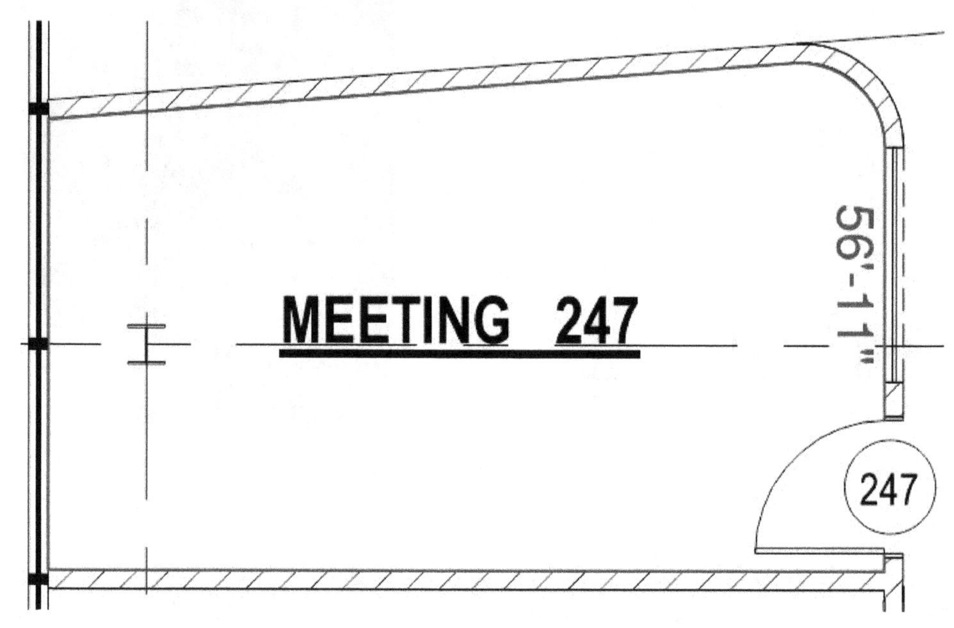

Figure 56 *The perimeter measurement in the Imperial units*

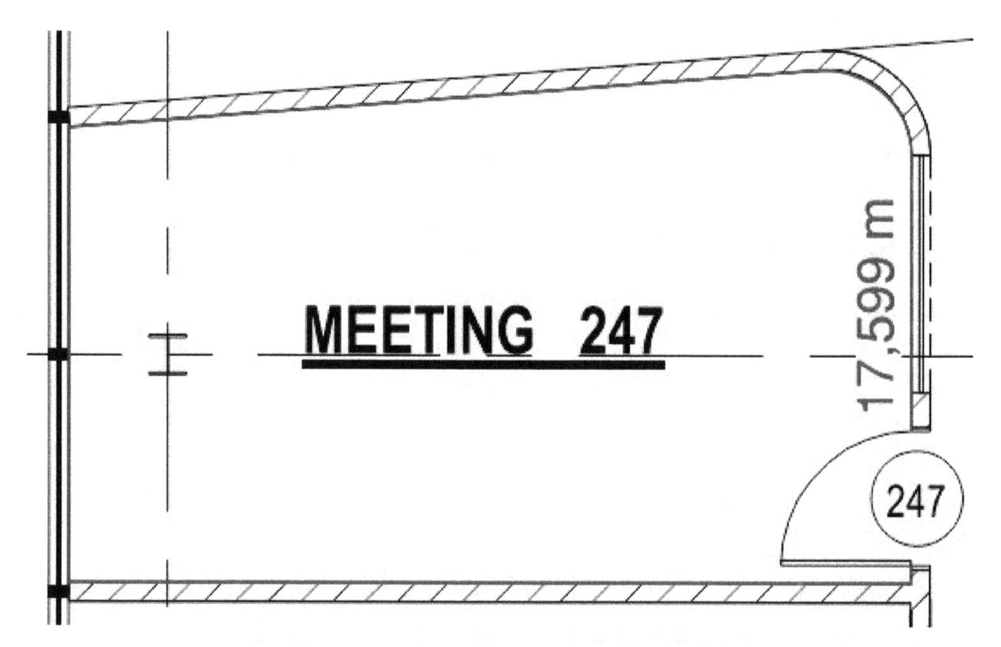

Figure 57 *The perimeter measurement in the Metric units*

14. Zoom to the extents of the PDF and then save it.

Section 6: Adding and Customizing the Polylength Measurement

In this section, you will navigate to the **OPEN OFFICE 200** room located at the top left of the plan and then add a polylength measurement between the columns. You will then customize this polylength measurement to measure beam lengths.

1. Navigate to the **OPEN OFFICE 200** room located at the top left of the plan, refer to Figure 58.
2. From the **Menu Bar**, click **Tools > Measure > Polylength**; the tool is invoked and you are prompted to select vertices of the perimeter to measure.

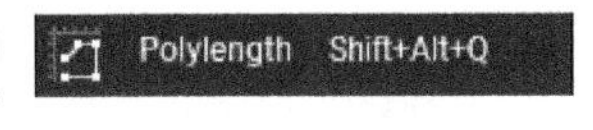

3. Zoom close to the vertex labeled as **1** in Figure 58 and snap to the midpoint of the column as the first vertex of the measurement.

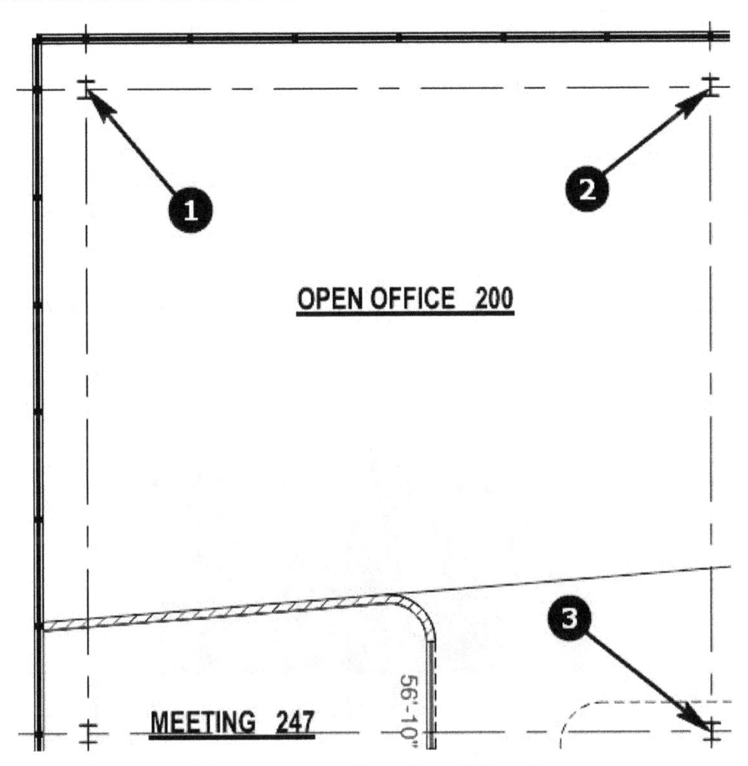

Figure 58 *The vertices to place the polylength measurement*

4. Zoom close to the vertex labeled as **2** in Figure 58 and then snap to the midpoint of the column as the second vertex of the measurement.

   Note that if you select an incorrect vertex, you can press the BACKSPACE key to deselect it.

5. Similarly, zoom close to the vertex labeled as **3** in Figure 58 and then snap to the midpoint of the column as the third vertex of the measurement.

6. After specifying the third vertex, press ENTER; the polylength measurement is placed.

   Next, you will change the subject of this measurement. Because this measurement represents beam lengths, you will specify the **Beam** type as the subject. However, because there are a number of different beam types, you will then specify the type in the **Label** field.

7. In the **Properties** panel > **Subject** field, enter **Beam** as the subject.
8. If you use the Imperial units, enter **S24X121** in the **Label** field and if you use the Metric units, enter **530 UB 82.0** in the **Label** field. On doing so, the label is also displayed on the measurement in the PDF file.
9. In the **Appearance** area, make the following changes:

Color:	**Blue**
Highlight:	**Not Checked**
Line Width:	**0.5**
Font Size:	**10**
Text Color:	**Blue**

   Next, you will turn off the visibility of the total length of the measurement. This can be done in the **Measurements** panel or the **Properties** panel. However, because you already have the **Properties** panel active, you can do it there.

10. In the **Properties** panel, click on the **Edit** button above the **Font** drop-down list; a cascading menu is displayed.
11. From the cascading menu, clear the **Length** check box, as shown in Figure 59.

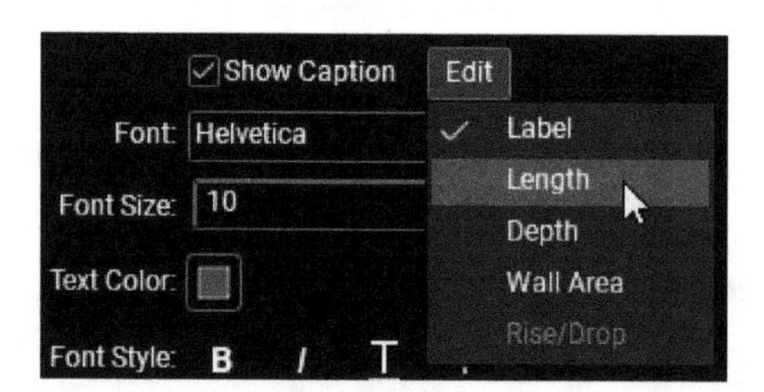

Figure 59 *Turning off the cumulative length caption for the measurement*

12. Press the ESC key to deselect the measurement. Figure 60 shows the polylength measurement with Imperial units and Figure 61 shows this measurement with Metric units.

Tip: *As mentioned earlier, because the polylength measurement can be assigned a rise and drop value, it is well suited to take off the length of mechanical ductwork.*

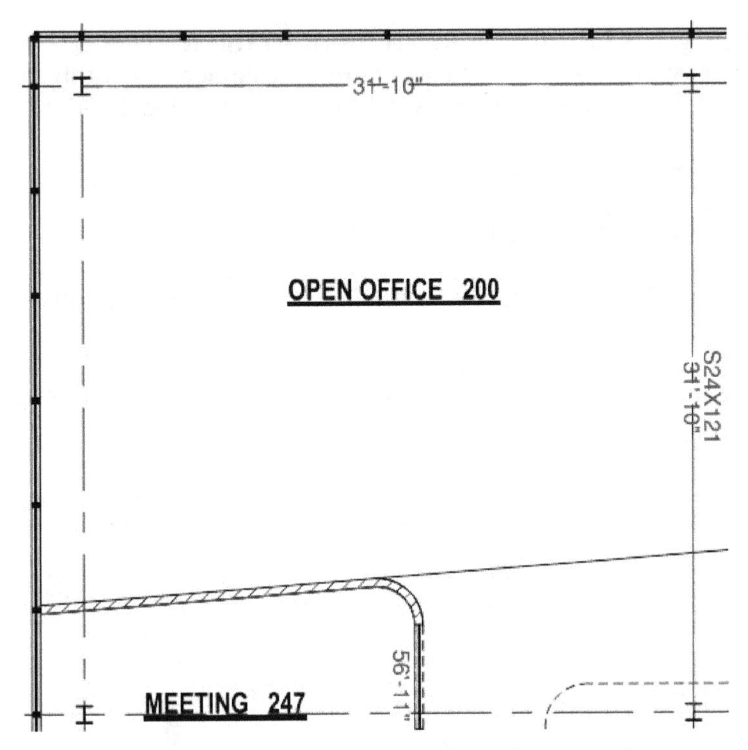

Figure 60 The polylength measurement in the Imperial units

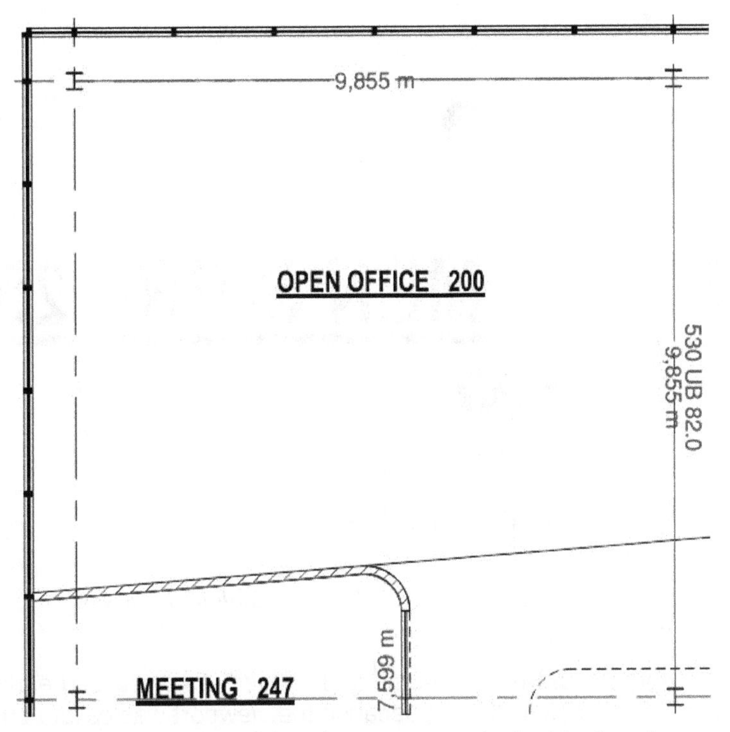

Figure 61 The polylength measurement in the Metric units

Section 7: Adding a Length Measurement to the MEN'S RR Viewport

In this section, you will use the customized **Project Manager's Length** measurement tool from the **Recent Tools** tool set to add a length measurement to the **MEN's RR** viewport.

1. In the **MEN'S RR** viewport on the lower left of the sheet.
2. Zoom close to the door tag **258B**, refer to Figure 62.
3. From the **Recent Tools** tool set, invoke the **Project Manager's Length** tool; the viewport is highlighted in Blue.
4. Snap to the vertices labeled as **1** and **2** in Figure 62 to place the customized length measurement.

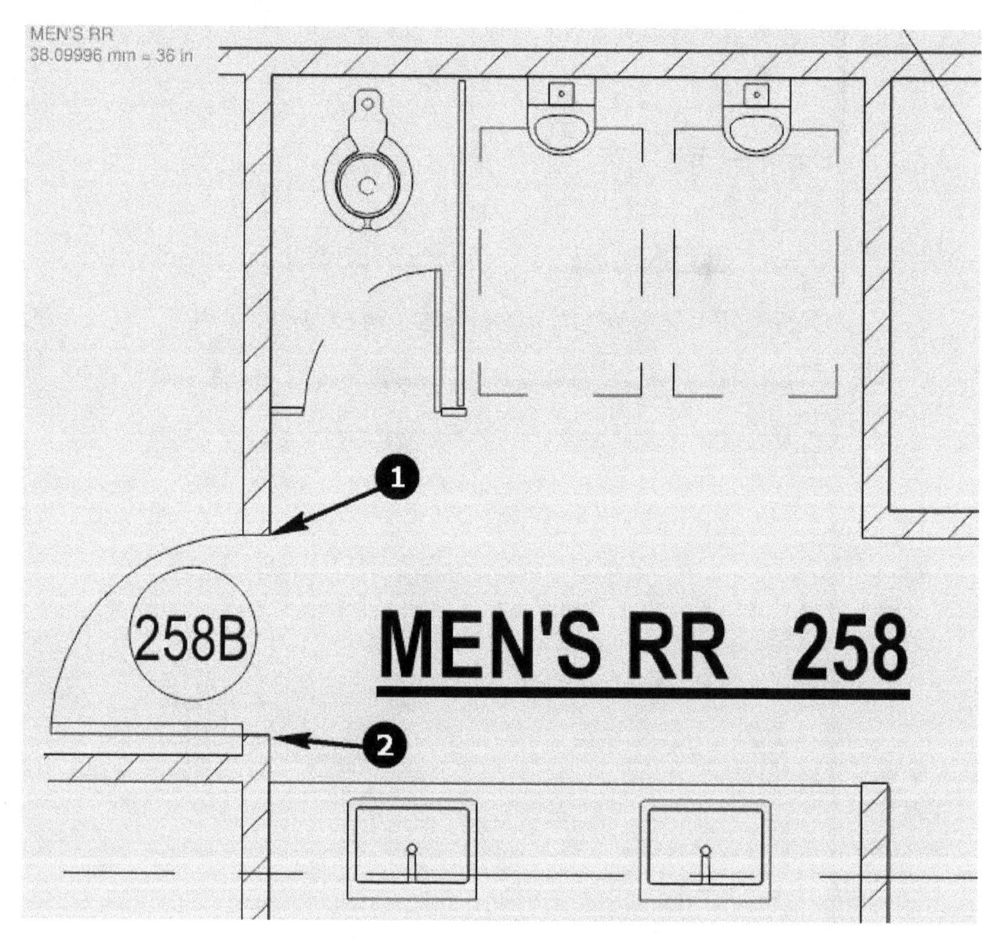

Figure 62 *The vertices to place the length measurement*

On doing so, the length of the door frame is displayed. As discussed earlier, the Imperial units show this value as **36 in**. This is because the viewport was calibrated with Inches as the units. You will now edit these units.

5. Activate the **Measurements** panel.

6. Scroll down to the **Viewports** area.

7. In the **Viewports** area, click on the **MEN'S RR** viewport; it is highlighted in Blue in the **Main Workspace**. Also, the scale options are displayed below the **Viewports** area.

8. In the field where it shows **36** as the value, change it to **3**.

9. From the units drop-down list on the right of the value list, select **ft' in"**, as shown in Figure 63.

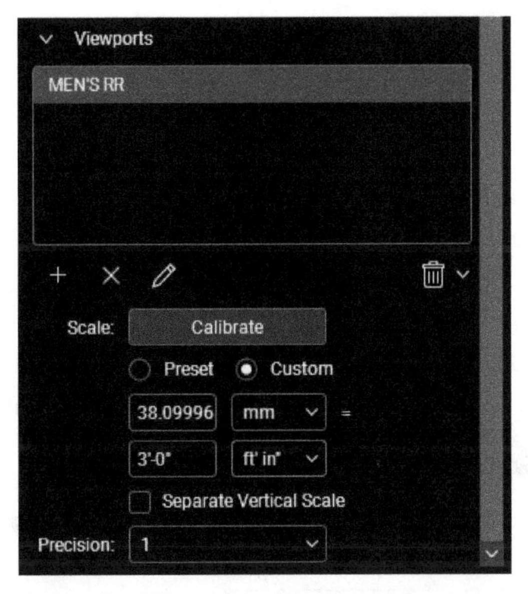

Figure 63 *Editing the calibration of the viewport*

10. Press the ESC key. Figure 64 shows the length measurement in the viewport with the Imperial units and Figure 65 shows this measurement in the Metric units.

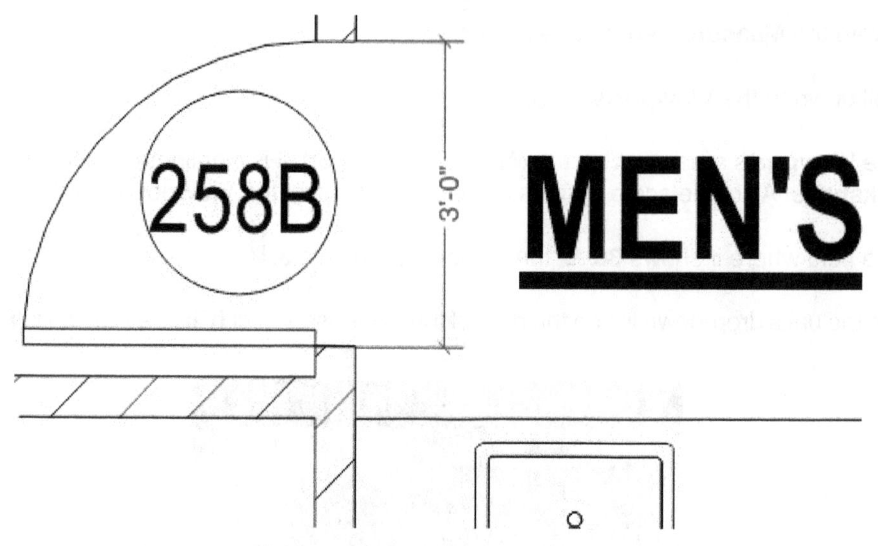

Figure 64 *The Imperial length measurement in the viewport*

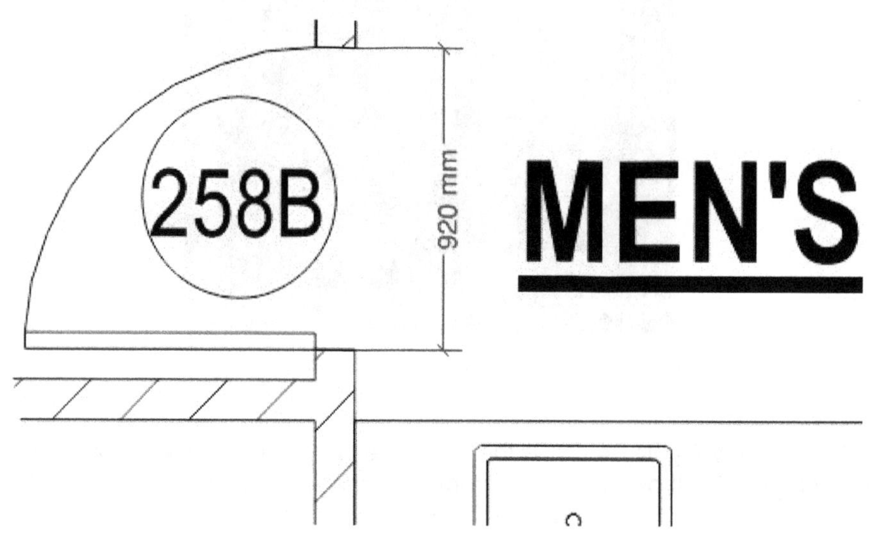

Figure 65 *The Metric length measurement in the viewport*

11. Zoom to the extents of the PDF.

12. Save the file and close Revu 2019.

Skill Evaluation

Evaluate your skills to see how many questions you can answer correctly. The answers to these questions are given at the end of the book.

1. In Revu 2019, you cannot set a standard scale for the sheet. (True/False)

2. The polylength measurement by default shows the length of the individual segments as well. (True/False)

3. The measurement tools cannot be customized. (True/False)

4. In Revu, you cannot turn off the line weights from the vector PDFs. (True/False)

5. You can specify different scale values for the detail views on the sheet. (True/False)

6. Which tool allows you to add a linear measurement between two selected points?

   (A) **Linear** (B) **Length**
   (C) **Points** (D) **Angle**

7. The regions on the sheet that can be defined their own individual scales are called:

   (A) Viewports (B) Viewpoints
   (C) Boxes (D) Windows

8. On what two bars at the bottom of the **Main Workspace** is the scale of the sheet displayed?

   (A) **Menu Bar** (B) **Status Bar**
   (C) **Toolbar** (D) **Navigation Bar**

9. Which option is used to turn off the visibility of line weights in the PDF file?

   (A) **View > No Line Weights** (B) **Tools > Line Weights**
   (C) **View > Disable Line Weights** (D) **Measure > Line Weights**

10. Which option allows you to specify the measurement scale by selecting two points and then define the distance between them?

    (A) Properties (B) Calibrate
    (C) Appearance (D) Measure

Class Test Questions

Answer the following questions:

1. Explain briefly the process of setting a standard sheet scale.
2. Explain briefly the process of calibrating a sheet.
3. Explain briefly how to create a viewport.
4. How will you turn off the line weights?
5. Explain briefly the process of placing a polylength measurement.

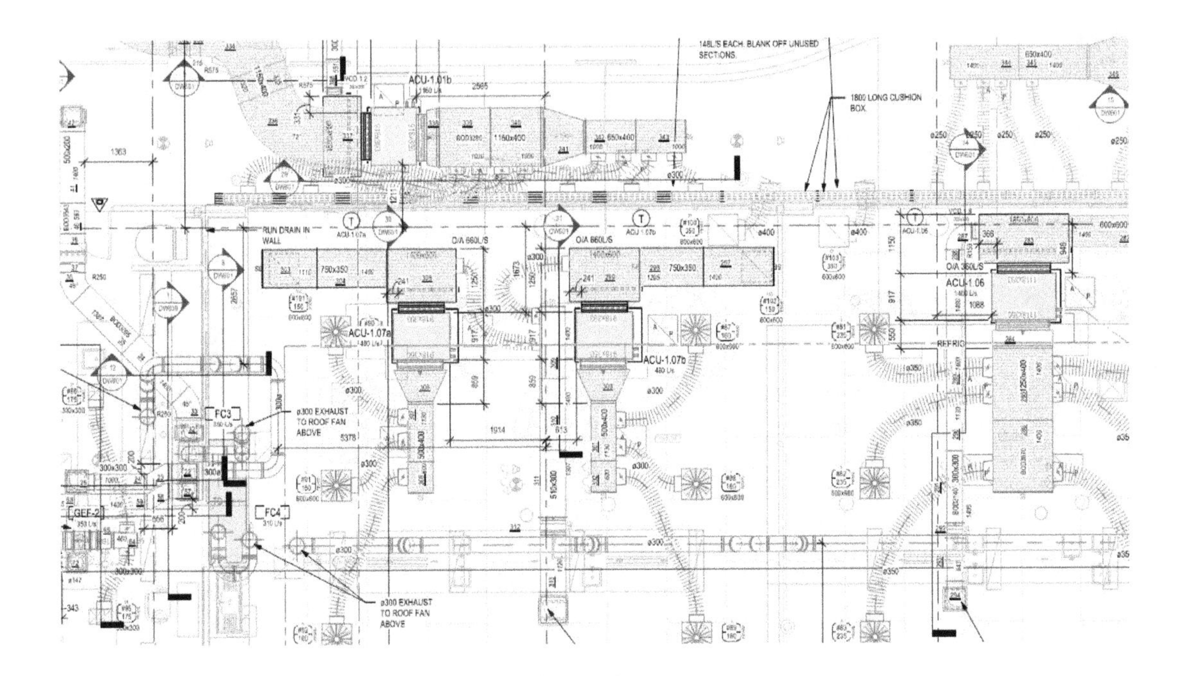

Chapter 7 - Working with the Measurement Tools - II

The objectives of this chapter are to:

- √ *Explain the process of adding and customizing the area measurement*
- √ *Explain the process of adding and customizing the volume measurement*
- √ *Explain the process of using the **Dynamic Fill** tool to fill a region and add multiple measurements*
- √ *Explain how to cutout regions from the area measurement*
- √ *Explain the process of adding and customizing the diameter measurements*
- √ *Explain how to use the various radii measurement tools*
- √ *Explain the process of adding and customizing an angle measurement*
- √ *Explain the process of creating a custom count measurement*

THE MEASUREMENT TOOLS - II

In this chapter, you will learn about some more measurement tools.

The Area Measurement

Menu: Tools > Measure > Area
Panel: Measurements > Area

The **Area** measurement tool is used to measure the area of a closed region that you define. This tool can be customized to takeoff the area measurements for items like carpets, tiles, stone, wooden floorings, and so on. When you invoke this tool, you will be prompted to select vertices of the area to measure. Similar to the polygon markup, you can drag two corners of the area measurement to measure a four-sided shape or you can snap to individual vertices to measure a multi-sided shape. Remember that while selecting the vertices, if you select an incorrect vertex, you can press the BACKSPACE key on the keyboard to deselect that point. Note that when you invoke this tool, the **Area Measurement Properties** area is displayed in the **Measurements** panel. In this area, you can specify the units of the area and volume measurements, along with the additional properties, such as depth and slope of the area measurement, as shown in Figure 1.

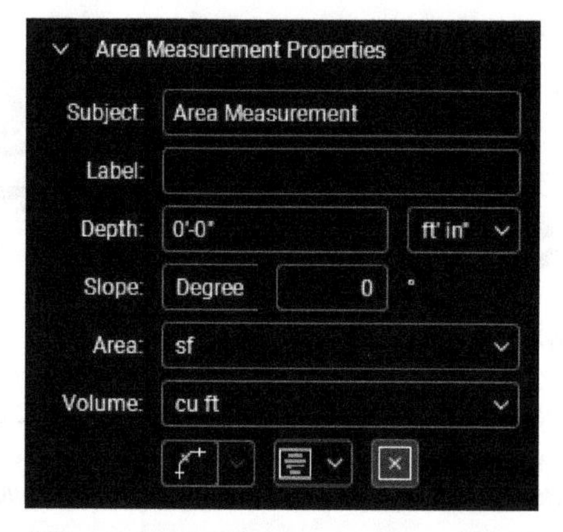

*Figure 1 The **Area Measurement Properties** area of the **Measurements** panel*

Once you have placed the area measurement, you can convert any of its linear segment into a curved segment, if they are not drawn straight or by holding down the SHIFT key. Having said that, in case of the areas that have non-linear segments, it is recommended to use the **Dynamic Fill** tool that will be discussed later in this chapter. Figure 2 shows the default area measurement in Imperial units and Figure 3 shows the same area measurement after customizing its appearance properties and changing the subject and label. Note that in both these figures, the area value caption is moved from its default location by holding down the SHIFT key.

***Note:** Once the area measurement is created, you can cutout regions from it using the **Cutout** tools that are discussed later in this chapter.*

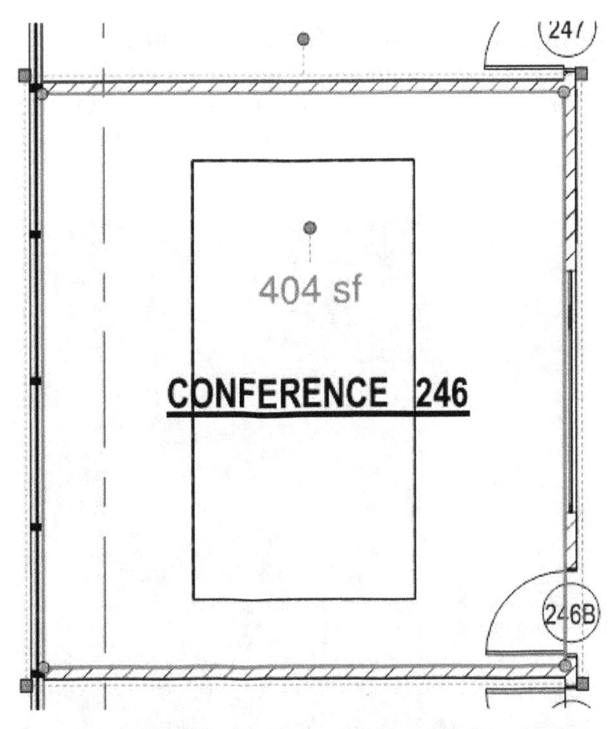

Figure 2 *The default area measurement in a four sided room*

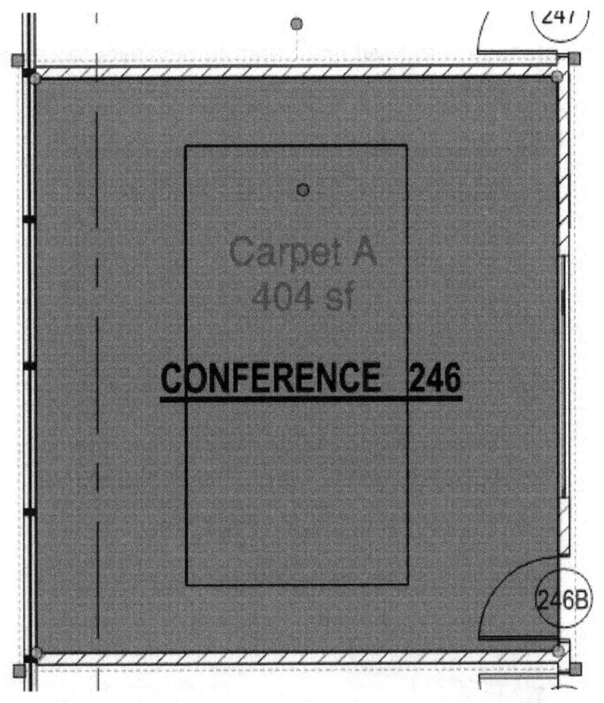

Figure 3 *The same area measurement after customizing it*

Displaying the Centroid of an Area Measurement

While measuring the area of a non-standard shape, it is sometimes required to display the centroid of that area. This can be done by selecting the **Show Centroid** option in the **Area Measurement Properties** area of the **Measurements** panel while creating the area, as shown in Figure 4, or in the **Properties** panel.

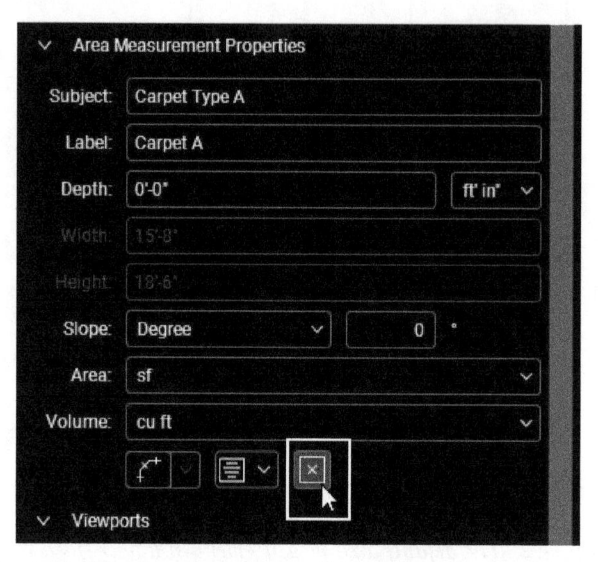

Figure 4 *Selecting the option to display the centroid of the area*

Figure 5 shows an area measurement with the centroid displayed as a cross.

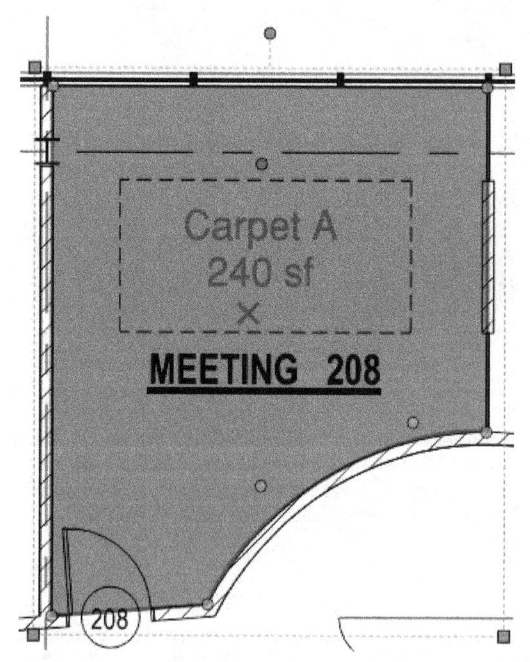

Figure 5 *The area measurement with the centroid turned on*

The Volume Measurement

Menu: Tools > Measure > Volume
Panel: Measurements > Volume

The **Volume** measurement tool works similar to the **Area** tool and is used to measure the volume of a closed region that you define. This tool can be customized to takeoff the volume measurements for items like concrete. When you invoke this tool, you will be prompted to select vertices of the area to measure. Before you select the vertices to create the volume measurement, you can specify the depth of the volume and its units in the **Volume Measurement Properties** area of the **Measurements** panel, as shown in Figure 6.

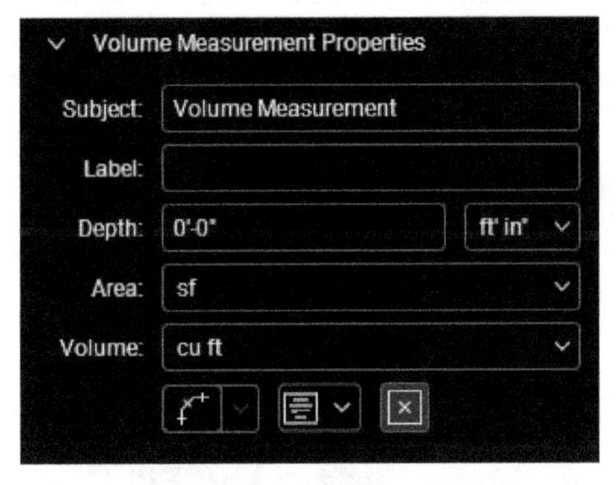

*Figure 6 The **Volume Measurement Properties** area of the **Measurements** panel*

Figure 7 shows the volume measurement of a slab customized as Concrete Grade M50. In this case, the measurement is also customized to display the **Depth** caption of **6 in**.

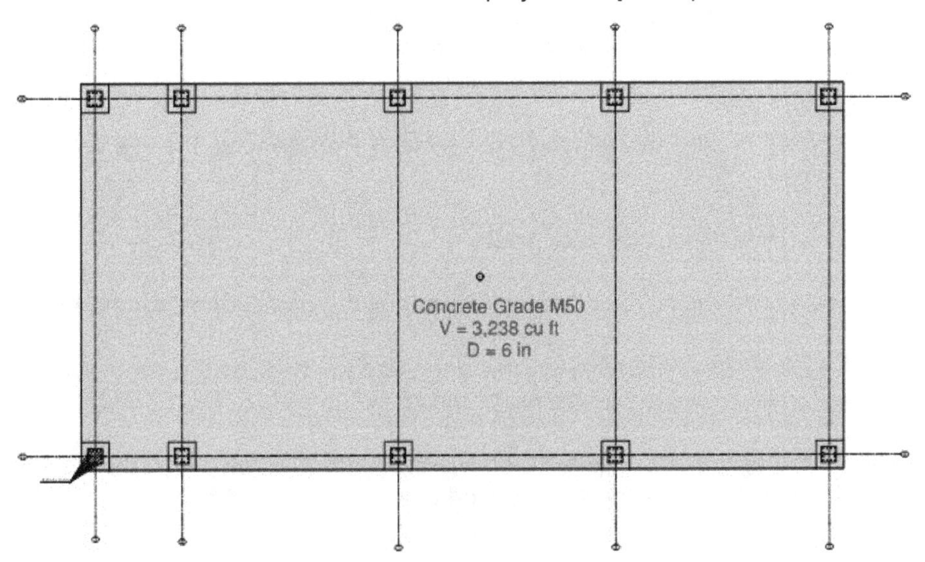

Figure 7 The volume measurement customized as Concrete Grade M50

The Dynamic Fill Measurement

Menu: Tools > Measure > Dynamic Fill
Panel: Measurements > Dynamic Fill

The **Dynamic Fill** measurement tool is an extremely versatile tool and is used to bucket fill a closed region of a vector PDF file. That region can then be added an area, volume, polylength, or a perimeter measurement, or any combination of these measurements. You can also create polygon markups or spaces to that filled region. You will learn more about spaces in later chapters.

For this tool to work effectively, it can only be used in closed regions of the PDFs. If the region you want to fill is not closed or has dashed lines around them, this tool also allows you to create boundaries to close the region you want to fill. When you invoke this tool, the **Dynamic Fill** toolbar, shown in Figure 8, is displayed at the top of the **Main Workspace**. The options in this toolbar are discussed next.

*Figure 8 The **Dynamic Fill** toolbar*

The various procedures for using the **Dynamic Fill** tool are discussed next.

Procedure for Reducing the Boundary and Fill Cursor Sizes

By default, the size of the circle that appears on the cursor to create a boundary or add a fill is really big because of which it gets hard to fill a smaller region. You can reduce the size for the current cursor instance of the **Dynamic Fill** tool by clicking on the **Dynamic Fill Settings** button available on the right of the **Delete** button on the **Dynamic Fill** toolbar. However, this value is restored to the original size when you invoke the **Dynamic Fill** tool the next time. Therefore, it is recommended to change the preferences to permanently reduce their size using Revu preferences. The procedure for doing this is discussed next.

1. From the top left of the Revu window, click **Revu > Preferences**; the **Preferences** dialog box is displayed.

2. From the left in this dialog box, click **Tools**.

3. From the area on the right, click **Measure** to display the measurement preferences.

4. Reduce the value of the **Fill Size** and **Boundary Size** spinners to **25%**, as shown in Figure 9. This is the lowest value you can define for these spinners.

5. Click **OK** in the dialog box. Figure 10 shows a boundary created with the default size of **100%** and Figure 11 shows the same boundary with a size of **25%**.

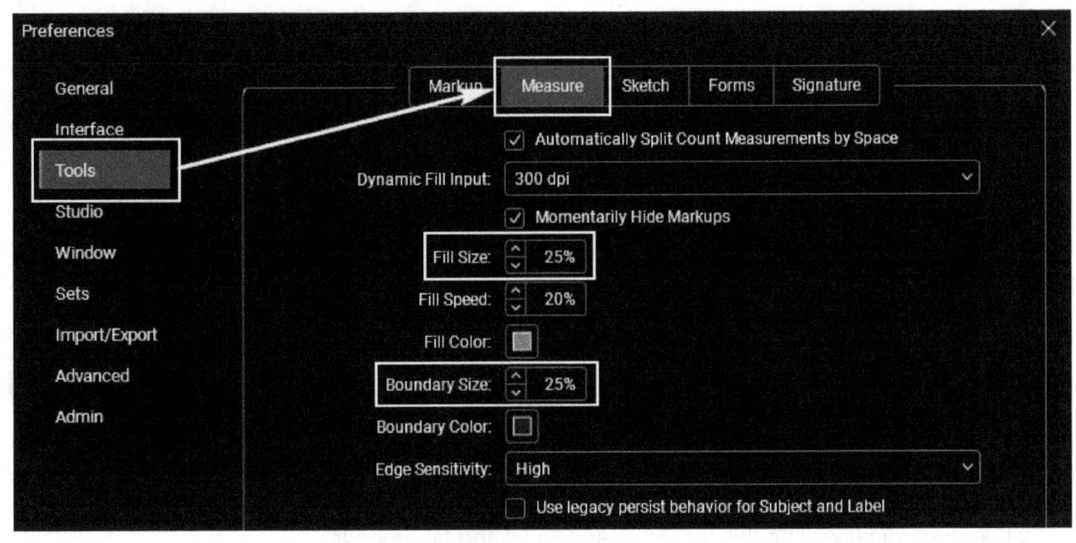

*Figure 9 Changing the **Fill Size** and **Boundary Size** preferences*

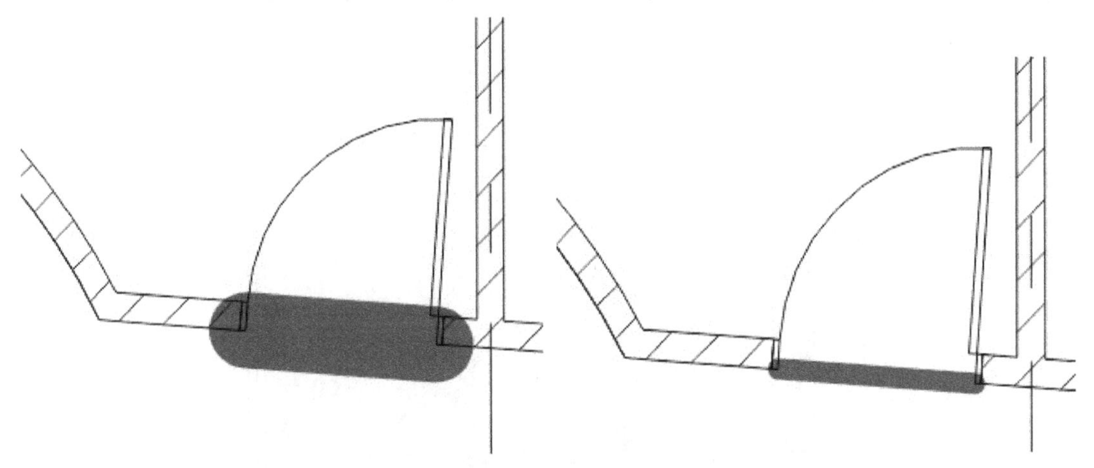

Figure 10 A boundary created on the door opening with a boundary size of 100%

Figure 11 A boundary created on the door opening with a boundary size of 25%

Procedure for Closing an Open Region by Creating a Boundary

In some vector PDFs the region that you want to fill may not be closed or it could have dashed lines instead of continuous lines as its boundaries. Another example is when you want to close a door opening so you can fill the room, including the door. In these cases, you can create a manual boundary. The following is the procedure for doing this.

1. From the **Status Bar**, make sure the **Snap to Content** option is turned on.

2. From the **Dynamic Fill** toolbar, invoke the **Add Boundary** tool; the cursor will change to a circle that allows you to define the endpoints of the boundary.

3. One by one, snap to the content of the PDF and define the boundary to close the region.

4. Once you have specified the last point of the boundary, press the ENTER or the C key to finish the process of boundary creation. Figure 9 shows a boundary created by defining two points to close the door opening.

Procedure for Dynamically Filling a Close Region

Once you have closed the region, you are ready to dynamically fill it. The following is the procedure for doing this.

1. From the **Dynamic Fill** toolbar, invoke the **Fill** tool; the cursor will change to a circle that allows you to fill the closed region. Note that the size of this circle will depend on the size you defined in the Revu preferences.

2. Press and hold down the Left Mouse Button in the closed region you want to fill; the process of filling the region starts, as shown in Figure 12. Note that you can move the cursor around while dynamically filling the region to speed up the fill process.

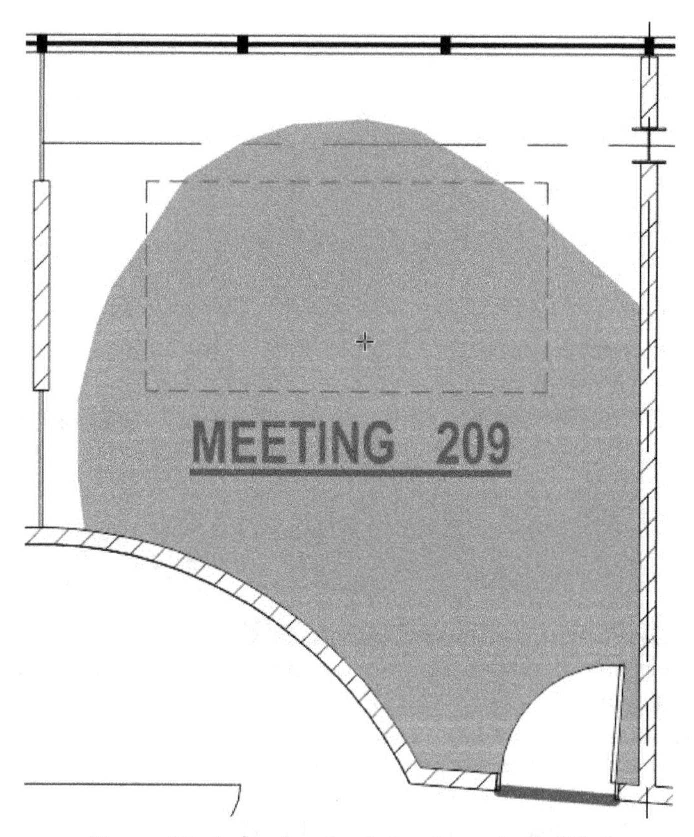

Figure 12 *A closed region being dynamically filled*

3. Make sure you fill all the closed regions that are isolated by continuous lines or boundaries. Figure 13 shows the room and the door space filled independently.

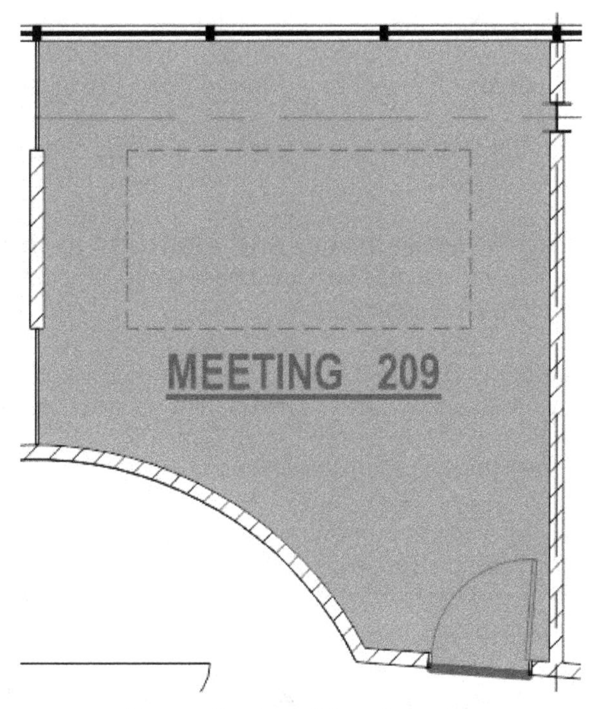

Figure 13 After filling the region

Procedure for Manually Controlling Fill and Boundary Sizes

If required, you can manually control the **Fill Size** and **Boundary** size using the **Dynamic Fill Settings** tool on the **Dynamic Fill** toolbar. The following is the procedure for doing this.

1. From the **Dynamic Fill** toolbar, invoke the **Dynamic Fill Settings** tool; the **Dynamic Fill Toolbar** will expand to show the settings, as shown in Figure 14.

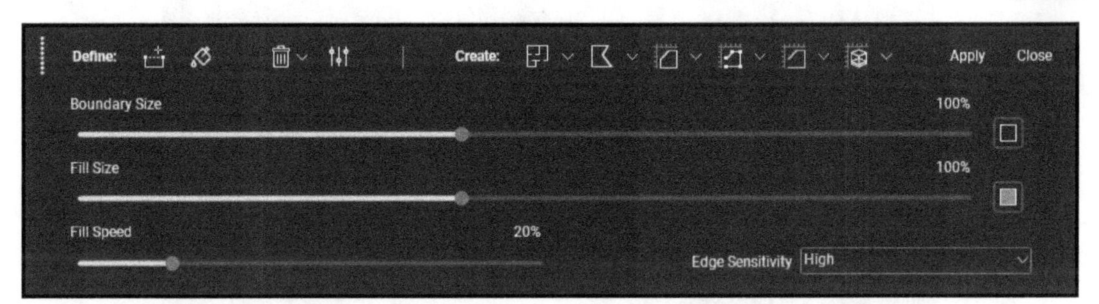

*Figure 14 The expanded **Dynamic Fill Toolbar** with the **Dynamic Fill** settings*

2. Use the **Boundary Size** or the **Fill Size** slider to change the size of the boundary or fill size cursor. Remember that this value will be reset to the value you defined in the Revu preferences when you invoke the **Dynamic Fill** tool the next time.

3. Use the **Fill Speed** slide to change the fill speed, if required.

4. Use the **Edge Sensitivity** to change it from **High** to **Low**, if required.

5. Click on the **Dynamic Fill Settings** button again to hide the expanded **Dynamic Fill Settings** area.

Procedure for Adding Measurements and Markups in the Filled Regions

Once you have filled a region, you can add various measurements or markups to those regions. The following is the procedure for doing this:

1. From the **Create** area of the **Dynamic Fill** toolbar, click on the button of the measurement or markup type you want to add. As mentioned earlier, you can add an area, a perimeter, a polylength, or a volume measurement, or any combination of these measurements. Additionally, you can add polygon markups and spaces.

2. If you have customized a measurement or a markup, click on the flyout next to that measurement or markup button and select that particular customized tool. Figure 15 shows the customized area measurement with the subject **Tile Type 1** being selected, which is saved on the **My Tools** toolbar.

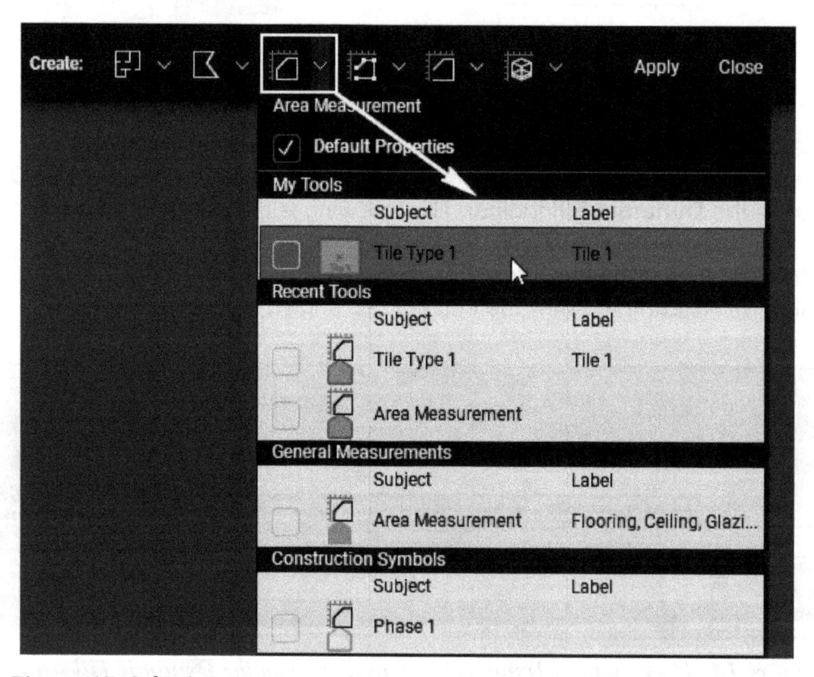

Figure 15 Selecting a customized area measurement to add to the filled region

3. Once you have selected all the measurements and markups to add, click the **Apply** button in the **Dynamic Fill** toolbar; the selected measurements and markups are added to the filled region. Figure 16 shows a room applied a customized area measurement listed as **Tile 1** and the perimeter measurement listed as **Skirting**.

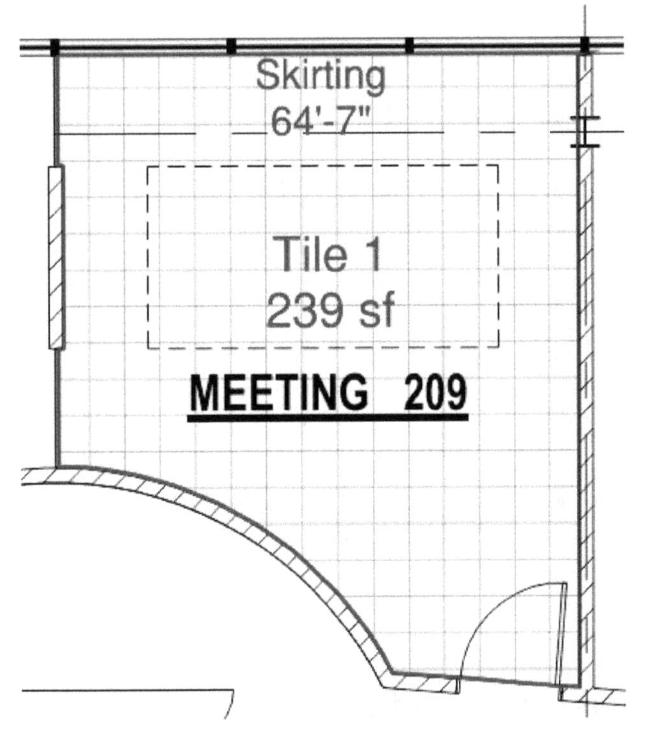

Figure 16 A customized area and perimeter measurement added to the room

***Tip**: Remember that if you select multiple measurements to add to a filled region, they will be placed overlapping on the PDF. In that case, you are better off using the **Markups List** if you need to select any one of those measurements.*

4. Continue dynamically filling additional regions and adding the measurements and markups.

5. Click the **Close** button on the **Dynamic Fill** toolbar to end the process and close the tool.

The Polygon Cutout and Ellipse Cutout Tools

Menu: Tools > Measure > Polygon Cutout / Ellipse Cutout
Panel: Measurements > Area > Polygon Cutout / Ellipse Cutout

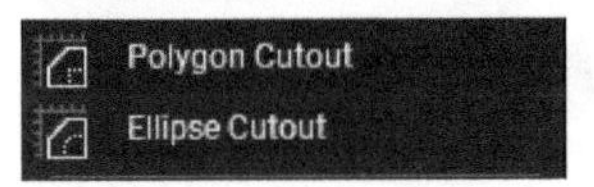

When using the **Dynamic Fill** tool, you can avoid the inner areas of a region, hence creating automatic cutouts. However, when you use **Area** or **Volume** tools, you cannot automatically cutout internal regions. In that case, you can use the **Polygon Cutout** or **Ellipse Cutout** tool to cutout a polygonal or a circular region from an existing area of volume measurements. When you invoke this tool and move the cursor over an area or a volume measurement, the cursor will change into a crosshair cursor and you can create the cutout. Once you create the cutout, the area or the volume measurement value will be automatically adjusted by subtracting the cutout value. Figure 17 shows a cutout created in the customized area measurement.

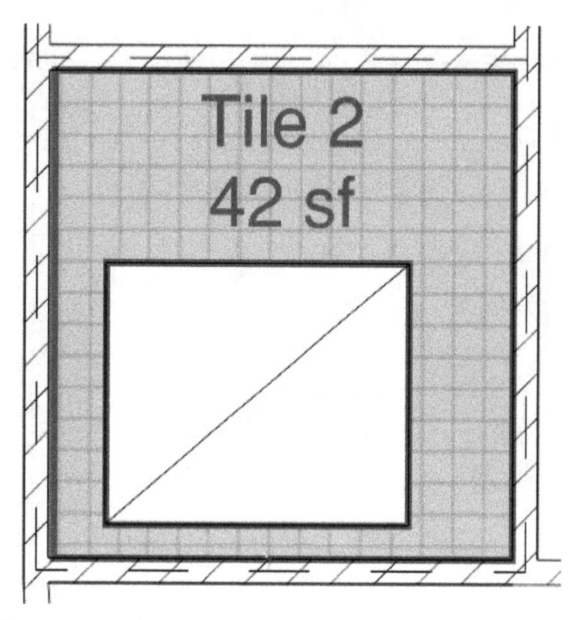

Figure 17 A rectangular region cutout from an existing area measurement

The Diameter Measurement Tool

Menu: Tools > Measure > Diameter
Panel: Measurements > Diameter

As the name suggests, the diameter measurement is used to measure the diameter of a circular region by specifying two points on the diameter of the circle. When you invoke this tool, you will be prompted to select the start point and drag to specify the endpoint. Similar to the other measurement tools, this tool can also be assigned a depth value in the **Diameter Measurement Properties** area of the **Measurement** panel. In this area, you can also turn on the additional captions, such as circumference, area, and so on. Figure 18 shows a customized diameter measurement.

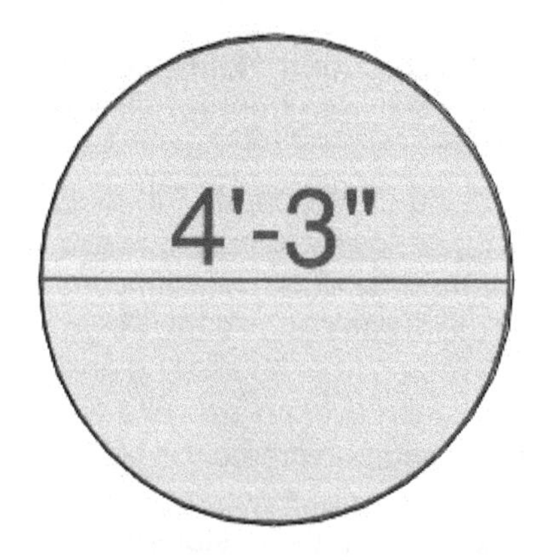

Figure 18 A customized diameter measurement

The Center Radius Measurement Tool

Menu: Tools > Measure > Center Radius
Panel: Measurements > Radius

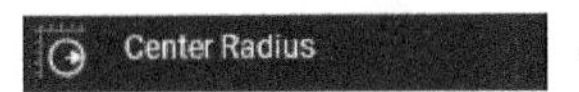

As the name suggests, the center radius measurement is used to measure the radius of a circular region by specifying the center of the circular region and a point on the circular region. Note that for this tool to work accurately, the center point of the circular region has to be known so it could be selected as the first point of the measurement. Similar to the other tools, you can display additional captions for this measurement as well, such as angle and arc length. Figure 19 shows a customized center radius dimension.

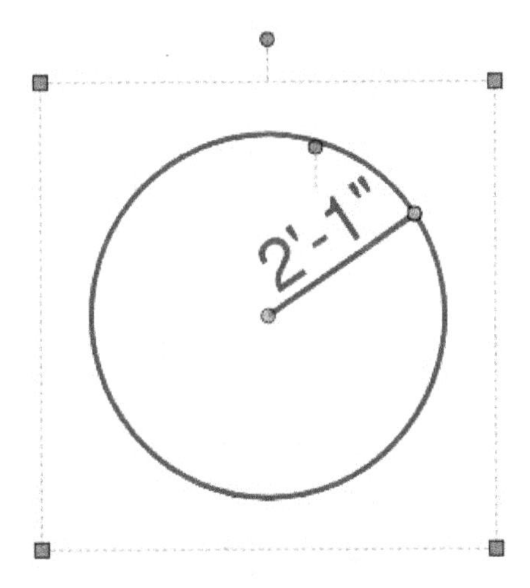

Figure 19 *A customized center radius measurement*

The 3-Point Radius Measurement Tool

Menu: Tools > Measure > 3-Point Radius
Panel: Measurements > Radius > 3-Point Radius

As the name suggests, the 3 point radius measurement is used to measure the radius of a circular region by specifying the first, second, and third point on that region. Figure 20 shows a customized 3 point radius measurement with the additional captions of angle and arc lengths displayed, along with the radius value.

The Angle Measurement Tool

Menu: Tools > Measure > Angle
Panel: Measurements > Angle

This tool is used to measure the angle between three points where the second point is the angle vertex point. When you invoke this tool, you

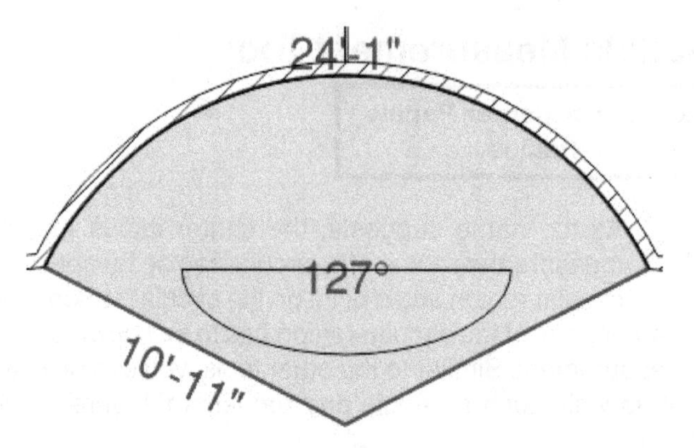

Figure 20 A customized 3 point radius measurement

will be prompted to select three points ABC to measure the angle between lines AB and BC. Once you specify the three points, the angle measurement will be displayed. Figure 21 shows a customized angle measurement where the second point was the corner of the room.

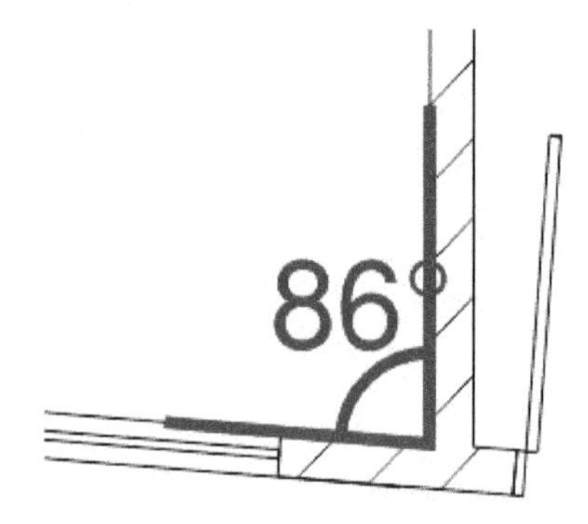

Figure 21 A customized angle measurement

The Count Measurement Tool

Menu: Tools > Measure > Count
Panel: Measurements > Count

The **Count** tool is used to count the number of items in the PDF file. This is an extremely versatile tool and can be customized to takeoff quantities of items such as electrical panels, VAVs, sinks, and so on. When you invoke this tool, a check mark is attached to the cursor. You can start the count by placing the first check mark on the item you want to count and then keep placing the check marks to continue the count. Figure 22 shows three items counted using the default count measurement tool.

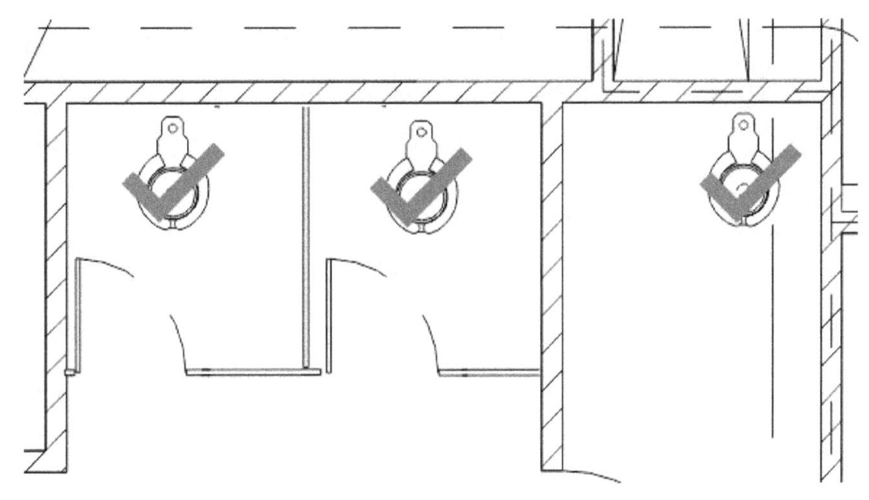

Figure 22 The default count measurement used to count items

Once you have placed the count measurement, you can right-click on any instance of it and select **Resume Count** to resume the count, if required. You can also split the selected count instance by right-clicking on it and selecting **Split Count** from the shortcut menu or split all of them by selecting **Split All** from the shortcut menu.

You can select any instance of the count from the **Main Workspace** and get the running count of the count in the **Comments** field of the **Properties** panel, which is available below the **Label** field. You can also change the symbol of the count in the **Appearance** area of the **Properties** panel, as shown in Figure 23. You can use the **Tool Chest** flyout in the **Appearance** area shown in Figure 23 to select a customized count from one of the custom tool sets.

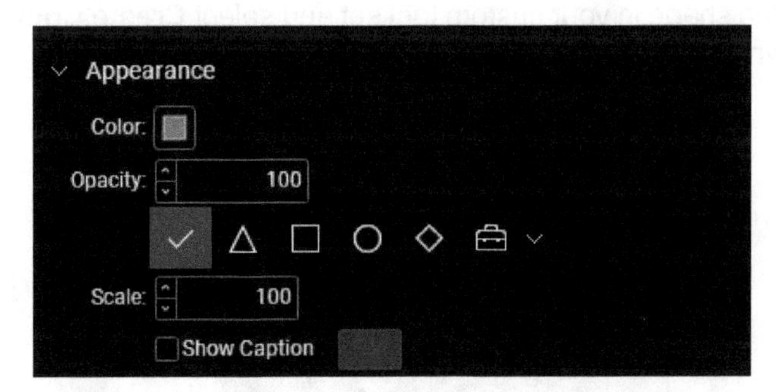

Figure 23 The options to change the count appearance

Procedure for Creating a Custom Count Measurement

As mentioned earlier, the **Count** tool can be customized to takeoff quantities of items such as electrical panels, VAVs, sinks, and so on. To do this, you first need to create a custom shape that you want to use as the count symbol and then customize the symbol as a count measurement. The following is the procedure for doing this:

1. Draw the shape you want to use as the count measurement. Note that if you have created a complex shape using multiple markups, you need to select all of them and right-click and select **Group** from the shortcut menu. Figure 24 shows a complex shape created using four different polygons and then grouped together to represent a VAV.

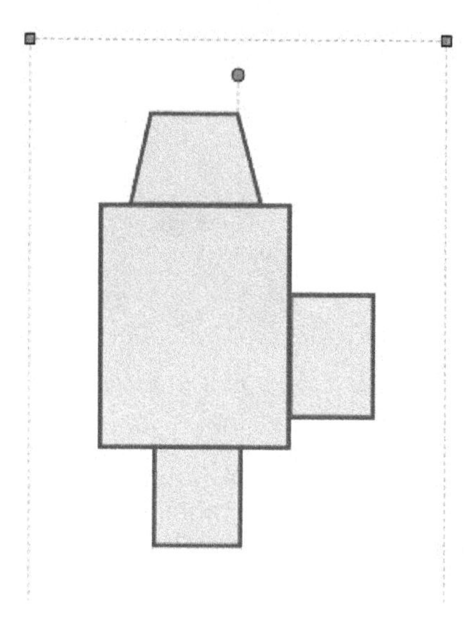

Figure 24 *The options to change the count appearance*

2. Right-click on the shape and add it to your custom tool set. At this stage, do not worry about changing the subject or label of this count.

3. Right-click on the shape in your custom tool set and select **Create Count** from the shortcut menu, as shown in Figure 25.

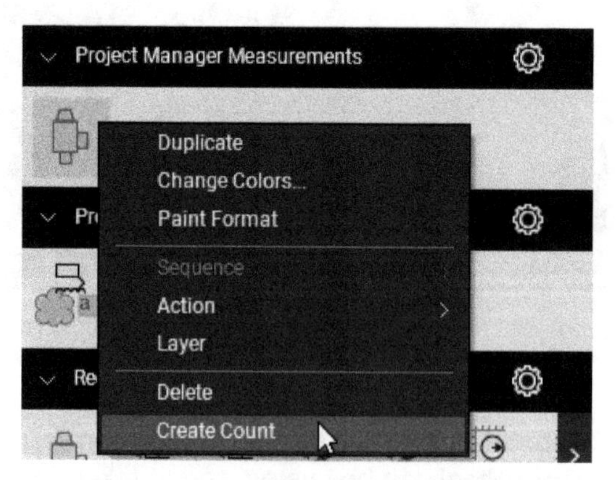

Figure 25 *Creating a count measurement from the custom tool*

On doing so, a custom count tool is added to the tool set. However, at this stage, this custom tool has a default subject and no label.

4. Place the custom count on the PDF. Notice that this count has a default subject and no label.

5. Press the ESC key to exit the count tool.

6. Select the count you placed on the PDF and change its subject and label in the **Properties** panel, as shown in Figure 26.

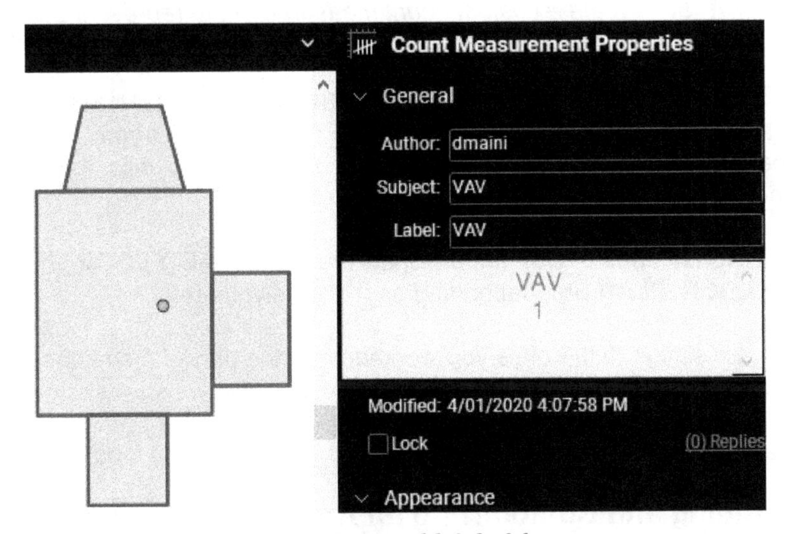

Figure 26 Changing the subject and label of the count measurement

7. From your custom tool set, delete the custom count and the custom symbol that got added.

8. Add the count whose subject and label you edited to your tool chest.

Tip: *If you want to get the cost value with your count measurements, you will have to first create custom columns and then link the count measurement to those custom columns before adding the count back to your tool set. You will learn more about custom columns in later chapters.*

Hands-on Tutorial	*In this tutorial, you will complete the following tasks:* *1. Add an area measurement and customize it.* *2. Cutout a region from the area measurement.* *3. Use the **Dynamic Fill** tool to fill various regions and create area measurements.* *4. Add a volume measurement and customize it.* *5. Create a custom count measurement to represent a VAV.*

Section 1: Opening the PDF File Calibrated in the Imperial or Metric Unit

For this tutorial, there are two different sets of PDF files provided to you. One set is calibrated in the Imperial units and the other one is calibrated in the Metric units. You need to open the file with the calibration units you prefer.

1. From the **Tutorial Files > C07** folder, open the **IMPERIAL_FLOOR_PLAN.pdf** or the **METRIC_FLOOR_PLAN.pdf**, depending on the units you prefer.

   These files are similar to the ones you worked on in the previous chapter. They also have the measurements from the previous chapter added to them.

2. Make sure you are currently in the **Training QTO** profile that you created earlier.

Section 2: Adding and Customizing an Area Measurement

In this section, you will navigate to the **CONFERENCE 246** room and add an area measurement. You will then customize this area measurement to represent the tile and carpet flooring types.

1. Navigate to the **CONFERENCE 246** room, which is available on the middle left of the plan.

2. From the **Menu Bar**, click **Tools > Measure > Area** to invoke this tool. Alternatively, you can invoke this tool from the **Measurements** panel. On doing so, you are prompted to select vertices of the area to measure.

   Before you specify the vertices to add the area measurement, you need to make sure the area units are configured to the preferred units.

3. In the **Measurements** panel, navigate to the **Area Measurement Properties** area; the properties of the default area measurement are listed there.

4. From the **Area** drop-down list, select **sf** if you are using the Imperial file, as shown in Figure 27. For the Metric file, select **sq m**.

   Although you will not add depth to this measurement, it is recommended that the volume units are also configured to the required units.

5. From the **Volume** drop-down list, select **cu ft** from the Imperial file, as shown in Figure 27, and **cu m** for the Metric file.

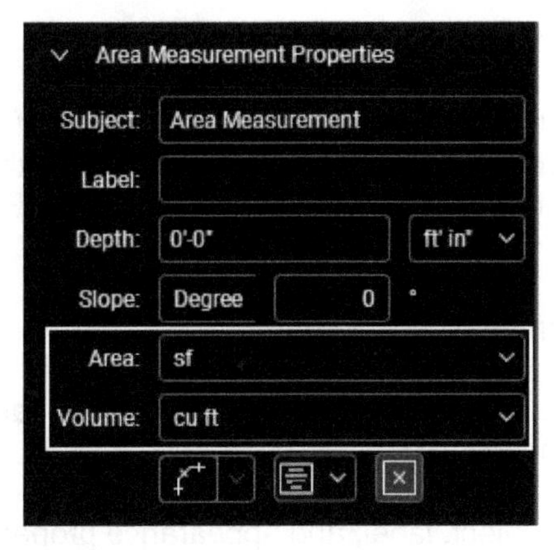

Figure 27 *Configuring the area and volume units for the Imperial file*

If you need to create an area measurement with more than four sides, you can snap to the vertices to define the area measurement segments. However, if you want to create a rectangular area measurement, then you can press and hold down the Left Mouse Button and drag the two opposite corners to create a rectangular area measurement. Because the **CONFERENCE 246** room is rectangular, you will press and drag the Left Mouse Button to define the two opposite vertices labeled as **1** and **2** in Figure 27.

6. Press and hold down the Left Mouse Button and drag from the vertex labeled as **1** to the vertex labeled as **2** in Figure 28 to add the area measurement.

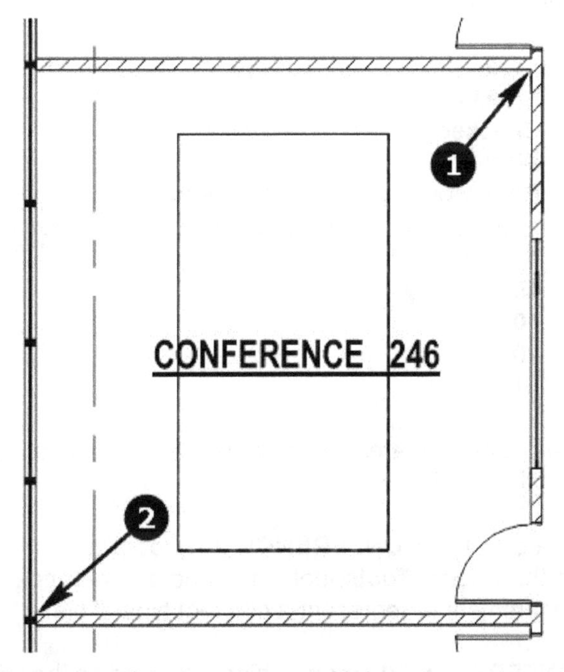

Figure 28 *The two vertices to add the area measurement*

7. Press the ESC key to exit the **Area** tool.

   You will now customize this area measurement. The default area measurement has the caption overlapping the room name. You will start by moving this area caption. Note that to move the caption, you need to first hold down the SHIFT key. Else, the entire measurement will move.

8. Select the area measurement you just added.

9. Press and hold down the SHIFT key and drag the area measurement caption above the room name so they are not overlapping.

   You will now modify the subject, label, and appearance properties of this measurement.

10. With the area measurement still selected, activate the **Properties** panel from the right side of the **Main Workspace**.

11. Change the subject to **Tile Type 1**.

12. Change the label to **Tile 1**; the label is now displayed with the area value caption.

13. If required, hold down the SHIFT key and move the caption further up so it is not overlapping with the room name.

14. Change the following appearance properties:

Color:	**Blue**
Fill Color:	**Blue Violet**
Highlight:	**Checked**
Hatch:	**Weave**
Hatch Color:	**Blue** *(available on the right of the **Hatch** drop-down list so you may need to drag the **Properties** panel and make it wide enough to dis-*
play	*this button)*
Scale:	**50**
Fill Opacity:	**40**
Font Size:	**10**
Text Color:	**Blue**

   Figure 29 shows the customized area measurement in the Imperial file and Figure 30 shows this measurement in the Metric file.

   You will now navigate to the **CONFERENCE 207** room and use the customized area measurement from the **Recent Tools** tool set to add an area measurement. You will then customize this measurement to represent a different type of tile.

15. Navigate to the **CONFERENCE 207** room, which is located on the left of the **MEETING 208** room at the top center of the plan.

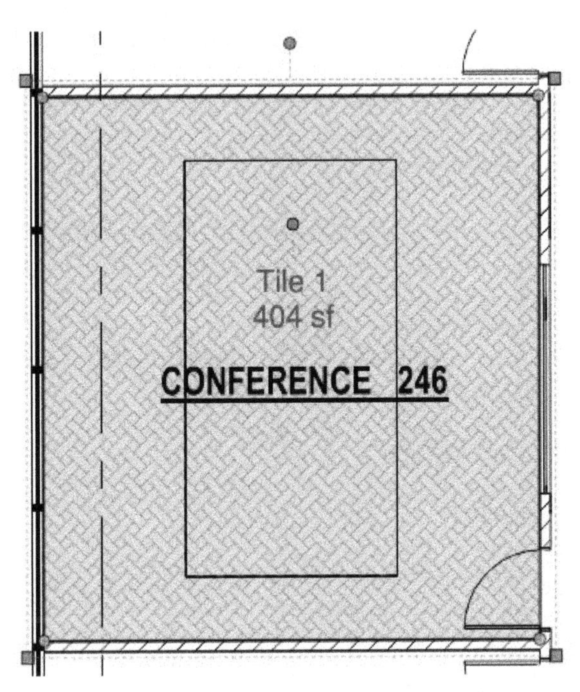

Figure 29 The customized area measurement in the Imperial file

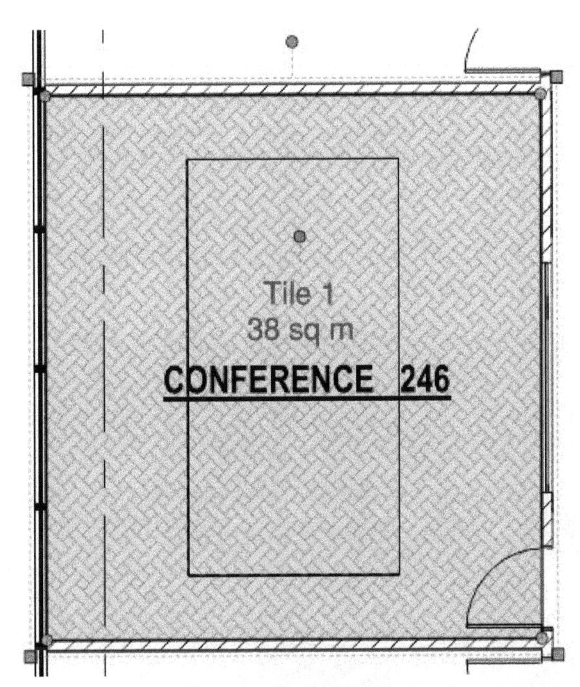

Figure 30 The customized area measurement in the Metric file

16. From the **Recent Tools** tool set, invoke the last **Area Measurement** tool, which is listed as **Tile Type 1**.

    Because this is not a rectangular area, you need to one by one select the four corners of the room to add the measurement.

17. One by one, select the four corners of the room to add the area measurement.

18. Press the ENTER key to finish selecting points; the area measurement is added.

    You will now change the subject, label, and appearance properties of this measurement to make it appear as a different type of tile.

19. Change the subject to **Carpet Type 1**.

20. Change the label to **Carpet 1**.

21. If required, hold down the SHIFT key and move the caption down so it is not overlapping with the room name, refer to Figures 31 or 32.

22. Change the following appearance properties:

Color:	**Violet**
Fill Color:	**Light Violet**
Highlight:	**Checked**
Hatch:	**Carpet** *(this is the hatch pattern you created in one of the previous chapters of this book and is listed under* ***Company ABC Patterns****)*
Hatch Color:	**Violet**
Text Color:	**Violet**

Figure 31 shows the customized area measurement in the Imperial file and Figure 32 shows this measurement in the Metric file.

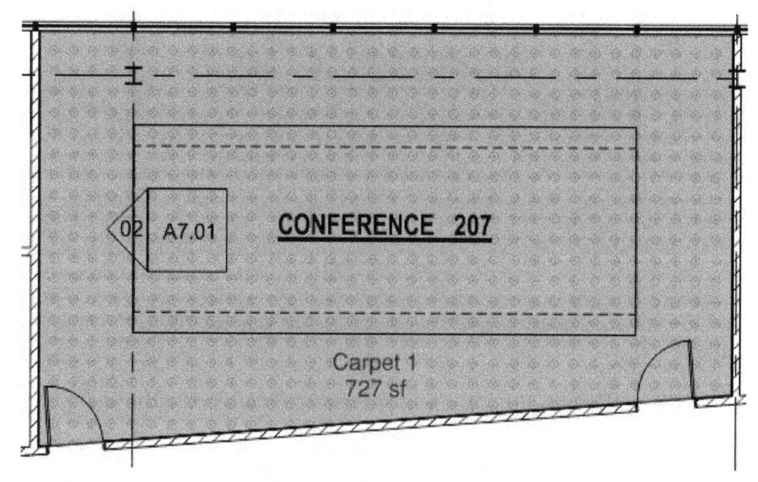

Figure 31 *The customized area measurement in the Imperial file*

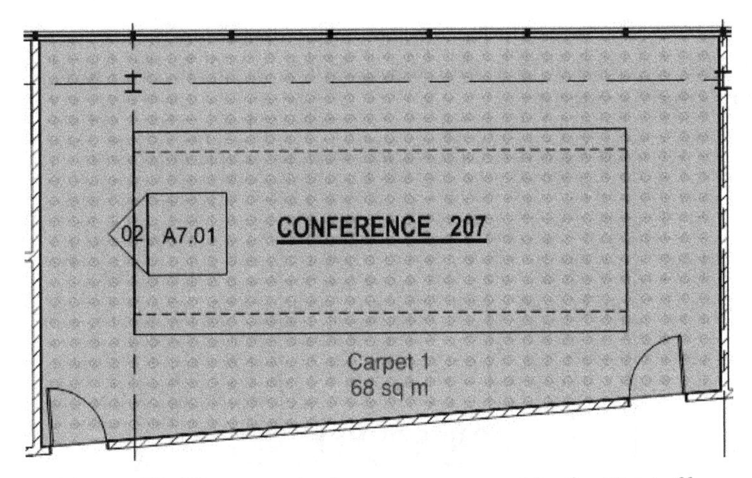

Figure 32 The customized area measurement in the Metric file

Section 3: Creating a Cutout from an Existing Area Measurement

In this section, you will cutout a region each from the two area measurements you created in the previous section.

1. From the **Menu Bar** or the **Measurements** panel, invoke the **Polygon Cutout** tool.

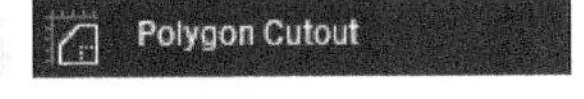

2. Hover the cursor over the last area measurement; the measurement is highlighted.
3. Press and hold down the Left Mouse Button and snap to the point labeled as **1** in Figure 33.

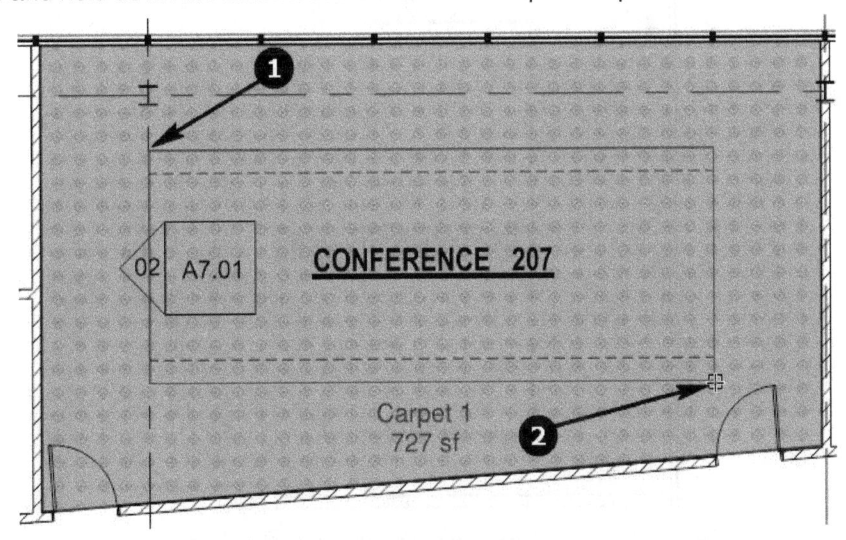

Figure 33 Snapping to the points to cutout a region

4. Drag the mouse to the point labeled as **2** in Figure 33; the selected region is removed and the area value is automatically adjusted.

5. Similarly, cutout a region from the Tile 1 measurement you created earlier. Figure 34 shows the Imperial Tile 1 measurement after the cutout and Figure 35 shows the Metric Tile 1 measurement after the cutout.

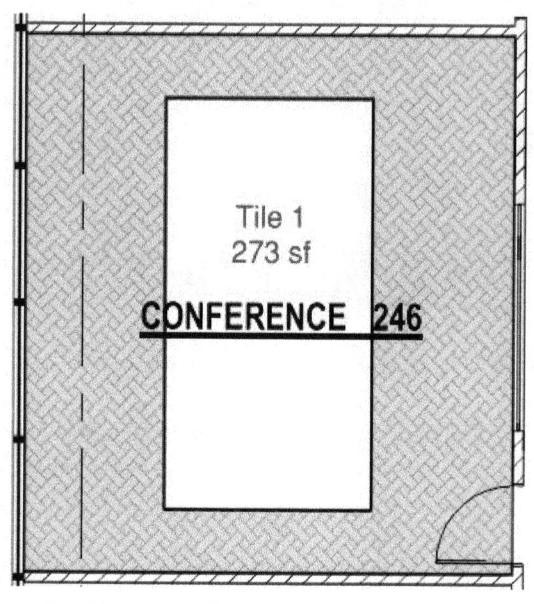

Figure 34 The Imperial area measurement after the cutout

Figure 35 The Metric area measurement after the cutout

6. Press the ESC key to exit the **Polygon Cutout** tool.

7. Save the file.

Section 4: Using the Dynamic Fill Tool to Create Multiple Measurements

In this section, you will use the **Dynamic Fill** tool to create various area and perimeter measurements together. But before you do that, you will change the default preferences of the fill boundary and fill size.

1. From the top left of the window, click **Revu > Preferences**; the **Preferences** dialog box is displayed.

2. In the dialog box, select **Tools > Measure**, refer to Figure 36 below.

   As discussed earlier in this chapter, the default sizes of the boundary and fill cursors in the **Dynamic Fill** tool is quite big. Instead of reducing these sizes every time you invoke the **Dynamic Fill** tool, you can configure the preferred sizes in this **Preferences** dialog box.

3. Set the values of the **Fill Size** and **Boundary Size** spinners to **25**, as shown in Figure 36.

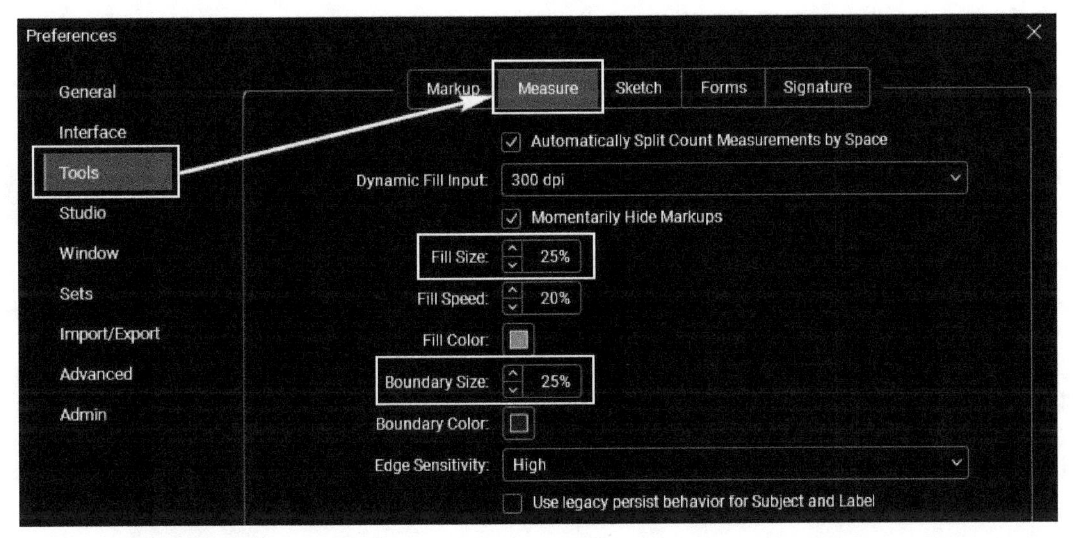

***Figure 36** Configuring the default preferences for the fill and boundary sizes*

4. Click **OK** in the dialog box.

5. Navigate to the **MEETING 208** room at the top center of the plan.

   You will now create the area measurement in the **MEETING 208** room using the **Dynamic Fill** tool. This tool can be invoked from the **Menu Bar** or from the **Measurements** panel. Alternatively, you can press the **J** key on the keyboard to invoke this tool.

6. Press the **J** key on the keyboard; this tool is invoked and the **Dynamic Fill** toolbar is displayed at the top of the screen.

   Notice that the **MEETING 208** room has a door that is not closed at the door frame. Therefore, you will first close the room at the door frame using the dynamic fill boundary.

7. In the **Dynamic Fill** toolbar, click the **Dynamic Fill Settings** button; the toolbar expands. Notice the **Boundary Size** and **Fill Size** spinners are set to the minimum value, as shown in Figure 37 because of the preferences you configured earlier.

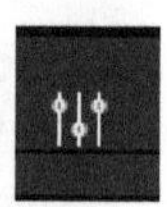

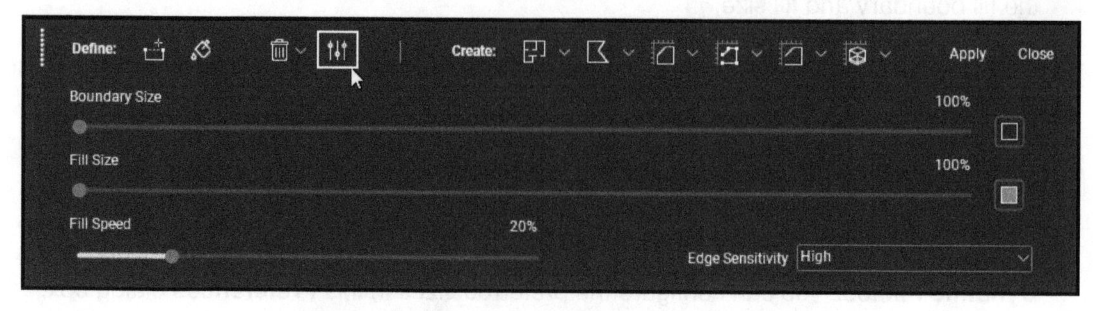

Figure 37 Reviewing the fill and boundary sizes

8. Click on the **Dynamic Fill Settings** button again to collapse the toolbar to the original size.

   You will now close the **MEETING 208** room at the door frame using the **Add Boundary** tool in the **Dynamic Fill** toolbar.

9. From the **Dynamic Fill** toolbar, click on the **Add Boundary** button; the cursor changes to the boundary cursor that is represented by a circle of the size you specified in the **Preferences** dialog box earlier.

   For you to be able to snap on the door frame vertices, you need to make sure the **Snap to Content** option is turned on from the **Status Bar**.

10. From the **Status Bar**, make sure the **Snap to Content** option is turned on.

11. Zoom close to the door frame of the **MEETING 208** room and snap to the two corner points of the door frame to close the boundary, as shown in Figure 38.

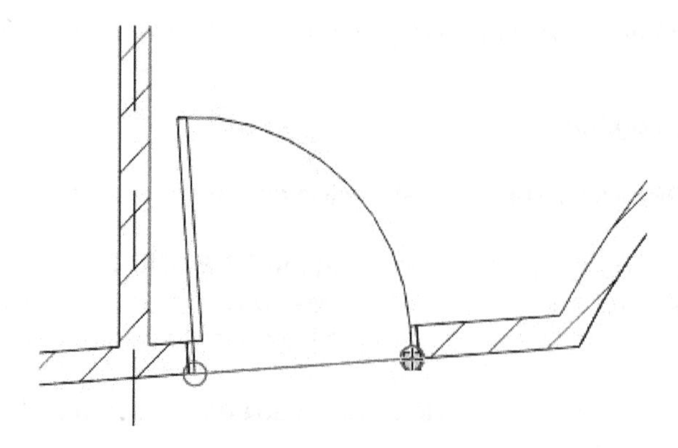

Figure 38 Closing the room boundary

12. Press the ENTER key to finish creating this boundary.

Next, you will fill the room using the **Fill** tool from the **Dynamic Fill** toolbar.

13. From the **Dynamic Fill** toolbar invoke the **Fill** tool; the cursor changes to the fill cursor represented by a circle of the size you specified in the **Preferences** dialog box earlier.

    To fill a closed region quicker, you can slowly drag the mouse while filling a region.

14. Press and hold down the Left Mouse Button in the **MEETING 208** room and slowly drag it around to fill the region.

    You will notice that the room is filled with Green color. However, the area closed by the door is not filled.

15. Press and hold down the Left Mouse Button to fill the door area as well, as shown in Figure 39.

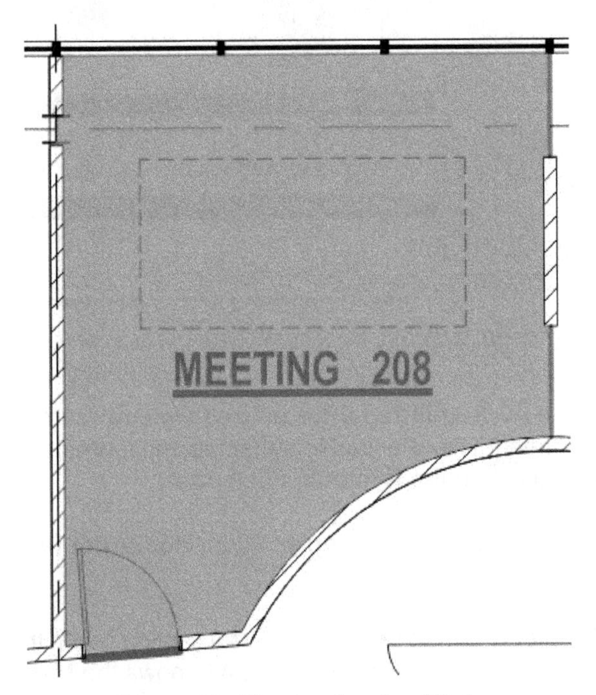

Figure 39 *The closed region filled*

Tip: *If the fill leaks out of the region you want to cover, you can press the CTRL+Z key to undo the last fill.*

Next, you will select the **Carpet Type 1** area measurement to fill this room. This area measurement is available in the **Recent Tools** section when you click on the flyout available on the right side of the **Area Measurement** tool on the **Dynamic Fill** toolbar.

16. On the **Dynamic Fill** toolbar > **Create** area, click on the flyout available on the right side of the **Area Measurement** tool (labeled as **1** in Figure 40) and select the **Carpet Type 1** area measurement, as shown in Figure 40.

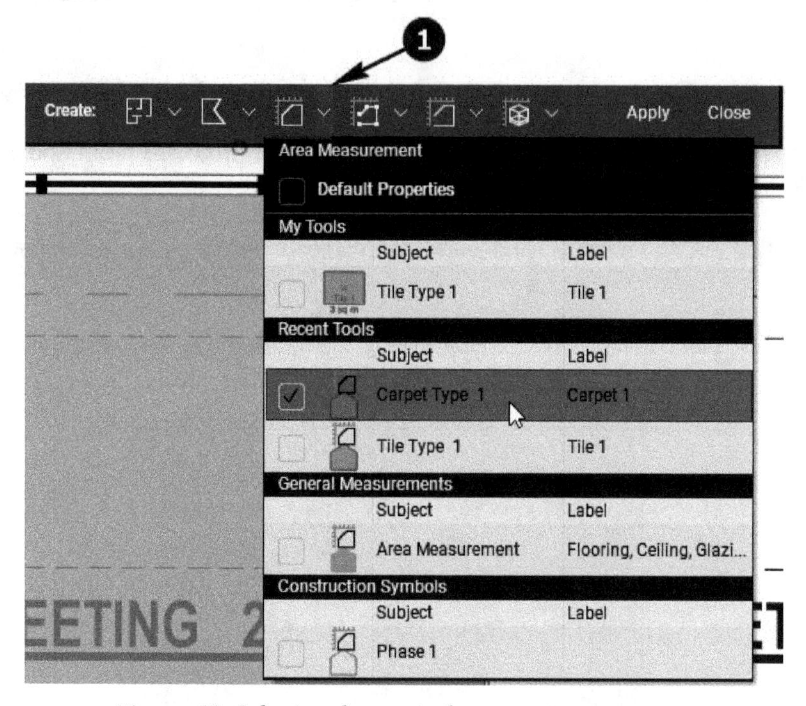

Figure 40 *Selecting the required area measurement type*

Note that although you have selected a customized area measurement, the fill regions will still be displayed in Green. This is because the actual measurement type is displayed after you click the **Apply** button on the **Dynamic Fill** toolbar.

17. On the **Dynamic Fill** toolbar, click the **Apply** button; the **Carpet Type 1** area measurement is created in the filled region.

18. Hold down the SHIFT key and drag the measurement caption above the room name. Figure 41 shows the Imperial measurement for and Figure 42 shows the Metric measurement.

    Notice that the **Dynamic Fill** toolbar is still active on top of the **Main Workspace.** You will now add the same area measurement type to the **MEETING 209** room.

19. From the **Dynamic Fill** toolbar, invoke the **Add Boundary** tool. Notice that when you invoke this tool, the previous area measurement temporarily disappears.

20. Close the door frame boundary of the **MEETING 209** room.

21. Press the ENTER key to exit the boundary creation process.

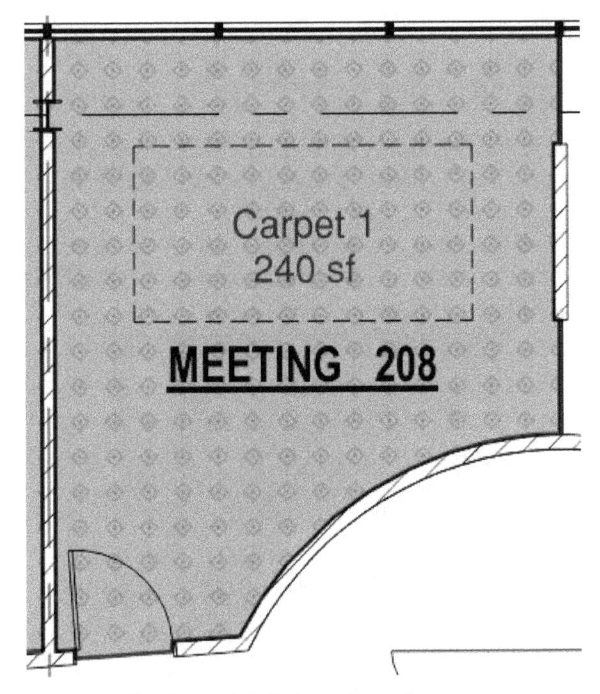

Figure 41 *The Imperial* ***Carpet Type 1*** *area measurement*

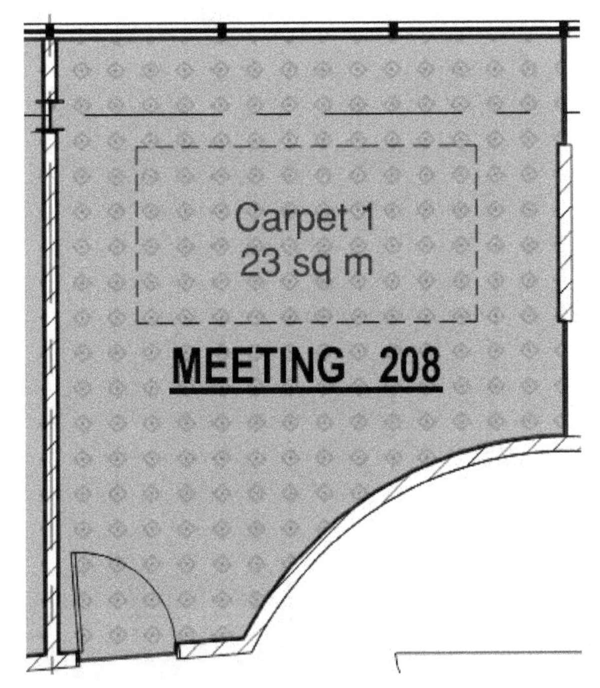

Figure 42 *The Metric* ***Carpet Type 1*** *area measurement*

22. From the **Dynamic Fill** toolbar, invoke the **Fill** tool.

23. Press and hold down the Left Mouse Button and fill the room and the door area.

24. Click the **Apply** button in the **Dynamic Fill** toolbar; because the required area measurement type is already selected in the **Dynamic Fill** toolbar, the room is filled with that measurement.

25. Hold down the SHIFT key and move the area caption above the room name. Figure 43 shows the Imperial measurement and Figure 44 shows the Metric measurement.

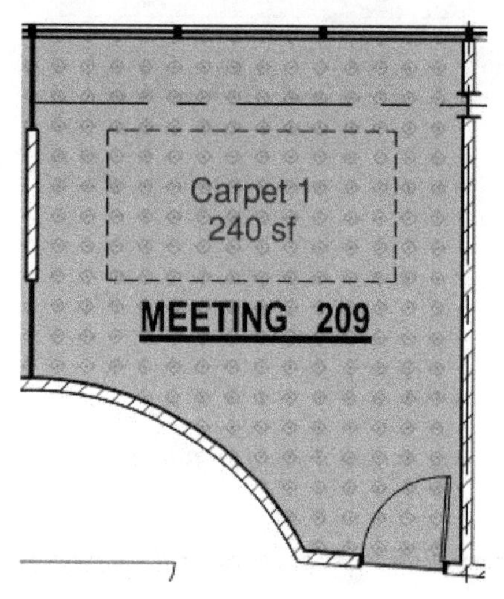

Figure 43 *The Imperial* ***Carpet Type 1*** *area measurement*

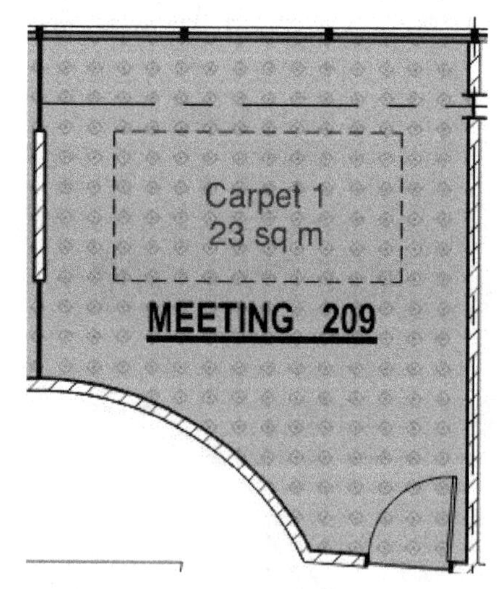

Figure 44 *The Metric* ***Carpet Type 1*** *area measurement*

26. Similarly, add this measurement to all the office and meeting rooms at the top of the plan. Note that in the **OFFICE 212**, **MEETING 214**, and **MEETING 216** rooms, you will have to close the top of the room to ensure the fill does not leak out of these rooms.

    Next, you will add the **Tile Type 1** area measurement to the **MEN'S RR** viewport on the lower left of the PDF file.

27. Navigate to the **MEN'S RR** viewpoint on the lower left of the PDF file.

28. Create the two boundaries labeled as **1** and **2** in Figure 45 to close this room.

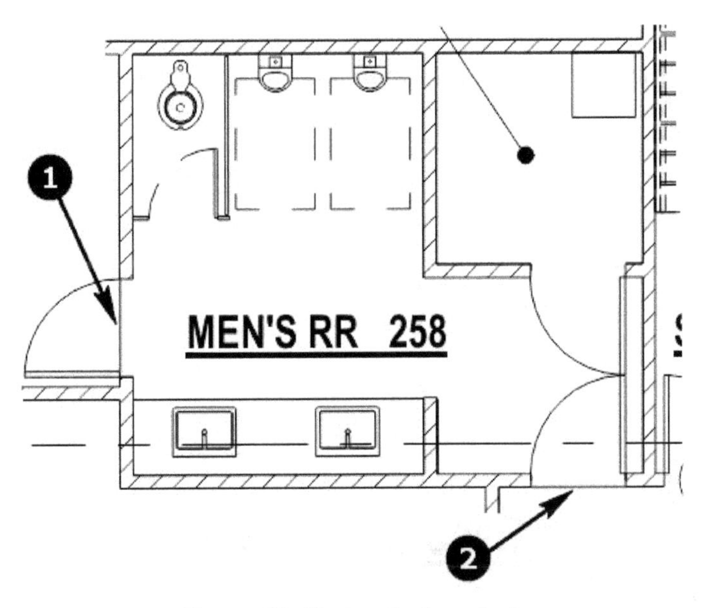

Figure 45 *Closing the boundaries*

29. Using the **Fill** tool, carefully fill the entire room avoiding the walls, as shown in Figure 46.

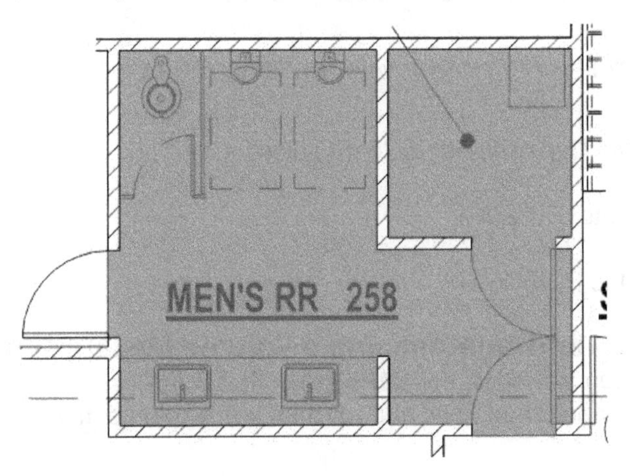

Figure 46 *Filling the room*

You need to apply the **Tile Type 1** measurement to this area. This measurement can be selected by clicking on the flyout on the right of the **Area Measurement** tool on the **Dynamic Fill** toolbar.

30. Click on the flyout on the right of the **Area Measurement** tool on the **Dynamic Fill** toolbar and select **Tile Type 1**.

31. Click the **Apply** button on the **Dynamic Fill** toolbar; the measurement is added to the filled region.

    Note that because of the scale of this viewpoint, the measurement text will appear small.

32. Hold down the SHIFT key and drag the area caption above the room name. Figure 47 shows the viewpoint with this area measurement applied.

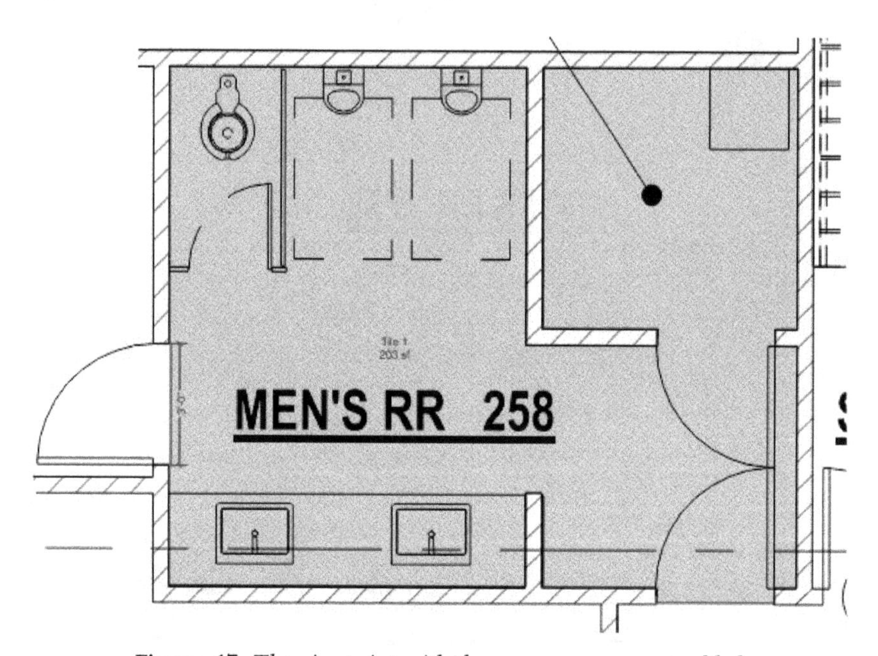

Figure 47 *The viewpoint with the area measurement added*

33. Click **Close** on the **Dynamic Fill** toolbar to close it.

34. Zoom to the extents of the file.

35. Save and close the PDF file.

Section 5: Adding and Customizing a Volume Measurement

In this section, you will open the **IMPERIAL_SLAB.pdf** or the **METRIC_SLAB.pdf** file, depending on your preferred units. You will then add a volume measurement and customize it.

1. From the **C07** folder, open the **IMPERIAL_SLAB.pdf** or the **METRIC_SLAB.pdf** file, depending on your preferred units.

   These files are calibrated with the Imperial or Metric scales. You will use the **Dynamic Fill** tool to add the volume measurement to the main slab.

2. Press the **J** key on the keyboard to invoke the **Dynamic Fill** tool; the **Dynamic Fill** toolbar is displayed.

   Before you start filling the slab, you will increase the size of the **Fill** cursor to **200**.

3. From the **Dynamic Fill** toolbar, click the **Dynamic Fill Settings** button; the toolbar expands and shows various sliders.

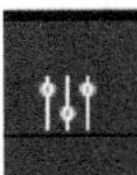

4. Drag the **Fill Size** slider all the way to the right so it is set to **200%.**

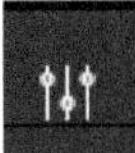

5. Click on the **Dynamic Fill Settings** button again to collapse the toolbar.

6. Fill the main slab ensuring the isolated footings are not filled, as shown in Figure 48.

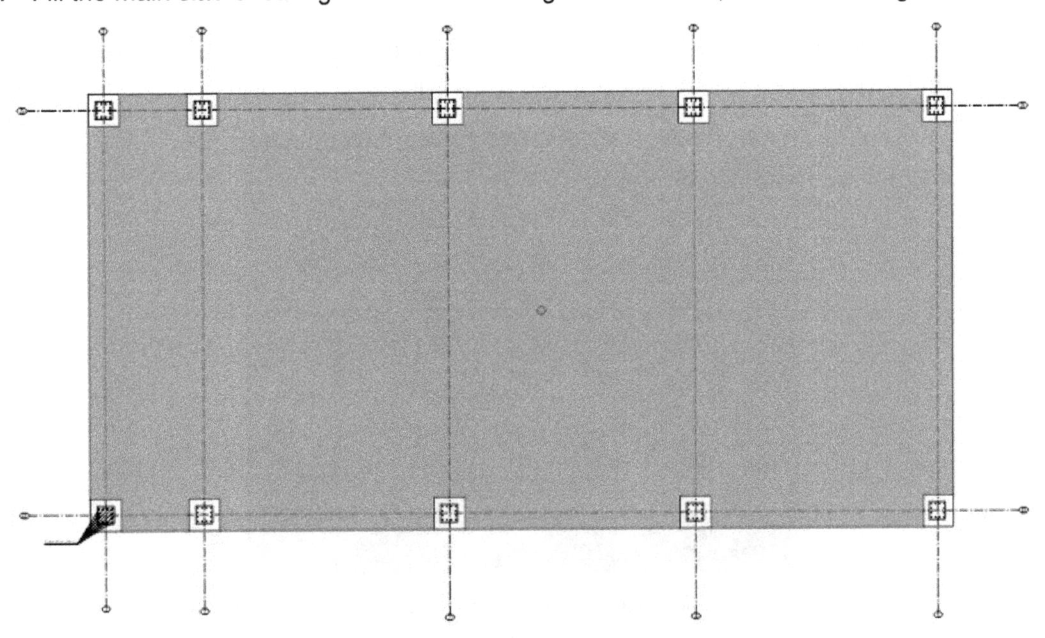

Figure 48 The main slab filled

   By default, the **Area Measurement** button is still on in the **Dynamic Fill** toolbar. You will turn off this button by clicking on it as you do not need to add an area measurement.

7. From the **Dynamic Fill** toolbar, click on the **Area Measurement** button to turn it off.

   Because you do not have a customized volume measurement in this file, you will add a default measurement and then customize it.

8. On the **Dynamic Fill** toolbar, click the **Volume Measurement** button to turn it on.

9. Click the **Apply** button and then click **Close** to close the **Dynamic Fill** toolbar.

   The default volume measurement will have incorrect measurement as you have not applied the required depth to it. Also, the measurement scale may not be correct. You will not customize this measurement.

10. With the volume measurement still selected, invoke the **Measurements** panel.

11. In the **Volume Measurement Properties** area > **Subject** field, enter **Concrete Grade M50** for the Imperial file and **Concrete Grade 200** for the Metric file.

12. In the **Label** field, enter **Concrete Grade M50** for the Imperial file and **Concrete Grade 200** for the Metric file.

13. In the **Depth** field, enter **6"** for the Imperial file and **200** for the Metric file.

14. From the **Area** drop-down list, select **sf** for the Imperial file and **sq m** for the Metric file.

15. From the **Volume** drop-down list, select **cu ft** for the Imperial file and **cu m** for the Metric file. Figure 49 shows these values for the Imperial file.

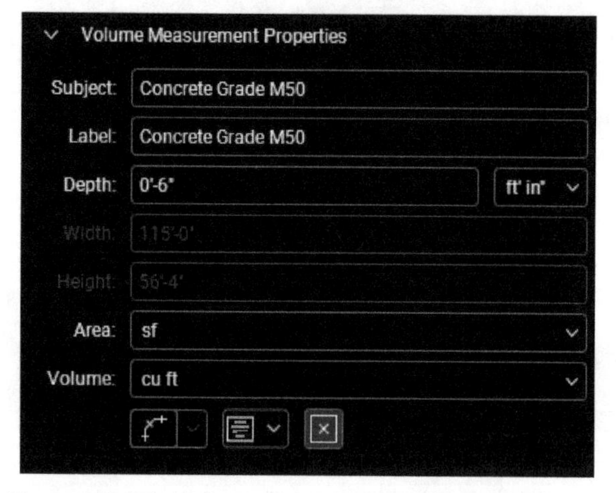

Figure 49 *The **Volume Measurement Properties** area*

16. In the **Properties** panel, change the following appearance properties:

Color:	**60% Gray** *(from the second from the bottom row in the color swatch)*
Fill Color:	**20% Gray** *(from the second from the bottom row in the color swatch)*
Highlight:	**Checked**
Font Size:	**36**
Text Color:	**Black**

Figure 50 shows the customized volume measurement in the Imperial file and Figure 51 shows this measurement in the Metric file.

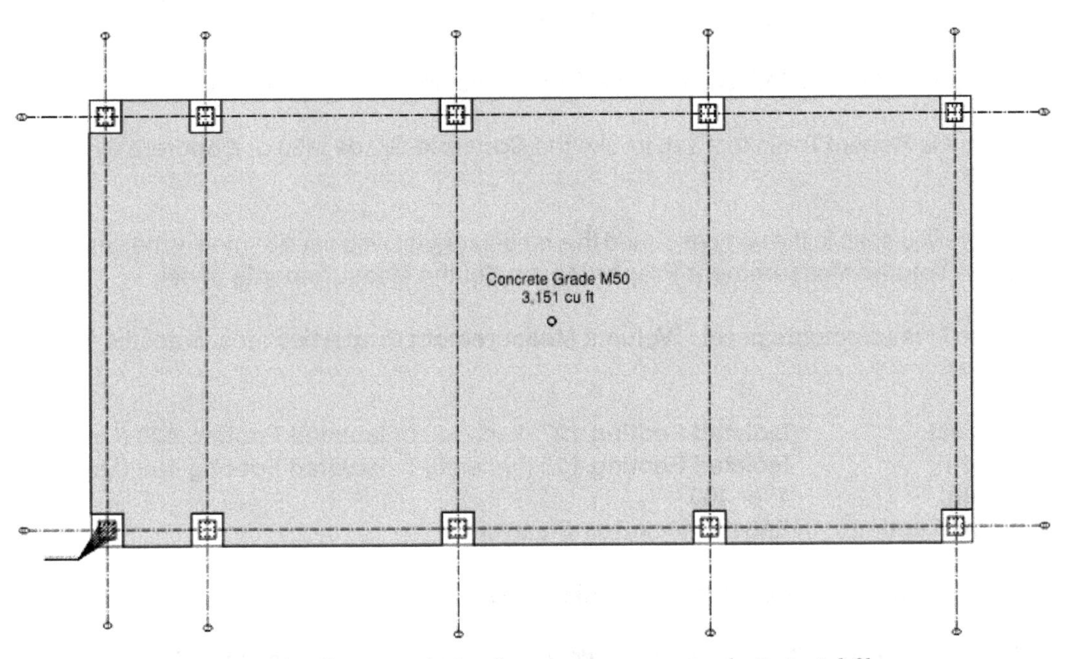

Figure 50 *The customized volume measurement in the Imperial file*

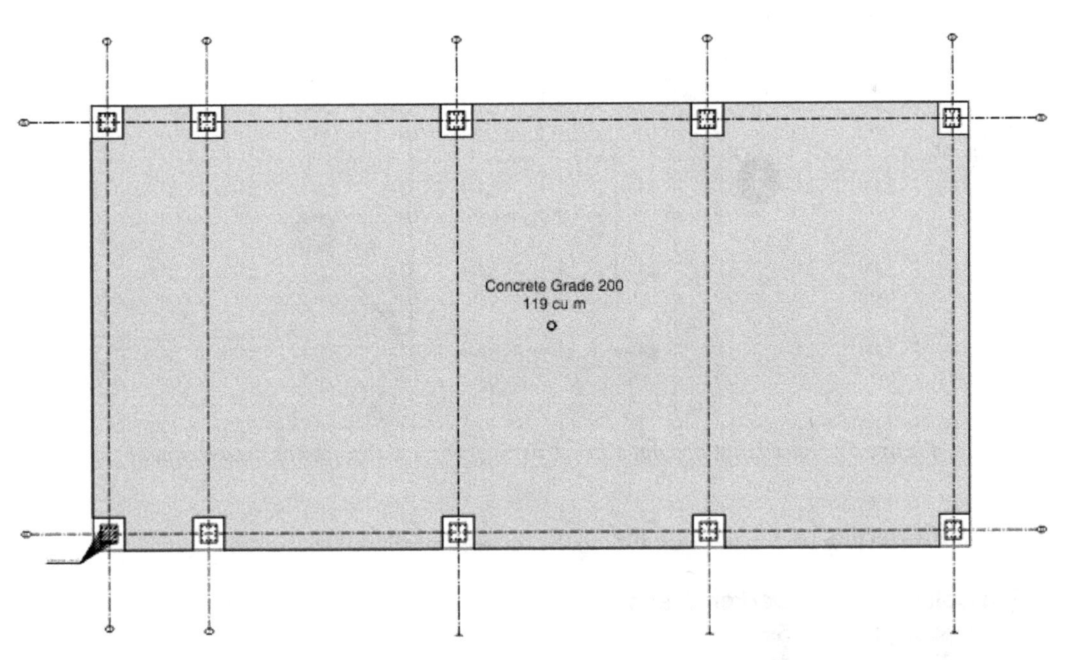

Figure 51 *The customized volume measurement in the Metric file*

Next, you will use the **Concrete Grade M50** or **Concrete Grade 200** tool from the **Recent Tools** tool set to add and volume measurement to one of the isolated footings. You will then customize this measurement for isolated footings.

17. Navigate to the isolated footing at the A1 grid intersection point on the top left of the plan.

18. From the **Recent Tools** tool set, invoke the **Concrete Grade M50** or **Concrete Grade 200** tool.

    Before you snap to the vertices to add this measurement, you need to modify the parameters in the **Volume Measurement Properties** area of the **Measurements** panel.

19. In the **Measurements** panel > **Volume Measurement Properties** area, enter the following values:

    Subject: **Isolated Footing 12"** (Imperial) or **Isolated Footing 400** (Metric)
    Label: **Isolated Footing 12"** (Imperial) or **Isolated Footing 400** (Metric)
    Depth: **1'** or **400**
    Show Caption: **Label Check Box Cleared**

20. In the Metric file only, change the **Precision** value to **0.1**.

21. Drag the two opposite corners on the isolated footings, as shown in Figure 52, to create the volume measurement.

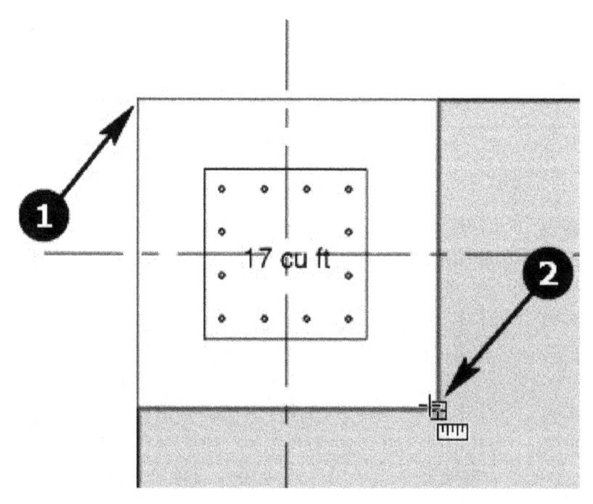

Figure 52 *Specifying two opposite corners for the isolated footing measurement*

22. In the **Properties** panel, change the following appearance properties:

    Fill Color: **Darker Green**
    Fill Opacity: **50**
    Font Size: **18**

23. If required, hold down the SHIFT key and drag the measurement caption so it is not interfering with the linework. Figure 53 shows this measurement for the Imperial plan and Figure 54 shows the same measurement for the Metric plan.

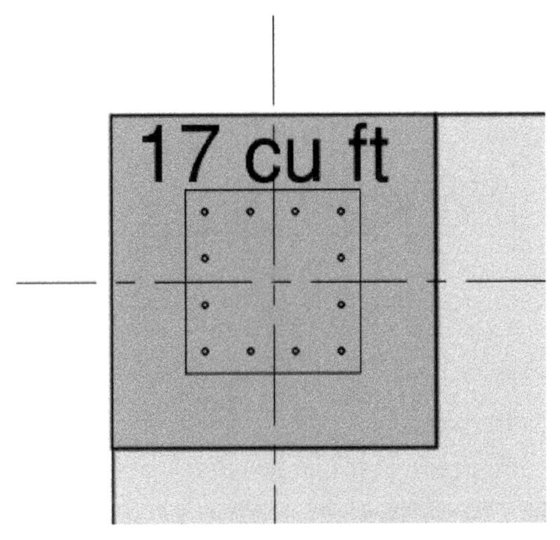

Figure 53 *The Imperial isolated footing measurement*

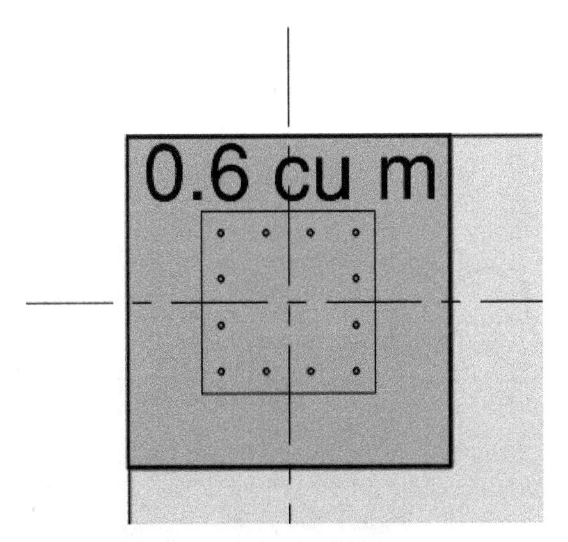

Figure 54 *The Metric isolated footing measurement*

24. Save and close the PDF file.

Section 6: Creating a Custom Count Measurement to Represent VAVs

In this section, you will use the **Polygon** tool to create multiple polygons and group them together to create a VAV symbol. You will then covert that symbol into a count measurement.

1. From the **C07** folder, open the **MECHANICAL-LAYOUT.pdf** file.

2. Navigate to **VAV-19** located near the top left of the layout.

3. From the **Project Manager Markups** tool set, invoke the **Project Manager's Polygon** tool.

4. Draw four polygons on top of the shape of **VAV-19**, as shown in Figure 55.

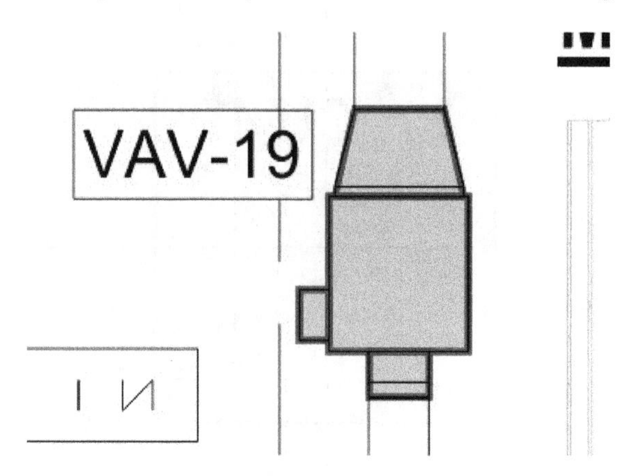

Figure 55 *Four custom polygons drawn on top of the existing VAV*

5. Hold down the SHIFT key and select the four polygons you drew.

6. Right-click on one of the selected polygons and select **Group** from the shortcut menu, as shown in Figure 56; the four polygons are grouped together.

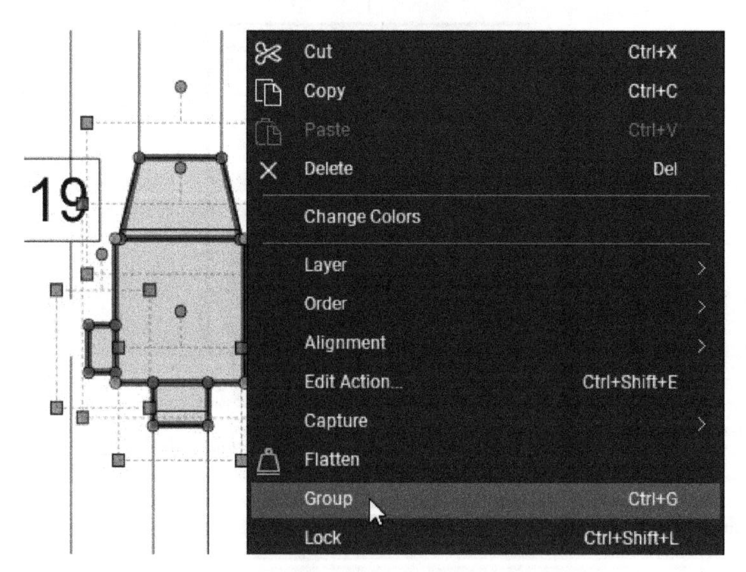

Figure 56 *Grouping the four polygons*

You will now add this group to the **Project Manager Measurements** tool set. You will not change the subject of this group at this stage.

7. Right-click on the group and add it to the **Project Manager Measurements** tool set.

   Next, you will convert this group of polygons into a count measurement.

8. On the **Project Manager Measurements** tool set, right-click on the group and select **Create Count**, as shown in Figure 57.

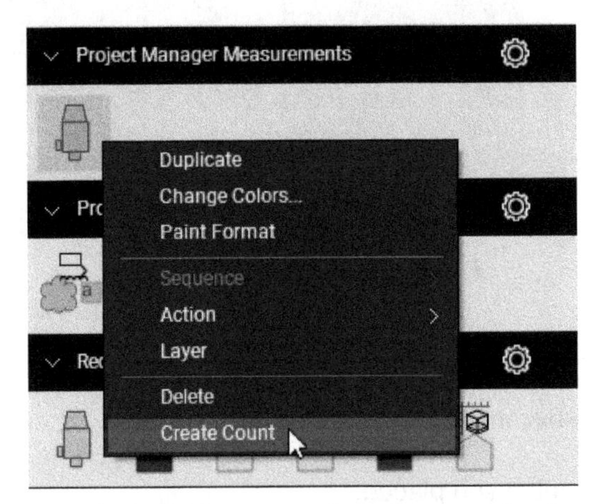

***Figure 57** Creating a count from the grouped polygons*

   On doing so, the tool for the custom count measurement is displayed on the right of the grouped polygons. The default subject of this tool is **Count Measurement**, as shown in Figure 58.

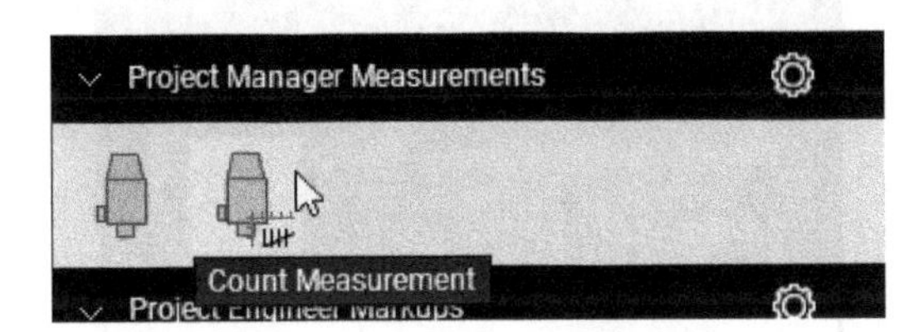

***Figure 58** The tool for the custom count measurement*

   You now need to delete the grouped polygons from the **Project Manager Measurements** tool set.

9. On the **Project Manager Measurements** ribbon panel, right-click on the grouped polygons and select **Delete** from the shortcut menu; the tool is deleted and the only tool available on this tool set now is the **Count Measurement** tool.

10. Using the **Project Manager Measurements** tool set > **Count Measurement** tool, place a VAV symbol anywhere on the PDF file.

11. Right-click on the VAV symbol count and select **Resume Count** from the shortcut menu, as shown in Figure 59.

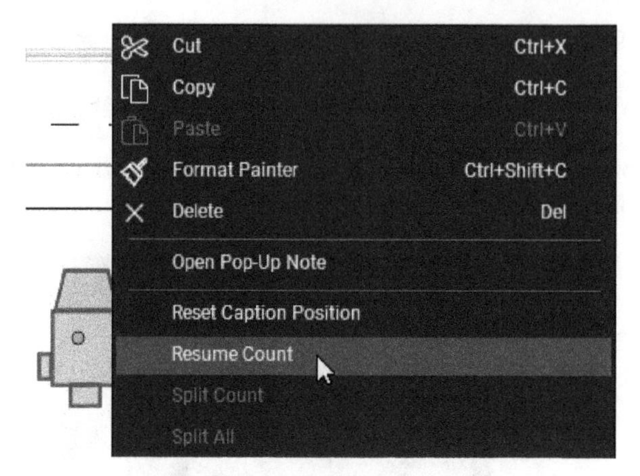

Figure 59 *Resuming the count*

12. Place a few more VAV symbols on the PDF file.

13. Press the ESC key to exit the tool.

14. Select the last VAV symbol you placed.

15. Activate the **Properties** panel and notice the count number displayed in the field below the **Label** field, as shown in Figure 60.

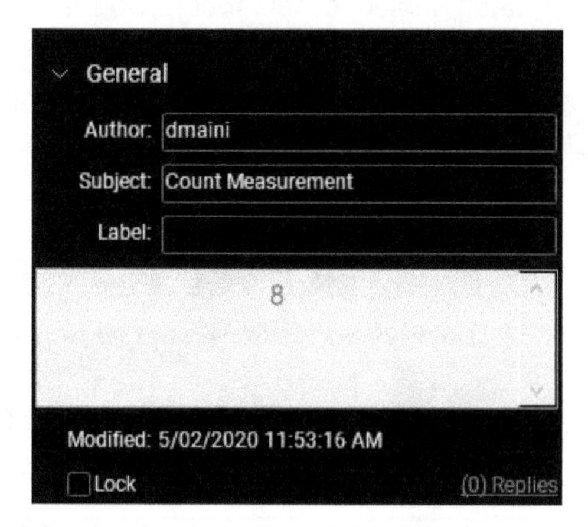

Figure 60 *The total count of the count measurement*

You will now close the **MECHANICAL-LAYOUT.pdf** file without saving changes. In the later chapters, you will calibrate this PDF file and search and count the VAVs using this count measurement you created.

16. Close the **MECHANICAL-LAYOUT.pdf** file without saving changes.

Skill Evaluation

Evaluate your skills to see how many questions you can answer correctly. The answers to these questions are given at the end of the book.

1. In Revu 2019, you cannot cutout a region from an area or a volume measurement. (True/False)

2. The fill size and the boundary size of the **Dynamic Fill** tool can be set in the **Preferences** dialog box. (True/False)

3. The count measurement tools cannot be customized. (True/False)

4. The label of the area measurement is automatically displayed when you place the measurement. (True/False)

5. In Revu, you cannot measure the radius of a circle. (True/False)

6. Which selection point is the angle vertex point in the **Angle Measurement** tool?

   (A) First (B) Fourth
   (C) Third (D) Second

7. Which tool allows you to fill a closed region and then add measurements to that region?

   (A) **Fill** (B) **Dynamic Fill**
   (C) **Area** (D) **Bucket Fill**

8. What are the two types of cutouts that can be removed from an existing area or volume measurement?

   (A) Polygonal (B) Elliptical
   (C) Polar (D) Angular

9. Which tool on the **Dynamic Fill** toolbar is used to close an open region?

   (A) **Add Boundary** (B) **Close**
   (C) **Cover** (D) **None**

10. Which type of measurement cannot be added using the **Dynamic Fill** tool?

    (A) Volume (B) Rectangle
    (C) Area (D) Polygon

Class Test Questions

Answer the following questions:

1. Explain briefly how to change the area measurement units.
2. Explain briefly the process of changing the precision.
3. What needs to be done to dynamically fill an open area?
4. How will you add an area measurement to a filled region?
5. Explain briefly the process of setting the fill size and boundary size preferences.

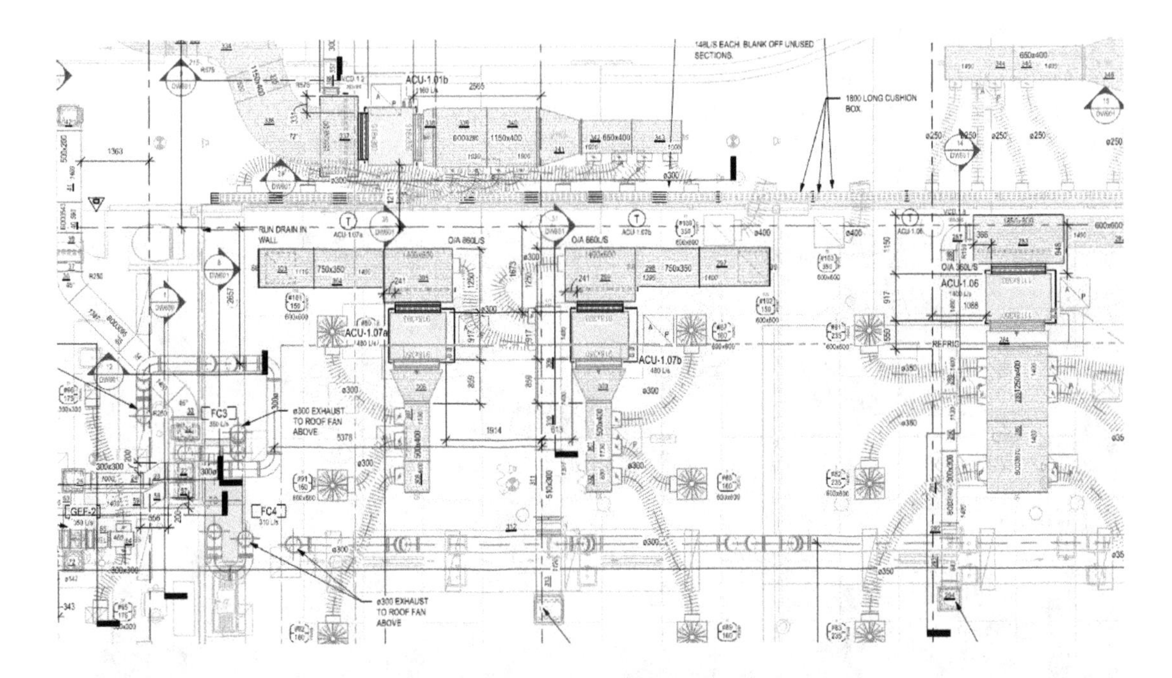

Chapter 8 - The Markups List

The objectives of this chapter are to:

- √ *Explain about the* ***Markups List***
- √ *Explain the process of sorting markups in the* ***Markups List***
- √ *Explain the process of filtering markups in the* ***Markups List***
- √ *Explain the process of exporting and importing markups from one PDF to the other*
- √ *Explain how to create various types of summary reports*
- √ *Explain the process of changing statuses of the markups*
- √ *Explain how to manage columns in the* ***Markups List***
- √ *Explain how to create custom columns and link them to the customized tools*

THE MARKUPS LIST

The **Markups List** is a tabular representation of all markups, including dimensions and stamps, added to all pages of the current PDF file. The various columns in this list show specific information about the markups, such as subject, page label, comments, author, and so on. Figure 1 shows an example of the **Markups List** displaying various types of markups added to the current file.

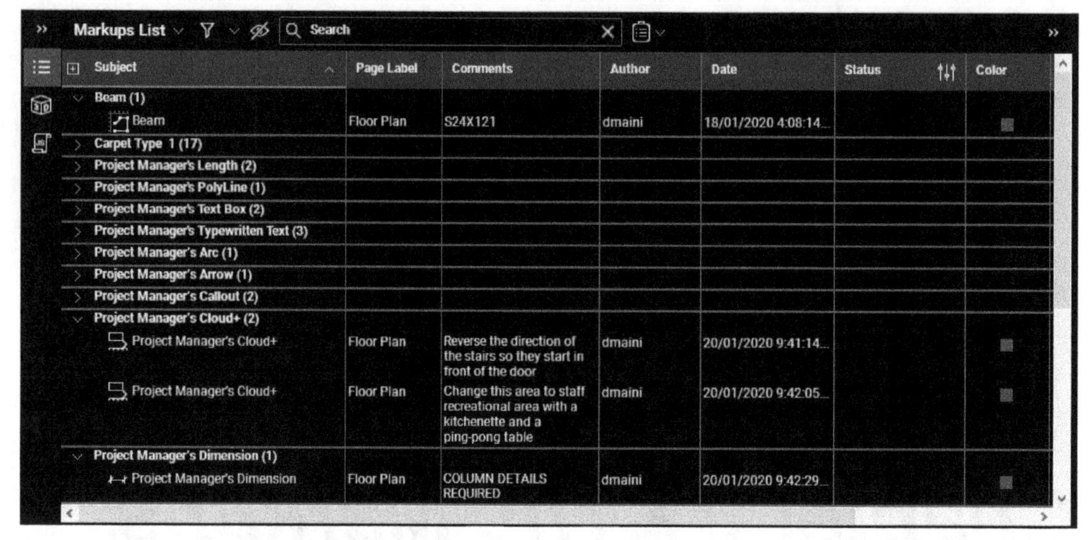

***Figure 1** The **Markups List** shows various markups added to the current PDF file*

The Procedure for Working with the Markups List

The following are the various procedures for working with the **Markups List**.

The Procedure for Controlling the Visibility of the Markups List

The following is the procedure for controlling the visibility of the **Markups List**.

1. From the bottom left of the Revu window, click the **Markups** button, as shown in Figure 2; the **Markups List** panel is displayed below the **Main Workspace**.

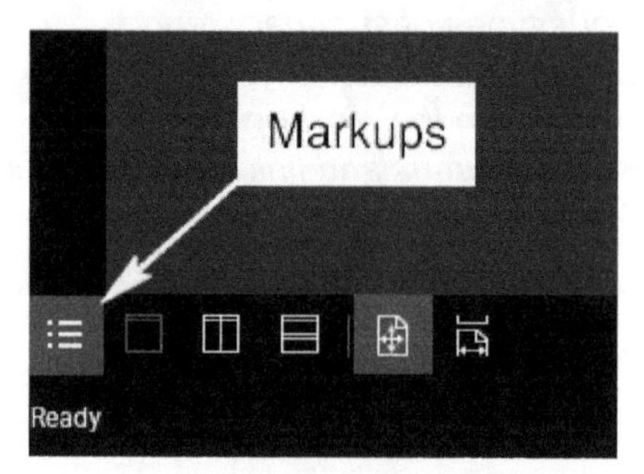

Figure 2** Turning on the visibility of the **Markups List

2. Click on the same button to hide the **Markups List**.

3. Right-click on the **Markups** button and select **Attach** to display the options to attach the **Markups List** to the other areas of the Revu window, as shown in Figure 3.

*Figure 3 Attaching the **Markups List** to the other areas of the Revu window*

4. Select the **Detach** option from the shortcut menu shown above to detach this panel and place it on a different monitor.

The Procedure for Sorting the Data in the Markups List

The following is the procedure for sorting the data in the **Markups List**.

1. By default, the data in the **Markups List** is sorted by the **Subject** column. This is evident by the arrow on this column name, which is labeled as **1** in Figure 4.

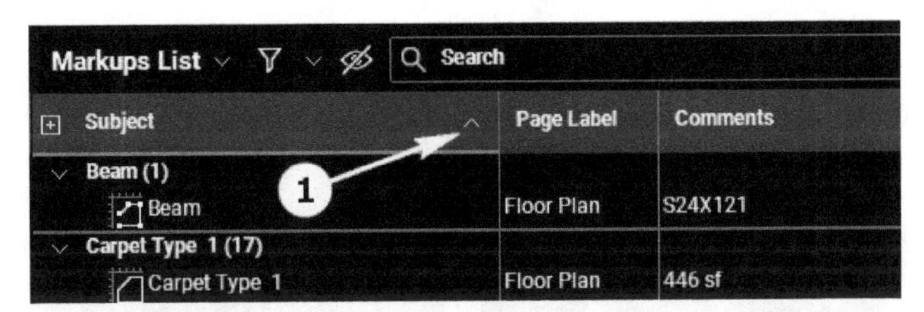

*Figure 4 The arrow indicating the data sorted by **Subject***

2. To sort the data by any other column, click on the name of that column. Figure 5 shows the data sorted by **Comments**, which is indicated by the arrow on that column.

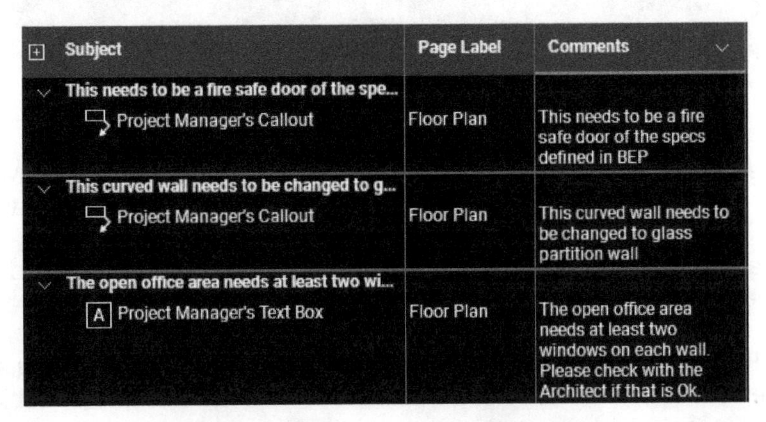

*Figure 5 The data sorted by **Comments***

The Procedure for Filtering the Data in the Markups List

By default, the **Markups List** shows all the markups added to the PDF file. However, you can easily filter the data and only display the content you want to review. The following is the procedure for doing this.

1. From the toolbar at the top in the **Markups List**, click the **Filter** button, as shown in Figure 6.

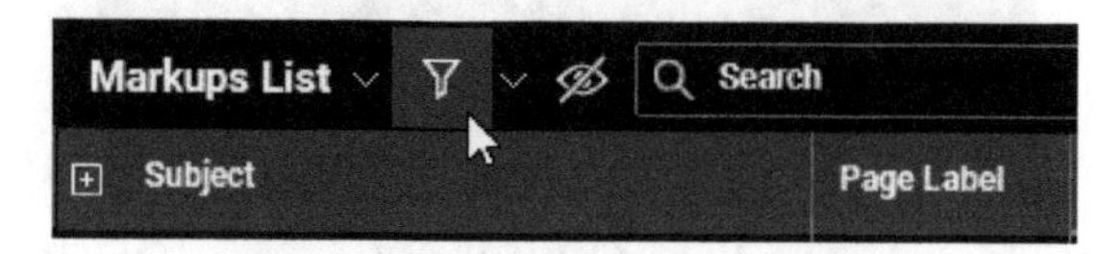

Figure 6 Turning the filtering on

On doing so, the **Filter** button is displayed on the name of all the columns, as shown in Figure 7.

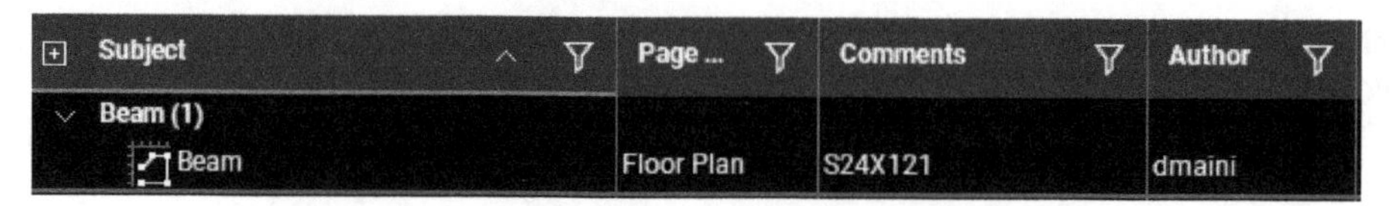

*Figure 7 The **Filter** buttons displayed on all the columns*

2. To filter the data, click on the **Filter** button on the name of that column and from the drop-down list that is displayed, select the value to filter based on. Figure 8 shows the data being filtered using **Subject > Carpet Type 1**.

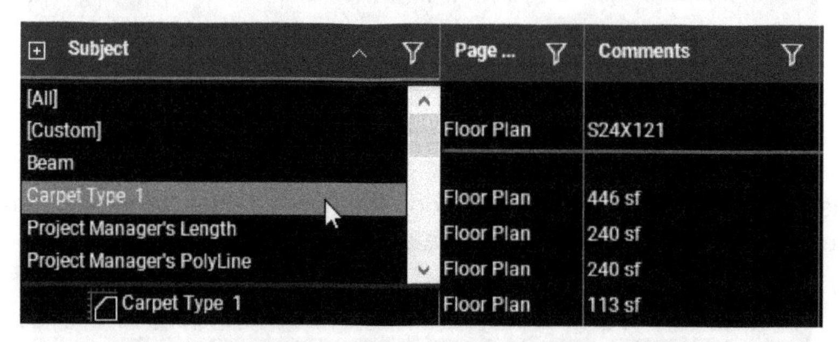

*Figure 8 Filtering the data by **Subject > Carpet Type 1***

As soon as you filter the data, the markups that are filtered out will be displayed grayed out in the PDF file and the filtered markups will be displayed with their original appearance.

3. To create a custom filter, click on the **[Custom]** option from the **Filter** drop-down list of the column; the **Custom Filter** dialog box will be displayed.

4. Using the fields in the **Custom Filter** dialog box, create the custom filter you want to use to display the data. Figure 9 shows this dialog box to filter the data whose subject is **Carpet Type 1** or **Tile Type 1**.

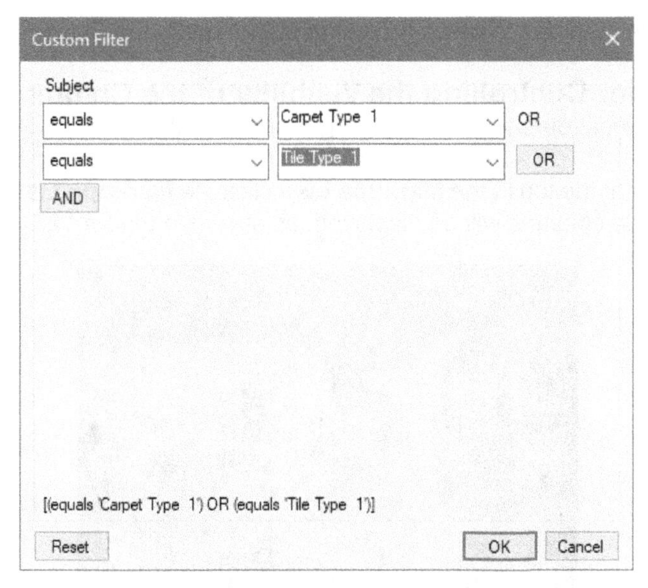

Figure 9 *Creating a custom filter*

5. Click **OK** in the dialog box; the markups with the subjects matching the custom filter will be displayed with their original appearances and the rest of the markups will be grayed out.

6. To clear the filters, click on the flyout on the right of the **Filter** button and select **Clear All**, as shown in Figure 10.

Figure 10 *Clearing all filters*

7. To turn off the filters, click on the **Filter** button on the toolbar at the top in the **Markups List**, as shown earlier in Figure 6.

The Procedure for Hiding/Displaying All the Markups

The following is the procedure for hiding all the markups from the PDF file.

1. From the toolbar at the top in the **Markups List**, click the **Hide Markups** button, as shown in Figure 11; all the markups are hidden from the PDF file.

Figure 11 *Hiding all the markups from the PDF file*

2. To display all the markups, click on the same button again.

The Procedure for Controlling the Visibility of the Columns

The following is the procedure for controlling the visibility of the columns in the **Markups List**.

1. From the toolbar at the top in the **Markups List**, click the **Markups List > Columns**; the list of all the available columns will be displayed, as shown in Figure 12.

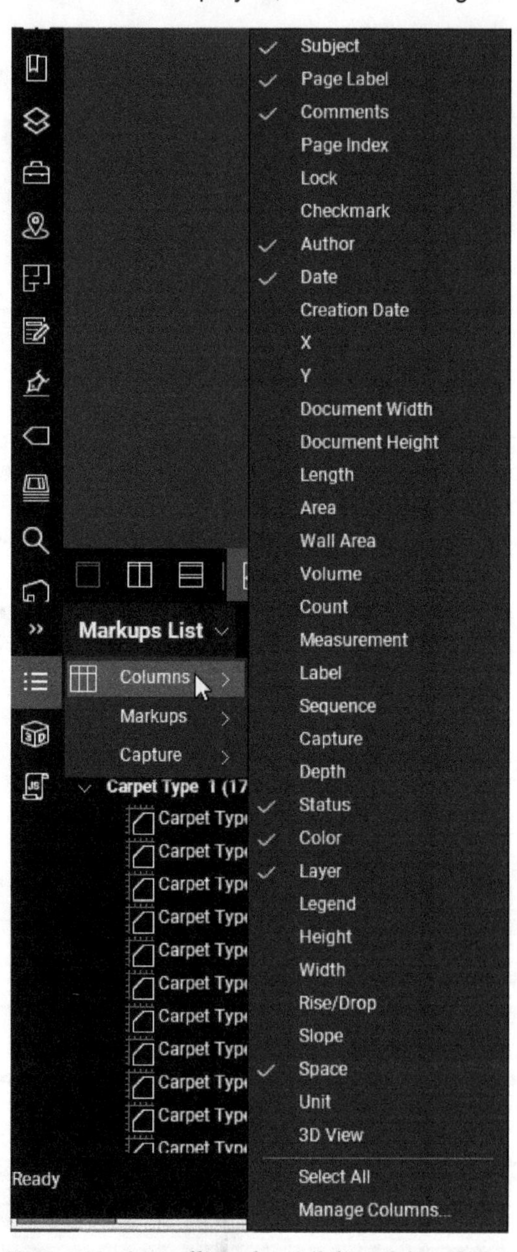

Figure 12 *Controlling the visibility of the columns*

The currently visible columns will be displayed with a check mark on the left of their names in this list.

2. Select any additional columns that you want to display in the **Markups List**.

3. To turn off the visibility of any of the columns currently turned on, click on their names again, the visibility of the columns will be turned off.

4. To change the order in which the columns are displayed in the **Markups List**, select and drag the column name to the left or right.

The Procedure for Creating Custom Columns

Revu is an extremely versatile program when it comes to design review and quantity takeoff. One of the reasons for this is that Revu allows you to create custom columns and assign additional information to the markups and takeoffs. The following is the procedure for creating various types of custom columns.

1. From the toolbar at the top in the **Markups List**, click the **Markups List > Columns**; the list of all the available columns will be displayed, as shown in Figure 12 earlier.

2. From the bottom of this list, click **Manage Columns**; the **Manage Columns** dialog box will be displayed, as shown in Figure 13.

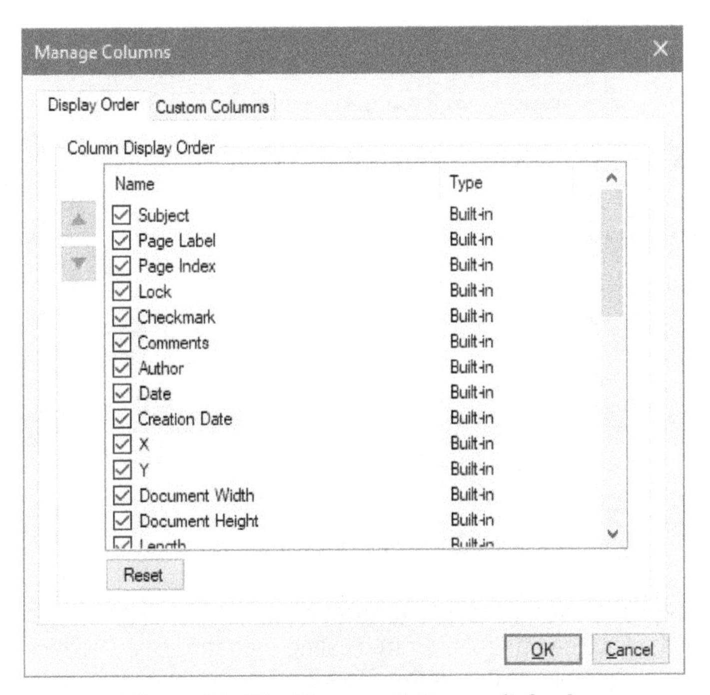

*Figure 13 The **Manage Columns** dialog box*

3. In this dialog box, select the **Custom Columns** tab.

4. On the **Custom Columns** tab, click the **Add** button; the **Add Column** dialog box will be displayed.

   Using this dialog box, you can create the following types of custom columns:

   Checkmark: This type is used to create a column with a check box for the data in the column. You can create a column that has the check box selected or cleared to start with. An example of this is the **Operations Manual** that can be selected for the equipment that have the operations manuals already supplied.

   Choice: This type of column is used to create a multiple-choice column. The choices you add can be assigned a numeric value representing the cost of that item. An example of this is the choice of selecting a custom status type for the markups. Alternatively, it could be used for selecting flooring types for the area measurement. Figure 14 shows a choice column called **Flooring Type** that allows you to select carpet or tile type flooring. This dialog box also shows the numeric value assigned to these columns that will allow you to calculate the cost of per square foot of that flooring type.

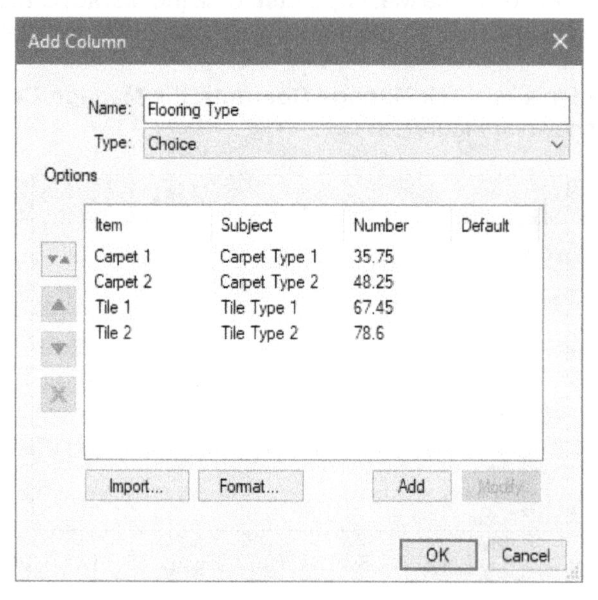

*Figure 14 The **Add Column** dialog box*

   Date: This type is used to create a custom date column.

   Formula: This type is used to create a formula column where you can add a formula to calculate values using the variables from existing standard or custom columns, constants, and functions. An example of this is the Flooring Cost column that calculates the flooring cost using the **Measurement** variable from the area measurement and multiplies it with the numeric value of the **Flooring Type** custom column you created earlier, as shown in Figure 15.

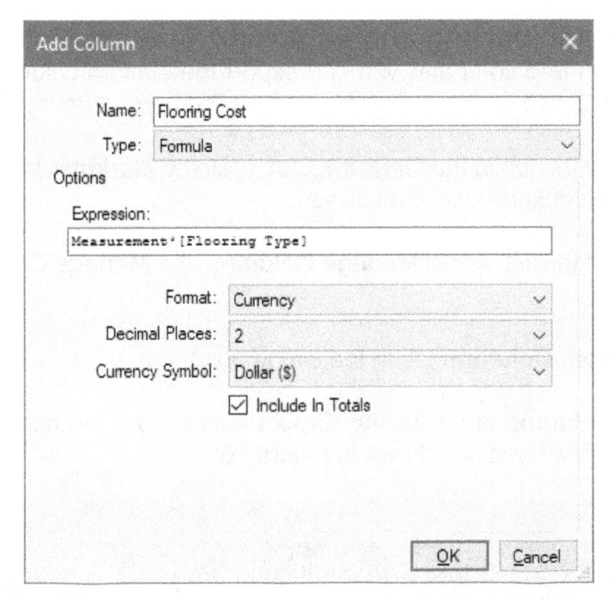

Figure 15 A custom column to calculate the flooring cost

What I do

*Once I create a custom column such as the **Flooring Type** column, I then assign the choice of that flooring type to the custom tool also. For example, I assign the choice of **Carpet Type 1** from the custom column to the custom area measurement tool that I create for **Carpet Type 1**. After assigning the choice, I add that custom tool to the custom tool set. This way, when I takeoff quantities using that custom tool, the selected choice is automatically assigned to that measurement. This results in the automatic calculation of the cost of that flooring type in the **Flooring Cost** column based on the formula. You will learn more about this later in this chapter.*

Number: This type is used to create a column with a normal, currency, or percentage numbers. You can also define a maximum and a minimum number for this column.

Text: This type is used to create a column that allows writing a single or multiline text.

5. Enter the name of the column in the **Name** field.
6. Select the type of the column from the **Type** field.
7. For the selected type, specify the required parameters in the **Add Column** dialog box.
8. Click **OK** in the **Add Column** dialog box; you will be returned to the **Manage Columns** dialog box and the custom column you created will be listed there.
9. Click **OK** in the **Manage Columns** dialog box; the custom columns you created will be listed in the **Markups List**.

The Procedure for Exporting and Importing Custom Columns

For the standardization in a company, you can export the custom columns and have all your team members import those columns. The following is the procedure for doing this.

1. From the toolbar at the top in the **Markups List**, click the **Markups List > Columns**; the list of all the available columns will be displayed.

2. From the bottom of the list, select **Manage Columns**; the **Manage Columns** dialog box will be displayed.

3. Activate the **Custom Columns** tab in the dialog box.

4. In the **Custom Columns** tab, click the **Export** button from the bottom left; the **Save As** dialog box will be displayed, as shown in Figure 16.

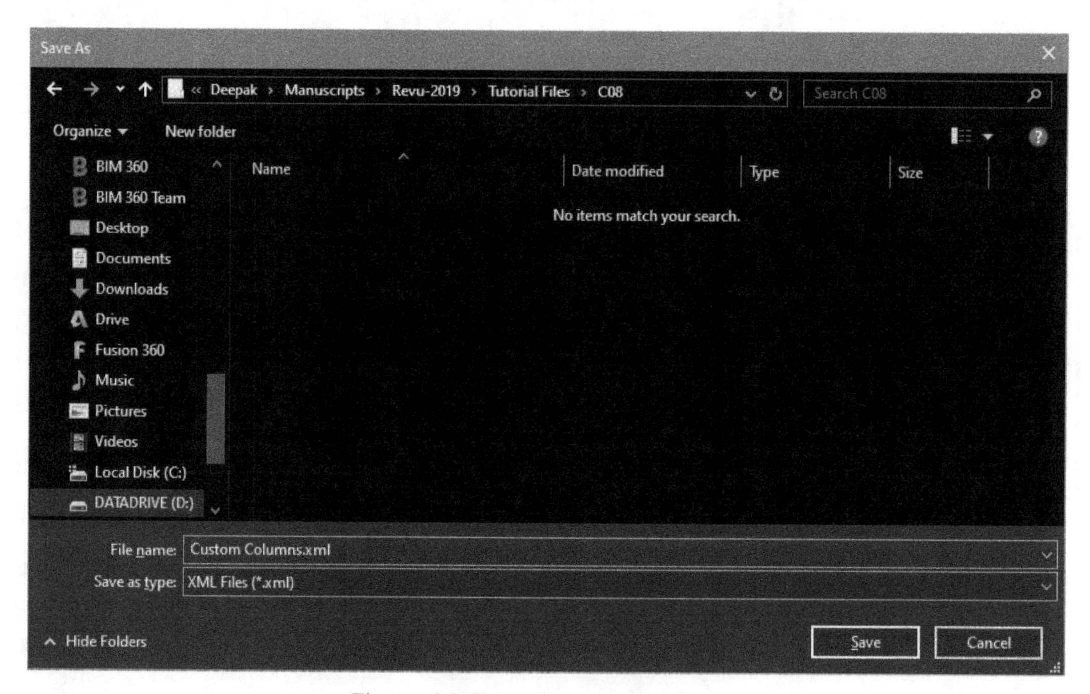

Figure 16 Exporting custom columns

5. Browse to the folder where you want to save the custom columns.

6. Specify the name of the custom columns XML file that will be saved and then click **Save** in the **Save As** dialog box.

   Note that all the custom columns you created will be saved in this XML file.

7. To import the custom columns, click the **Import** button on the lower left in the **Custom Columns** tab of the **Manage Columns** dialog box; the **Open** dialog box will be displayed.

8. Browse to the XML file of the custom columns and then double-click on it; the **Import Custom Columns** warning box will be displayed informing you that the process will replace any existing custom columns and will be immediately applied to the active document.

9. Click **Yes** in the warning box; the custom columns will be imported.

10. Click **OK** in the **Manage Columns** dialog box; the imported custom columns will be displayed in the **Markups List**.

The Procedure for Setting Status for the Markups in the Markups List

Once the markups in the PDF file are actioned as needed, they need to have their status changed to ensure it reflects what action has been taken. There is a default **Status** column available from which you can select the required status for the markup. The following is the procedure for doing this.

1. In the **Markups List**, scroll to the markup that has been actioned and requires its status to be changed.

2. In the **Status** column, double-click on the field of that markup; a shortcut menu will be displayed, as shown in Figure 17.

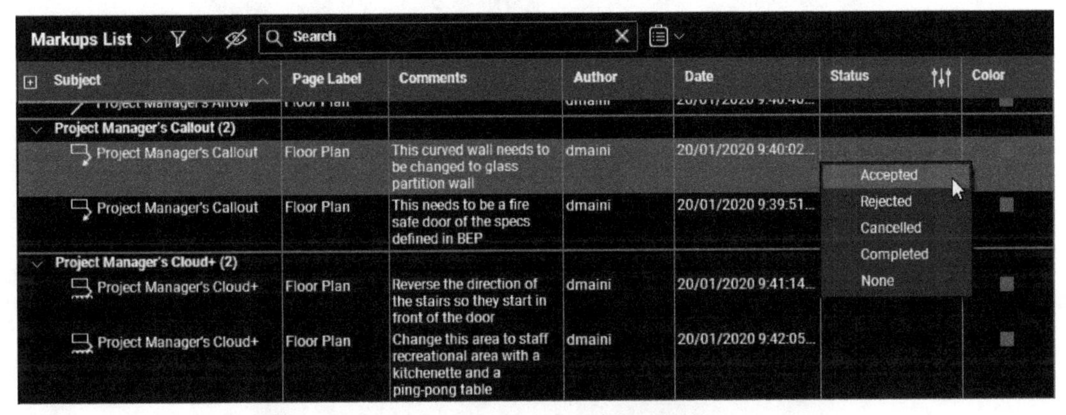

Figure 17 Selecting the required status from the markup

3. Select the required status for the markup from the menu shown in Figure 17; the **Status** column will show the details of the status applied, as shown in Figure 18.

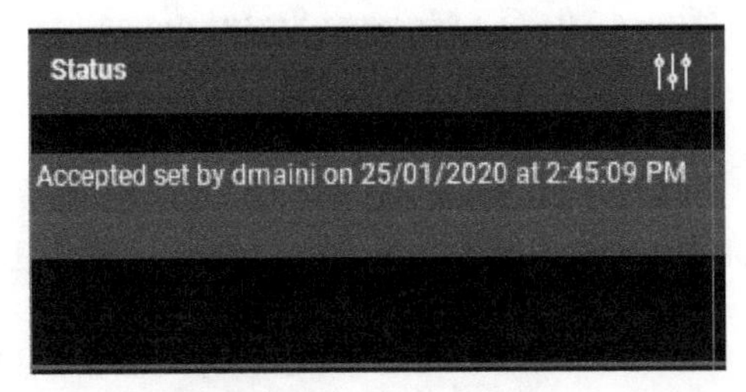

*Figure 18 The selected status reflected in the **Status** column*

The Procedure for Creating Custom Statuses

If your company has some custom statuses, you can create those and then assign them to the markups. The following is the procedure for doing this.

1. In the **Markups List > Status** column, click on the icon enclosed in the rectangle in Figure 19 and then click **Manage Status**.

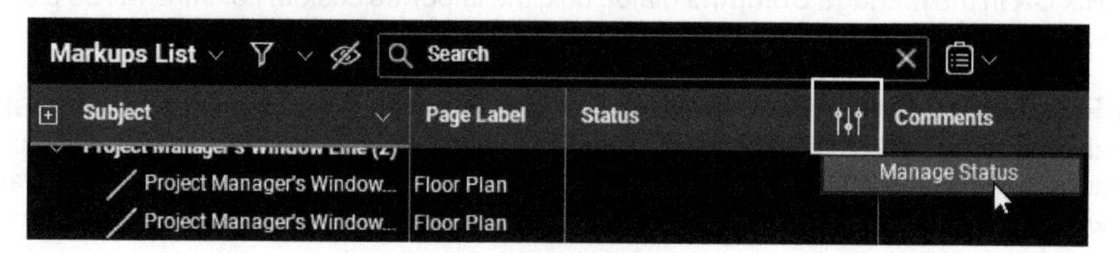

Figure 19 *Creating custom statuses*

On doing so, the **Manage Status** dialog box will be displayed, as shown in Figure 20. This dialog box displays the default state models and their respective statuses.

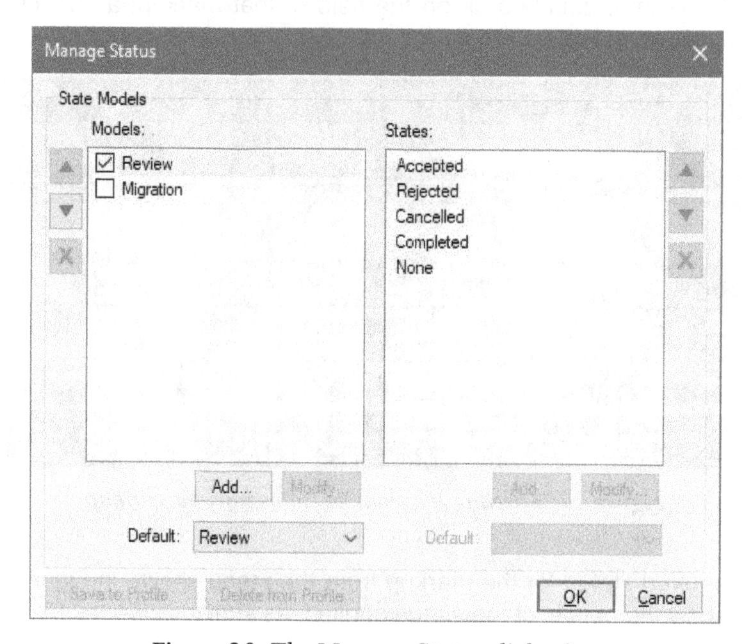

Figure 20 *The* ***Manage Status*** *dialog box*

To create custom statuses, you first need to create the model.

2. Click the **Add** button below the **Models** area; the **Model** dialog box will be displayed.
3. Enter the name of the model in the **Model** dialog box, as shown in Figure 21.
4. Click **OK** in the **Model** dialog box; the model name will be displayed in the **Models** area.

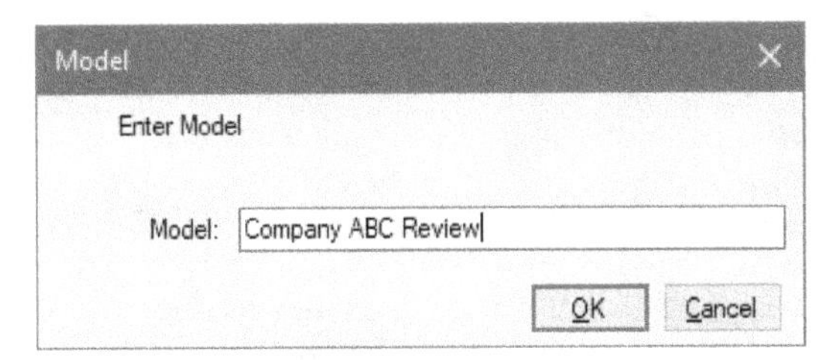

Figure 21 Creating the status model

5. With the custom model name selected in the **Models** area, click the **Add** button below the **States** area; the **State Properties** dialog box will be displayed.

6. Enter the name of the status in the **Name** field.

7. If required, add additional text in the **Text** field. The text added in this field will appear in the **Comments** column of the markup.

   With these custom statues, you can change the color of the markup when the status is applied by selecting the **Color** check box. However, note that once the color is changed, you will not be able to restore the original appearance color of the markup easily. You will have to either use **Match Properties** or manually change the color in the **Properties** window.

8. Select the **Color** check box and then select the color to be used for the status. Figure 22 shows the **State Properties** dialog box with a state created.

*Figure 22 The **State Properties** dialog box*

9. Click **OK** in the dialog box; you will be returned to the **Manage Status** dialog box.

10. Similarly, create additional states. Figure 23 shows the **Manage Status** dialog box with the various states created for the custom status.

11. From the **Models** area, clear the check boxes of the status models you do not want to use. Remember that if you have multiple models selected, they will be available to assign while changing statuses. Figure 23 shows the two models selected. So both these models will be available to be used while changing statues of the markup.

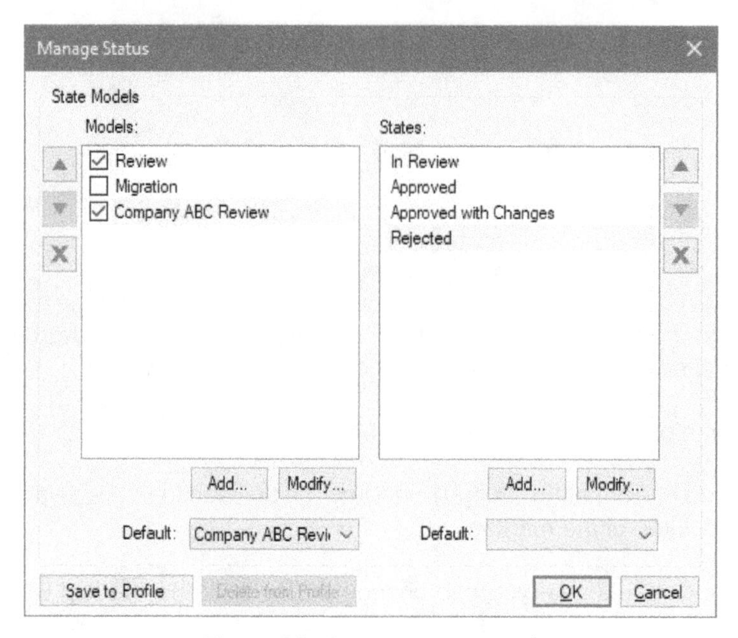

Figure 23 Custom states created

12. To save custom status to the current profile, click the **Save to Profile** button on the lower left of the **Manage Status** dialog box.

13. Click **OK** in the **Manage Status** dialog box.

14. In the **Markups List**, double-click on the **Status** field of the markup you want to apply the status to and then select your custom status, as shown in Figure 24.

Figure 24 Applying custom status to a markup

The Procedure for Creating Markup Summary

Creating a summary of all the markups added to the PDF file is an important part of the design review process. Revu allows you to create summaries in the CSV, XML, and PDF format. You can also print the summary directly. The following is the procedure for doing this creating the markup summary.

1. From the toolbar at the top in the **Markups List**, select the **Summary** button enclosed in the rectangle in Figure 25, and then select the summary type.

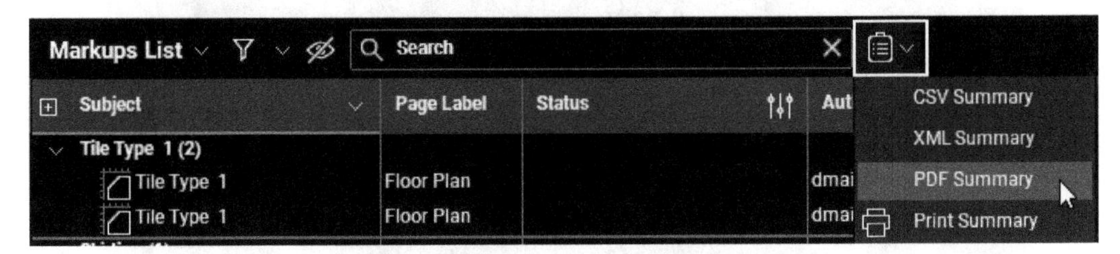

Figure 25 *Selecting the summary type*

   On doing so, the **Markup Summary** dialog box will be displayed.

2. On the **Columns** tab of the **Markup Summary** dialog box, select the columns to be included in the summary.

3. On the **Filter and Sort** tab, filter and sort the markups the way you want them to appear in the summary report.

4. On the **Output** tab, configure the settings for the summary.

> ***What I do***
>
> *I normally create the summary in the PDF format. This format allows me to append the summary pages to the current PDF and also create hyperlinks from the summary pages to the original pages of the PDF. That way I can easily navigate to the original markup from the summary pages.*

   Once you configure the settings in the **Markup Summary** dialog box, Revu allows you to save the configuration to be reused.

5. To save the configuration, click the **Save Config** button on the lower right of the **Markup Summary** dialog box and save the BCF file of the configuration.

   Figure 26 shows the **Markup Summary** dialog box after configuring all the settings.

6. Click **OK** in the **Markup Summary** dialog box.

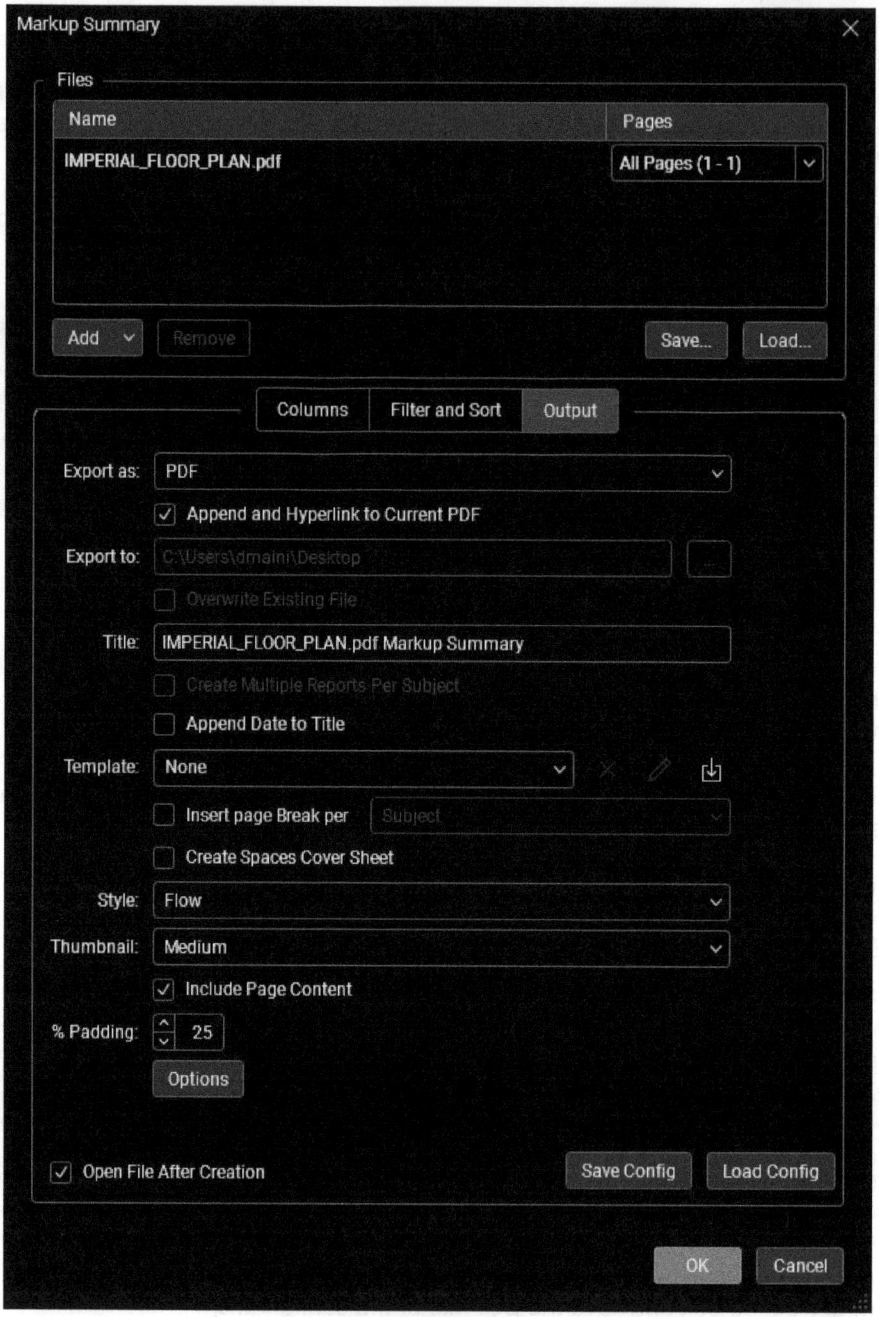

Figure 26 *The* ***Markup Summary*** *dialog box after configuring the settings*

Depending on the type of summary you configured, the summary report will be created. Figure 27 shows the **Thumbnails** panel with the pages of PDF summary report created and appended to the current file with the hyperlinks created between the summary pages and the markups on the original page.

Figure 27 *The* ***Thumbnails*** *tab showing various summary pages*

Hands-on Tutorial	*In this tutorial, you will complete the following tasks:* *1. Sort the markups in the PDF file using various columns.* *2. Filter markups using standard and custom filters.* *3. Drag various columns to reorder them in the **Markups List**.* *4. Create custom statuses.* *5. Export markups summary.* *6. Create custom columns.* *7. Link customized measurement tools to the custom columns.* *8. Create a legend using the tools from a custom tool set.*

Section 1: Sorting Markups using Various Markups List Columns

In this section, you will open the **IMPERIAL_FLOOR_PLAN.pdf** or **METRIC_FLOOR_PLAN.pdf** file from the **C08** folder. You will then display the **Markups List** and then sort the markups using various columns in the **Markups List**.

1. From the **C08** folder, open the **IMPERIAL_FLOOR_PLAN.pdf** or **METRIC_FLOOR_PLAN.pdf** file. Notice that the file has various markups and measurements already added.

   Before proceeding any further, you need to change the profile to the **Revu Training** profile you created in Chapter 1 of this book.

2. Activate the **Revu Training** profile and then hide any panels displayed on the left or right of the **Main Workspace**.

3. From the lower left corner of the Revu window, click the **Markups** button, as shown in Figure 28; the **Markups List** is displayed below the **Main Workspace**.

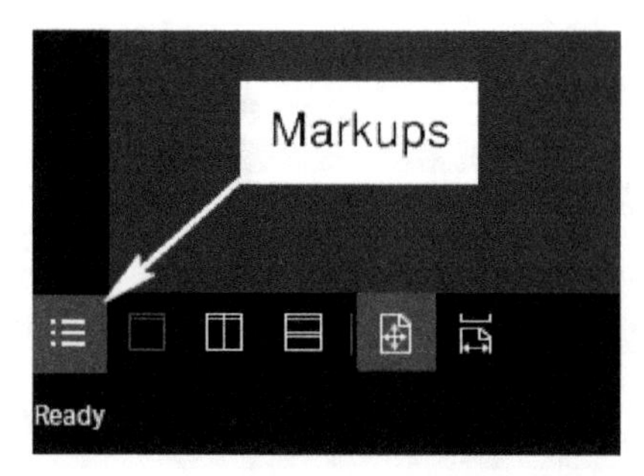

Figure 28 *Turning on the visibility of the **Markups List***

   By default, the data in the **Markups List** is sorted by their subject in the alphabetic order. This is indicated by the arrow displayed on the **Subject** column, as shown in Figure 29.

4. Click on the **Subject** column name; the markups are sorted in the reverse alphabetic order, as shown in Figure 30.

5. Click on the name of the **Comments** column; the markups are sorted by their comments and an arrow is displayed on the right of this column name.

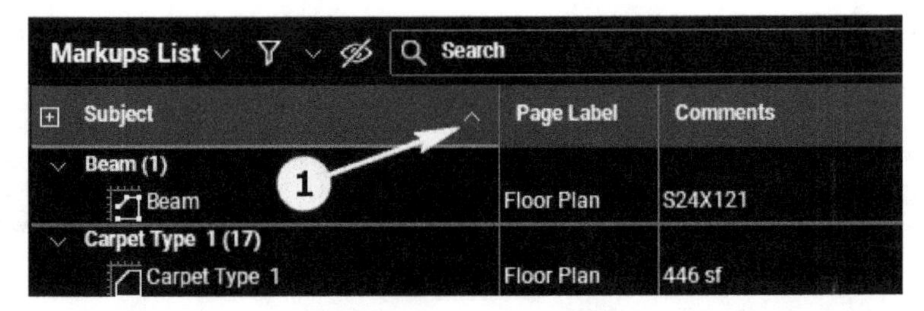

Figure 29 *The arrow indicating the data sorted by* ***Subject***

Markups List | Search

Subject	Page Label	Comments	Date	Status	Author
Tile Type 1 (2)					
Tile Type 1	Floor Plan	273 sf	18/01/2020 4:08:14...		dmaini
Tile Type 1	Floor Plan	203 sf	18/01/2020 4:08:25...		dmaini
Skirting (1)					
Skirting	Floor Plan	56'-10"	18/01/2020 4:08:14...		dmaini
Project Manager's Window Line (2)					
Project Manager's Window...	Floor Plan		20/01/2020 9:38:32...		dmaini
Project Manager's Window...	Floor Plan		20/01/2020 9:39:17...		dmaini
Project Manager's Wall Socket Lin...					
Project Manager's Wall So...	Floor Plan		20/01/2020 9:38:32...		dmaini
Project Manager's Wall So...	Floor Plan		20/01/2020 9:39:17...		dmaini
Project Manager's Note (1)					
Project Manager's Note	Floor Plan	Sec. 7-1809. Field Testing. Only when inspection indicates that the construction is not in accordance with the	27/12/2019 3:57:59...		dmaini

Figure 30 *Sorting the markups in the* ***Markups List***

6. Click on the name of the **Comments** column again and notice how the comments have resorted.

7. Similarly, sort the data in the **Markups List** by their color by clicking on the **Color** column.

   You will now restore the original sort settings by sorting the data by their subject.

8. Sort the markups by their subject by clicking on the **Subject** column name.

Section 2: Filtering Markups using the Standard and Custom Filters

In this section, you will filter the markups using the standard filters available. You will then create custom filters to filter the markups.

1. From the toolbar at the top in the **Markups List**, click the **Filter** button, as shown in Figure 31.

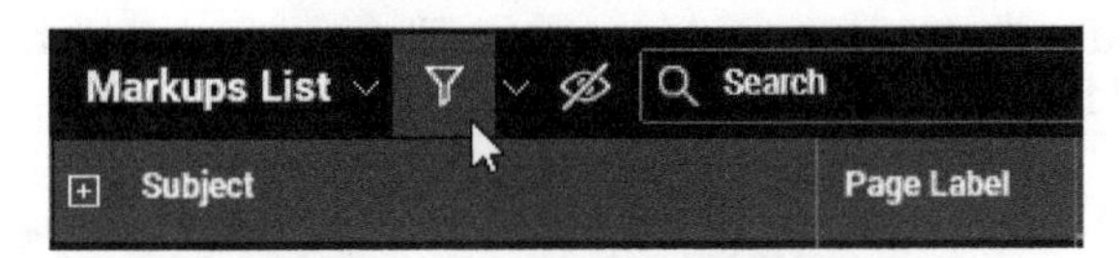

Figure 31 *Turning the filtering on*

On doing so, the **Filter** buttons are displayed on top of all the columns, as shown in Figure 32.

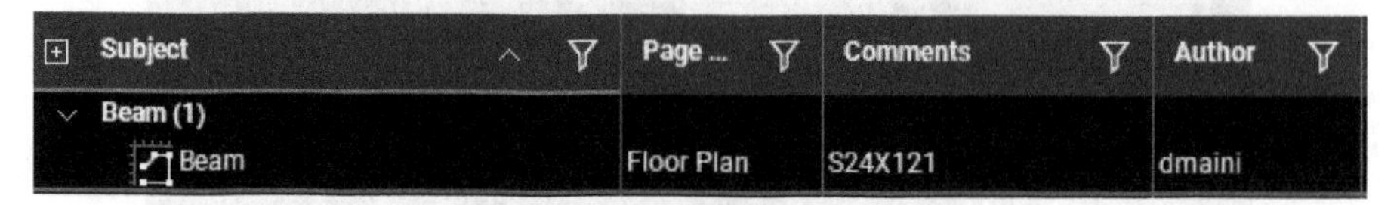

*Figure 32 The **Filter** buttons displayed on all the columns*

You will now filter the data using the **Carpet Type 1** subject.

2. Click on the **Filter** button on the right of the **Subject** column; the drop-down list showing various subjects is displayed, refer to Figure 33.

3. From the drop-down list, select **Carpet Type 1**; as shown in Figure 33.

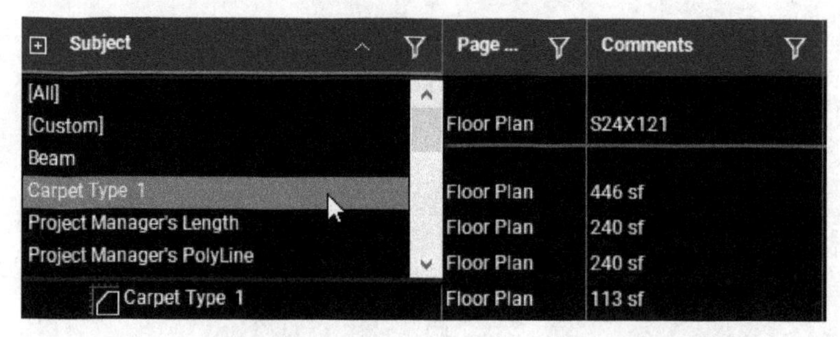

*Figure 33 Filtering the data by **Subject > Carpet Type 1***

Notice the markups that are filtered out are displayed in Gray in the PDF file and the filtered markups will be displayed with their original appearance, as shown in Figure 34. Also, the filtered out data has disappeared from the **Markups List**.

You will now filter the data based on another subject.

4. Click on the **Filter** button on the right of the **Subject** column and select **Project Manager's Text Box** from the drop-down list; the data is filtered based on this subject on the PDF file as well as the **Markups List**.

   Next, you will create a custom filter to display the data based on multiple subjects.

5. Click on the **Filter** button on the right of the **Subject** column and select **[Custom]** from the drop-down list, as shown in Figure 35; the **Custom Filter** dialog box is displayed.

   By default, this dialog box displays the current filter applied, which is **Project Manager's Text Box**. You will first reset this filter.

6. From the lower left in the **Custom Filter** dialog box, click **Reset**; all the filters are removed in the **Custom Filter** dialog box.

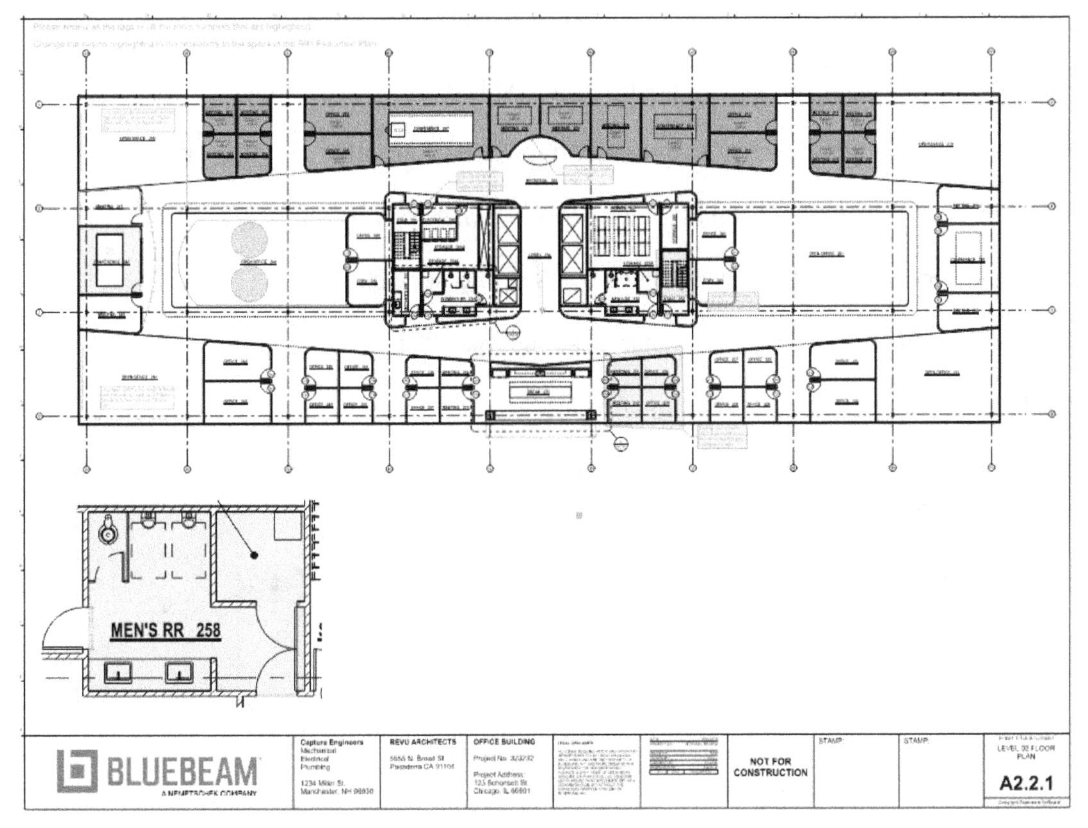

Figure 34 The data filtered by ***Subject > Carpet Type 1***

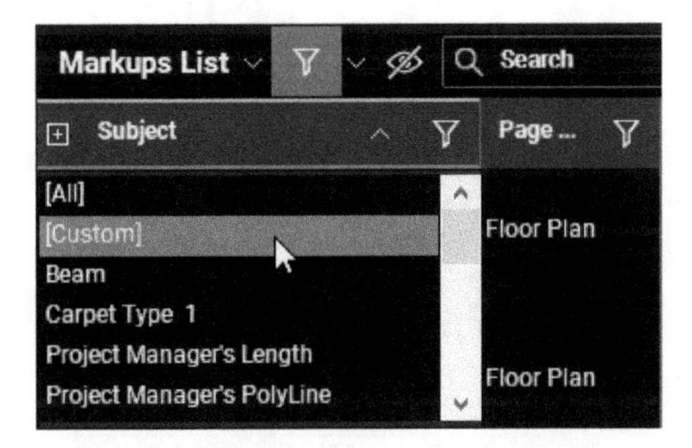

Figure 35 Creating a custom filter

You will now create a custom filter to display the markups that have their subject as **Carpet Type 1**, **Tile Type 1**, or **Skirting**.

7. From the first drop-down list in the **Subject** area, select **equals**.

8. From the second drop-down list in the **Subject** area, select **Carpet Type 1**.

Next, you need to add an additional row to add the next subject.

9. Click the **OR** button on the right of the first row; another row is added.
10. From the first drop-down list in the second row, select **equals**.
11. From the second drop-down list in the second row, scroll down and select **Tile Type 1**.

Finally, you will add the row to include the **Skirting** subject.

12. Click the **OR** button on the right of the second row; another row is added.
13. From the first drop-down list in the third row, select **equals**.
14. From the second drop-down list in the third row, scroll down and select **Skirting**. Figure 36 shows the **Custom Filter** dialog box with these three rows.

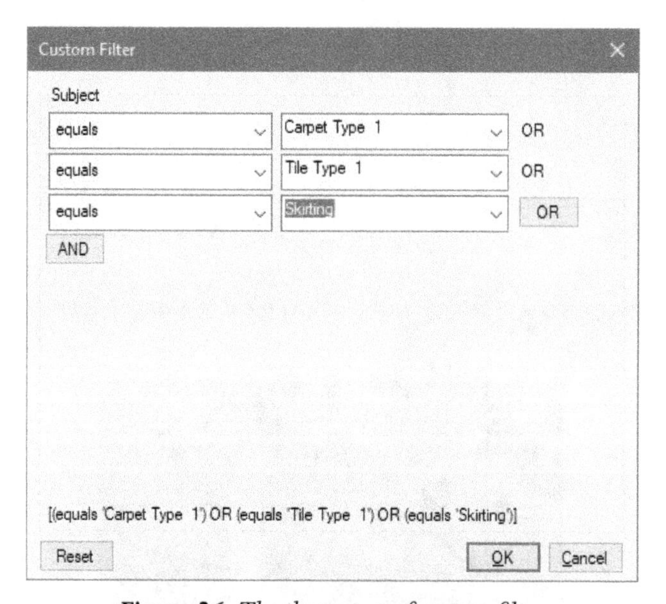

Figure 36 *The three rows of custom filter*

15. Click **OK** in the dialog box; the data in the **Markups List** and the PDF file is filtered based on the custom filter you created.

You will now clear the filter to redisplay all the data.

16. From the toolbar at the top in the **Markups List** click the **Filter** button; the filter is removed and all the data is displayed in the **Markups List** as well as the PDF file.

Section 3: Relocating Columns in the Markups List

In this section, you will drag various columns to reorder them in the **Markups List**.

1. Press and hold down the Left Mouse Button on the name of the **Comments** column and drag it to the right of the **Date** column.

2. Release the Left Mouse Button when you see an arrow on top of the right edge of the **Date** column; the **Comments** column is now located on the right of the **Date** column.

3. Drag and drop the **Color** column on the left of the **Author** column.

4. Drag and drop the **Page Label** column on the right of the **Comments** column. Figure 37 shows the **Markups List** after reordering the columns.

Markups List | Search

Subject	Date	Comments	Page Label	Status	Color	Author
Beam (1)						
Beam	18/01/2020 4:08:14...	S24X121	Floor Plan			dmaini
Carpet Type 1 (17)						
Carpet Type 1	18/01/2020 4:08:14...	446 sf	Floor Plan			dmaini
Carpet Type 1	18/01/2020 4:08:14...	240 sf	Floor Plan			dmaini
Carpet Type 1	18/01/2020 4:08:14...	240 sf	Floor Plan			dmaini
Carpet Type 1	18/01/2020 4:08:14...	113 sf	Floor Plan			dmaini
Carpet Type 1	18/01/2020 4:08:14...	136 sf	Floor Plan			dmaini
Carpet Type 1	18/01/2020 4:08:14...	113 sf	Floor Plan			dmaini
Carpet Type 1	18/01/2020 4:08:14...	128 sf	Floor Plan			dmaini
Carpet Type 1	18/01/2020 4:08:14...	230 sf	Floor Plan			dmaini
Carpet Type 1	18/01/2020 4:08:14...	217 sf	Floor Plan			dmaini

*Figure 37 The **Markups List** after reordering the columns*

Because you made a few changes to the profile, it is better to save it now.

5. Save the current profile by clicking on **Revu > Profiles > Save Profile**.

Section 4: Creating and Applying Custom Statuses

In this section, you will create custom statuses and then apply those to the markups. You will make sure the markups change color based on their statuses.

1. In the **Markups List > Status** column, click on the icon enclosed in the rectangle in Figure 38 and then click **Manage Status**.

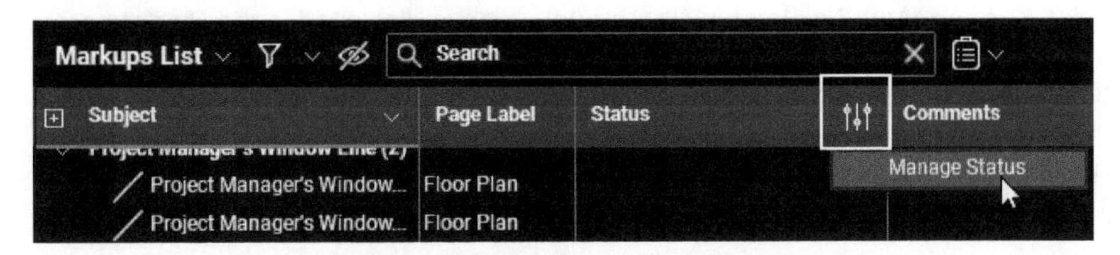

Figure 38 Creating custom statuses

On doing so, the **Manage Status** dialog box is displayed, as shown in Figure 39. This dialog box displays the default state models and their respective statuses.

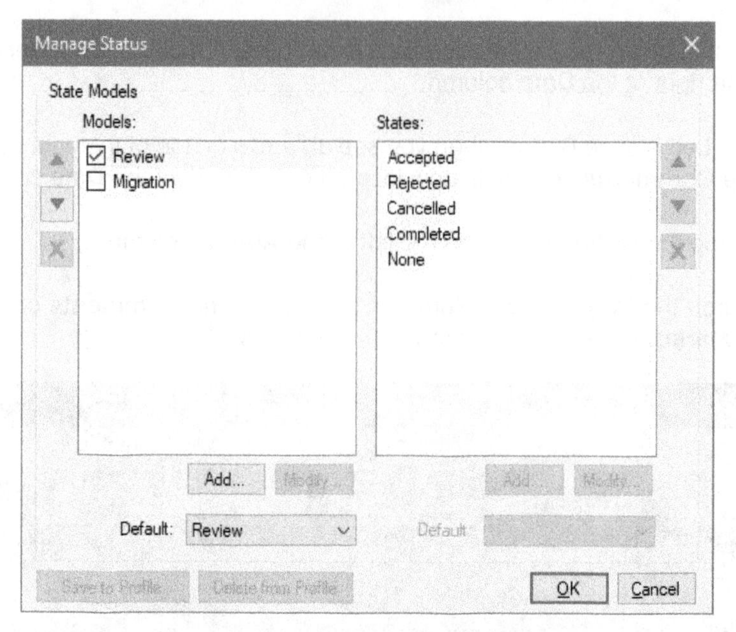

Figure 39 *The **Manage Status** dialog box*

To create custom statuses, you first need to create the model.

2. Click the **Add** button below the **Models** area; the **Model** dialog box is displayed.

3. Enter **Company ABC Review** as the name of the model in the **Model** dialog box, as shown in Figure 40.

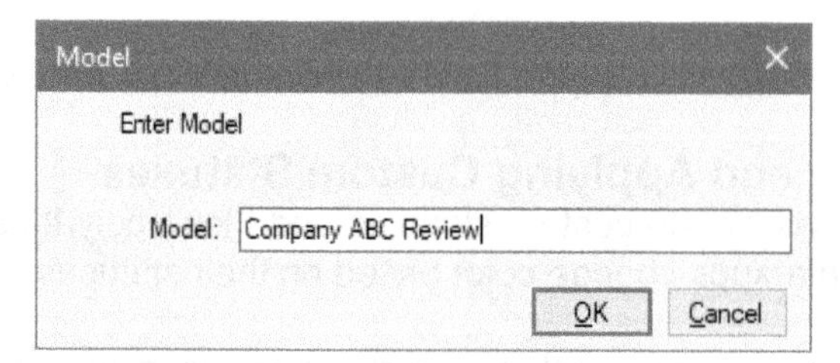

Figure 40 *Creating the status model*

4. Click **OK** in the **Model** dialog box; the model name is displayed in the **Models** area.

   You are now ready to add various statuses for this model.

5. With the custom model name selected in the **Models** area, click the **Add** button below the **States** area; the **State Properties** dialog box will be displayed.

6. Enter **In Review** as the name of the state in the **Name** field, refer to Figure 41.

The text you enter in the **Text** field is displayed in the **Comments** column in the **Markups List**. However, note that this text will override any text or comments of the existing markups. Ideally, it is NOT RECOMMENDED to enter any text in the **Text** field. The following steps show what happens when you enter the text in this field.

7. In the **Text** field, enter the following:

   This markup is being reviewed at this stage. Further action will be taken once the review is completed.

   With these custom statues, you can change the color of the markups when the status is applied by selecting the **Color** check box. Note that once the color is changed by the status, to restore it you will have to either use **Match Properties** or manually change the color in the **Properties** window.

8. Select the **Color** check box; the color swatch is activated on the right of this check box.

9. From the **Color** swatch, select the **Yellow** color. Figure 41 shows the **State Properties** dialog box with this state created.

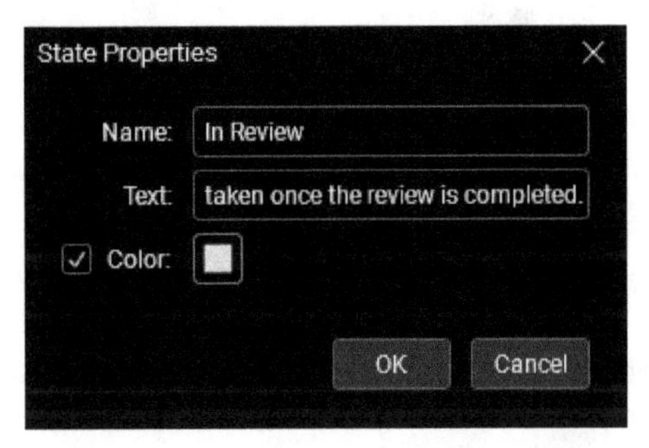

*Figure 41 The **State Properties** dialog box*

10. Click **OK** in the dialog box; you are returned to the **Manage Status** dialog box and the **In Review** state is displayed in the **States** area.

11. Similarly, create the following additional states.

Name	**Approved**
Text	**This markups is approved.**
Color	**Blue**
Name	**Approved with Conditions**
Text	**This markups is approved provided all the relevant conditions are met.**
Color	**Green**
Name	**Rejected**
Text	**The team does not think this change should be made.**
Color	**Red**

Figure 42 shows the **Manage Status** dialog box with the various states created for the custom status.

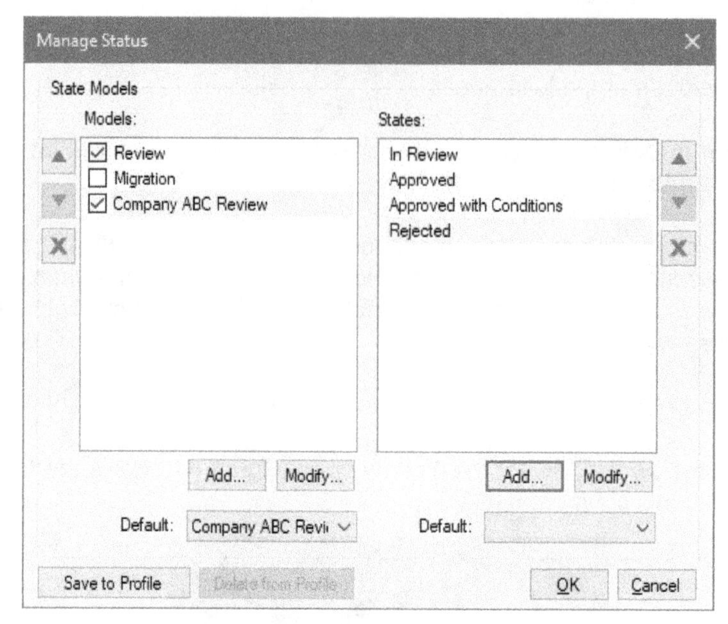

***Figure 42** The **Manage Status** dialog box with the custom states created*

You will now make sure these custom statuses are saved in your current profile.

12. From the bottom left in the **Manage Status** dialog box, click the **Save to Profile** button; the custom statues are saved in the current profile and will be available in the **Markups List** whenever you activate this profile.

    To ensure the default **Review** model is also available to apply default statuses to the markups, you need to make sure the check box on the left of this model is selected.

13. From the **Models** area of the dialog box, make sure the check box on the left of the **Review** and **Company ABC Review** models are selected.
14. Click **OK** in the **Manage Status** dialog box, you are returned to the Revu window.

    You are now ready to change the statuses of the existing markups.

15. In the **Markups List**, scroll to the two **Project Manager's Text Box** markups.
16. Hold down the SHIFT key and select both the **Project Manager's Text Box** markups.
17. Right-click on any of the two selected markups; a shortcut menu is displayed.

18. From the shortcut menu, select **Set Status > Company ABC Review > In Review**, as shown in Figure 43.

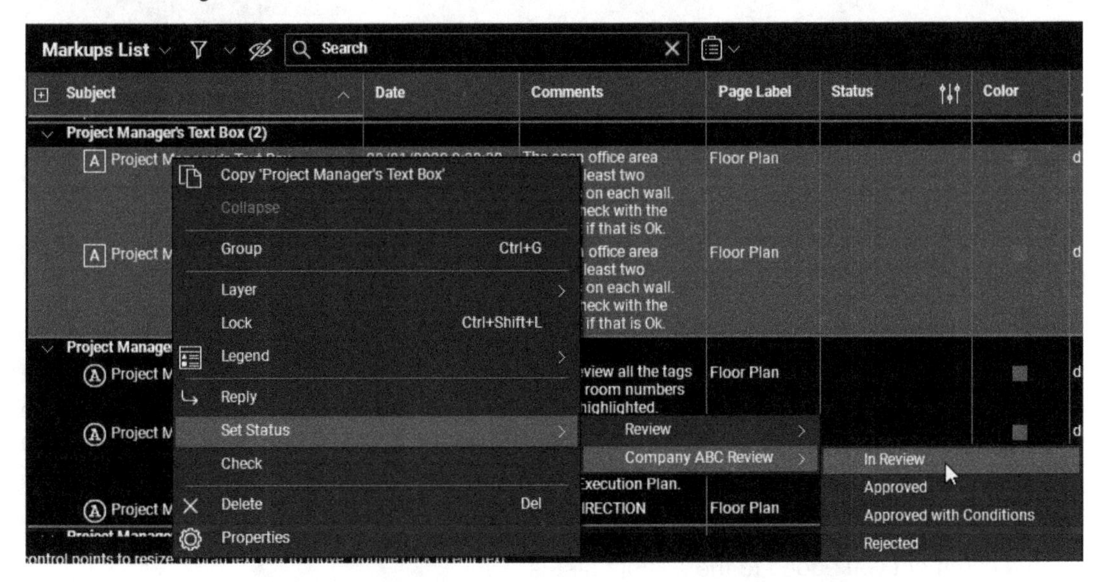

Figure 43 Applying custom status to the selected markups

On doing so, the **Status** column shows the custom status applied, along with the information about the user who applied the status and the date and time, as shown in Figure 44. Also, the color of these two markups on the PDF file is changed to the status color of **Yellow**.

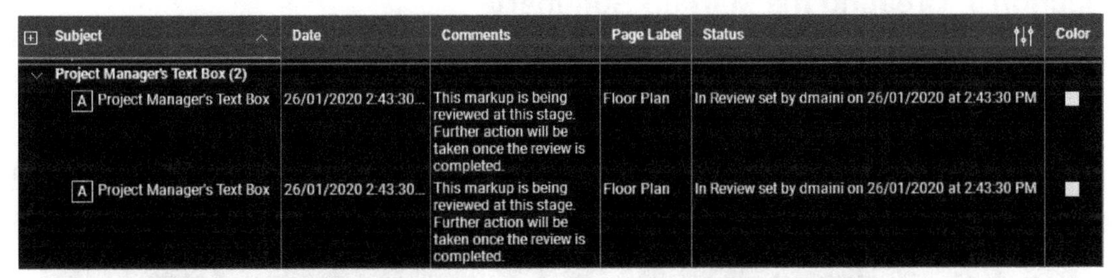

*Figure 44 The custom status shown in the **Markups List***

19. Navigate to one of these markups in the **Main Workspace** and notice the content of the text box markup is changed to match the status comment.

20. In the **Markups List**, select the four **Project Manager's Highlight** markup that highlight the basins in the PDF file.

21. Right-click on any of the selected markups and from the shortcut menu, select **Set Status > Company ABC Review > Approved**; the status is applied to the markups in the **Markups List**, as shown in Figure 45, and the **Comments** column shows the comment associated with this status. Also, notice that the color of these four markups is changed in the PDF file.

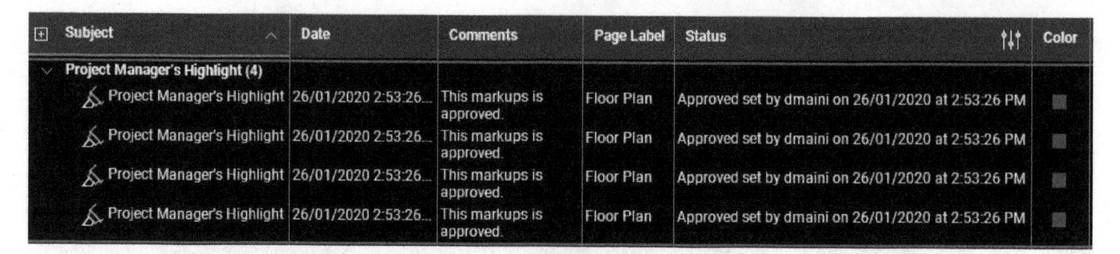

Subject	Date	Comments	Page Label	Status	Color
Project Manager's Highlight (4)					
Project Manager's Highlight	26/01/2020 2:53:26...	This markups is approved.	Floor Plan	Approved set by dmaini on 26/01/2020 at 2:53:26 PM	
Project Manager's Highlight	26/01/2020 2:53:26...	This markups is approved.	Floor Plan	Approved set by dmaini on 26/01/2020 at 2:53:26 PM	
Project Manager's Highlight	26/01/2020 2:53:26...	This markups is approved.	Floor Plan	Approved set by dmaini on 26/01/2020 at 2:53:26 PM	
Project Manager's Highlight	26/01/2020 2:53:26...	This markups is approved.	Floor Plan	Approved set by dmaini on 26/01/2020 at 2:53:26 PM	

***Figure 45** The custom status applied to the highlight markups*

22. In the **Markups List**, select the second **Project Manager's Cloud+** markup that talks about changing the area to the staff recreational area.

23. Right-click on the selected markup and from the shortcut menu, select **Set Status > Company ABC Review > Rejected**; the status is applied to the markup in the **Markups List** and the **Comments** column shows the comment associated with this status. Also, notice that the color of this markup is changed to Red in the PDF file.

24. Navigate to this markup in the **Main Workspace**; notice that the text is changed to reflect the comment you added to the custom status.

    As mentioned earlier, if you do not want the text in the markup to change, you need to make sure you do not add any text in the **Text** field while creating the custom state for the status.

25. Save the current PDF file.

Section 5: Creating the Markup Summary

In this section, you will create the markup summary in the PDF format. You will append the summary pages to the current PDF file and create hyperlinks between the summary pages and the original PDF page.

1. From the toolbar at the top in the **Markups List**, select the **Summary** button enclosed in the rectangle in Figure 46; a shortcut menu is displayed, as shown in Figure 46.

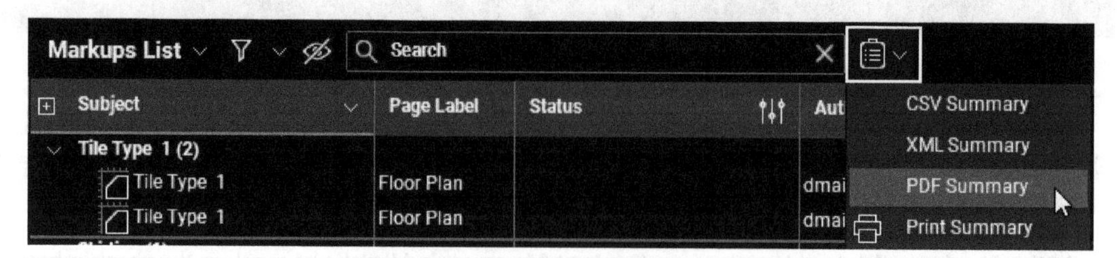

***Figure 46** Creating a markup summary*

2. From the shortcut menu, click **PDF Summary**, as shown in Figure 46; the **Markup Summary** dialog box is displayed.

The markup summary you need to create should not include any of the measurements. So you need to filter out those measurements in the **Filter and Sort** tab of the **Markup Summary** dialog box.

3. Activate the **Filter and Sort** tab in the **Markup Summary** dialog box.

4. From the **Subject** drop-down list, deselect the **Beam**, **Carpet Type 1**, **Skirting**, and **Tile Type 1** subjects and make sure the rest of them are selected, as shown in Figure 47.

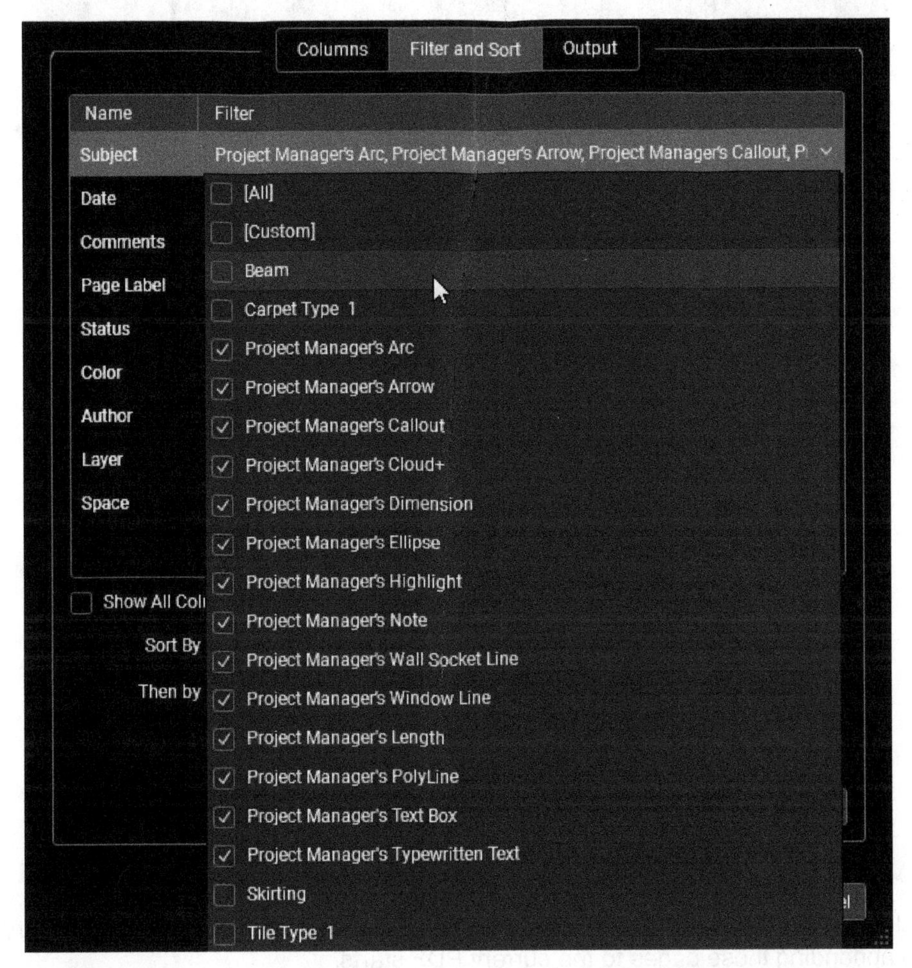

Figure 47 Deselecting the subject types to be excluded from the summary report

5. Activate the **Output** tab in the dialog box.

6. From the **Export as** drop-down list, make sure **PDF** is selected.

7. Select the **Append and Hyperlink to Current PDF** check box available below the **Export as** drop-down list.

8. From the **Style** drop-down list, make sure **Flow** is selected.

9. From the **Thumbnail** drop-down list, select **Medium.** Figure 48 shows the **Output** tab after configuring these settings.

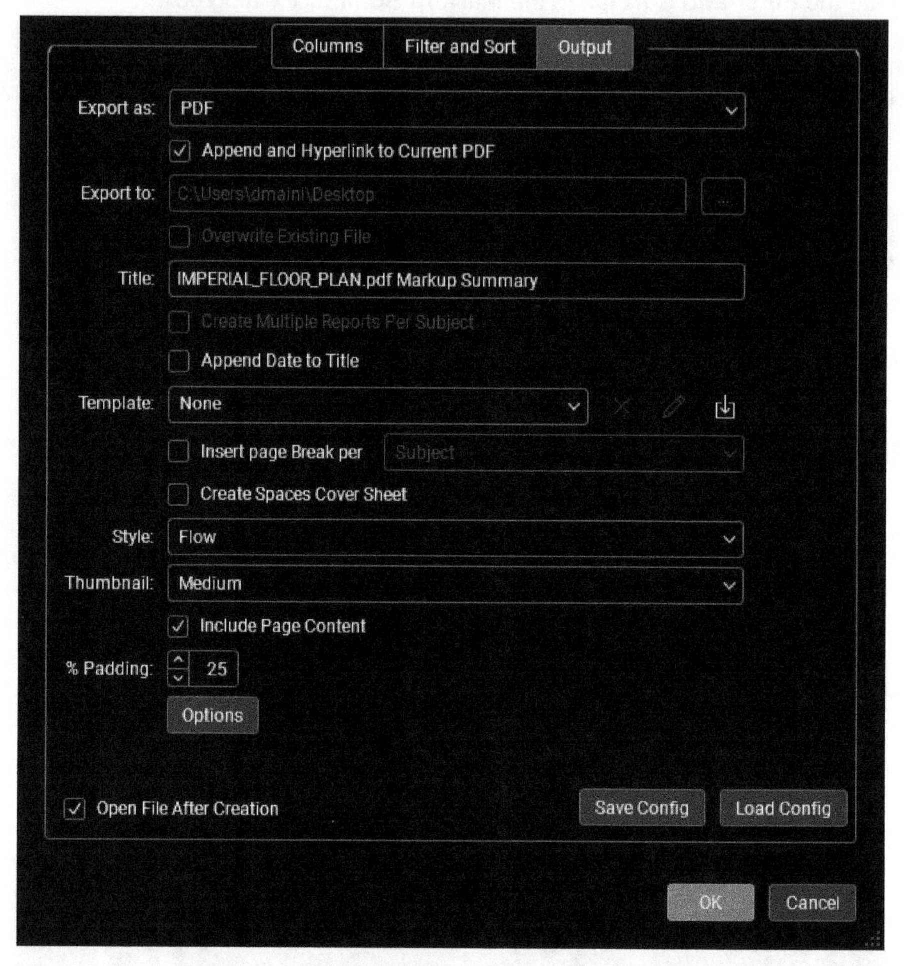

Figure 48 *Configuring the summary output settings*

10. Click **OK** in the **Markup Summary** dialog box; the process of writing the summary report and appending those pages to the current PDF starts.

    Once the summary is created, the first page of the report is displayed in the **Main Workspace**.

11. From the **Panel Bar**, invoke the **Thumbnail** panel; the summary pages are appended with the original page of the PDF, as shown in Figure 49.

12. Zoom close to one of the markups on the summary pages.

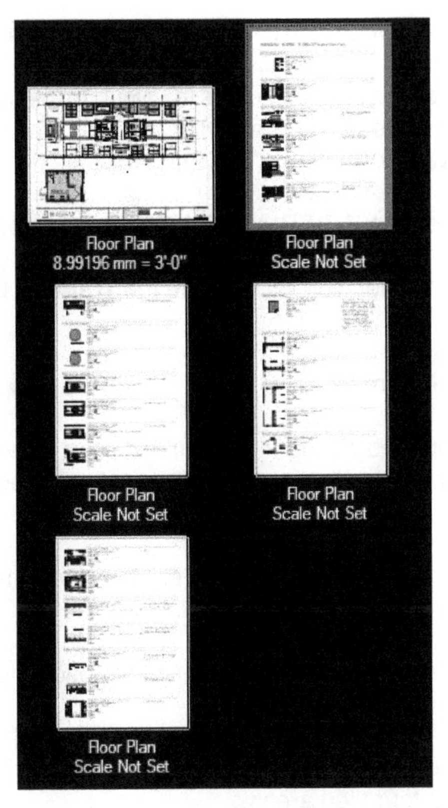

Figure 49 *The summary pages appended with the original page of the PDF*

13. Click on the thumbnail of the markup on the summary page; you are taken to that particular markup on the main page because there is a hyperlink created between the thumbnails on the summary pages and the original markup.

14. Using the **Thumbnails** panel, navigate to another summary page.

15. Click on the summary thumbnail of a markup to navigate to that markup on the original PDF file.

16. Similarly, repeat this process to review various markups from the summary pages.

17. Navigate to the original page of the PDF file.

18. Zoom to the extents of the page.

19. Save and close the PDF file.

Section 6: Creating Custom Columns to Calculate the Cost of the Area Takeoffs

In this section, you will create custom columns to calculate the cost of the various customized area takeoff tools that you created in the earlier chapters. To do that, you will first open the **IMPERIAL_TAKEOFF.pdf** or **METRIC_TAKEOFF.pdf** file from the **C08** folder.

1. From the **C08** folder, open the **IMPERIAL_TAKEOFF.pdf** or **METRIC_TAKEOFF.pdf** file.

   Before you proceed any further, you will activate the **Training QTO** profile that you created in the earlier chapters.

2. Activate the **Training QTO** profile.

   The file that you have opened has four customized area measurements to takeoff carpet areas and four customized area measurements to takeoff tiled areas. You will now review these customized measurements.

3. Navigate to the top left of the plan and notice the four area measurements customized to takeoff the four carpet types.

4. One by one click on the four area measurements and notice their subjects and labels in the **Measurements** or **Properties** panel.

5. Similarly, navigate to the right side of the plan and review the four area measurements customized to takeoff the four tiled areas.

6. Hide any panels on the left or right of the **Main Workspace**.

7. Activate the **Markups List**.

   You are now ready to create custom columns. You will create two custom columns. The first one is the choice column that will be used to select one of the eight flooring types. You will assign a numeric value to each flooring type choice that will represent the cost of per square foot or square meter of the flooring type. The second custom column is a formula column for multiplying the area measurement takeoff value with the numeric value assigned to the flooring type choices.

8. From the toolbar at the top in the **Markups List**, click the **Markups List > Columns**; the list of all the available columns is displayed, as shown in Figure 50.

9. From the bottom of this list, select **Manage Columns**; the **Manage Columns** dialog box is displayed.

10. In the **Manage Columns** dialog box, activate the **Custom Columns** tab.

11. From the bottom right in the **Custom Columns** area, click the **Add** button; the **Add Column** dialog box is displayed.

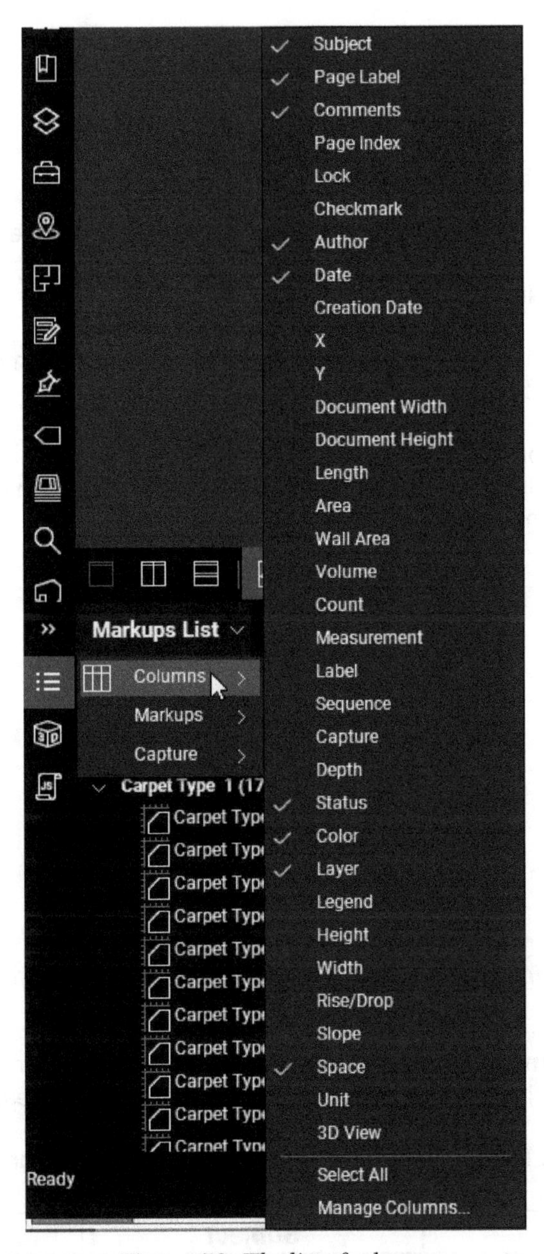

Figure 50 The list of columns

12. In the **Name** field, enter **Flooring Type** as the name of the custom column.

13. From the **Type** drop-down list, select **Choice**; the parameters in the **Options** area of this dialog box are adjusted to allow you to create multiple choices for the flooring type.

14. Click on the **Add** button on the lower right of the **Options** area; the **Manage Choice Item** dialog box is displayed.

To make sure you are able to map this flooring type to the custom tool, you need to make sure you enter the label of the customized flooring tool in the **Item** field and the subject in the **Subject** field.

15. In the **Item** field, type **Carpet 1**.

16. In the **Subject** field, type **Carpet Type 1**. Be careful not to add any extra spaces in the name.

17. Select the **Assign Numeric Value** check box.

    The numeric value that you assign here represents the cost of the per square foot or per square meter of the flooring type.

18. In the numeric field on the right, enter **3.75** as the value for the Imperial file and **37.75** as the value for the Metric file. Figure 51 shows the **Manage Choice Item** dialog box after creating this choice for the Imperial file.

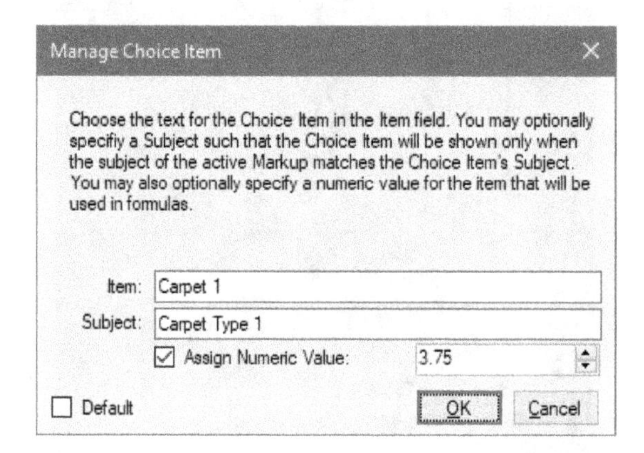

Figure 51 *The choice item created for the Imperial file*

19. Click **OK** in the **Manage Choice Item** dialog box; you are returned to the **Add Column** dialog box and the choice you created is listed in the **Options** area.

20. Similarly, create the following choices. Note the numeric values for Imperial and Metric.

Item	Subject	Numeric Value
Carpet 2	Carpet Type 2	4.5 (I) or 45.50 (M)
Carpet 3	Carpet Type 3	4.8 (I) or 48.75 (M)
Carpet 4	Carpet Type 4	5.5 (I) or 55.85 (M)
Tile 1	Tile Type 1	6.5 (I) or 65.50 (M)
Tile 2	Tile Type 2	7.75 (I) or 70.75 (M)
Tile 3	Tile Type 3	8.5 (I) or 85.45 (M)
Tile 4	Tile Type 4	10.35 (I) or 100.35 (M)

Figure 52 shows the **Add Column** dialog box with all these choices created for the Imperial file.

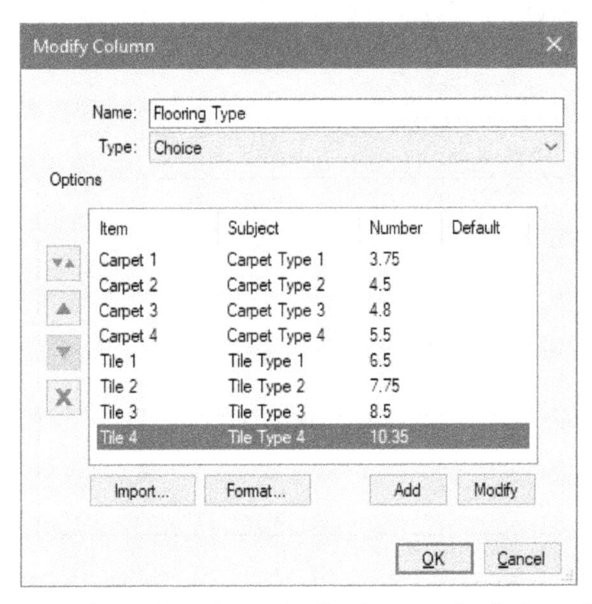

Figure 52 *The various choices for flooring type for the Imperial file*

21. Click **OK** in the **Add Column** dialog box; you are returned to the **Manage Columns** dialog box with the custom column you created listed in the **Custom Columns** tab.

    Next, you will create a custom column to calculate the total cost of each flooring type.

22. On the lower right in the **Custom Columns** area of the **Manage Columns** dialog box, click the **Add** button; the **Add Column** dialog box is displayed.

23. In the **Name** field, type **Flooring Cost** as the name of the custom column.

24. From the **Type** drop-down list, select **Formula**; the parameters in the **Options** area are modified to allow you to create the formula based on the expression.

    The **Expression** field is not a drop-down list. This means that there is nothing to select from in this field. However, as soon as you start typing, this field will display the parameters to select based on what you type. In your case, you need to create an expression to multiply the measurement with the **Flooring Type** column you created earlier. To select these two parameters, you can simply type the first letter and the **Expression** field will be populated with the parameters that you can select.

25. In the **Expression** field, type **M**; a drop-down list of parameters is displayed, as shown in Figure 53.

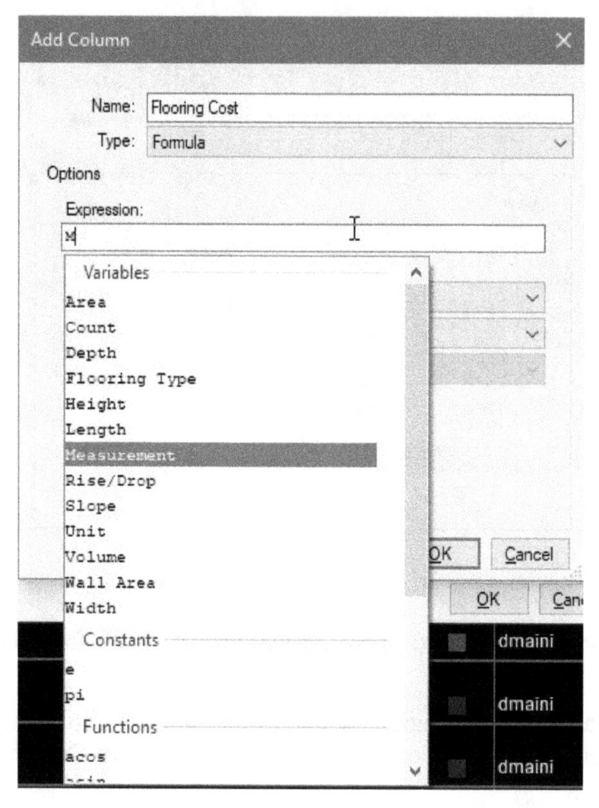

Figure 53 Selecting the parameters to create the expression

26. Double-click on the **Measurement** variable; it is listed in the **Expression** field.

27. Type * in the **Expression** field for multiplication symbol.

    Next, you need to select the **Flooring Type** column. For that, you can simply type **F** in the **Expression** field and then select this column name from the list displayed in this field.

28. In the **Expression** field, type **F**; a drop-down list is displayed with various parameters that you can select.

29. Double-click on the **Flooring Type** parameter; it is listed in the **Expression** field.

    Because this column represents a monetary value, you will change the format of this column to currency.

30. From the **Format** drop-down list, select **Currency**.

    You will use the default values from the **Decimal Places** and **Currency Symbol** lists. Figure 54 shows the **Add Column** dialog box after configuring the parameters for this column

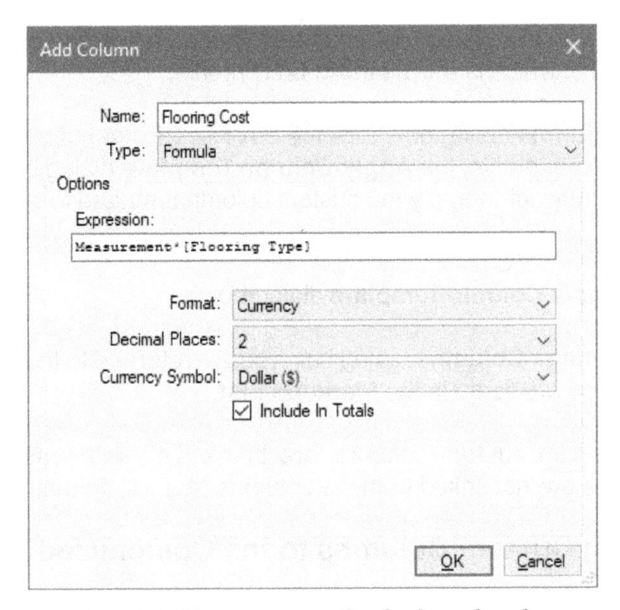

Figure 54 The parameters for the formula column

31. Click **OK** in the **Add Column** dialog box; you are returned to the **Custom Columns** dialog box with the two custom columns listed in the **Custom Columns** tab, as shown in Figure 55.

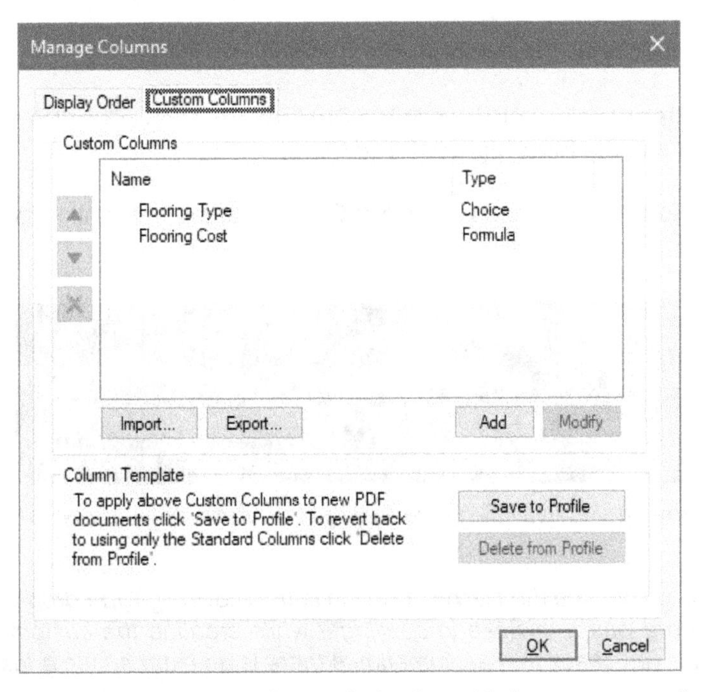

*Figure 55 The **Manage Columns** dialog box with the two custom columns*

Before you click **OK** in this dialog box, you need to ensure you save these custom columns to the current profile, which is the **Training QTO** profile.

32. In the **Manage Columns** dialog box, click the **Save to Profile** button to save these custom columns to the current profile; the **Apply Column Template** dialog box is displayed asking you to confirm if you want to apply the custom column template to the current PDF and all new PDFs.

33. Click **OK** in the **Apply Column Template** dialog box.

34. Click **OK** in the **Manage Columns** dialog box; you are returned to the Revu window and the two custom columns are listed in the **Markups List**.

    At this stage, the two custom columns are blank. This is because the eight custom measurement tools are not linked to these columns. You will do that in the next section.

Section 7: Linking Custom Columns to the Customized Measurement Tools

In this section, you will link the custom columns created in the previous section to the customized area measurement tools in the current PDF file. You will then add these tools to the custom toolset you created in the earlier chapters.

1. Navigate to the top left of the plan where the four carpet measurements are created.

2. Select the **Carpet Type 1** measurement.

3. In the **Properties** panel, scroll down to the **Custom** area; you will notice the **Flooring Type** drop-down list and the **Flooring Cost** field there.

4. Click on the **Flooring Type** drop-down list; **Carpet 1** is the only option available in this list, as shown in Figure 56.

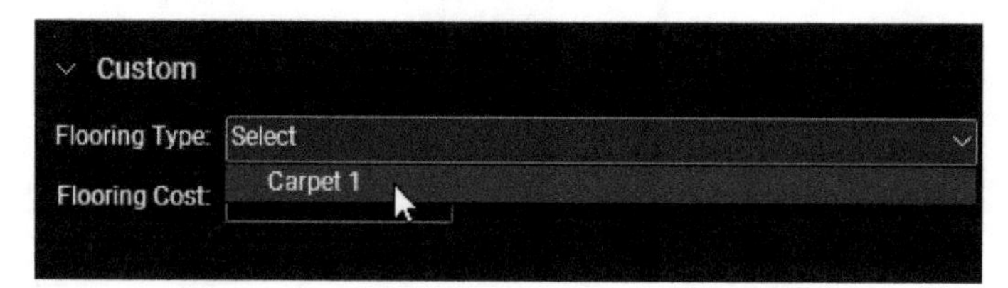

Figure 56 *Mapping the flooring type to the customized measurement tool*

***Tip**: If you do not see the **Carpet 1** option in the **Flooring Type** drop-down list then that means there was a mistake in spelling it while creating the custom column choices. Make sure you check that as sometimes there is an extra space added to the names.*

The reason you only see the **Carpet 1** option for the selected measurement is because this is automatically mapped based on the subject and label of the customized tool and the custom column.

5. From the **Flooring Type** drop-down list, select **Carpet 1**; the value in the **Flooring Cost** field is automatically adjusted to reflect the total cost of the carpet type 1 flooring. Also, notice that the **Markups List** shows these two columns automatically populated now for the selected measurement.

6. From the **Main Workspace**, select the **Carpet Type 2** measurement.

7. From the **Properties** panel > **Custom** area > **Flooring Type** drop-down list, select **Carpet 2**; the flooring cost is updated in the **Properties** panel and also in the **Markups List**.

8. Similarly, select the remaining carpet and tile types and map them to their respective flooring types in the **Properties** panel.

   Figure 57 shows the **Markups List** with the two custom columns populated for the measurement types in the Imperial file. Note that in this figure, the two custom columns are reordered to sit next to the **Comments** column.

Subject	Date	Comments	Flooring Type	Flooring Cost
Carpet Type 1 (1)				**$423.75**
Carpet Type 1	26/01/2020 5:58:11 PM	113 sf	Carpet 1 3.75	$423.75
Carpet Type 2 (1)				**$508.50**
Carpet Type 2	26/01/2020 6:02:06 PM	113 sf	Carpet 2 4.50	$508.50
Carpet Type 3 (1)				**$652.80**
Carpet Type 3	26/01/2020 6:03:04 PM	136 sf	Carpet 3 4.80	$652.80
Carpet Type 4 (1)				**$704.00**
Carpet Type 4	26/01/2020 6:03:06 PM	128 sf	Carpet 4 5.50	$704.00
Tile Type 1 (1)				**$734.50**
Tile Type 1	26/01/2020 6:03:13 PM	113 sf	Tile 1 6.50	$734.50
Tile Type 2 (1)				**$984.25**
Tile Type 2	26/01/2020 6:03:16 PM	127 sf	Tile 2 7.75	$984.25
Tile Type 3 (1)				**$960.50**
Tile Type 3	26/01/2020 6:03:19 PM	113 sf	Tile 3 8.50	$960.50
Tile Type 4 (1)				**$1,397.25**
Tile Type 4	26/01/2020 6:18:06 PM	135 sf	Tile 4 10.35	$1,397.25

*Figure 57 The **Markups List** with the custom columns populated*

You are now ready to add these customized tools to the **Project Manager Measurements** tool set that you created in the earlier chapters.

9. Add the eight customized area measurements to the **Project Manager Measurements** tool set.

10. Double-click on all the tools to change them to the **Properties** mode, as shown in Figure 58.

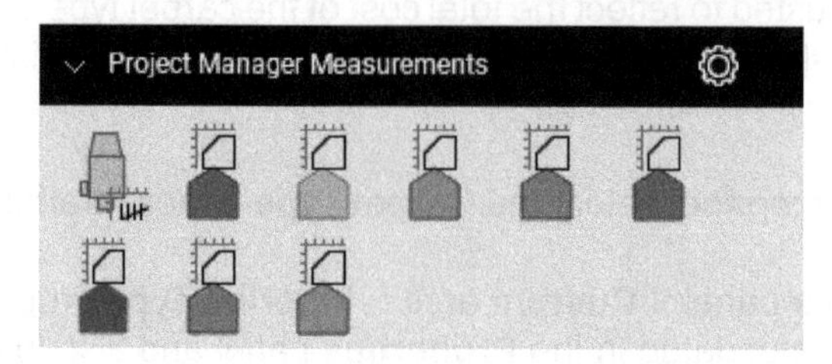

***Figure 58** The customized measurement tools added to the custom tool set*

To ensure these customized tools work the way they are supposed to, it is important to test them.

11. Using the customized tools in the **Project Manager Measurements** tool set, add an area measurement of each type on the PDF and notice how the flooring cost updates in the **Markups List**.

12. Delete all the recent area measurements.

13. Zoom to the extents of the PDF.

14. Save the current file.

Section 8: Creating a Legend Using the Tools in the Custom Tool Set

One of the main advantages of adding the takeoff tools to your custom tool set is to be able to create a legend using that tool set. In this section, you will create a legend using the tools in the **Project Manager Measurements** tool set.

1. In the **Tool Chest** panel, click on the cogwheel on the top right of the **Project Manager Measurements** tool set and select **Legend > Create New Legend**, as shown in Figure 59.

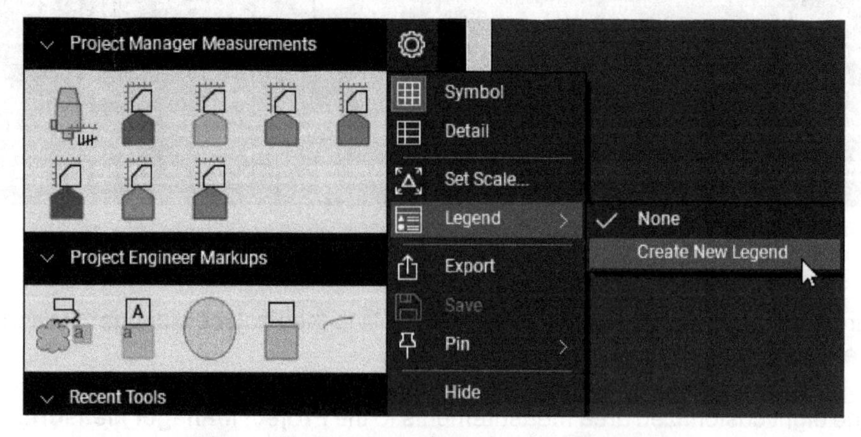

***Figure 59** Creating a legend using the tool set*

On doing so, the preview of the legend is attached to the cursor.

2. Place the legend near the bottom right of the plan; the default legend is placed. Figure 60 shows the legend in the Imperial file.

Project Manager Measurements

Description	Quantity	Unit
Carpet Type 1	113	sf
Carpet Type 2	113	sf
Carpet Type 3	136	sf
Carpet Type 4	128	sf
Tile Type 1	113	sf
Tile Type 2	127	sf
Tile Type 3	113	sf
Tile Type 4	135	sf

Figure 60 *The default legend in the Imperial file*

You will now customize this legend by changing its title. You will then include the **Flooring Cost** column and also displaying the grid lines of the legend.

3. With the legend still selected, display the **Properties** panel.

4. In the **Properties** panel > **Title** area > **Title** field, enter **FLOORING TAKEOFFS**; the title of the legend is changed.

5. In the **Properties** panel > **Columns** area, click on **Edit Columns**; the **Edit Legend Columns** dialog box is displayed.

6. Scroll down in the dialog box and select the **Flooring Cost** check box.

7. Click **OK** in the **Edit Legend Columns** dialog box; the **Flooring Cost** column is displayed in the legend and the cost of each flooring type is displayed.

8. In the **Properties** panel > **Table** area, change the color to **Blue**.

9. Change the **Line Width** to **0.5**.

10. From the **Table Style** drop-down list, select **Gridlines**; the grid lines of the legend are displayed.

11. In the **Symbol Size** spinner, enter **200** as the value; the legend table size is increased.

12. If required, move the legend back to the bottom right of the plan. Figure 61 shows the legend in the Imperial file and Figure 62 shows the legend in the Metric file.

FLOORING TAKEOFFS				
	Description	Quantity	Unit	Flooring Cost
	Carpet Type 1	113	sf	$423.75
	Carpet Type 2	113	sf	$508.50
	Carpet Type 3	136	sf	$652.80
	Carpet Type 4	128	sf	$704.00
	Tile Type 1	113	sf	$734.50
	Tile Type 2	127	sf	$984.25
	Tile Type 3	113	sf	$960.50
	Tile Type 4	135	sf	$1,397.25

Figure 61 *The legend in the Imperial file*

FLOORING TAKEOFFS				
	Description	Quantity	Unit	Flooring Cost
	Carpet Type 1	11	sq m	$415.25
	Carpet Type 2	13	sq m	$591.50
	Carpet Type 3	11	sq m	$536.25
	Carpet Type 4	12	sq m	$670.20
	Tile Type 1	11	sq m	$720.50
	Tile Type 2	12	sq m	$849.00
	Tile Type 3	11	sq m	$939.95
	Tile Type 4	13	sq m	$1,304.55

Figure 61 *The legend in the Metric file*

13. Add the legend to the **Project Manager Measurements** tool set.

14. Save and close the current file.

Skill Evaluation

Evaluate your skills to see how many questions you can answer correctly. The answers to these questions are given at the end of the book.

1. The **Markups List** is a tabular representation of all markups, including dimensions and stamps, added to all pages of the current PDF file. (True/False)

2. The **Markups List** can only be displayed at the bottom of the **Main Workspace**. (True/False)

3. The **Markups List** allows you to sort and filter markups. (True/False)

4. You can create custom columns using the **Markups List**. (True/False)

5. The columns in the **Markups List** cannot be reordered. (True/False)

6. Which one of the following is not a format to generate the summary report?

   (A) Word (B) PDF
   (C) CSV (D) XML

7. While generating the summary report, which one of the following formats allows you to append and hyperlink summary pages to the original pages of the PDF?

   (A) None (B) PDF
   (C) CSV (D) XML

8. Which tool on the toolbar available on the top in the **Markups List** allows you to hide all markups?

   (A) **Close** (B) **Turn Off**
   (C) **No Markups** (D) **Hide Markups**

9. Which one is not a type of custom column that can be created?

   (A) Notes (B) Checkmark
   (C) Choice (D) Formula

10. Which type of custom column is used to create a multiple-choice column?

    (A) Formula (B) Date
    (C) Text (D) Choice

Class Test Questions

Answer the following questions:

1. Explain briefly the process of creating custom filters.
2. Explain briefly how to hide all the markups on the PDF file.
3. Explain the process of creating a custom column.
4. What is the procedure for reordering the columns?
5. Explain briefly the process of linking a custom measurement tool to a custom column.

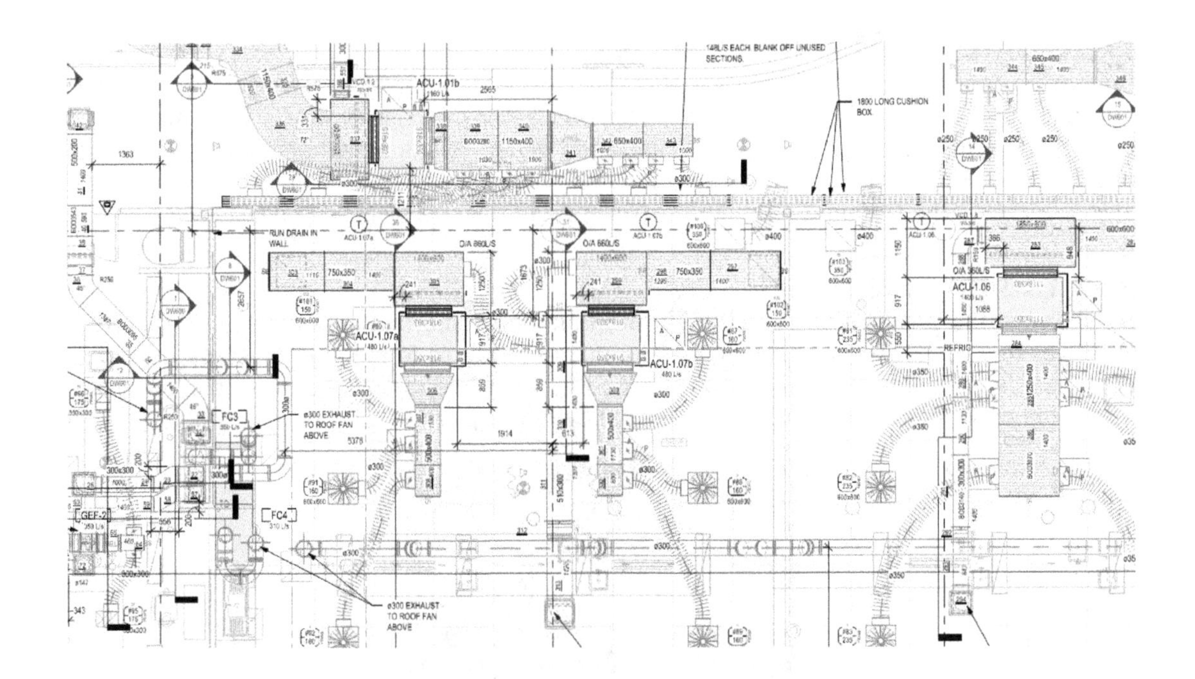

Chapter 9 - Document Management and Hyperlinks

The objectives of this chapter are to:

- √ *Explain about the **Thumbnails** panel*
- √ *Explain how to insert blank pages in the PDF*
- √ *Explain the process of deleting and reordering pages from a PDF file*
- √ *Explain how to insert pages from an existing PDF file*
- √ *Explain how to extract pages of an existing PDF file into individual PDFs*
- √ *Explain the taking snapshots of a part of PDF and paste that snapshot*
- √ *Explain how to create various types of hyperlinks in the PDF*

DOCUMENT MANAGEMENT

Performing document management operations such as adding pages, removing pages, reordering pages, and so on has been an extremely tedious task in the past. However, Revu makes this process extremely easy by using the **Thumbnails** panel. This is an extremely versatile panel and provides you all the document management tools in one single location. These tools are also available in the **Document** menu. The **Thumbnail** panel is discussed next.

The Thumbnails Panel

As shown in the previous chapter, the **Thumbnails** panel shows all the pages of the current PDF file. This panel also provides you with a number of tools to manage these pages of the document. These tools are available by clicking the **Thumbnails** option at the top in the panel or by right-clicking anywhere in this panel, as shown in Figure 1.

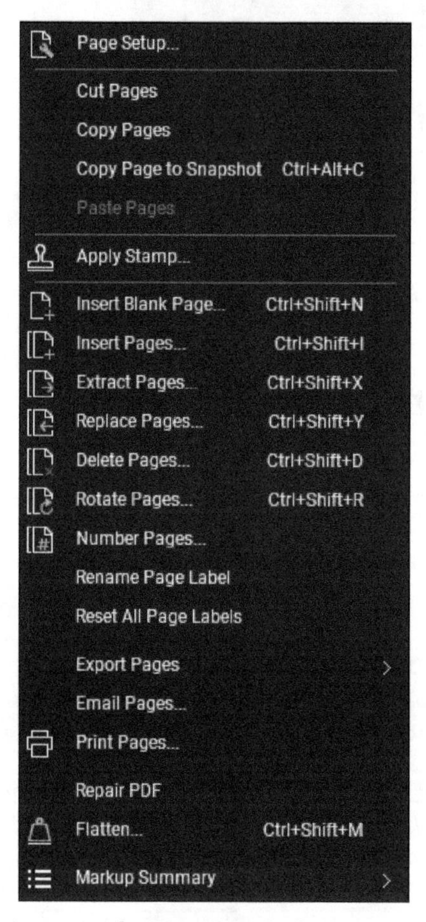

*Figure 1 The **Thumbnails** shortcut menu*

The procedures for managing documents using various tools available in the **Thumbnails** shortcut menu are discussed next.

The Procedure for Reordering Pages of a PDF File

Reordering pages of a PDF file is one of the most commonly used functionality. This can be easily done using the **Thumbnails** panel. The following is the procedure for doing this:

1. From the **Panel Access Bar**, turn on the **Thumbnails** panel, as shown in Figure 2.

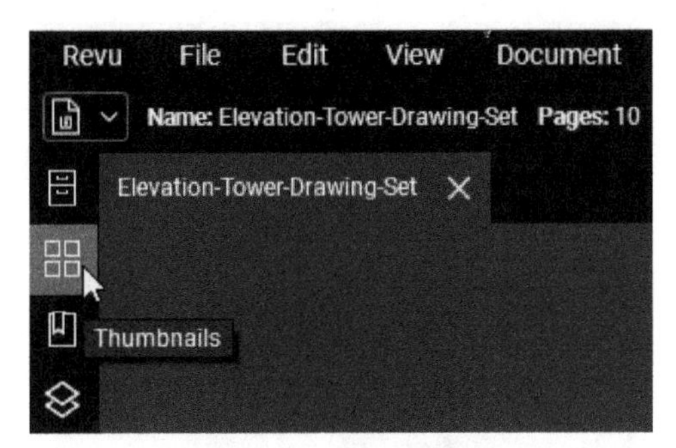

*Figure 2 Turning the **Thumbnails** panel on*

2. In the **Thumbnails** panel, hold down the SHIFT or CTRL key and select the pages you want to reorder.

3. Drag and drop these pages before or after any other pages of the document. Figure 3 shows the **GA No: 1**, **GA No: 2**, and the **GA No: 3** pages of the PDF being placed before the **DWG No: 100** page.

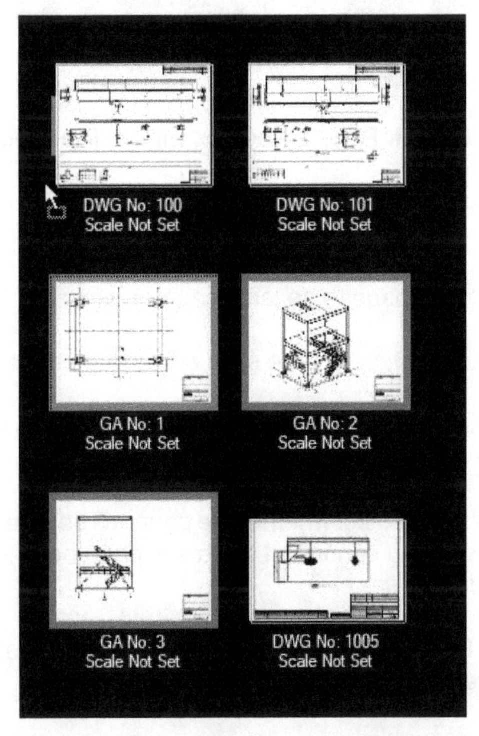

Figure 3 Reordering the pages of a PDF file

The Procedure for Inserting a Blank Page in the PDF File

The following is the procedure for inserting a blank page in the PDF:

1. Right-click on any page of the PDF in the **Thumbnails** panel and select **Insert Blank Page** from the shortcut menu, as shown in Figure 4.

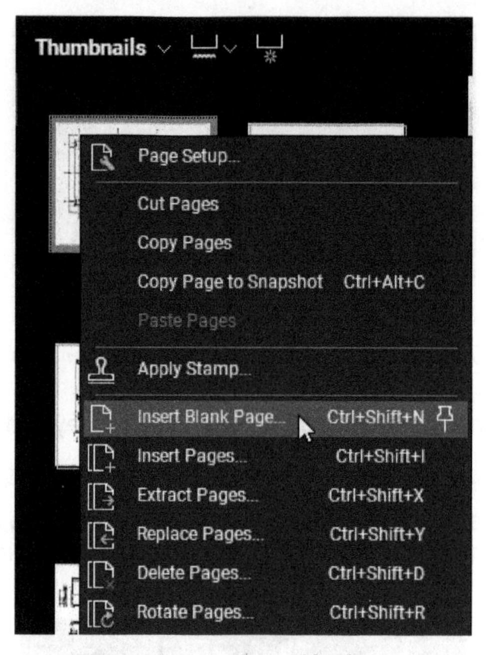

Figure 4 *Inserting a blank page*

On doing so, the **Insert Blank Page** dialog box will be displayed, as shown in Figure 5.

2. From the **Template** drop-down list, select the size of the page you want to insert. Leaving **Custom** in this drop-down list allows you to manually enter the width and height of the page.

3. Select the orientation of the page to be landscape or portrait.

4. From the **Style** drop-down list, select the style of the page to be inserted.

5. In the **Page Count** spinner, select the number of blank pages to be inserted. By default, one blank page will be inserted.

6. From the **Into XXX.pdf** area, select whether the page will be inserted before or after the first or last page. Alternatively, you can insert the blank page before or after the page number entered in the **Page** edit box.

7. Click **OK** in the **Insert Blank Page** dialog box; the blank page(s) will be inserted before or after the specified page.

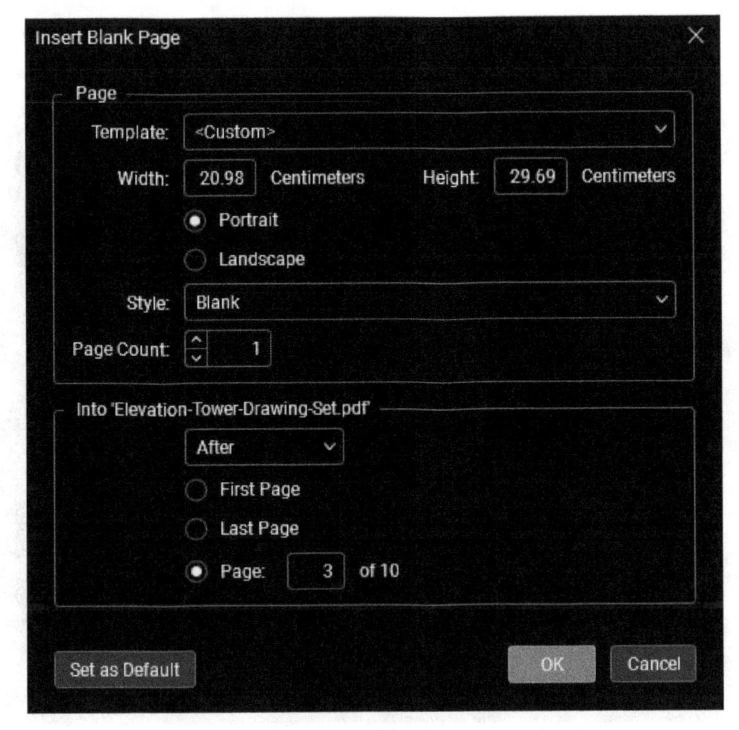

*Figure 5 The **Insert Blank Page** dialog box*

The Procedure for Inserting Pages from an Existing PDF File

The following is the procedure for inserting pages from an existing PDF:

1. Right-click on any page of the PDF in the **Thumbnails** panel and select **Insert Pages** from the shortcut menu, as shown in Figure 6.

Figure 6 Inserting pages from an existing PDF file

On doing so, the **Select Files to Insert** dialog box will be displayed, as shown in Figure 7.

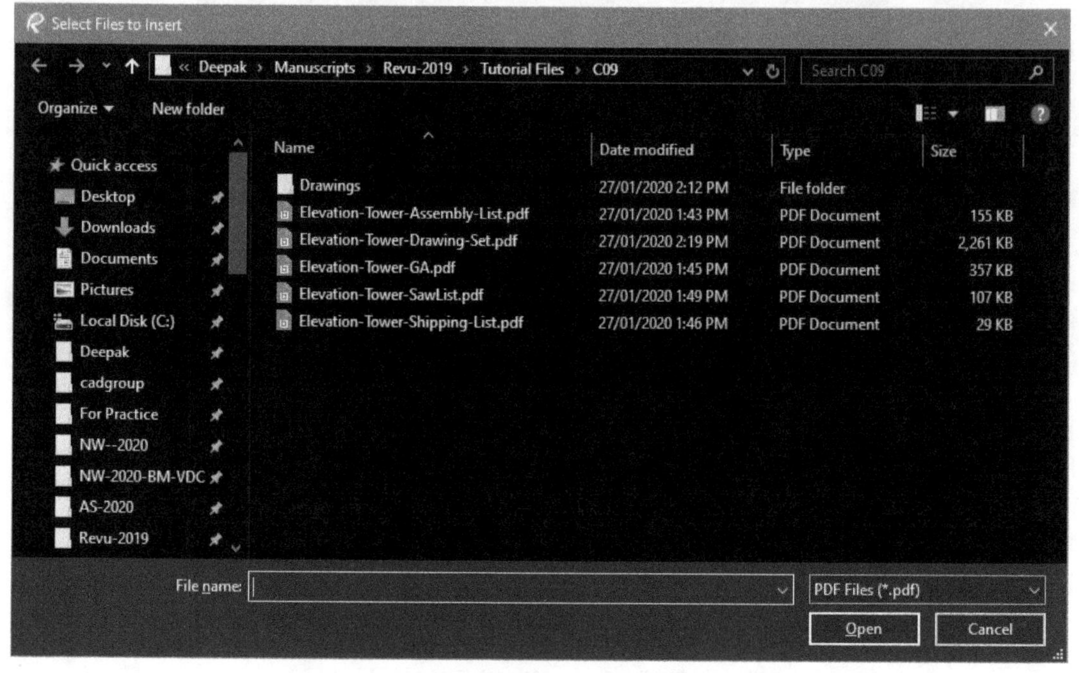

*Figure 7 The **Select Files to Insert** dialog box*

2. Browse and select the files you want to insert pages from.

3. Click **Open** in the **Select Files to Insert** dialog box; the **Insert Pages** dialog box will be displayed, as shown in Figure 8.

4. In the **Insert Files** area, select the PDFs and reorder them using the up and down arrow buttons available on the left side of this area.

5. In the same area, double-click on the **Pages** field of the selected files to display the **Select Pages** dialog box.

6. In **Select Pages** dialog box, select the pages from the selected PDF file to insert.

7. In the **Options** area of the **Insert Pages** dialog box, select the check box of the option you want to turn on.

8. In the **Into XXX.pdf** area, select the option to insert the pages before or after the specified page.

9. Click **OK** in the dialog box; the selected pages of the PDF files will be inserted in the current PDF file.

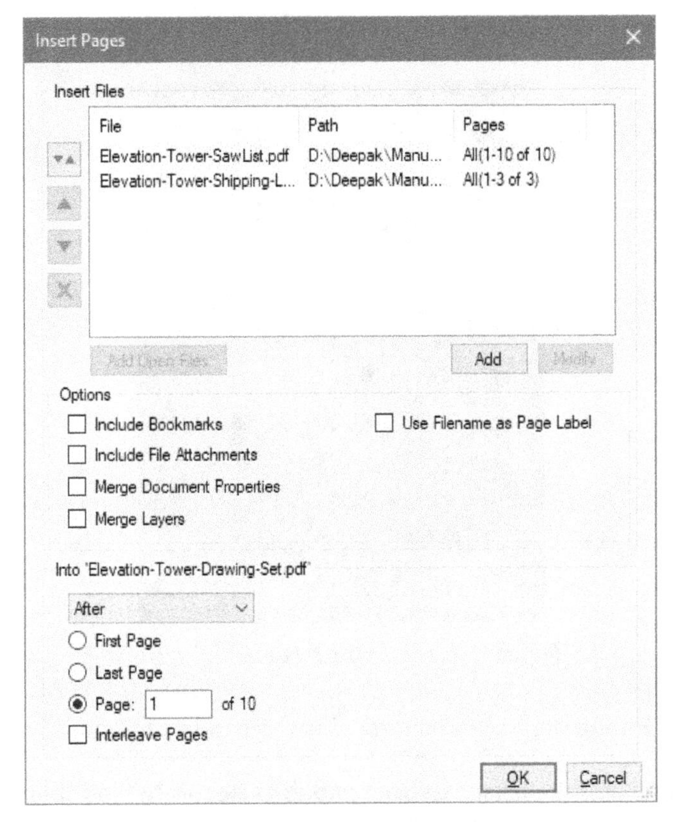

*Figure 8 The **Insert Pages** dialog box*

The Procedure for Extracting Pages of the Current PDF File

The following is the procedure for extracting pages of the current PDF file as individual files:

1. Right-click on any page of the PDF in the **Thumbnails** panel and select **Extract Pages** from the shortcut menu, as shown in Figure 9.

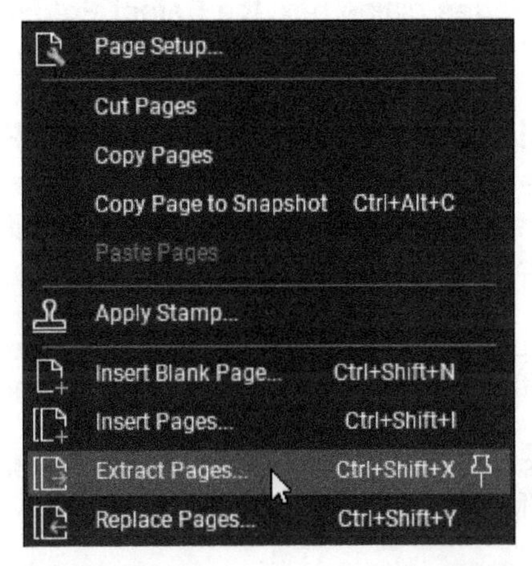

Figure 9 Extracting pages of the current PDF file

On doing so, the **Extract Pages** dialog box will be displayed, as shown in Figure 10.

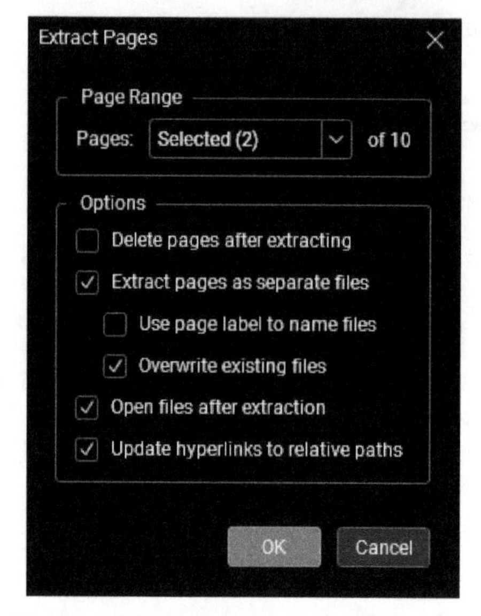

Figure 10 *The* ***Extract Pages*** *dialog box*

2. From the **Page Range** area > **Pages** drop-down list, select the pages to be extracted.

3. From the **Options** area, select the **Extract pages as separate files** check box; the additional two check boxes are enabled below this check box.

4. Select the **Use page labels to name files** check box. This will ensure the new PDF files have the same name as the page labels.

5. Select the **Open files after extraction** check box to open the extracted files. Note that selecting this check box is not a good idea if you are extracting a large number of pages.

6. Click **OK** in the **Extract Pages** dialog box; the **Select Folder** dialog box will be displayed that allows you to select the folder in which you want to extract the PDF files.

The Procedure for Replacing Pages of the Current PDF File with the Pages of an Existing PDF File

The following is the procedure for replacing the selected pages of the current PDF file with the pages of an existing PDF file:

1. Right-click on any page of the PDF in the **Thumbnails** panel and select **Replace Pages** from the shortcut menu; the **Open** dialog box will be displayed.

2. Open the PDF file whose pages you want to select; the **Replace Pages** dialog box will be displayed, as shown in Figure 11.

3. In the upper area of the **Replace Pages** dialog box, select the pages to be replaced.

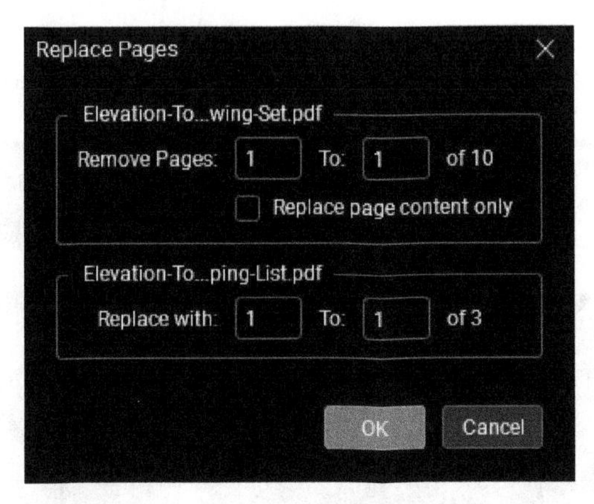

*Figure 11 The **Replace Pages** dialog box*

4. Select the **Replace page content only** check box to only replace the content of the pages and not the actual page.

5. In the lower area of the dialog box, select the pages to be replaced with.

6. Click **OK** in the dialog box; the selected pages of the current PDF file will be replaced with the selected pages of the other PDF file.

The Procedure for Deleting Pages of the Current PDF File

The following is the procedure for deleting the selected pages of the current PDF file:

1. Right-click on any page of the PDF in the **Thumbnails** panel and select **Delete Pages** from the shortcut menu; the **Delete Pages** dialog box will be displayed, as shown in Figure 12.

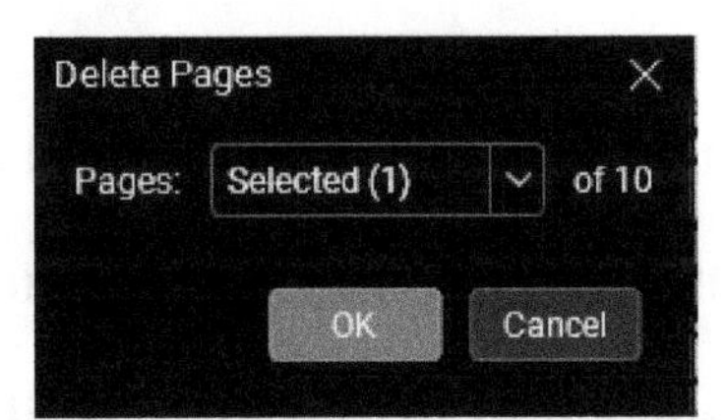

*Figure 12 The **Delete Pages** dialog box*

2. Select the pages you want to delete or create a custom range of pages to be deleted.

3. Click **OK** in the **Delete Pages** dialog box; the specified pages will be deleted from the current PDF file.

The Procedure for Rotating Pages of the Current PDF File

A number of times when the PDF files are created, the pages in those files are not oriented correctly. Revu allows you to easily rotate the selected pages of the current PDF file. The procedure for doing that is discussed next.

1. Right-click on any page of the PDF in the **Thumbnails** panel and select **Rotate Pages** from the shortcut menu; the **Rotate Pages** dialog box will be displayed, as shown in Figure 13.

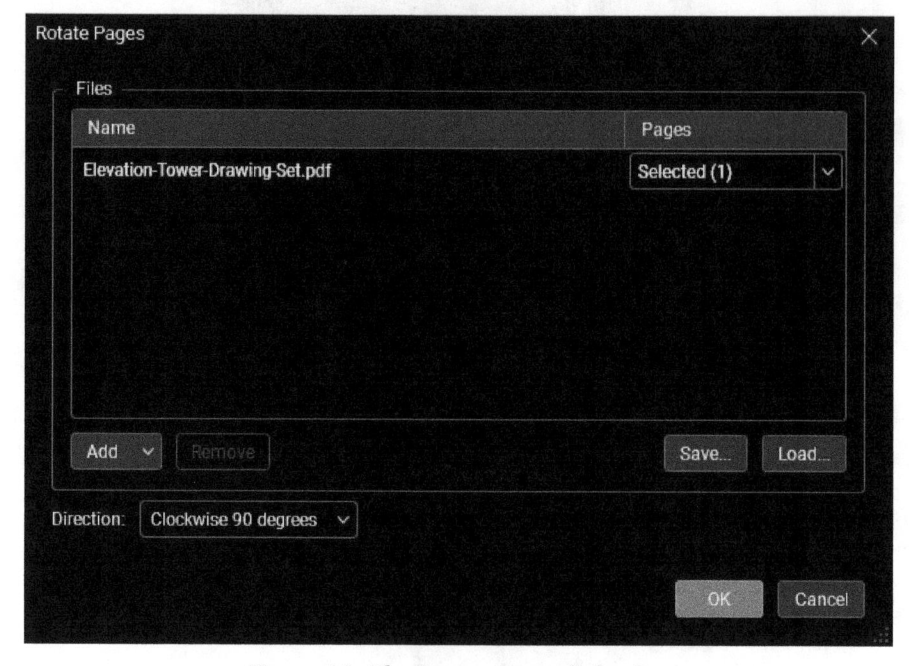

Figure 13 The ***Rotate Pages*** *dialog box*

2. In the **Pages** field, select the pages to be rotated or create a custom range.

3. If required, click on the **Add** button on the bottom left of the **Files** area to add more files to rotate pages.

4. In the **Direction** list, select the direction in which you want to rotate the selected pages.

5. Click **OK** in the dialog box; the selected pages will be rotated in the specified direction.

The Procedure for Manually Numbering and Labeling Pages of the Current PDF File

The following is the procedure for manually numbering and labeling pages of the current PDF file:

1. Right-click on any page of the PDF in the **Thumbnails** panel and select **Number Pages** from the shortcut menu; the **Page Numbering and Labeling** dialog box will be displayed, as shown in Figure 14.

2. In the **Numbering** area > **Style** drop-down list, select the style of numbering to be used.

3. In the **Prefix** field, type the prefix to be used for the numbering and labeling.

4. In the **Start** spinner, enter the start number to be used.

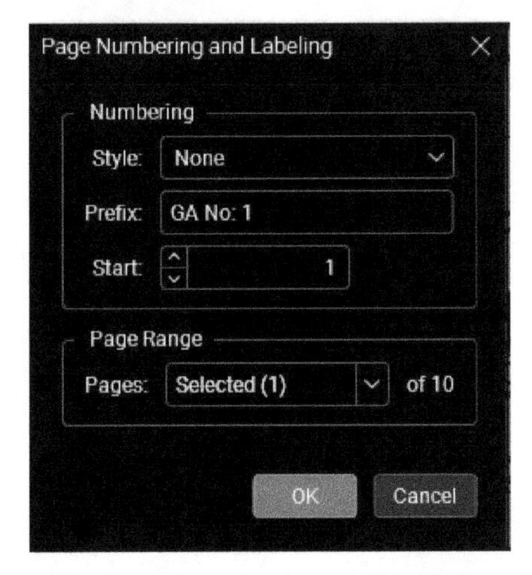

*Figure 14 The **Page Numbering and Labeling** dialog box*

5. In the **Page Range** area > **Pages** drop-down list, select the pages to be numbered and labeled. Alternatively, you can create a custom page range.

6. Click **OK** in the dialog box; the pages are numbered and labeled.

The Procedure for Automatic Extraction of Page Labels of the Current PDF File

Revu allows you to automatically extract the page labels from an existing area of the current PDF file. This is generally used when you want the information available in the titleblock of a drawing to be used as the page label. The following is the procedure for doing this:

1. Click on the **Create Page Labels** button available at the top in the **Thumbnails** panel, as shown in Figure 15.

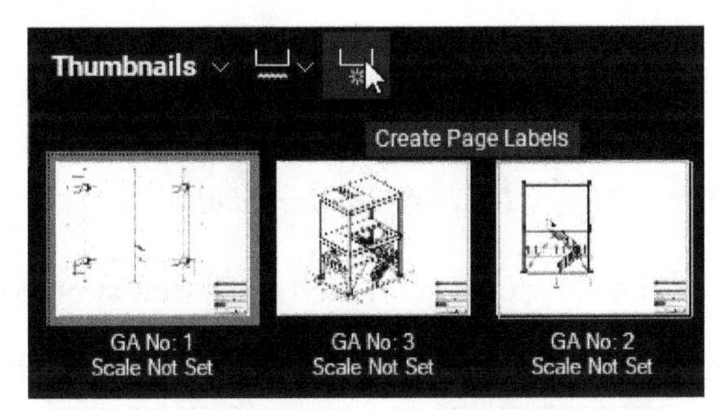

Figure 15 Selecting the tool to create page labels

On doing so, the **Create Page Labels** dialog box will be displayed, as shown in Figure 16.

2. If you have bookmarks created in the current PDF file, you can select the **Bookmarks** radio button from the **Options** area.

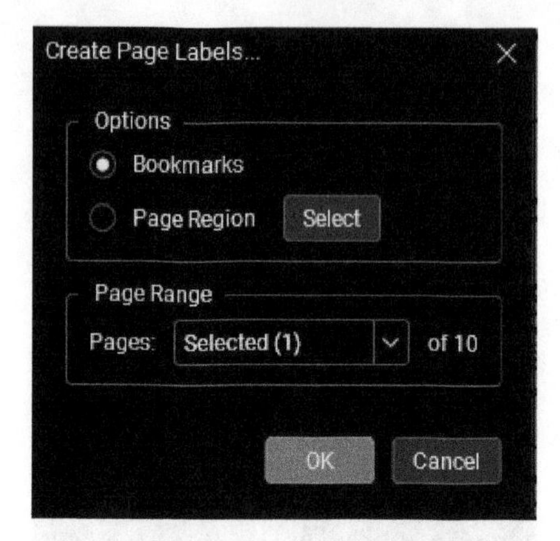

Figure 16 The ***Create Page Labels*** *dialog box*

3. To extract the page label information from the titleblock, select the **Page Region** radio button from the **Options** area.

4. Click the **Select** button on the right of the **Page Region** radio button; you will be returned to the PDF file.

5. Create a box around the region from where you want to extract the page label information; the **AutoMark** dialog box will be displayed, as shown in Figure 17.

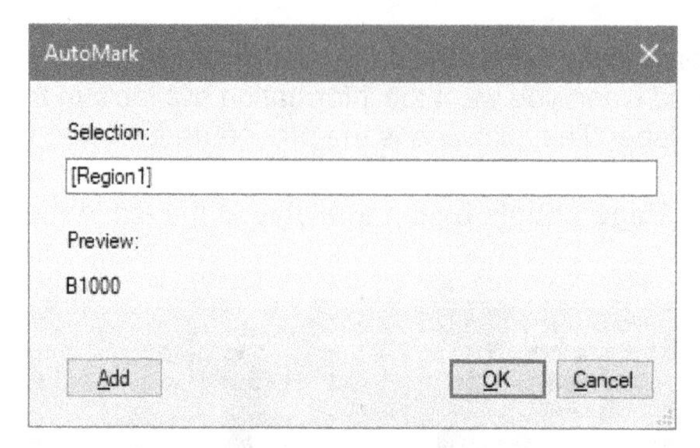

Figure 17 The ***AutoMark*** *dialog box*

6. To add a prefix to the page label, click at the start of the **Selection** field and enter the prefix.

7. To extract additional content from the titleblock to be included in the page label, click the **Add** button and drag a box around the new region to extract the information.

***Tip**: If the current drawing set has different size titleblocks, then it is recommended to use this method to apply the page labels to only those pages that match the current titleblock. You can repeat this process to create the page labels for the pages that have a different size titleblock.*

8. Click **OK** in the **AutoMark** dialog box to return to the **Create Page Labels** dialog box.

9. From the **Page Range** area, select the pages to apply the page labels to. Alternatively, you can create a custom page range to apply the page label to.

10. Click **OK** in the **Create Page Labels** dialog box; the page labels will be applied to the specified number of pages.

The Procedure for Exporting Pages of the Current PDF File

Revu allows you to export pages of the current PDF file in the various image formats. Alternatively, you can export pages in Word, Excel, or PowerPoint formats. The following is the procedure for doing this:

1. Select the pages you want to export in the **Thumbnails** panel and then right-click on any page and select **Export Pages**; the cascading menu will be displayed with the various formats in which you can export these pages, as shown in Figure 18.

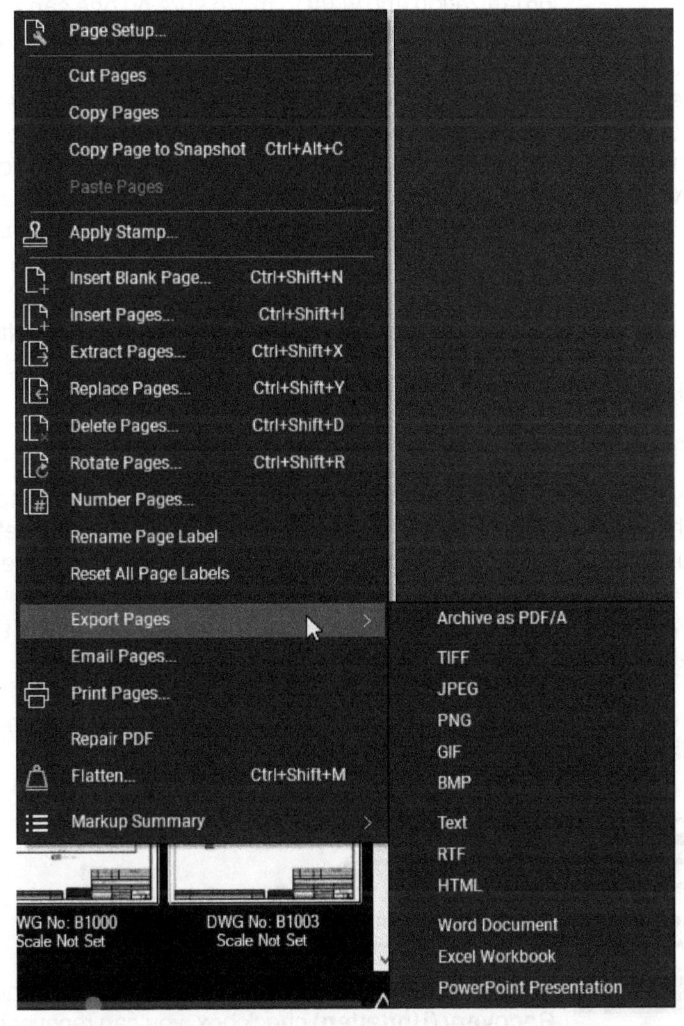

***Figure 18** Exporting pages of the current PDF file*

2. From the cascading menu, select the format in which you want to export the pages you selected in the **Thumbnails** panel; the **Save As** dialog box will be displayed.

3. Browse and select the folder in which you want the exported file to be saved.

4. Click **Save** in the **Save As** dialog box; the selected pages will be exported in the format you selected and then opened on your machine.

Flattening Markups

Menu: Document > Flatten
Thumbnail Panel: Right-click > Flatten

Whenever you want to send your marked up PDF file outside your organization and want to make sure no one can alter your markups, you are better off flattening those markups. The flattening process converts the markups into PDF content that are then protected from being edited. You can flatten the entire document so all the markups are flattened or flatten only the selected markups. To flatten the entire document, you can invoke this tool from the **Document** menu or the **Thumbnails** panel. To flatten the selected markups, you can select them and then right-click and invoke this tool. When you invoke this tool, the **Flatten Markups** dialog box will be displayed, as shown in Figure 19. The various options available in this dialog box are discussed next.

Files Area

This area shows the current document and you can select the pages to flatten. You can also create a page range to flatten. Using the **Add** button in this area, you can add more PDF files to flatten the markups.

Options Area

This area allows you to select the markup types you want to flatten. Selecting the **Type** check box at the top of this area will select all the markup types to flatten. Alternatively, you can select individual markup types to flatten. Remember that if you select the flatten the form fields, they cannot be unflattened. Selecting the **Exclude Filtered Markups** radio button will not flatten the markups that are currently filtered out. Selecting the **Allow Markup Recovery (Unflatten)** check box will allow another Revu user to unflatten your flattened markups.

Once you have selected the required options, click the **Flatten** button in the dialog box, the markups will be flattened and will not be selectable anymore.

Unflattening Flattened Markups

Menu: Document > Unflatten
Thumbnail Panel: Right-click > Unflatten

While flattening the markups, if you selected the **Allow Markup Recovery (Unflatten)** check box, you can recover those markups by unflattening them. This is done using the **Unflatten** tool. When you

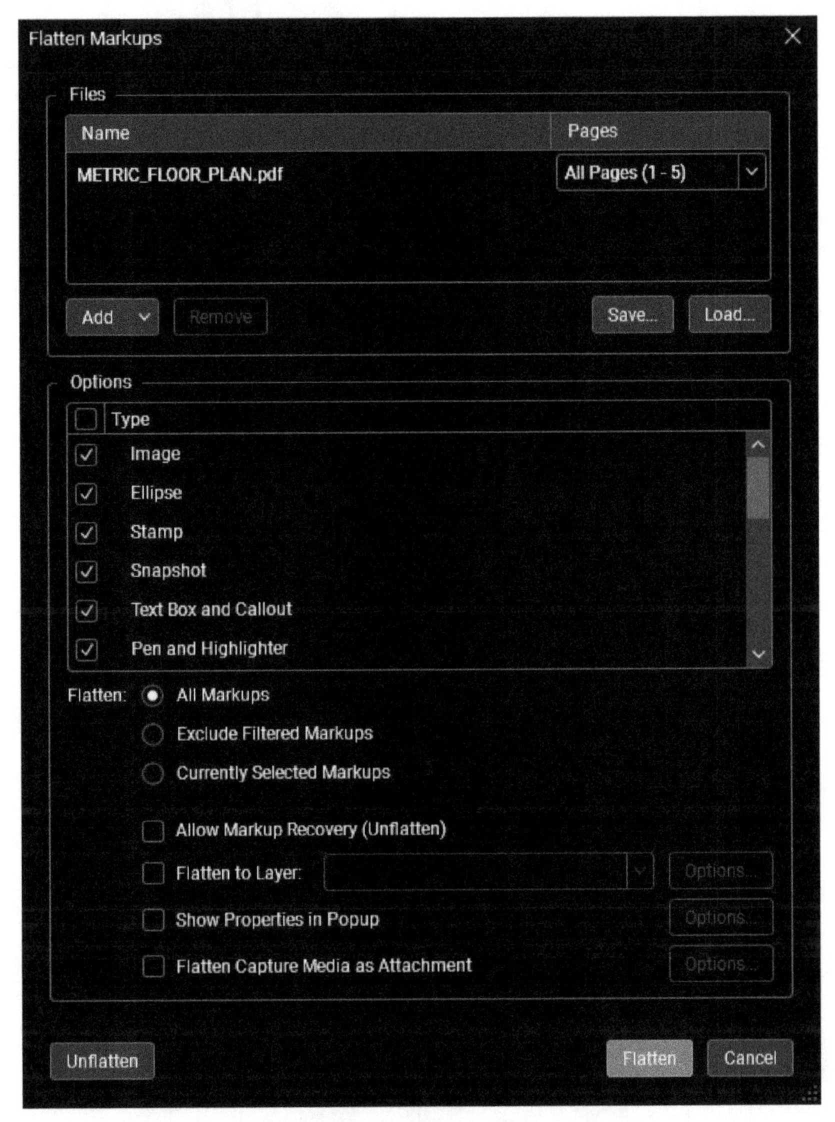

*Figure 19 The **Flatten Markups** dialog box*

invoke this tool, all the markups that can be recovered will be unflattened and become editable.

Taking Snapshots of the Sections of the PDF File

Menu: Edit > Snapshot
Shortcut: G

The **Snapshot** tool allows you to copy the content of the PDF so you can paste it at a different location. This tool maintains the vector integrity of the content that you capture and allows you to snap on to that content, provided the **Snap to Markup** option is turned on from the **Status Bar**. When you invoke this tool, you will be prompted to select a region to take the snapshot. You can drag two opposite corners to create a rectangular snapshot. Alternatively, you can select multiple points to create a multi-sided snapshot shape. Once you have specified all the vertices of the snapshot, press the ENTER key to finish the process. On doing so, the content will be copied to the clipboard and you can paste it wherever you want.

Once you have pasted the snapshot content, you can right-click on it and select the **Change Colors** option from the shortcut menu to display the **Color Processing** dialog box, as shown in Figure 20. Using this dialog box, you can change the source color of the snapshot content to any required color.

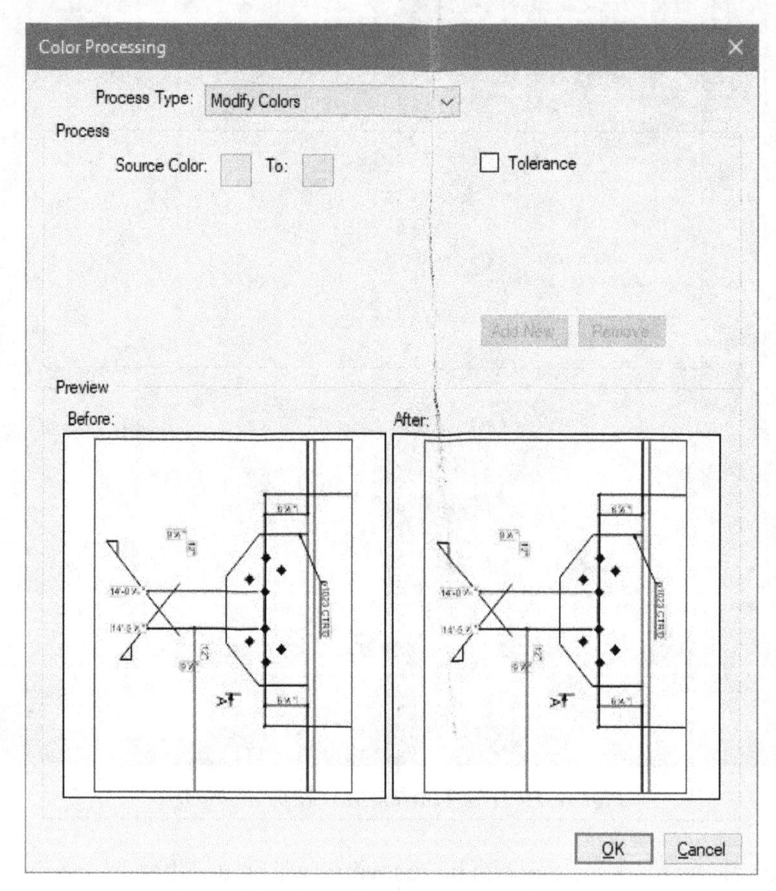

*Figure 20 The **Color Processing** dialog box*

What I do

*I normally use the **Snapshot** tool to copy a smaller region from the PDF file and paste it on a new page as a scaled-up content. That allows me to add markups about that region easily and with more details. I also create hyperlinks between the original region and the copied region on another page. You will learn more about hyperlinks later in this chapter.*

Cutting and Pasting Content of the PDF File

Menu: Edit > PDF Content > Cut Content

The **Cut Content** tool allows you to cut the content of the PDF file and make it available to be pasted somewhere else. You can cut the content using a box created by dragging two opposite corners or click points to create a multi- sided shape. Once the content is cut, you can use the **Paste** tool from the **Edit** menu or the CTRL + V key to paste the content wherever you want.

Erasing Content of the PDF File

Menu: Edit > PDF Content > Erase Content

The **Erase Content** tool allows you to permanently erase the content of the PDF file and make it available to be pasted somewhere else. Similarly to the **Cut Content** tool, this tool also lets you drag two opposite corners to create a box or click points to create a multisided shape to erase the content of the PDF file.

ADDING HYPERLINKS

Menu: Tools > Hyperlink

The **Hyperlink** tool is an extremely versatile tool used to perform various actions, such as navigating to other locations in the current document, opening another document or folder, or navigating to the web location. When you invoke this tool, you will be prompted to select a region to place a hyperlink. You can specify the region by clicking two opposite corners. Alternatively, if you want to hyperlink a text, you can simply highlight the text. Once you have specified the region or highlighted the text to hyperlink, the **Action** dialog box will be displayed. Using this dialog box, you can add the hyperlink actions mentioned earlier.

Using the Hyperlink Tool to Perform Various Actions

The following are the procedures for performing various actions using the **Hyperlink** tool.

The Procedure for Using Hyperlink to Navigate to Another Page in the Current Document

The following is the procedure for navigating to another page in the current document using the **Hyperlink** tool.

1. From the **Tools** menu, invoke the **Hyperlink** tool; you will be prompted to select a region to place the hyperlink.

2. Click the two opposite corners to define a rectangular region or highlight the text to place the hyperlink; the **Action** dialog box will be displayed, as shown in Figure 21.

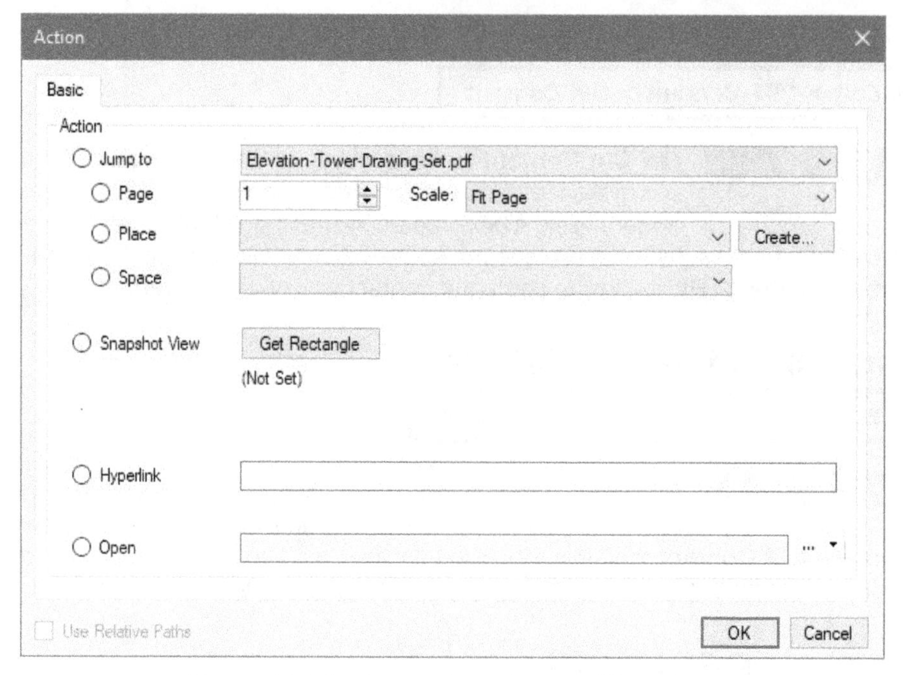

***Figure 21** The **Action** dialog box to add hyperlink*

3. In the **Action** area of the dialog box, select the **Page** radio button.

4. In the spinner on the right of the **Page** radio button, enter the page number you want to navigate to.

5. From the **Scale** drop-down list, specify the option to define how the page will be displayed when you navigate to it.

6. Click **OK** in the dialog box; the hyperlink will be created and highlighted on the PDF file.

7. Press the ESC key to exit the **Hyperlink** tool.

The Procedure for Using Hyperlink to Create and Navigate to a Place in the Current Document

Places are named regions that can be created in the current document and then navigated to using the **Hyperlink** tool. The following is the procedure for doing this.

1. From the **Tools** menu, invoke the **Hyperlink** tool; you will be prompted to select a region to place the hyperlink.

2. Click the two opposite corners to define a rectangular region or highlight the text to place the hyperlink; the **Action** dialog box will be displayed.

3. In the **Action** area of the dialog box, select the **Place** radio button.

4. Click the **Create** button on the right of the **Place** drop-down list; the **Place** dialog box will be displayed, as shown in Figure 22.

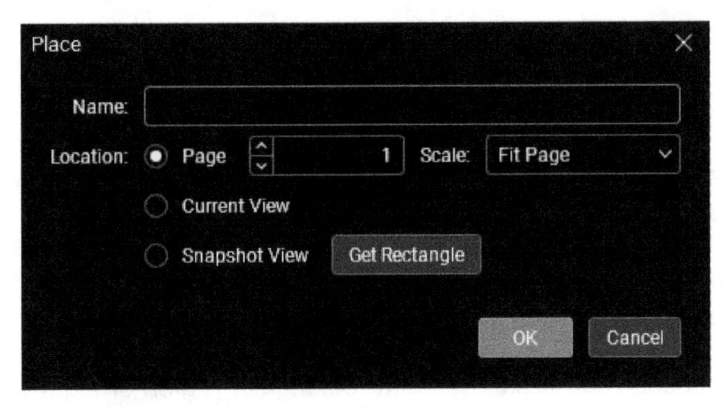

*Figure 22 The **Place** dialog box to create a place*

5. In the **Name** field, enter the name of the place.

6. Click the **Get Rectangle** button to create the place; you will be returned to the PDF file.

7. Using the **Thumbnails** panel, navigate to the page you want to create the place on.

8. Specify the two opposite corners to create the place; you will be returned to the **Place** dialog box.

9. Click **OK** in the **Place** dialog box; you will be returned to the **Action** dialog box and the name of the place you created will be available in the **Place** drop-down list.

10. Click **OK** in the **Action** dialog box.

11. Press the ESC key to exit the **Hyperlink** tool.

The Procedure for Using Hyperlink to Create and Navigate to a Snapshot View

The following is the procedure for creating a snapshot view and navigating to it using a hyperlink.

1. From the **Tools** menu, invoke the **Hyperlink** tool; you will be prompted to select a region to place the hyperlink.

2. Click the two opposite corners to define a rectangular region or highlight the text to place the hyperlink; the **Action** dialog box will be displayed.

3. In the **Action** area of the dialog box, select the **Snapshot View** radio button.

4. Click the **Get Rectangle** button; you will be returned to the PDF file.

5. Using the **Thumbnails** panel, navigate to the page to create the snapshot view on.

6. Specify the two opposite corners to create the snapshot view; you will be returned to the **Action** dialog box.

7. Click **OK** in the **Action** dialog box.

8. Press the ESC key to exit the **Hyperlink** tool.

The Procedure for Using Hyperlink to Navigate to a Website

The following is the procedure for navigating to a website using a hyperlink.

1. From the **Tools** menu, invoke the **Hyperlink** tool; you will be prompted to select a region to place the hyperlink.

2. Click the two opposite corners to define a rectangular region or highlight the text to place the hyperlink; the **Action** dialog box will be displayed.

3. In the **Action** area of the dialog box, select the **Hyperlink** radio button.

4. Type or paste the website you want to navigate to in the field on the right of the **Hyperlink** radio button.

5. Click **OK** in the **Action** dialog box.

6. Press the ESC key to exit the **Hyperlink** tool.

The Procedure for Using Hyperlink to Open a Folder or a File

The following is the procedure for using a hyperlink to open a folder or a file.

1. From the **Tools** menu, invoke the **Hyperlink** tool; you will be prompted to select a region to place the hyperlink.

2. Click the two opposite corners to define a rectangular region or highlight the text to place the hyperlink; the **Action** dialog box will be displayed.

3. In the **Action** area of the dialog box, select the **Open** radio button.

4. Click on the flyout on the right of the **Browse** button to specify whether you want to open a file or a folder, as shown in Figure 23.

5. In the **Select File** or **Select Folder** dialog box, browse and select the file or folder to open.

6. Make sure the **Use Relative Path** check box is selected on the lower left of the **Action** dialog box.

7. Click **OK** in the **Action** dialog box.

8. Press the ESC key to exit the **Hyperlink** tool.

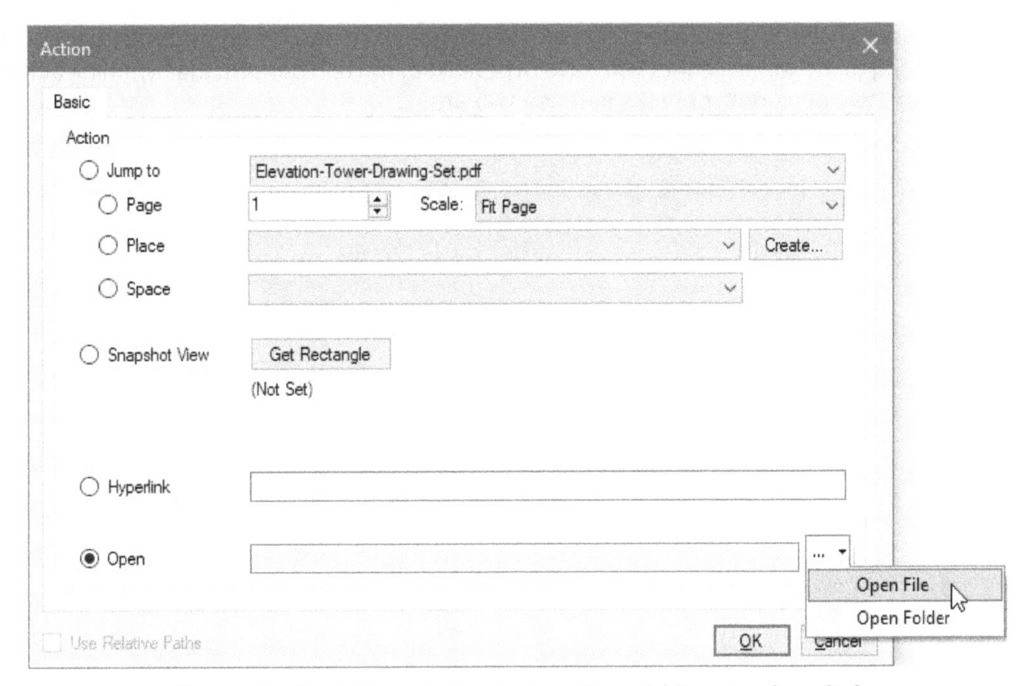

Figure 23 Specifying whether to open file or folder using hyperlink

The Links Panel

The **Links** panel is your one-stop shop to review and access all hyperlinks added to the current document. This panel is divided into two areas: **Places** and **Hyperlinks**, as shown in Figure 24.

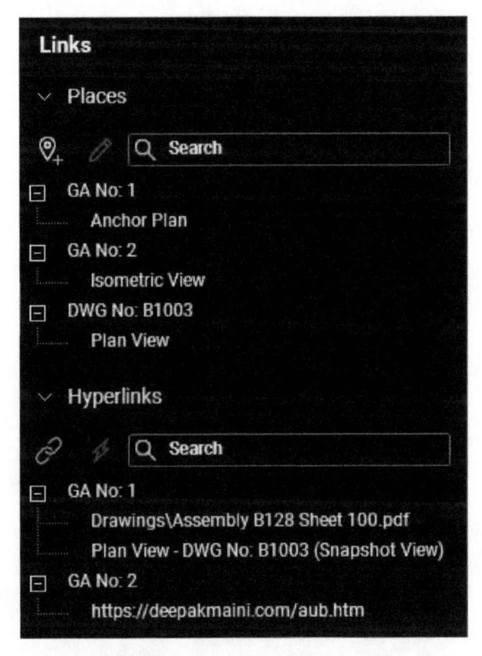

*Figure 24 The **Links** panel*

The **Places** area shows all the places that you have hyperlinked and the **Hyperlinks** area shows the files, folders, or the websites that you have hyperlinked. You can perform the hyperlink actions by clicking on the places or hyperlinks in these two areas.

Hands-on Tutorial

In this tutorial, you will complete the following tasks:
1. *Open a multi-page PDF file and reorder pages in that file.*
2. *Create automated page labels for the pages in the PDF file.*
3. *Extract pages of the current PDF file.*
4. *Take snapshot of a region of the PDF file and paste it on a new sheet.*
5. *Create hyperlinks in the current document.*
6. *Flatten the content of the PDF.*

Section 1: Opening a Multi-page PDF File and Reordering Pages

In this section, you will open the **Elevation-Tower-Drawing-Set.pdf** file from the **C09** folder. You will then use the **Thumbnails** panel to reorder pages in this PDF file.

1. From the **C09** folder, open the **Elevation-Tower-Drawing-Set.pdf** file.

2. From the **Panel Access Bar**, invoke the **Thumbnails** panel; the pages in the current PDF file are displayed in this panel, as shown in Figure 25.

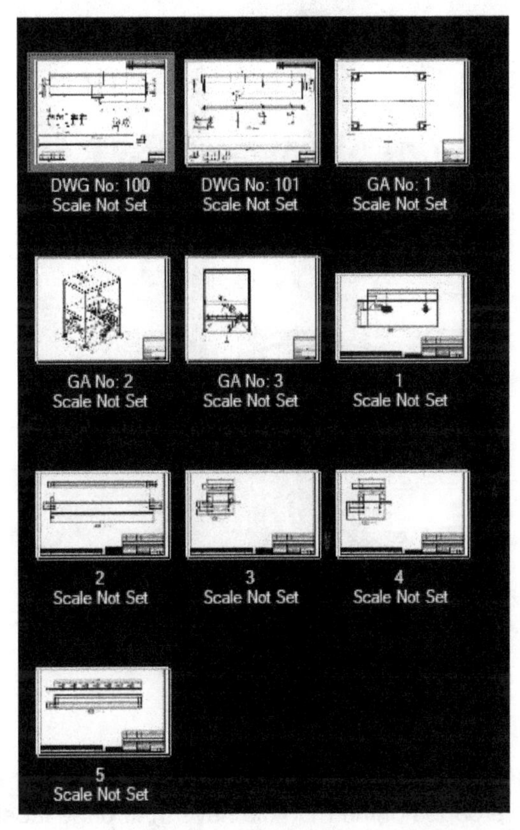

*Figure 25 The **Thumbnails** panel showing the pages in the PDF file*

***Tip**: You can use the slider bar at the bottom of the **Thumbnails** panel to resize the thumbnails in this panel.*

The first two pages in this PDF file are the assembly drawings and the next three pages are the General Arrangement (GA) drawings. You will now reorder the assembly drawing page so they are placed after the GA pages by dragging and dropping them.

3. Hold down the shift key and select the first two pages of the PDF in the **Thumbnails** panel.

4. Drag the two selected pages and drop them after the GA pages, as shown in Figure 26; the pages are reordered.

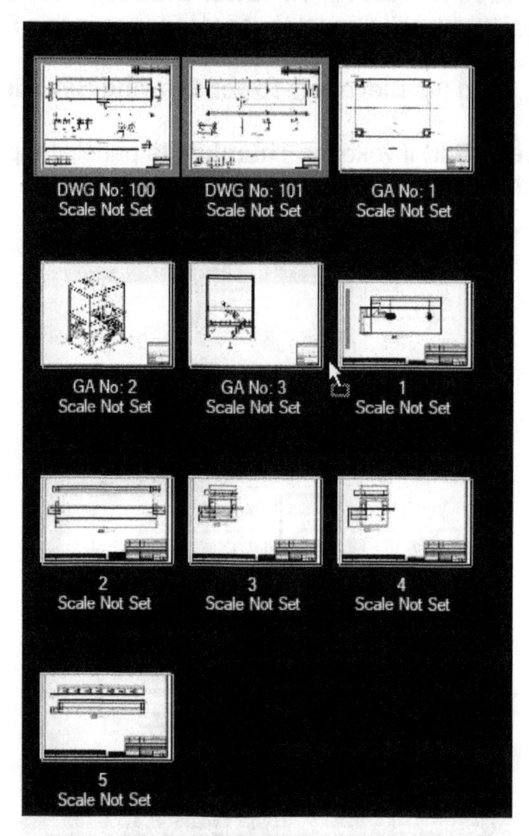

Figure 26 Reordering pages of the PDF file

5. Save the PDF file.

Section 2: Creating Page Labels by Extracting the Page Information

Notice that the first five pages of the PDF file have their page labels defined. However, the last five pages are labeled as 1 to 5. In this section, you will extract the page labels for these pages from the titleblocks on these pages.

1. From the toolbar available at the top in the **Thumbnails** panel, click the **Create Page Labels** button, as shown in Figure 27.

Figure 27 Invoking the tool to create page labels

On doing so, the **Create Page Labels** dialog box is displayed, as shown in Figure 28.

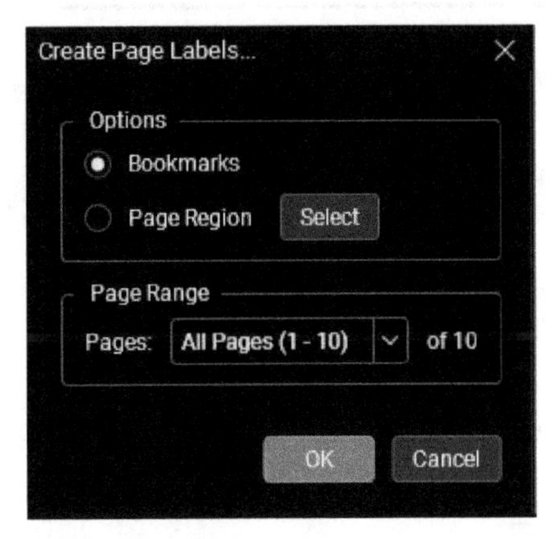

*Figure 28 The **Create Page Labels** dialog box*

2. In the **Options** area of the dialog box, click the **Page Region** radio button.

   The page label will be extracted by selecting the region where the text you want to use is located in the title block of pages 5-10 of this PDF file.

3. Click the **Select** button on the right of the **Page Region** radio button; you are returned to the PDF file and are prompted to select a rectangle.

4. From the **Thumbnails** panel, click on the sixth sheet of the PDF, which is labeled as **1** in this panel.

5. Navigate to the bottom right of the titleblock where the **Drawing No.** text appears.

6. In the field where the **Drawing No.** text appears, click two opposite corners to create a rectangle, as shown in Figure 29.

   On doing so, the region to extract the page label is defined and the **AutoMark** dialog box is displayed, as shown in Figure 30. In this dialog box, you will add a prefix to the page label.

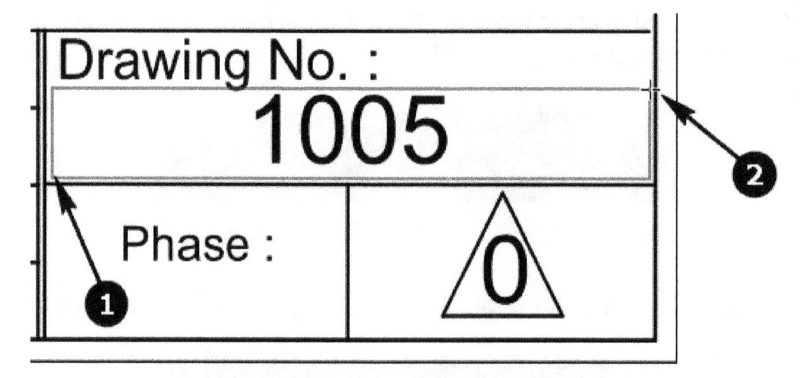

Figure 29 Specifying the two points to extract the page label

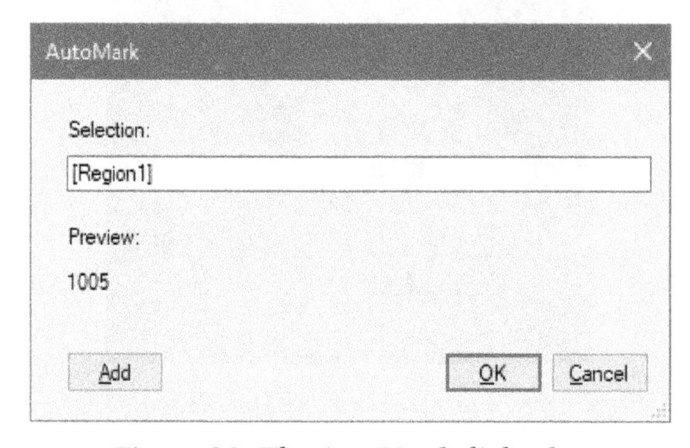

*Figure 30 The **AutoMark** dialog box*

7. In the **AutoMark** dialog box, click on the left of the **[Region1]** text to move the cursor there.

8. Type **DWG No:** as the prefix so the text in the **Selection** field appears as **DWG No: [Region1]**.

9. Click **OK** in the **AutoMark** dialog box; you are returned to the **Create Page Labels** dialog box.

   Next, you need to define the pages to which this page label will be assigned.

10. From the **Page Range** area > **Pages** drop-down list, select **Custom Range**.

11. Type **6-10** as the custom range.

12. Click **OK** in the **Create Page Labels** dialog box; the page labels are assigned to pages 6 to 10, as shown in Figure 31.

13. Save the PDF file.

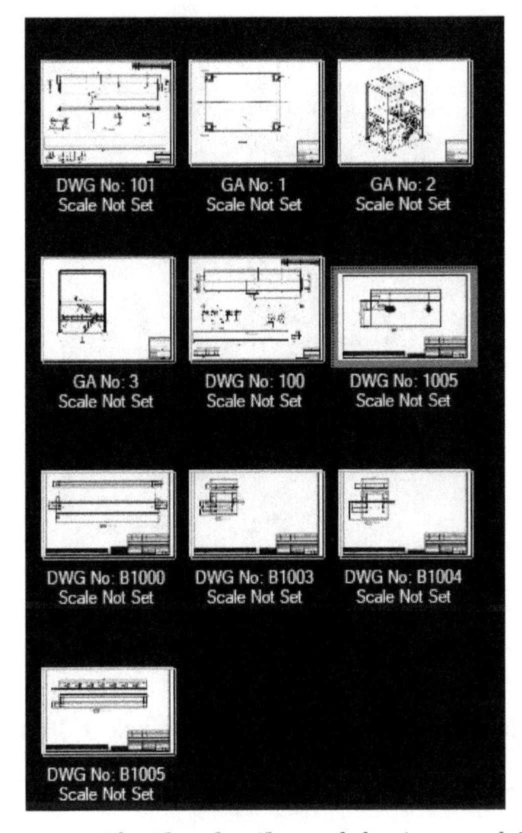

Figure 31 *The* ***Thumbnails*** *panel showing page labels*

Section 3: Extracting Pages of the Current PDF File

In this section, you will extract the last six pages of the current PDF file as individual PDFs.

1. Right-click anywhere in the **Thumbnails** panel and select **Extract Pages** from the shortcut menu, as shown in Figure 32.

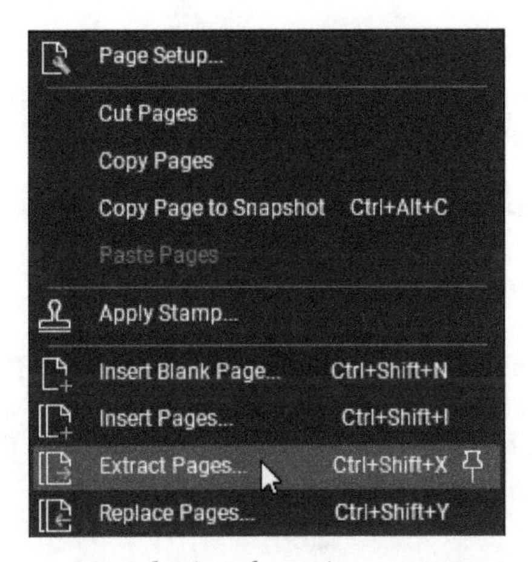

Figure 32 *Selecting the option to extract pages*

On doing so, the **Extract Pages** dialog box is displayed, as shown in Figure 33.

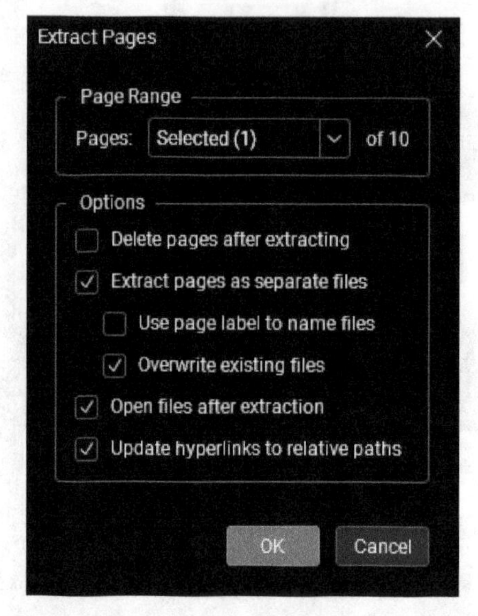

Figure 33 The ***Extract Pages*** *dialog box*

2. From the **Page Range** area > **Pages** drop-down list, select **Custom Range**.
3. Enter **6-10** as the page range to extract in the same field.
4. In the **Options** area, make sure the **Delete pages after extracting** is not selected.
5. Select the **Extract pages as separate files** check box, if not already selected.
6. Select the **Use page label to name files** check box.
7. Make sure the **Open files after extraction** check box is selected to ensure the PDF files are opened after extraction.

Note: *It is not recommended to have the* ***Open files after extraction*** *check box selected when you are extracting a large number of pages.*

8. Click **OK** in the **Extract Pages** dialog box; the **Select Folder** dialog box is displayed.
9. Browse and select the **C09 > Extracted Drawings** folder.
10. Click the **Select Folder** button on the lower right of the **Select Folder** dialog box; the **Extract Pages has completed** message box is displayed informing you that some drawings have special characters that are replaced by an underscore in the file name.
11. Click **OK** in the message box; the drawings are extracted and opened in the Revu window.

12. Review the five extracted files that are opened in the Revu window and notice their page labels in the **Thumbnails** panel that match with their file names.

13. Close the five extracted drawings.

14. Save the original PDF file and then close it.

Section 4: Inserting a Blank Page in a PDF File and Cutting and Pasting the Content

In this section, you will open the **Elevation-Tower-Assembly-List.pdf** file and then insert a blank page at the start of this PDF file. You will then cut some content from the first original page of this document and paste it on the new page.

1. From the **C09** folder, open the **Elevation-Tower-Assembly-List.pdf** file.

   Notice in the **Thumbnails** panel that this file has 19 pages. You will now insert a blank page at the start of this document.

2. In the **Thumbnails** panel, right-click on the first page and select **Insert Blank Page** from the shortcut menu, as shown in Figure 34.

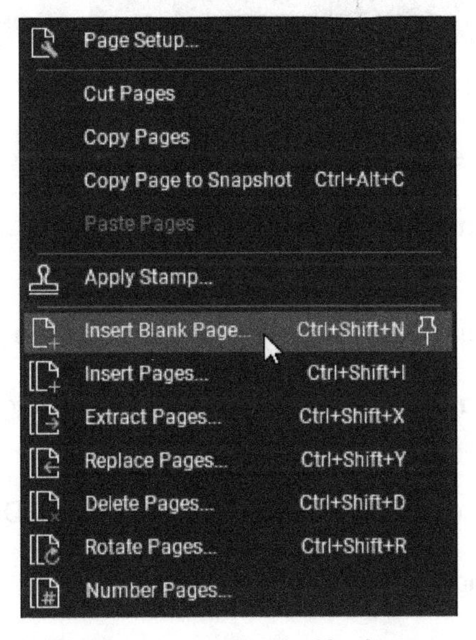

Figure 34 *Inserting a blank page*

   On doing so, the **Insert Blank Page** dialog box is displayed, as shown in Figure 35. The default template in this dialog box is set to the page size of the original PDF. Therefore, you do not need to change the page template.

3. In the **Page** area of the dialog box, select **Landscape** to set the orientation of the new page.

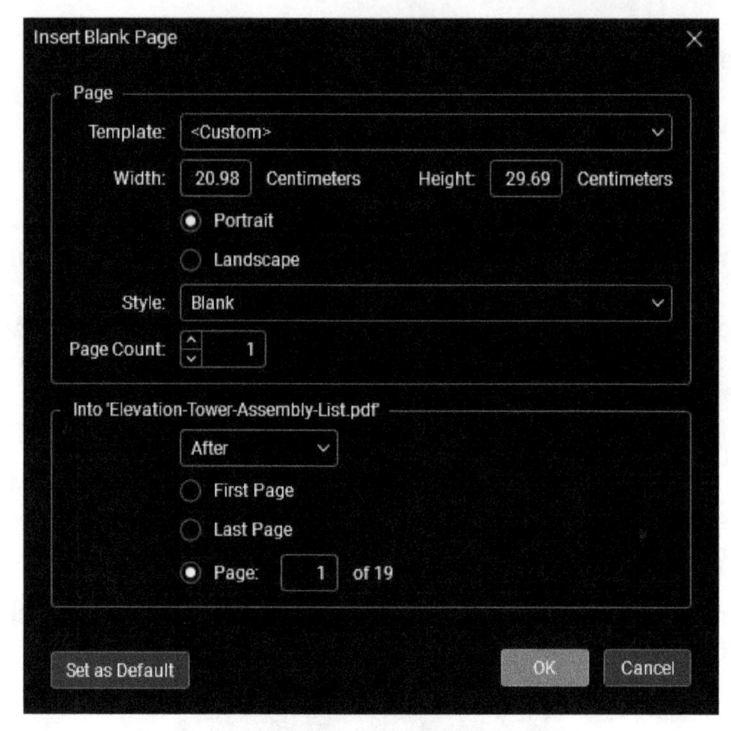

*Figure 35 The **Insert Blank Page** dialog box*

4. In the **Page Count** spinner, make sure **1** is selected.

5. From the **Into 'Elevation-Tower-Assembly-List.pdf'** area, select **Before** from the drop-down list at the top.

6. Select the **First Page** radio button. This ensures the new page is added before the first page.

7. Click **OK** in the **Insert Blank Page** dialog box; a new blank page is inserted as the first page of this PDF file.

   Next, you will cut the top description area of page 2 of the PDF and will copy it on page 1.

8. Navigate to the top of page 2 of the PDF.

9. From the **Menu Bar**, click **Edit > PDF Content > Cut Content**; the cursor changes to a crosshair so you can define the region to cut.

10. Pres and hold down the Left Mouse Button and drag two opposite corners labeled as **1** and **2** in Figure 36 to cut the region; the PDF content is cut and copied to the clipboard.

    Next, you will paste this content on the first page of this document.

11. Navigate to the top of page 1 of the document, which is a blank page.

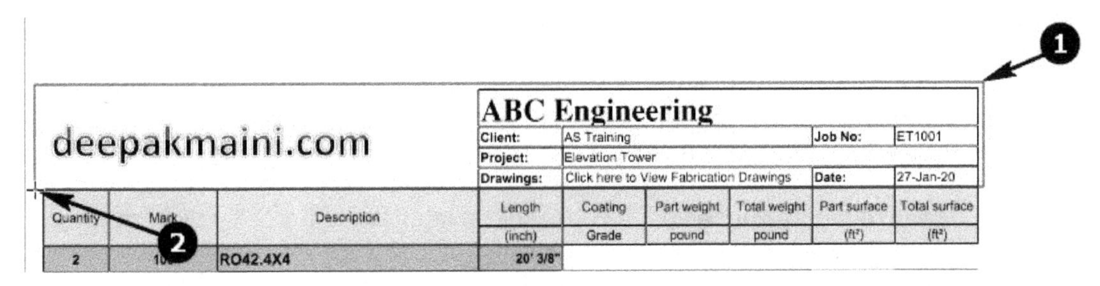

Figure 36 Selecting the region to cut content

12. Right-click on the blank page in the **Main Workspace** and select **Paste** from the shortcut menu; the content is pasted.

13. Move the content so it is located centrally on the top of the page, as shown in Figure 37.

deepakmaini.com

ABC Engineering			
Client:	AS Training	Job No:	ET1001
Project:	Elevation Tower		
Drawings:	Click here to View Fabrication Drawings	Date:	27-Jan-20

Figure 37 The content copied on the first page

Notice that the line at the bottom of the table is missing. You will now draw a line manually at the bottom of the table. You need to make sure you turn on the **Snap to Markup** option from the **Status Bar** to be able to snap on the lines of the copied content.

14. Turn on the **Snap to Markup** option from the **Status Bar**.

15. Using the **Line** markup tool, draw a black line of 0.5 width to close the table at the bottom.

16. Save the PDF file.

Section 5: Taking Snapshot and Pasting in the PDF

In this section, you will open the **Elevation-Tower-GA.pdf** file and take a snapshot of the view in that PDF file. You will then paste that view in the currently opened PDF file.

1. From the **C09** folder, open the **Elevation-Tower-GA.pdf** file.

   This file has one page that shows the Isometric view of the elevation tower. You will now take the snapshot of this view and paste it in the other PDF you have currently opened.

2. Type the **G** key on the keyboard to invoke the **Snapshot** tool; the cursor changes to a crosshair and you are prompted to select a region to take the snapshot.

Because you need to create a four-sided box to snapshot, you can press and drag the mouse to define two opposite corners.

3. Press and hold down the Left Mouse Button and drag two opposite corners around the Isometric view to create the snapshot of the view.

4. Close the **Elevation-Tower-GA.pdf** file; you are returned to page 1 of the other document.

5. Right-click on the first page and select **Paste** from the shortcut menu; the snapshot view is pasted.

   Notice that the size of the view is too big for the current sheet. You will now reduce the scale of this snapshot view to 20% size in the **Properties** panel.

6. With the snapshot view still selected, reduce its scale to **20%** in the **Properties** panel.

7. Move the view so it is located at the center of the page, as shown in Figure 38.

ABC Engineering			
Client:	AS Training	Job No:	ET1001
Project:	Elevation Tower		
Drawings:	Click here to View Fabrication Drawings	Date:	27-Jan-20

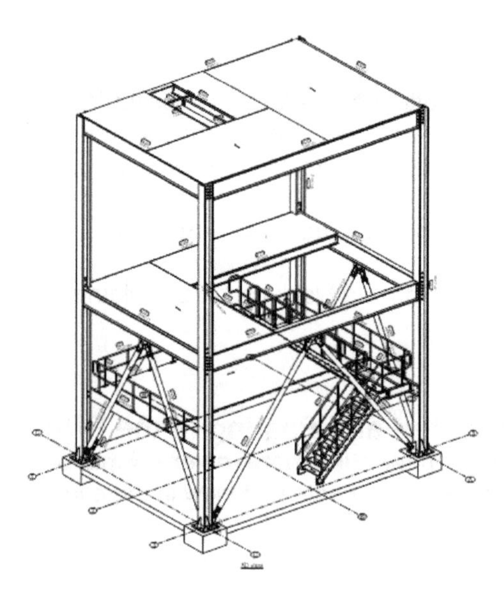

Figure 38 The content copied on the first page

You will now change the color of the linework in this Isometric view. Because the **Snapshot** tool maintains the vector integrity of the content you capture, you can right-click on it and select the option to change its color.

8. Right-click on the snapshot view in the **Main Workspace** and select **Change Colors** from the shortcut menu; the **Color Processing** dialog box is displayed, as shown in Figure 39.

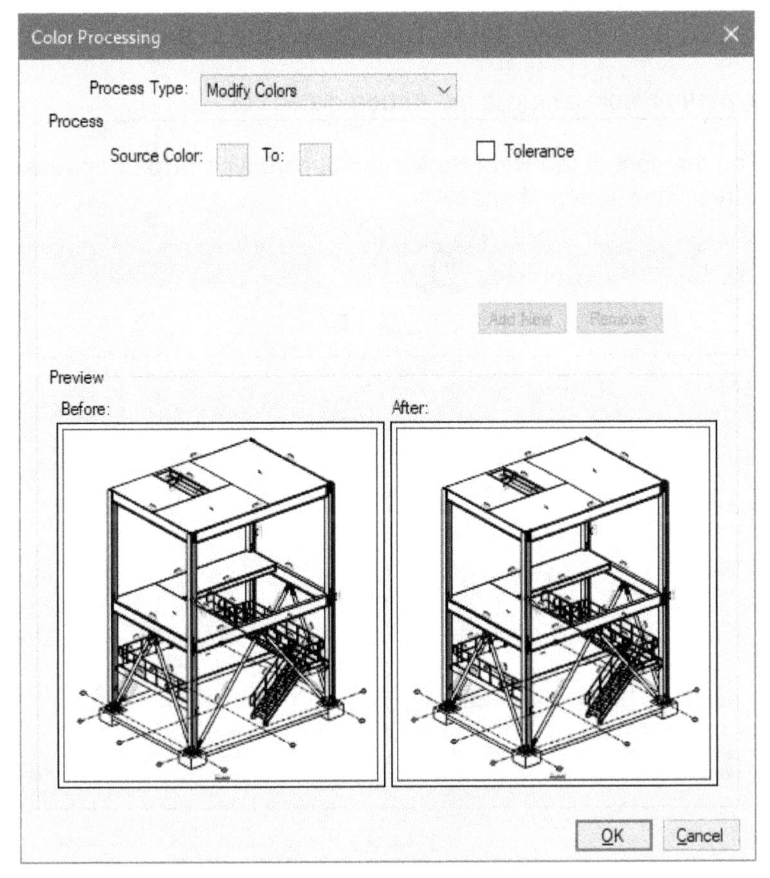

*Figure 39 The **Color Processing** dialog box*

9. Click the button on the right of **Source Color** and select the Black color from the list.

10. Click the button on the right of **To** and select the Blue color from the list; the preview of the Isometric view in the After area changes to Blue lines.

11. Click **OK** in the **Color Processing** dialog box; the color change is applied to the Isometric view.

12. Save the file.

Section 6: Creating Hyperlinks

In this section, you will create various hyperlinks to the PDF file you have currently opened. The first hyperlink you will create will be around the logo on the top left of the first page.

1. From the **Menu Bar**, click **Tools > Hyperlink** to invoke this tool; you are prompted to select a region to place a hyperlink.

2. Specify two points around the **deepakmaini.com** logo to define the region to place hyperlink; the **Action** dialog box is displayed.

   For this hyperlink, you will enter a website to open in the **Action** dialog box.

3. Click the **Hyperlink** radio button in the **Action** dialog box.

4. In the field on the right of the **Hyperlink** radio button, type **https://deepakmaini.com/** as the web address, as shown in Figure 40.

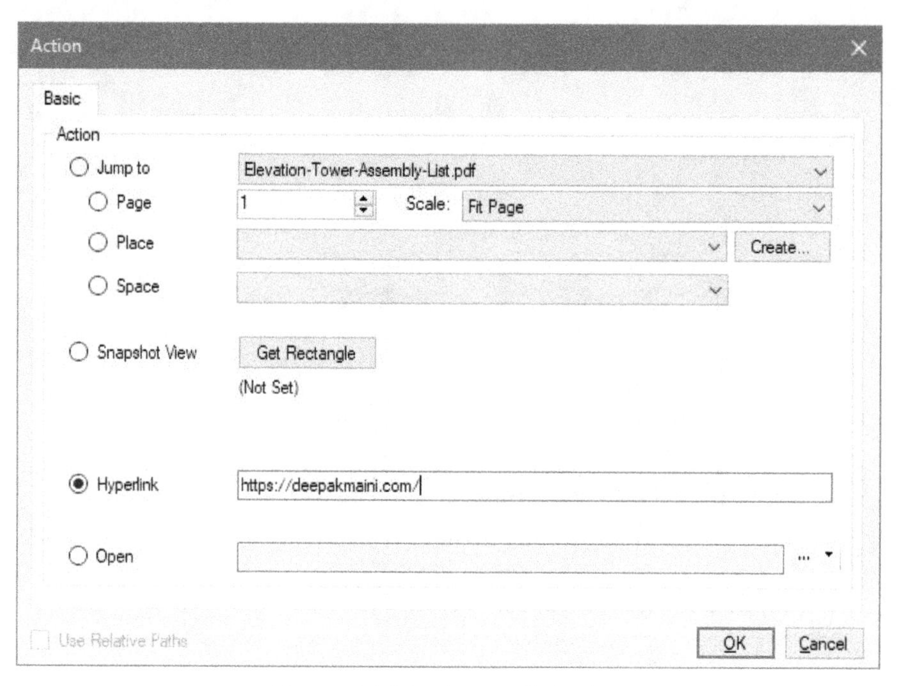

***Figure 40** The **Action** dialog box*

5. Click **OK** in the **Action** dialog box; the hyperlink is added and the region around the logo is displayed in Blue.

6. Press the ESC key to exit the **Hyperlink** tool; the region around the logo is no more displayed in Blue.

***Tip**: To edit the hyperlink, you can invoke the **Links** panel and then double-click on the link in that panel. You will learn more about the **Links** panel later in this section of the tutorial.*

The next hyperlink will be added to the **Click here to View Fabrication Drawings** text in the table on the top right. Because you cut this content and pasted it here, you cannot directly select the text. You will have to draw a box around the text, similar to the way you did for the logo.

7. From the **Menu Bar**, click **Tools > Hyperlink** to invoke this tool; you are prompted to select a region to place a hyperlink.

8. Click two opposite corners around the **Click here to View Fabrication Drawings** text in the table on the top right to define the region to place the hyperlink; the **Action** dialog box is displayed.

9. In the **Action** dialog box, click the **Open** radio button.

10. Click the flyout on the right of the **Open** field and select **Open File**, as shown in Figure 41.

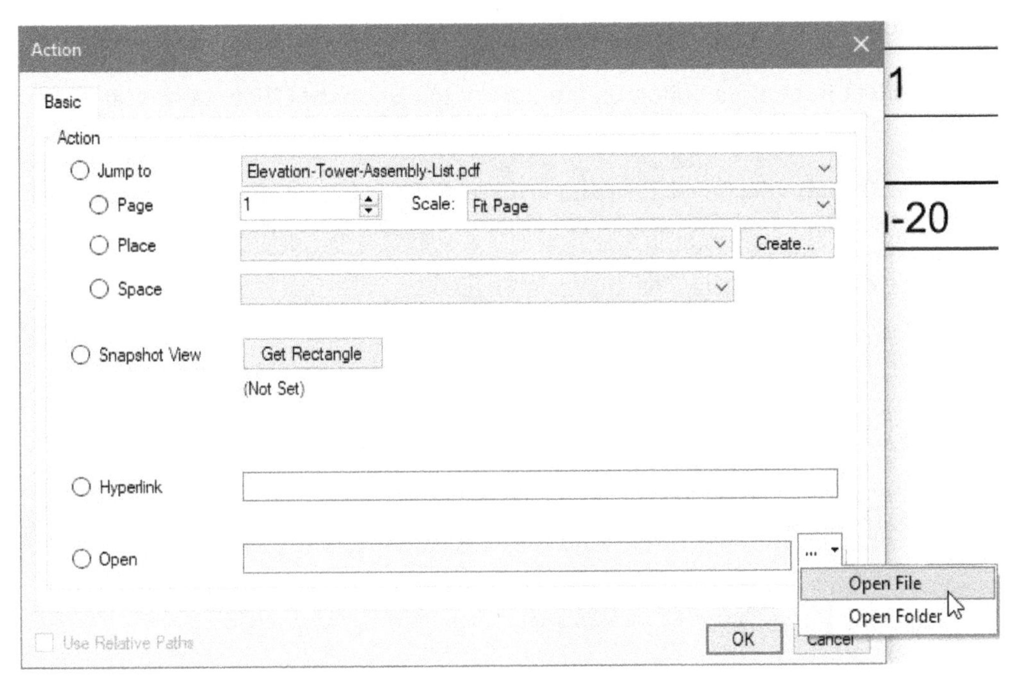

Figure 41 Selecting the option to open a file using hyperlink

On doing so, the **Open File** dialog box is displayed.

11. Browse to the **C09** folder and double-click on the **Elevation-Tower-Drawing-Set.pdf** file; you are returned to the **Action** dialog box and the name of the file is displayed in the field on the right of the **Open** radio button.

12. Make sure the **Use Relative Paths** check box is selected on the lower left in the **Action** dialog box.

13. Click **OK** in the **Action** dialog box; the region around the text you selected earlier is displayed in Blue suggesting that the hyperlink is created around that text.

The last hyperlink will be to the total weight of the design on the last page. But first, you will write a text using the **Typewriter** tool on the right of the project name.

14. Using the **Typewriter** tool, type **Total Weight = 109,196.46 Pounds** on the right of the **Elevation Tower** text in the **Project:** field.

15. Change the font size to 10 and use Helvetica or Arial font.

16. Invoke the **Hyperlink** tool and draw a box around the text you wrote to specify the region to place hyperlink; the **Action** dialog box is displayed.

17. In the **Action** dialog box, click the **Create** button on the right of the **Place** drop-down list; the **Place** dialog box is displayed.

18. In the **Name** field, enter **Weight Value** as the name of the place.

19. Click the **Get Rectangle** button on the right of the **Snapshot View** radio button; you are returned to the PDF file.

20. Using the **Thumbnails** panel, navigate to the last page of the PDF and then navigate to the table at the bottom that shows the total values, refer to Figure 42.

21. Draw a box around the table, as shown in Figure 42

TOTAL QUANTITY	127	
TOTAL WEIGHT	109,196.46	pound
TOTAL PAINT AREA	7,336.46 ^{2}	ft^2

Figure 42 *Selecting the region to create the space*

On doing so, you are returned to the **Place** dialog box and the information about the place is available in this dialog box, as shown in Figure 43.

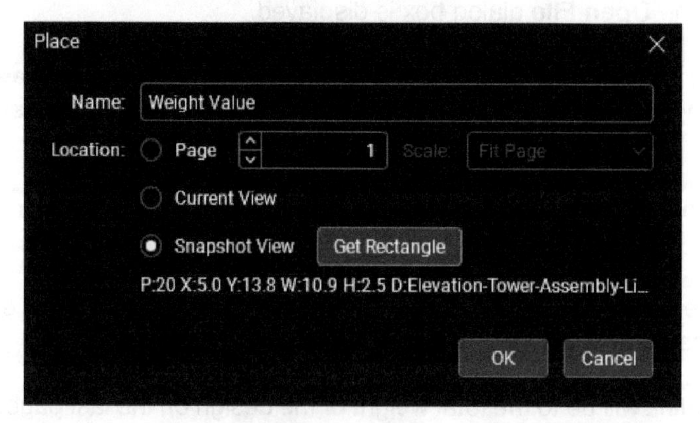

Figure 43 *The **Place** dialog box showing the details of the place you created*

22. Click **OK** in the **Place** dialog box; you are returned to the **Action** dialog box and the place you created is listed in this dialog box.

23. Click **OK** in the **Action** dialog box.

24. Press the ESC key to exit the **Hyperlink** tool.

    Notice that you are currently on the last page of the PDF file. This is because the place you created will bring you to this page. If you want to return to the first page of the PDF, you can use the **Thumbnails** panel. Alternatively, you can create a hyperlink on this page to return to page 1. This is what you will do now.

25. Below the weight table on the last page, write a text **Return to Page 1**.

26. Invoke the **Hyperlink** tool and draw a box around the text you wrote to specify the region to place hyperlink, as shown in Figure 44.

TOTAL QUANTITY	127	
TOTAL WEIGHT	109,196.46	pound
TOTAL PAINT AREA	7,336.46 ²	ft²

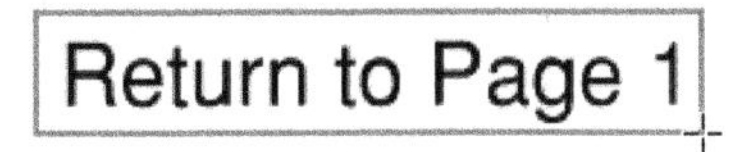

Figure 44 *Specifying the region around the text to place the hyperlink*

27. In the **Action** dialog box, select **Jump to** radio button; the **Page** radio button is automatically selected and the spinner on the right of this button is set to page 1.

28. From the **Scale** drop-down list, make sure **Fit Page** is selected.

29. Click **OK** in the **Action** dialog box; you are returned to the PDF file and the region around the text is highlighted in Blue.

30. Press the ESC key.

    You will now test the various hyperlinks you created.

31. With the current view still at the last page, click on the **Return to Page 1** text; the hyperlink navigates you to page 1 of the PDF and the view is set to the full page view. This is because you selected the **Fit Page** option while creating this hyperlink.

32. Hover the cursor over the **deepakmaini.com** logo on the top left of the page; the hyperlinked website is displayed as a tooltip.

33. Click on the logo; the website is opened in the Revu window.

34. Close the website window to return to the original PDF.

35. Still on the first page of the PDF, hover the cursor over the text **Click here to View Fabrication Drawings**; the name of the PDF file is displayed as the tool tip.

36. Click on the text and the linked PDF file opens in the Revu window.

37. Close the **Elevation-Tower-Drawing-Set.pdf** file and return to the original PDF.

38. Hover the cursor over the **Total Weight = 109,196.46 Pounds** text; the hyperlink associated with that region is displayed as the tooltip.

    Notice that in this tooltip, the name of the space and the page number is also displayed.

39. Click on the **Total Weight = 109,196.46 Pounds** text to navigate to the table on the last page that shows the total weight.

40. Click on the **Return to Page 1** text to return to the first page of the PDF.

41. Save the PDF file.

    Next, you will activate the **Links** panel to review the various hyperlinks added to the current PDF file.

42. From the **Panel Access Bar**, invoke the **Links** panel; the hyperlinks added to the current PDF file are displayed in the **Places** and **Hyperlinks** area, as shown in Figure 45.

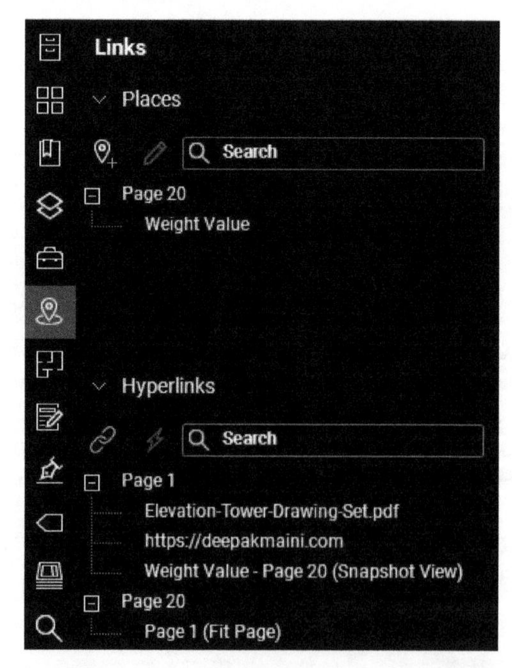

***Figure 45** The **Links** panel with the hyperlinks added to the current PDF*

43. In the **Places** area of the **Links** panel, click on the **Weight Value** place; you are navigated to the weight value table on the last page of the PDF.

44. In the **Hyperlinks** area of the **Links** panel, click on the **Elevation-Tower-Drawing-Set.pdf** link; you are navigated to page 1 of the PDF where this hyperlink was added. However, the linked PDF file is not opened.

    This is because the **Links** panel is used to navigate to the location of the hyperlink in the PDF and not action the hyperlink.

45. Zoom to the extents of page 1.

46. Save the current PDF file.

Section 7: Flattening the Markups in the PDF File

You added some text and also a snapshot view in the current PDF file. To make sure all this newly added content is locked from editing when this document leaves your organization, you need to make sure you flatten the markups in this PDF file. In this section, you will do this.

1. From the **Menu Bar**, click **Document > Flatten**; the **Flatten Markups** dialog box is displayed.

   The currently opened document is already listed in the **Files** area. Also, by default, all pages of the document are selected in the **Pages** field of the **Files** area. But you need to make sure that is the case.

2. From the **Files** area > **Pages** field, make sure **All Pages (1-20)** is selected, as shown in Figure 46.

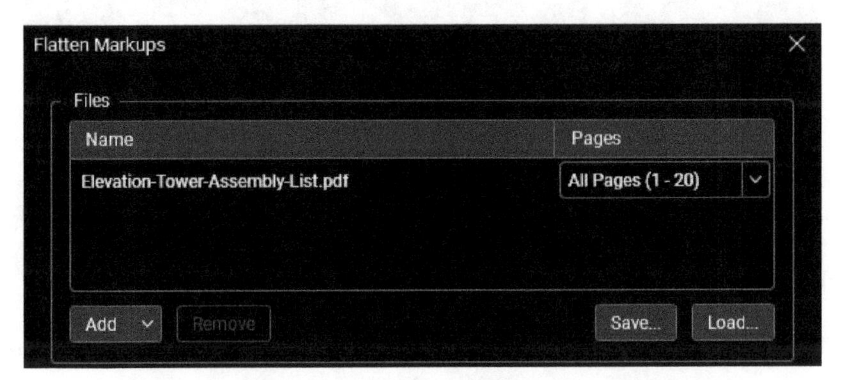

*Figure 46 The **Flatten Markups** dialog box > **Files** area*

3. In the **Options** area, make sure the **Snapshot** check box is selected, refer to Figure 47.

4. In the **Flatten** area, make sure the **All Markups** radio button is selected, as shown in Figure 47. This ensures if there are any filtered out markups, they are flattened as well.

The **Allow Markup Recovery (Unflatten)** check box allows you to recover the markup, if required. However, this also means that someone else outside your organization will be able to recover these markups if they have Bluebeam Revu. Therefore, it is not recommended to turn this check box on.

5. Make sure the **Allow Markup Recovery (Unflatten)** check box is not selected, as shown in Figure 47.

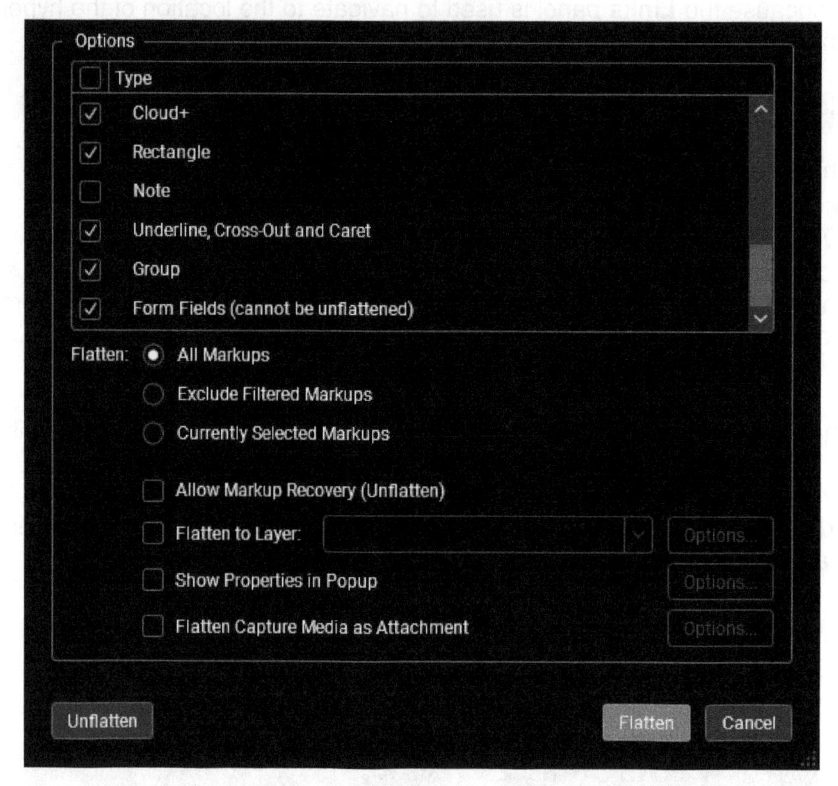

Figure 47 *The **Flatten Markups** dialog box > **Options** area*

6. Click the **Flatten** button on the lower right of the dialog box; the markups are flattened.

7. Hover the cursor over the Isometric view and notice that you cannot select it.

8. Hover the cursor over any of the text where you added a hyperlink and notice that it still works.

9. Save the current file.

10. Close the file.

Skill Evaluation

Evaluate your skills to see how many questions you can answer correctly. The answers to these questions are given at the end of the book.

1. The **Thumbnails** panel shows all the pages of the current PDF file. (True/False)

2. You can insert blank pages in the current PDF file using the **Thumbnails** panel. (True/False)

3. The markups once flattened cannot be unflattened. (True/False)

4. You cannot insert hyperlinks on the text in Revu. (True/False)

5. The **Thumbnails** panel allows you to reorder pages. (True/False)

6. Which tool is used to insert a hyperlink to open a website?

   (A) **Link** (B) **Open**
   (C) **Hyperlink** (D) **Insert**

7. While creating hyperlinks, which option allows you to navigate to a named region that you create?

   (A) **Place** (B) **Regions**
   (C) **Pages** (D) **Areas**

8. Which tool allows you to copy and paste a region of the PDF to another page or another PDF, still maintaining the vector integrity of the markup?

   (A) **Snapshot** (B) **Camera**
   (C) **Picture** (D) **None**

9. Which tool in the **Thumbnails** panel allows you to automatically extract page labels from a defined region on the PDF?

   (A) **Labels** (B) **Create Page Labels**
   (C) **Choose** (D) **Extract Info**

10. Which option in the **Action** dialog box allows you to open a file using hyperlinks?

    (A) **Page** (B) **Open**
    (C) **Place** (D) **Space**

Class Test Questions

Answer the following questions:

1. Explain briefly the process of inserting a blank page in the current PDF.
2. Explain briefly how to flatten all markups.
3. Explain the process of adding a hyperlink to a named place.
4. Explain the process of reordering pages.
5. Explain the process of creating a snapshot.

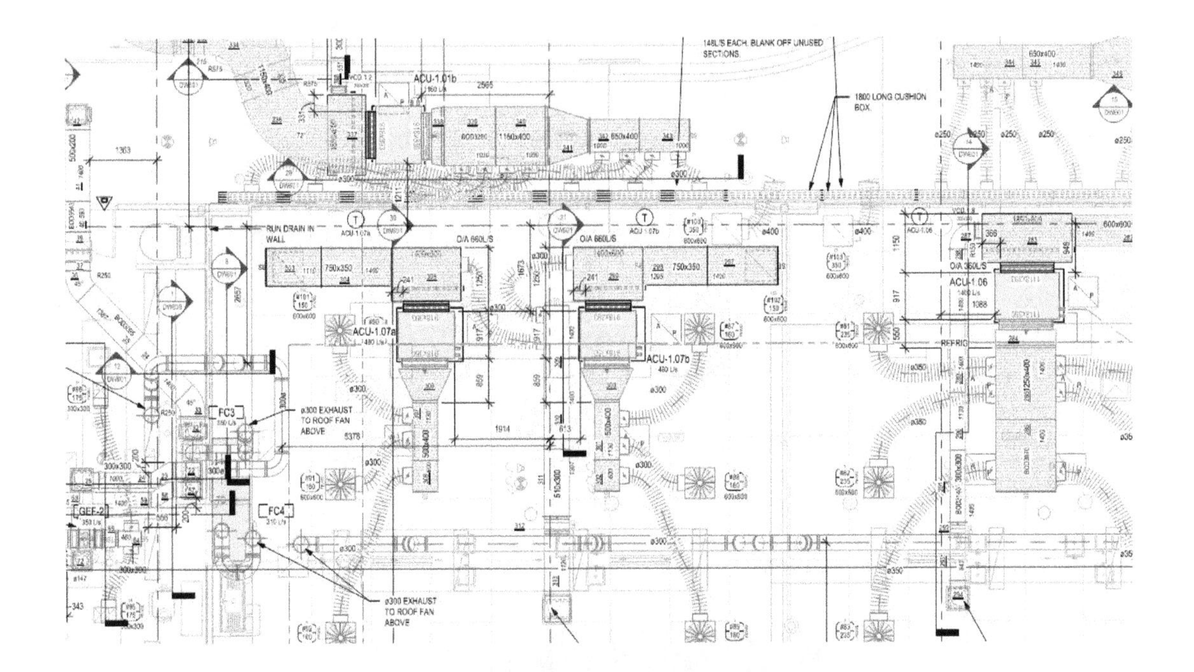

Chapter 10 - Searching and Comparing PDFs and Inserting Images

The objectives of this chapter are to:

- √ *Explain the types of searches that can be performed in the PDF files*
- √ *Explain how to perform text and visual searches*
- √ *Explain the process of comparing documents to find the differences*
- √ *Explain how to overlay pages*
- √ *Explain how to insert images*

PERFORMING TEXT AND VISUAL SEARCHES IN PDFs

Revu is an extremely powerful program when it comes to performing searches. It not only allows you to search for texts in vector PDFs, but also lets you perform visual searches for shapes and symbols in the drawing documents. Performing visual searches is a technique widely used in taking off quantities in the Mechanical or Electrical documents when you have to search for symbols representing items like lighting fixtures, switchboards, VAVs, and so on. Both these types of searches are performed using the **Search** panel. The following are the procedures for performing both these types of searches.

The Procedure for Searching Texts in PDFs

The following is the procedure for searching texts in the PDF files:

1. Invoke the **Search** panel.

2. From the top of the **Search** panel, make sure the **Text** button is selected, refer to Figure 1.

3. In the search field, type the text you want to search. Figure 1 shows **B128** as the text being searched.

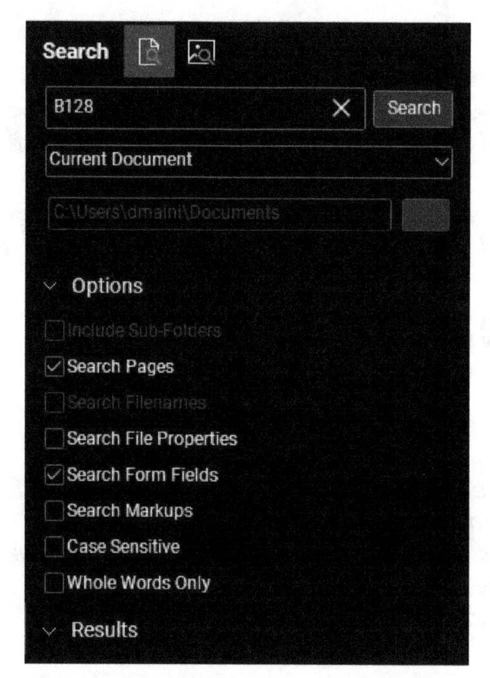

Figure 1 The ***Search*** *panel to perform text search*

4. In the drop-down list below the search field, select the scope of the search. You can select the current page, the current document, all open documents, recent documents, or the specified folder.

5. In the **Options** area, select the options to refine your search. If you want to search the text in the markups you added, make sure you select the **Search Markups** check box.

6. Click the **Search** button; the process of searching the text will start. Once the process finishes, the search results will be displayed in the **Search** panel, as shown in Figure 2.

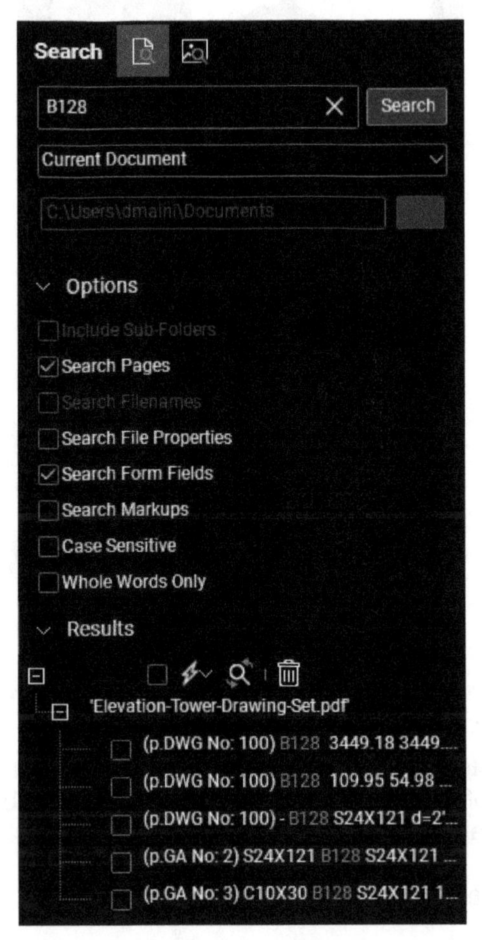

*Figure 2 The search results displayed in the **Results** area of the **Search** panel*

7. One by one, click on the search results to navigate to that text element in the PDF file.

***Note:** It is important to note that the scanned documents cannot be used to perform a text search. This is because the scanned documents are actually images.*

The Procedure for Performing Visual Searches in PDFs

The following is the procedure for performing visual searches in the PDF files:

1. Invoke the **Search** panel.

2. From the top of the **Search** panel, make sure the **Text** button is selected, refer to Figure 3.

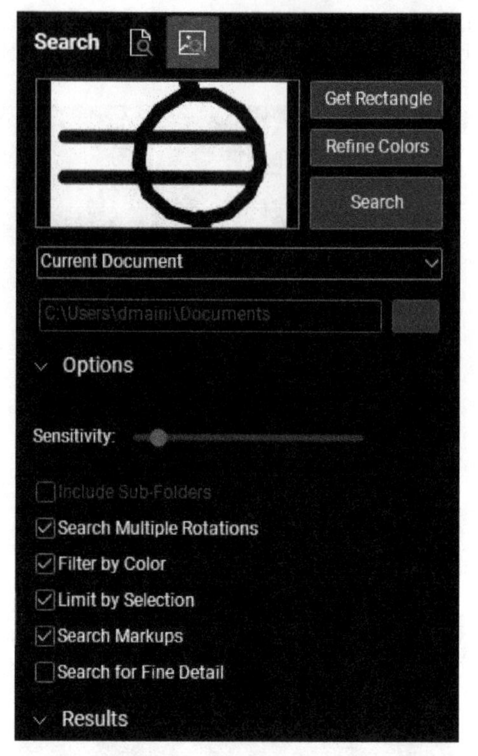

*Figure 3 The **Search** panel to perform visual search*

3. Click the **Get Rectangle** button and drag a box around the element in the PDF that you want to search; the image of the item enclosed in the rectangle will be displayed in the preview area. Figure 3 shows a receptacle that was enclosed in the rectangle to be searched.

4. If required, click the **Refine Colors** button to refine the color of the item to be searched.

5. In the drop-down list below the search field, select the scope of the search. You can select the current page, the current document, all open documents, recent documents, or the specified folder.

6. In the **Options** area, use the **Sensitivity** slider to control the sensitivity of the search.

7. Select the check boxes of the other options in this area to refine your search.

8. Click the **Search** button; the process of searching the item will start. Once the process finishes, the search results will be displayed in the **Search** panel. Figure 4 shows the **Results** area of the **Search** panel with some of the search results.

9. One by one, click on the search results to navigate to that searched items in the PDF file.

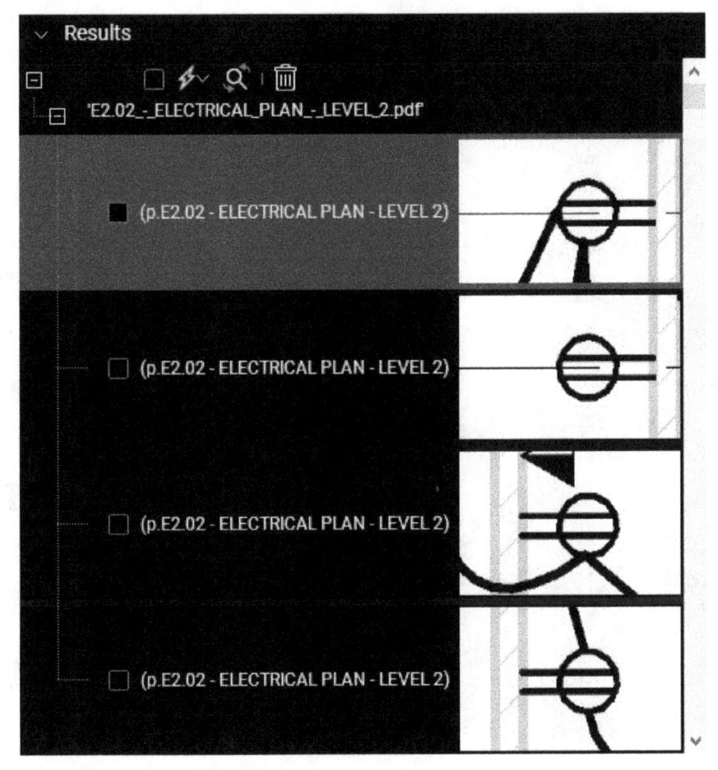

Figure 4 The results of the visual search

The Procedure for Applying the Count Measurement to the Searched Items

While performing quantity takeoffs, it is important to apply the count measurement to the searched elements so they are included in the takeoff counts. The following is the procedure for doing this:

1. Once you have performed the text or visual search, click the **Select All** check box from the toolbar available at the top in the **Results** area, as shown in Figure 5.

Figure 5 Selecting all the search results

2. Scroll down in the results list to make sure the unwanted items are not selected in the search results. If you see any unwanted items, clear the check box on the left of that item.

3. Click the **Check Options** flyout on the right of the **Select All** button and select **Apply Count Measurement to Checked**; the cascading menu will be displayed with all the count measurements that you can apply, as shown in Figure 6.

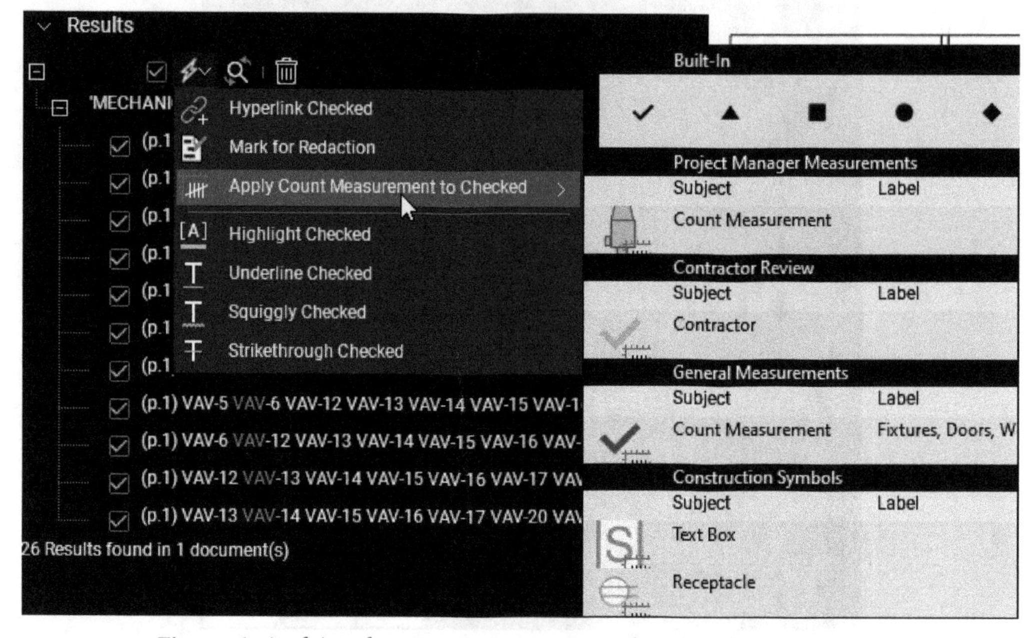

Figure 6 Applying the count measurement to the selected searched items

4. Select the required count measurement to apply. If the count measurement is associated with the custom columns to provide the cost, that cost information will be displayed in the **Markups List**.

> **What I do**
>
> *I use this method of applying count measurements extensively to takeoff quantities in the Mechanical, Electrical, and Plumbing discipline PDFs. It allows me to perform text or visual searches in those documents and then apply my custom count measurements to get the total cost of the takeoffs.*

COMPARING DIFFERENCES BETWEEN TWO VERSIONS OF THE SAME DOCUMENT

The AEC industry is known for making regular changes to the design and the related documents. As a result, you may see regular version changes in the PDF document set that you are provided. This also means that you need to regularly find differences between the two versions of documents that you are provided. Revu provides you with a couple of extremely powerful tools to compare the versions of the documents. Both these tools are discussed next.

The Compare Documents Tool

Menu: Document > Compare Documents

This tool allows you to select two versions of the same document and compare them. The resulting changes are enclosed in revision clouds for you to visually identify the changes between the documents. When you invoke this tool, the **Compare Documents** dialog box will be displayed, as shown in Figure 7.

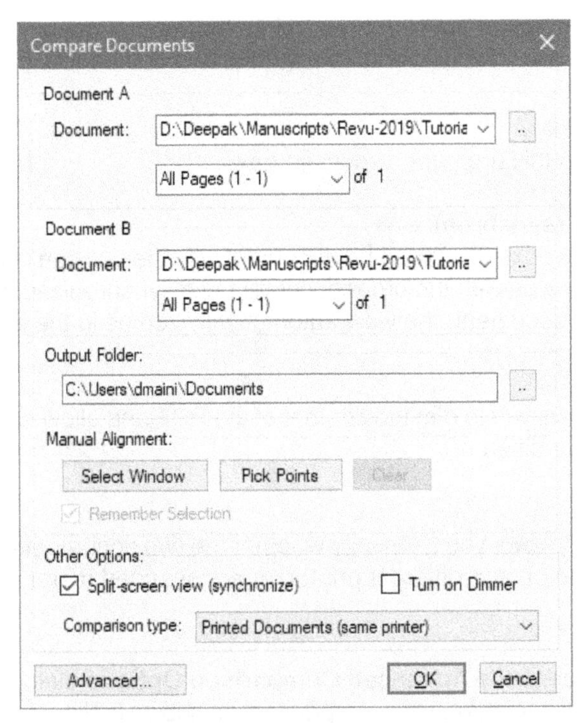

*Figure 7 The **Compare Documents** dialog box*

The options available in this dialog box are discussed next.

Document A / Document B Area

These areas allow you to select the two documents to compare and their sheets you want to compare. The documents currently opened in the Revu sessions are available in the lists available in these areas. If the documents you want to use for comparison is not available in the lists, you can click the **Browse** buttons on the right of the lists and browse and select the documents.

Output Folder Area

When you compare two documents, you get the option to save the compare results in a separate document. This area allows you to select the folder where the output document with the comparison results will be saved.

Manual Alignment Area

This area provides the options to manually aligned documents in cases where the two documents being compared are not printed at the same size sheets. The two options available in this area are discussed next.

Select Window

This method allows you to select a region of the document to compare by defining a window in the document and specify the region to compare.

Pick Points

This method allows you to specify three points on the first document and then the same three points on the second document to align the two documents manually.

Other Options Area

The options available in this area are discussed next.

Split-screen view (synchronize)

Selecting this check box will ensure the document with the revision clouds of the differences is displayed side by side with the other document in a synchronized view. As a result, when you zoom into one document, the view automatically zooms in the other document as well.

Turn on Dimmer

This check box allows you to dim the content of the PDF and allow easier visual inspections of the areas that are different.

Comparison type

This drop-down list allows you to specify whether the two documents being compared were printed on the same printer, different printer, or are scanned raster documents.

Advanced

Clicking this button displays the **Advanced Comparison Options** dialog box that allows you to specify advanced settings related to the comparison, as shown in Figure 8. The options available in this dialog box are discussed next.

Grid Size

This slider allows you to define the grid size in which the documents will be divided as regions to perform the comparison.

Pixel Density

This slider allows you to define the pixel density for each grid region to perform the comparison between the documents.

Color Sensitivity

This slider controls the color sensitivity threshold for the comparison.

Margin

This field allows you to set the margin around the border to ignore the areas for comparison.

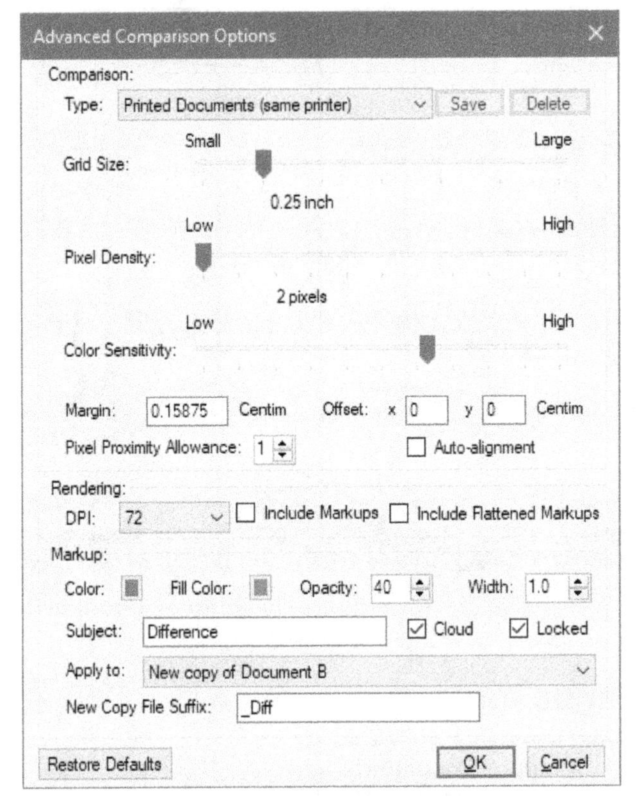

Figure 8 The **Advanced Comparison Options** dialog box

Rendering Area

This area provides the options for specifying the value for the document rendering and whether or not to include unflattened and flattened markups in comparison.

Markup Area

This area provides the options to define the color, fill color, opacity, and width of the revision clouds that will enclose the changes. You can also specify the document whose copy will be created with the changes and the suffix to be applied to the name of the new document.

Once you have specified the settings in the **Advanced Comparison Options** dialog box, click **OK** in this dialog box to return to the **Compare Documents** dialog box. Click **OK** in this dialog box as well to start the comparison process. Figure 9 shows the Revu screen displaying the two compared documents in a synchronized view.

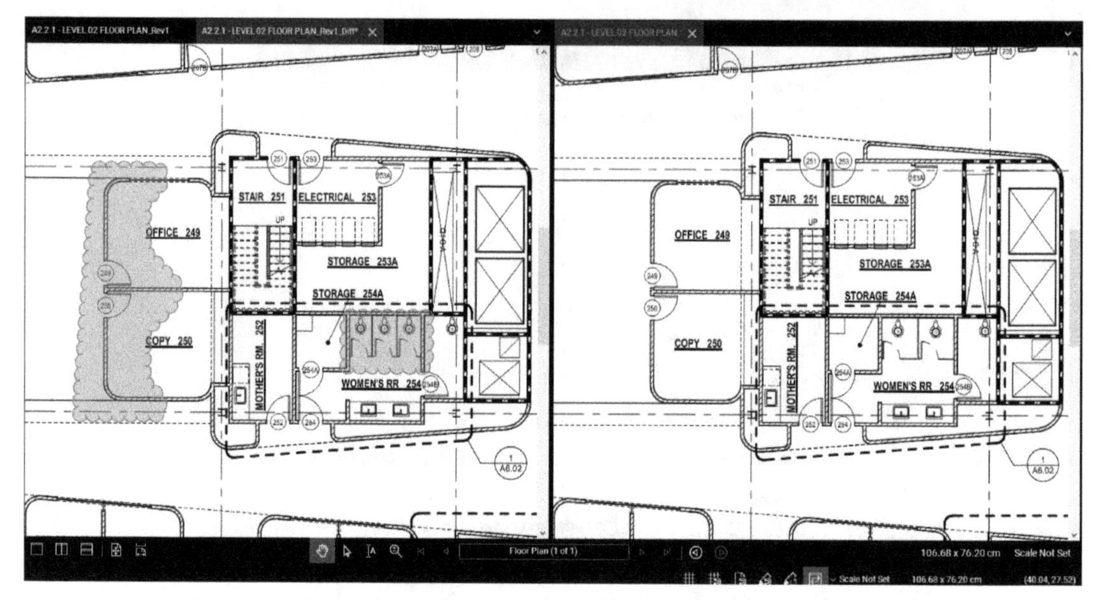

Figure 9 The two compared documents displayed in a synchronized view

The Overlay Pages Tool

Menu: Document > Compare Documents

Overlay Pages...

This tool allows you to overlay the two documents to compare on top of each other with different color linework. This allows you to easily identify the changes between the documents due to the change in the color of that linework. When you invoke this tool, the **Overlay Pages** dialog box will be displayed, as shown in Figure 10.

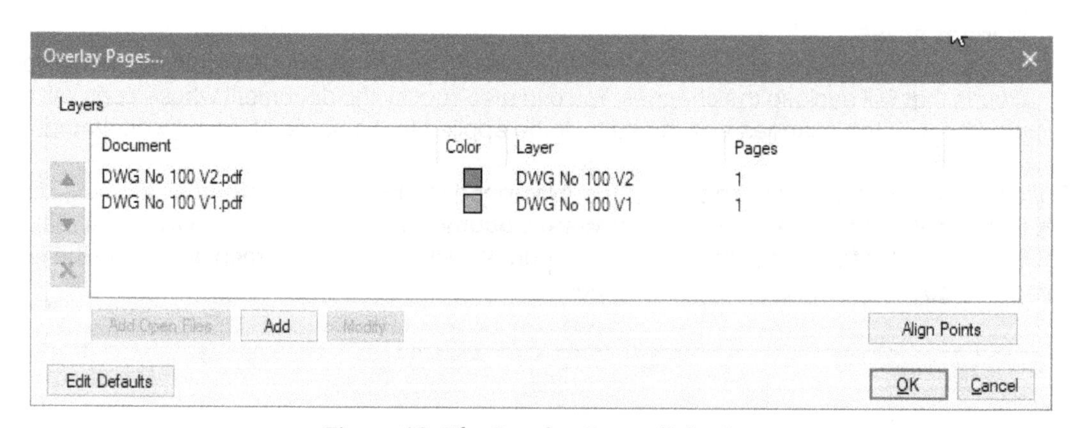

*Figure 10 The **Overlay Pages** dialog box*

The currently opened documents will be displayed in this dialog box. Alternatively, you can use the **Add** button to load the documents to overlay. Selecting a document in the **Layers** area and clicking the **Modify** button allows you to modify the color and file related settings of the selected document.

The **Align Points** button is used to manually align the documents to compare.

> ***What I do***
>
> *No matter how much we talk about BIM and 3D, we get PDF files from the MEP consultants on regular basis. I use this **Overlay Pages** tool to overlay all the services PDF together to create a single federated PDF and review if the services line up correctly or not.*

Figure 11 shows the zoomed in view of the two documents overlayed and showing the differences.

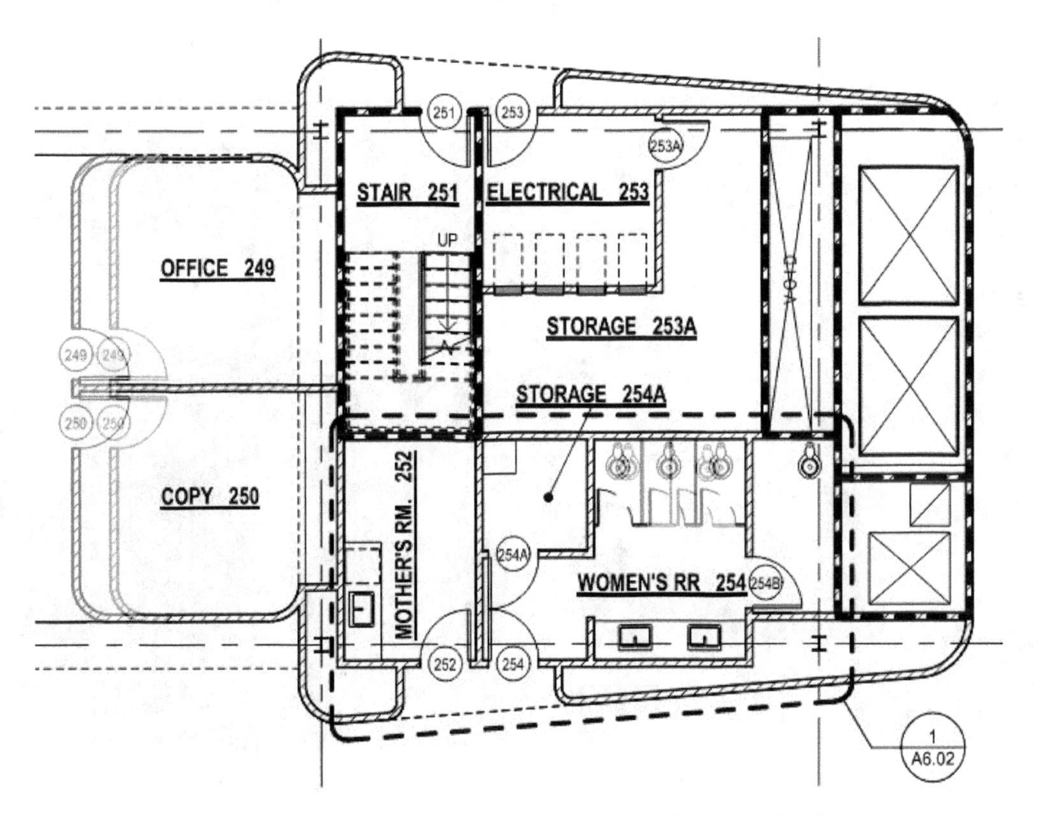

Figure 11 *The two compared documents overlayed on top of each other*

INSERTING IMAGES

Revu allows you to insert images from various sources, such as an existing image from a file, a new image from the camera of the current device, or an image from the scanner. Note that an image once inserted in the PDF is then embedded in it. This means that you do not separately provide the image file along with the PDF document. The tools to insert images using all these sources are discussed next.

Inserting an Image from an Existing File

Menu: Tools > Markup > Image > From File
Shortcut: I

This tool allows you to insert an existing image saved locally on your computer or on an accessible network drive. When you invoke this tool, the **Open** dialog box will be displayed, as shown in Figure 12.

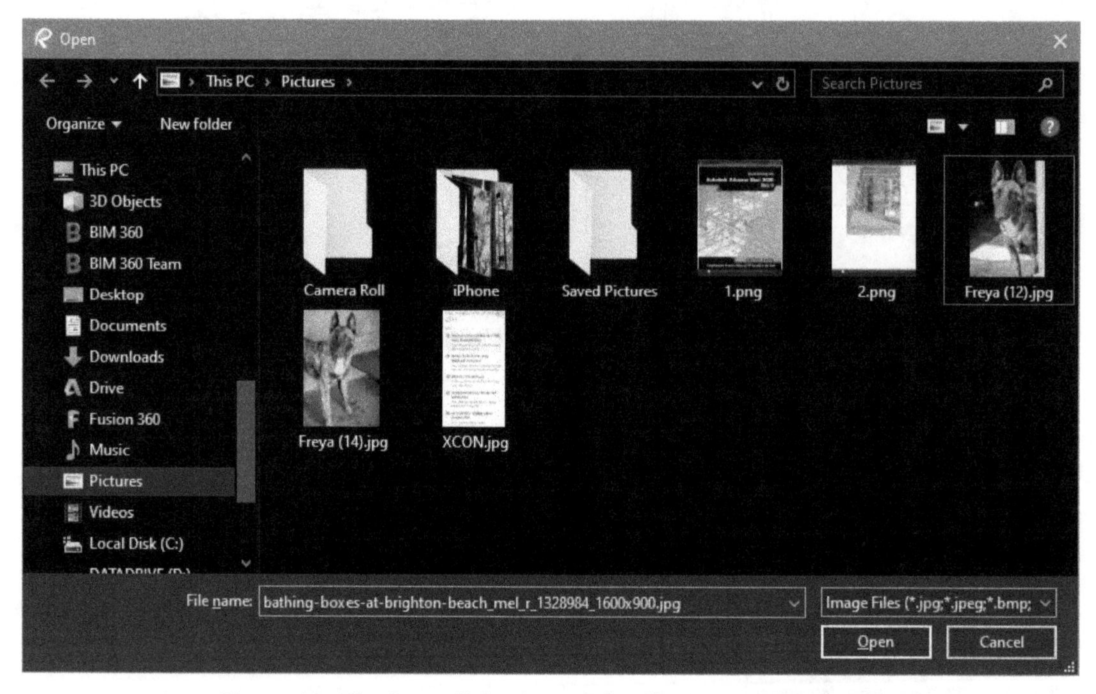

***Figure 12** The **Open** dialog box to select the image to be inserted*

Using this dialog box, you can browse and select the image to be inserted. Once you select the image and click **Open** in the dialog box, you will be returned to the PDF file and will be prompted to select a region to insert the image. You can drag two opposite corners to insert the image in the specified size box or click on a location to place the original size image. Note that holding down the SHIFT key allows you to insert a distorted scale image.

What I do

I normally insert a smaller size image on the main page of the PDF and also insert a bigger size image on a new page of the PDF so I have more area to add comments related to that image. I then add hyperlinks to take the user from the smaller image to the bigger image on a different page and then back to the smaller image on the original page.

Inserting an Image from the Device Camera

Menu: Tools > Markup > Image > From Camera

From Camera... Ctrl+Alt+I

This tool allows you to use the camera of your device to take a picture and then insert it as an image in the current PDF file. When you invoke this tool, the camera of the current device is opened and the camera view is displayed in the preview window. You can take the picture using this window and then insert it similar to the method discussed earlier.

Inserting an Image from Attached Scanner

Menu: Tools > Markup > Image > From Scanner

From Scanner... Shift+I

This tool allows you to use the scanner attached to the current device and select the scanned image to be inserted in the PDF file. When you invoke this tool, the **Select Source** dialog box will be displayed, as shown in Figure 13.

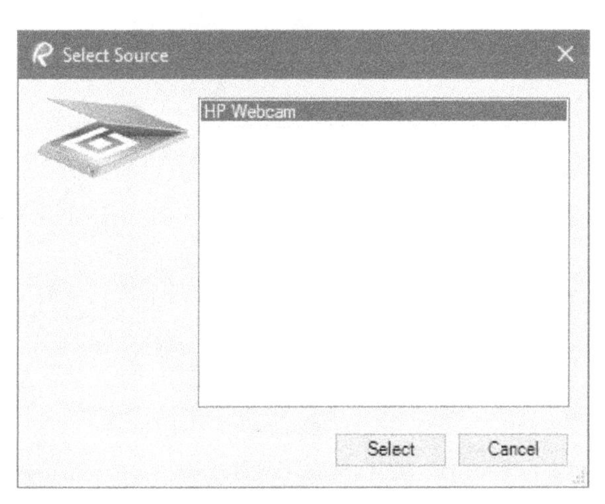

Figure 13 *The* ***Select Source*** *dialog box to select the scanner*

Once you select the scanner, the dialog box related to the attached scanner will be displayed using which you can accept the current scan or scan another image to be used.

Cropping the Inserted Image

Menu: Tools > Markup > Image > Crop Image

This tool allows you to crop an existing image. When you invoke this tool, you will be prompted to select the region on an image to crop. As you move the cursor over an image, it changes to a crosshair and you can drag two opposite corners to crop the image.

Hands-on Tutorial

In this tutorial, you will complete the following tasks:
1. *Perform text search in a PDF file.*
2. *Perform visual search in a PDF file.*
3. *Compare two different versions of a document and review changes.*
4. *Use the **Overlay Pages** tool to overlay the Electrical and Mechanical files.*
5. *Insert an existing image in the current PDF file.*
6. *Create hyperlinks between images on different pages of the current PDF file.*

Section 1: Performing Text Search in the PDF File

In this section, you will open the **Architectural-Plan-10.pdf** file from the **C10** folder. You will then use the **Search** panel to search for the text "Office". You will then add a default count measurement to all the searched instances.

1. From the **C10** folder, open the **Architectural-Plan-10.pdf** file.
2. From the **Panel Access Bar**, invoke the **Search** panel.
3. From the top in this panel, make sure the **Text** option is selected to perform the text search, refer to Figure 14.
4. In the text field, enter **office** as the text to search, as shown in Figure 14.
5. From the drop-down list below the text field, select the **Current Page (Floor Plan)** option, if not already selected, as shown in Figure 14.

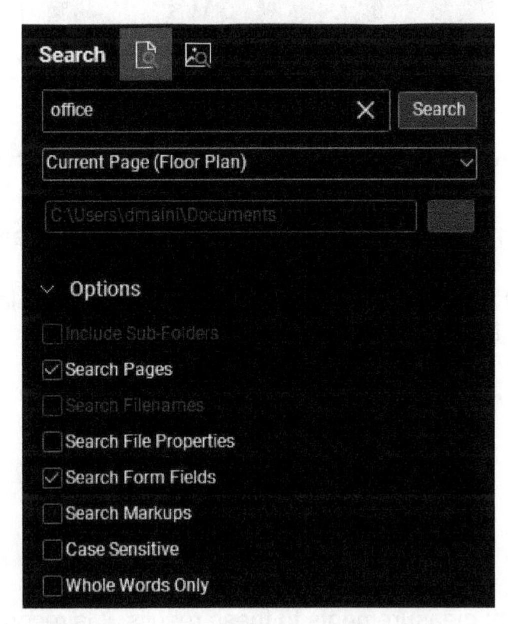

Figure 14 Configuring settings to perform the text search

6. In the **Options** area, clear the **Search Markups** check box, if selected.

7. Clear the **Case Sensitive** check box, if selected.

   You are now ready to perform the search.

8. Click the **Search** button available on the right of the text field.

   Once the search is performed and the results are displayed in the **Results** area. Also, the lower left corner of the **Results** area informs you about the number of results, which in this case is 29, as shown in Figure 15.

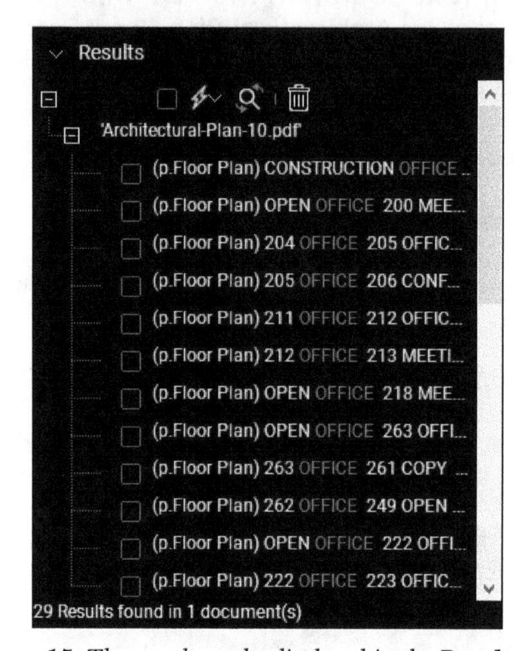

*Figure 15 The search results displayed in the **Results** area*

You will now select all the search results and apply the count measurement them.

9. Select the **Select All** check box at the top in the **Results** area, labeled as **1** in Figure 16; all the search results are selected.

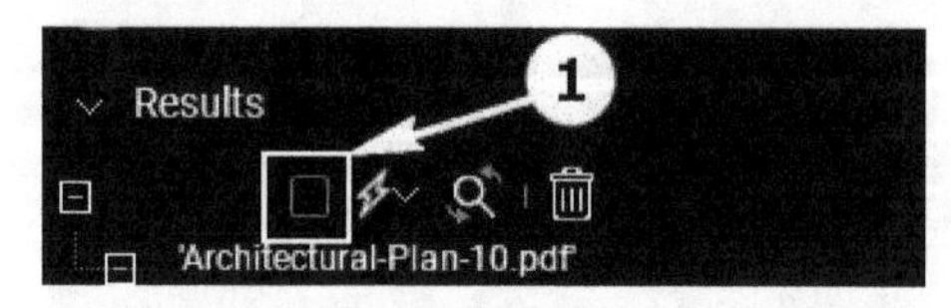

Figure 16 Selecting all the search results

Before you apply count measurements to these results, it is recommended to review these results to ensure no unwanted text is selected.

10. Zoom to the top left of the PDF file. This ensures that when you click on a search result, you are automatically navigated to its location.

11. Click on the first search result; the text is highlighted near the bottom right of the window and you are navigated to the area where this text is located.

    Notice that this text is located in the titleblock. You do not need to count this text.

12. From the **Search** panel > **Results** area, clear the check box on the left of the first search result.

13. Similarly, click on the other search results and notice that they are the offices in the plan.

    You are now ready to apply the count measurement to the selected search results. In this case, you will apply the default count measurement.

14. Scroll to the top of the search results.

15. From the toolbar at the top of the search results, click **Check Options** > **Apply Count Measurement to Checked**; a cascading menu is displayed, as shown in Figure 17.

16. From the **Built-In** area, select the default check box, as shown in Figure 17; all the selected search results are applied a count measurement and check marks are displayed on them in the **Main Workspace**.

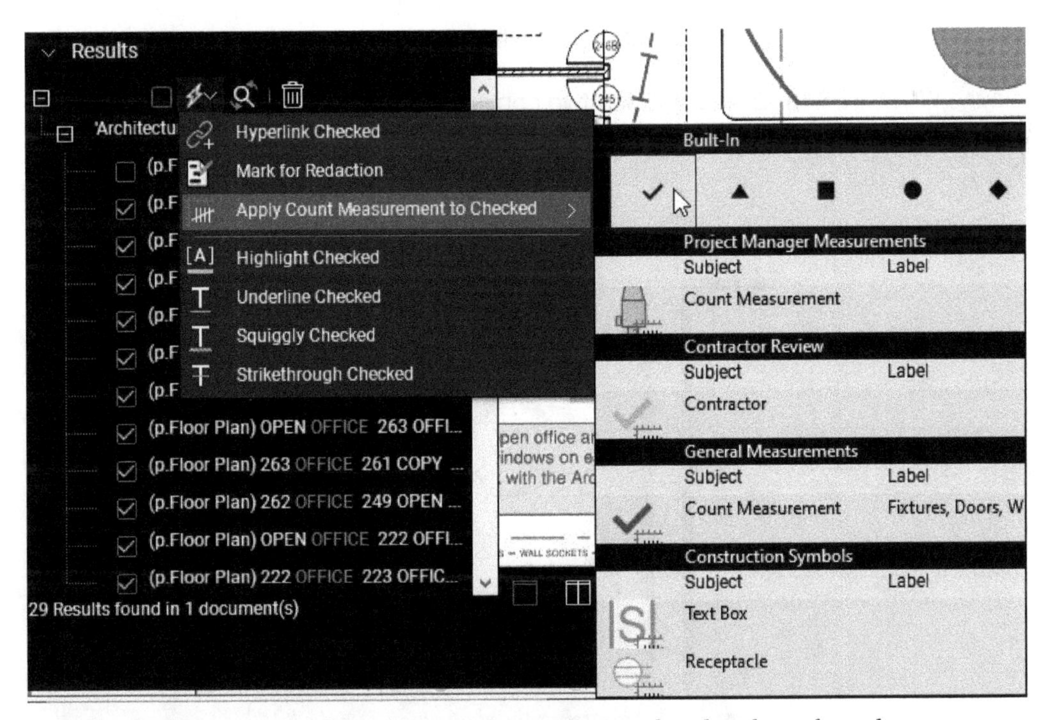

Figure 17 Applying the count measurement to the selected search results

17. Invoke the **Markups List** and notice that there are 28 count measurements added to the document, as shown in Figure 18.

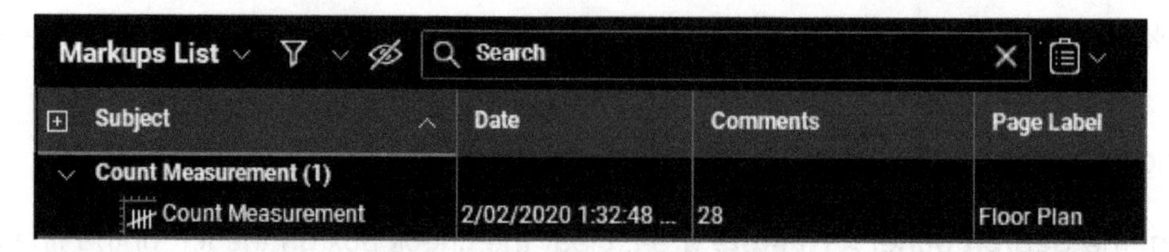

***Figure 18** The **Markups List** shows the number of count measurements added*

18. Hide the **Markups List**.

Section 2: Performing Visual Search in the PDF File

In this section, you will perform a visual search in the current PDF file.

1. Navigate to the **WOMEN'S RR 254** room around the center of the plan.

2. From the top in the **Search** panel, click the **Visual** button to perform the visual search; the options in the **Search** panel are modified to the visual search options.

   You will now use the **Get Rectangle** button and draw a box around a block in the **WOMEN'S RR 254** room that you want to search.

3. Click the **Get Rectangle** button; the cursor changes to a crosshair.

4. In the **WOMEN'S RR 254** room, click two opposite corners labeled as **1** and **2** in Figure 19 to select the shape you want to search.

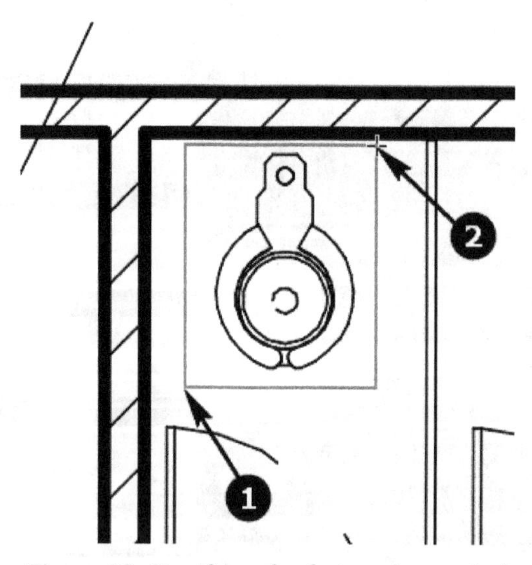

***Figure 19** Specifying the shape to be searched*

On doing so, that shape is displayed in the preview area of the **Search** panel, as shown in Figure 20.

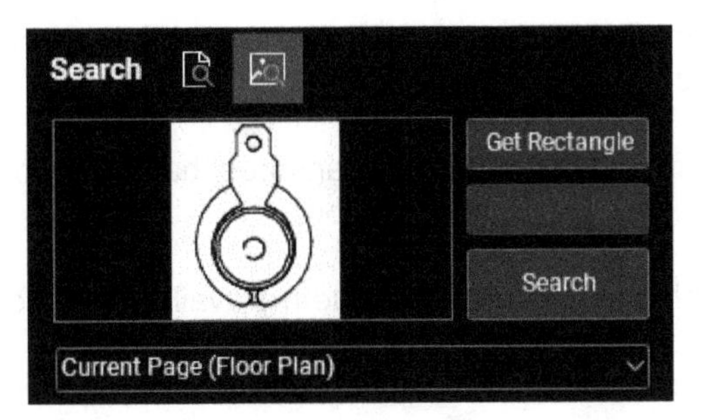

*Figure 20 The specified shape displayed in the **Search** panel*

5. In the **Options** area of the **Search** panel, drag the **Sensitivity** slider to around a quarter of the distance.

6. Select the **Search Multiple Rotations** check box, if not already selected.

   You are now ready to perform the search.

7. Click the **Search** button; the process of performing visual search starts.

   Once the search process is completed, the search results are displayed in the **Results** area and the total number of items found are listed at the bottom left of the **Results** area, as shown in Figure 21.

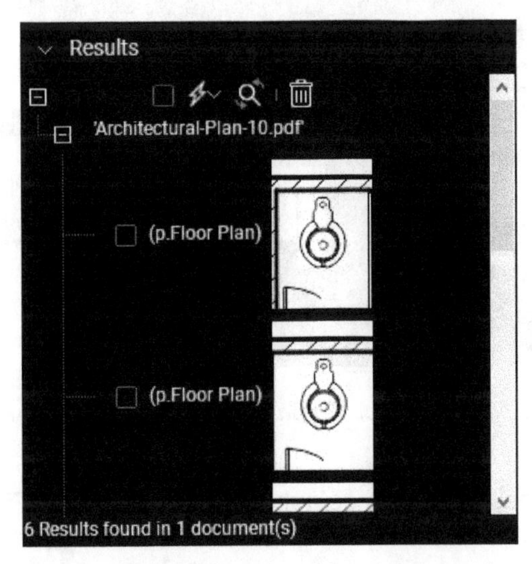

Figure 21 The results of the visual search

You will now select all the search results and apply the count measurement them.

8. Select the **Select All** check box at the top in the **Results** area, refer to Figure 16 above; all the search results are selected.

   Before you apply count measurements to these results, it is recommended to review these results to ensure no unwanted text is selected.

9. Navigate to the top left of the PDF file. This ensures that when you click on a search result, you are automatically navigated to its location.

10. Click on the first search result; the searched item is available on the lower right of the window.

11. Similarly, click on the other search results and review their locations.

    You are now ready to apply the count measurement to the selected search results. In this case, you will apply the default count measurement.

12. Scroll to the top of the search results.

13. From the toolbar at the top of the search results, click **Check Options** > **Apply Count Measurement to Checked**; a cascading menu is displayed, as shown in Figure 22.

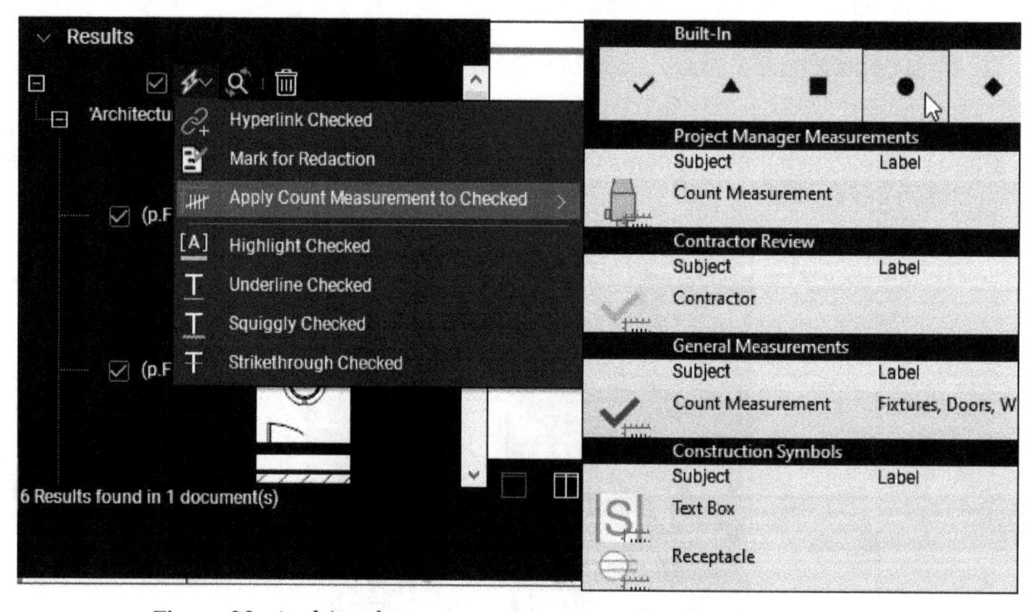

Figure 22 Applying the count measurement to the selected search results

14. From the **Built-In** area, select the default circle, as shown in Figure 22; all the selected search results are applied a count measurement and Red circles are displayed on them in the **Main Workspace**.

15. Invoke the **Markups List** and notice that there are 6 count measurements added to the document.

16. Hide the **Markups List**.

17. Save and close the current PDF file.

Section 3: Comparing Two Versions of a Document

In this section, you will open the **1st Floor Plan.pdf** and **1st Floor Plan_Rev 1.pdf** files from the **C10** folder and compare the differences between the two documents.

1. From the C**10** folder, open the **1st Floor Plan.pdf** and **1st Floor Plan_Rev 1.pdf** files.

   Before proceeding further, make a note of whether the **1st Floor Plan.pdf** file is listed as the first tab above the **Main Workspace** or the second tab.

2. Hide all the panels that are turned on to make more space for the **Main Workspace**.

3. From the **Menu Bar**, click **Document > Compare Documents**; the **Compare Documents** dialog box is displayed, as shown in Figure 23.

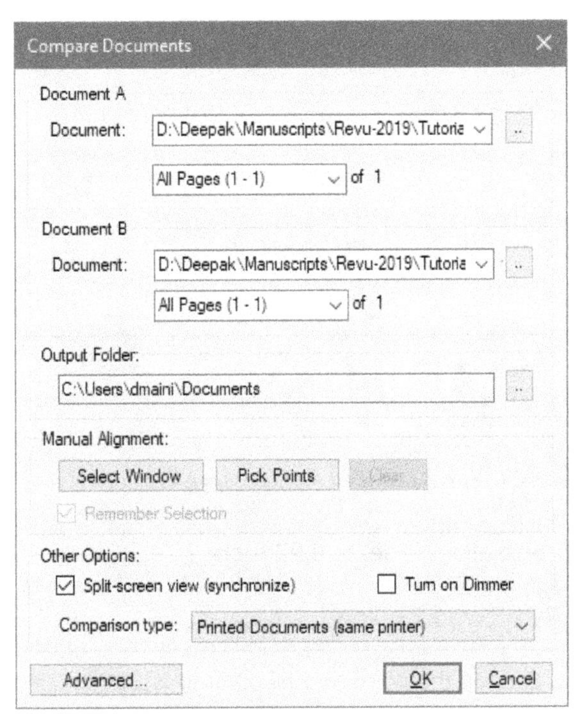

*Figure 23 The **Compare Documents** dialog box*

The **Document A** and **Document B** areas already show the two documents that are currently opened to be compared. You will now change the output folder.

4. From the **Output Folder** area, click the **Browse** button and browse and select the **C10** folder.

5. In the **Other Options** area, make sure the **Split-screen view (synchronize)** check box is selected.

   You will now configure some advanced settings for the comparison.

6. Click the **Advanced** button on the lower left of the dialog box; the **Advanced Comparison Options** dialog box is displayed, as shown in Figure 24.

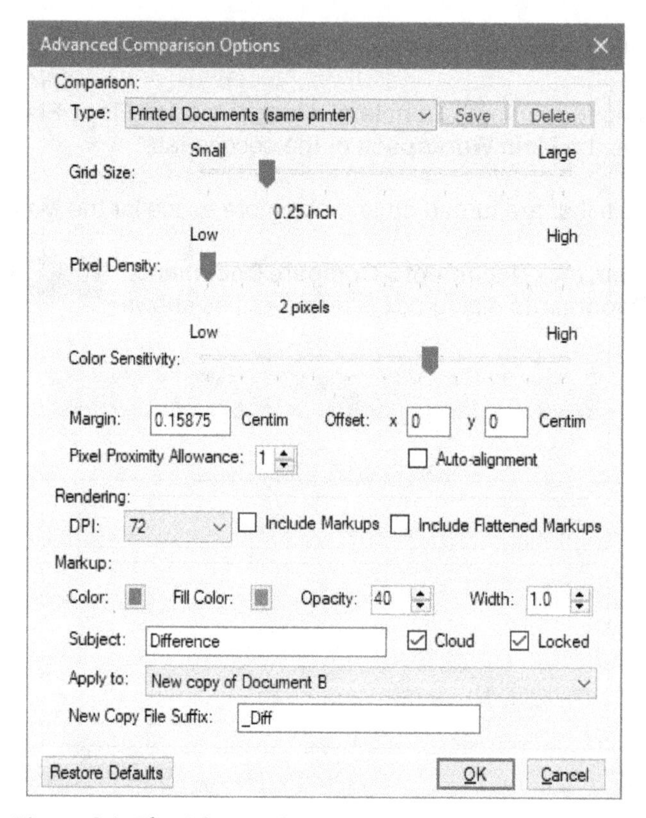

*Figure 24 The **Advanced Comparison Options** dialog box*

7. Make sure the **Grid Size** slider is set to **0.25 inch**.

8. Set the **Pixel Density** slider to **2 pixels**.

9. In the **Markup** area, change the color and fill color to **Blue**.

   You will now select the option to create a new document with the revision clouds around the changes. However, the option you select from this list will depend on whether the **1st Floor Plan.pdf** document is listed as the first tab or the second tab above the **Main Workspace**.

10. From the **Apply to** drop-down list, select **New copy of Document B** if the **1st Floor Plan.pdf** file is listed as the first tab at the top of the **Main Workspace**. Else, select **New copy of Document A** from this drop-down list.

The new document will have the same name and **_Diff** as the suffix because that is listed in the **New Copy File Suffix** edit box. You will accept this as the value.

11. Click **OK** in the **Advanced Comparison Options** dialog box; you are returned to the **Compare Documents** dialog box.

    You are now ready to compare the two documents.

12. Click **OK** in the **Compare Documents** dialog box; the two documents are compared and a new PDF file, which is the copy of the **1st Floor Plan_Rev_1.pdf** is created with a suffix of **_Diff**. Also, the new and **1st Floor Plan.pdf** files are displayed in the synchronized view in the **Main Workspace**, as shown in Figure 25.

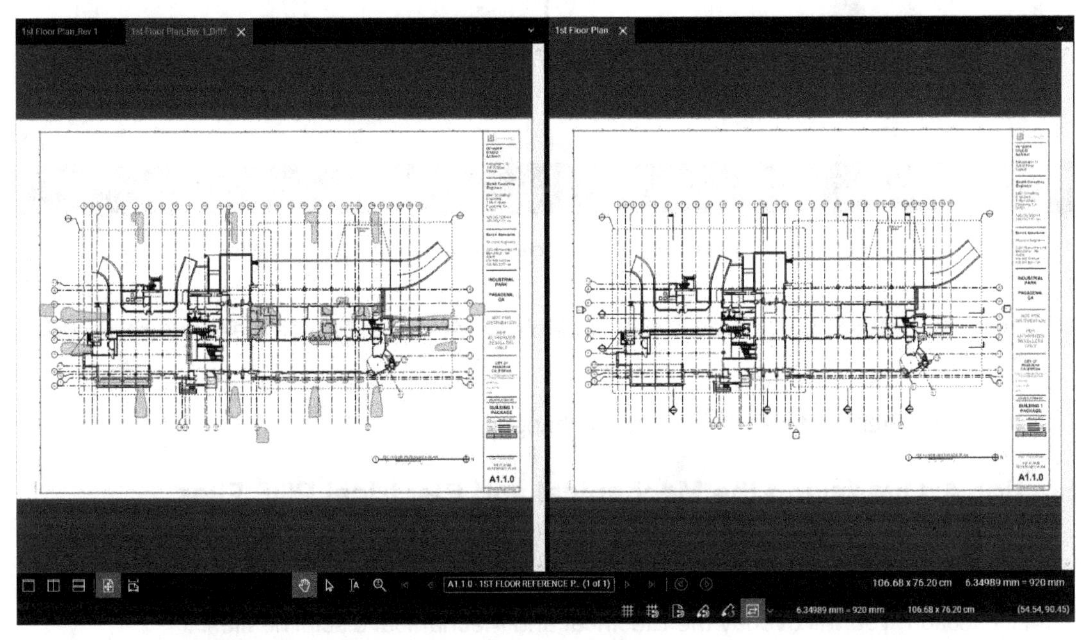

Figure 25 *The compared documents displayed in a synchronized view*

Notice that every change is enclosed in a revision cloud in the new document. You will now review these changes.

13. Zoom close to one of the revision clouds in the new document; the view automatically zooms in the other document as well for you to review the changes between the original document and the revised document, refer to Figure 26.

14. Similarly, navigate to the other revision clouds in the new document and notice how the view automatically updates in the other document as well.

15. Save the changes in the **1st Floor Plan_Rev 1_Diff.pdf** and close it.

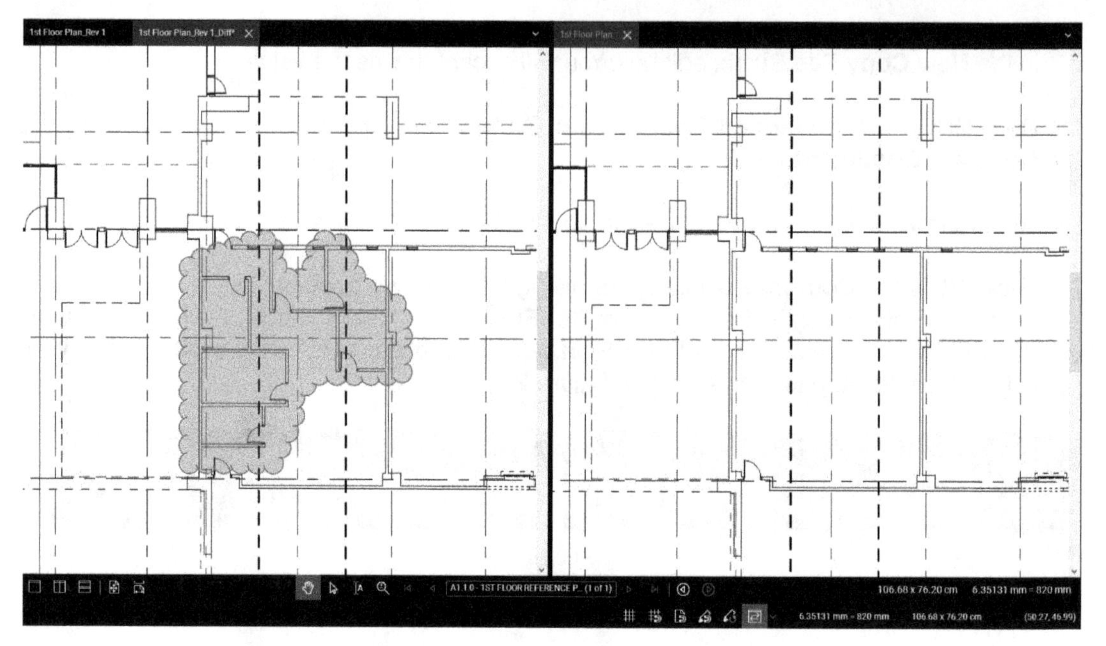

Figure 26 Reviewing the differences between the two PDF files

16. From the **Status Bar** at the bottom of the **Main Workspace**, turn off the **Synchronize View** button.

17. Close the other two files without saving changes.

Section 4: Overlaying the Mechanical and Electrical PDF Files

As mentioned earlier, no matter how much we talk about BIM and 3D, there are always PDF files provided to us for various services. To ensure all the services line up properly, you can use the **Overlay Pages** tool and overlay the services PDFs together and review them. In this section, you will overlay the Electrical and Mechanical discipline files.

1. From the C**10** folder, open the **Electrical-Layout.pdf** and **Mechanical-Layout.pdf** files.

2. One by one review the two PDF files and notice that they do not appear to be printed with the same scale. Also, notice on the right of the **Status Bar** that the two PDF files are created on different size sheets.

   Because of the differences between the two files, you will manually align them to ensure they sit correctly when overlayed together.

3. From the **Menu Bar**, click **Document > Overlay Pages**; the **Overlay Pages** dialog box is displayed, as shown in Figure 27.

   You will use the default color options for the two files. However, you need to manually align the two documents.

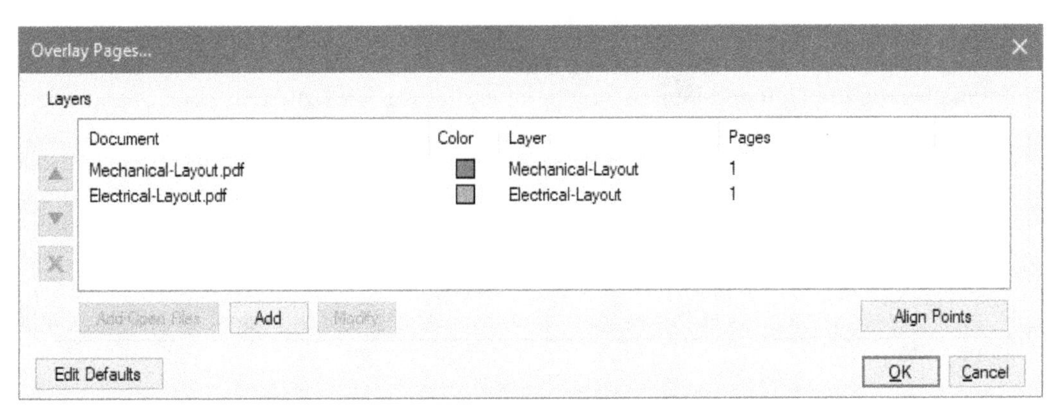

*Figure 27 The **Overlay Pages** dialog box*

4. Click the **Align Points** button; the **Bluebeam Revu** information box is displayed informing you that you need to choose 3 points on both documents for manual alignment.

5. Click **OK** in the information box; you are returned to the PDF files.

   To align the documents, you will select the grid intersection points **A01**, **A10**, and **C10**.

6. Turn on the **Snap to Content** option from the **Status Bar**.

7. From the **View** menu, select the **Disable Line Weights** option to turn off the line weights.

8. In the first document, zoom close to the **A01** grid intersection point and select it as the first point of alignment, as shown in Figure 28. Note that this figure shows the first point in the Mechanical file. In your case, this could be in the Electrical file, depending on which file was opened first.

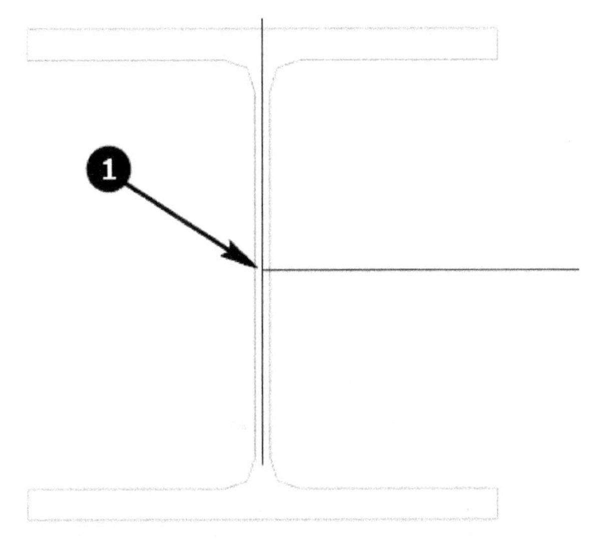

Figure 28 Selecting the first alignment point

9. Zoom close to the **A10** grid intersection point and select it as the second point of alignment.

10. Zoom close to the **C10** grid intersection point and select it as the third point of alignment.

    As soon as you specify the third point of alignment, the second document is opened. Notice that in the second document, the locations of the three points from the first document are shown in Gray. As you can see, these points are way off from the location in the second document. As a result, it was important to manually align the two documents.

11. Repeat the same process in the second document and select the same three points in the same order.

    As soon as you select the third point of alignment, the **Overlay Pages** dialog box is redisplayed.

12. Click **OK** in the **Overlay Pages** dialog box; a new document is created and the two overlayed documents are combined in this file and displayed on the screen, as shown in Figure 29.

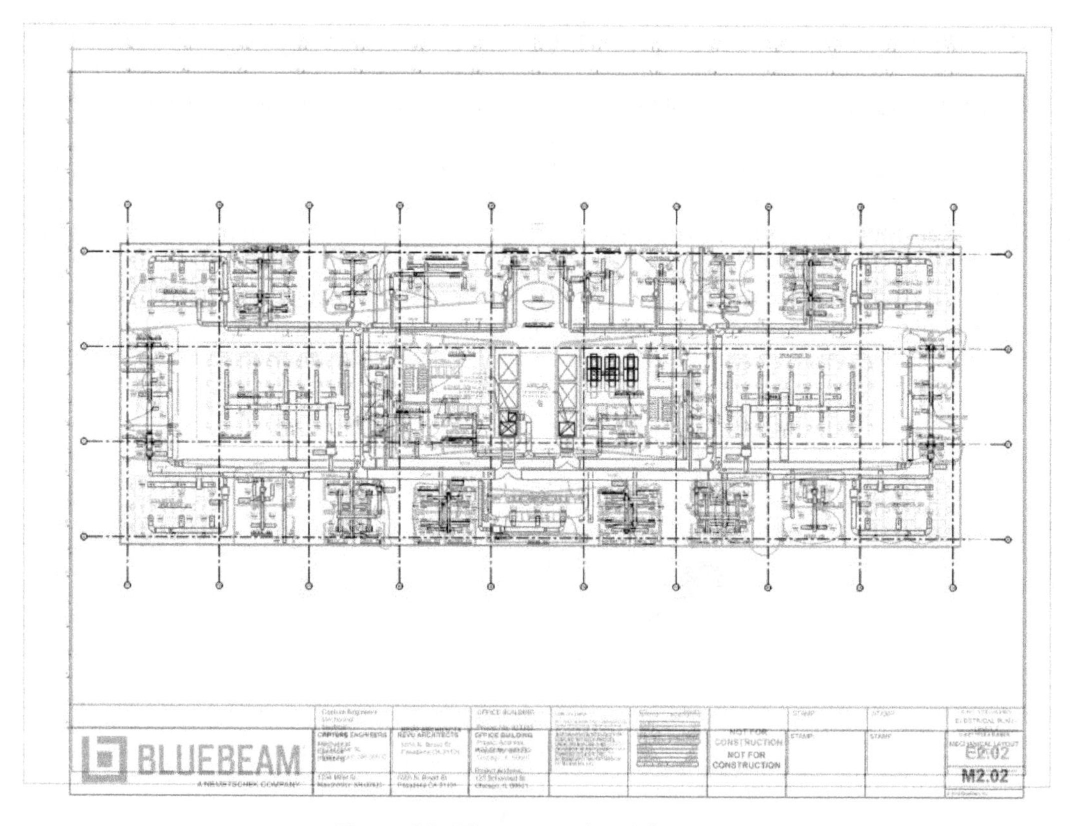

Figure 29 The two overlayed documents

13. Zoom in and notice how the Electrical and Mechanical discipline items are located with respect to each other. The Mechanical items are displayed in Red and the Electrical items are displayed in Green as these are the default colors.

*Tip: You can open the **Layers** panel and control the visibility of the **Electrical-Layout** and **Mechanical-Layout** layers.*

14. Save the **Overlay** file in the **C10** folder.

15. Close all the files.

Section 5: Inserting an Existing Image

In this section, you will open the **Duct-Issue.pdf** file and then insert an existing image in this file close to the markup in the file. You will then insert a new blank page and then insert a bigger copy of the same image. Finally, you will create hyperlinks between the two pages of the PDF.

1. From the C**10** folder, open the **Duct-Issue.pdf** file.

2. Navigate to the cloud+ markup.

   You will now insert an existing image to the right of this cloud. The image you will use is available in the **C10** folder.

3. From the **Menu Bar**, click **Tools > Markup > Image > From File**. Alternatively, you can press the **I** key on the keyboard. On doing so, the **Open** dialog box is displayed, as shown in Figure 30.

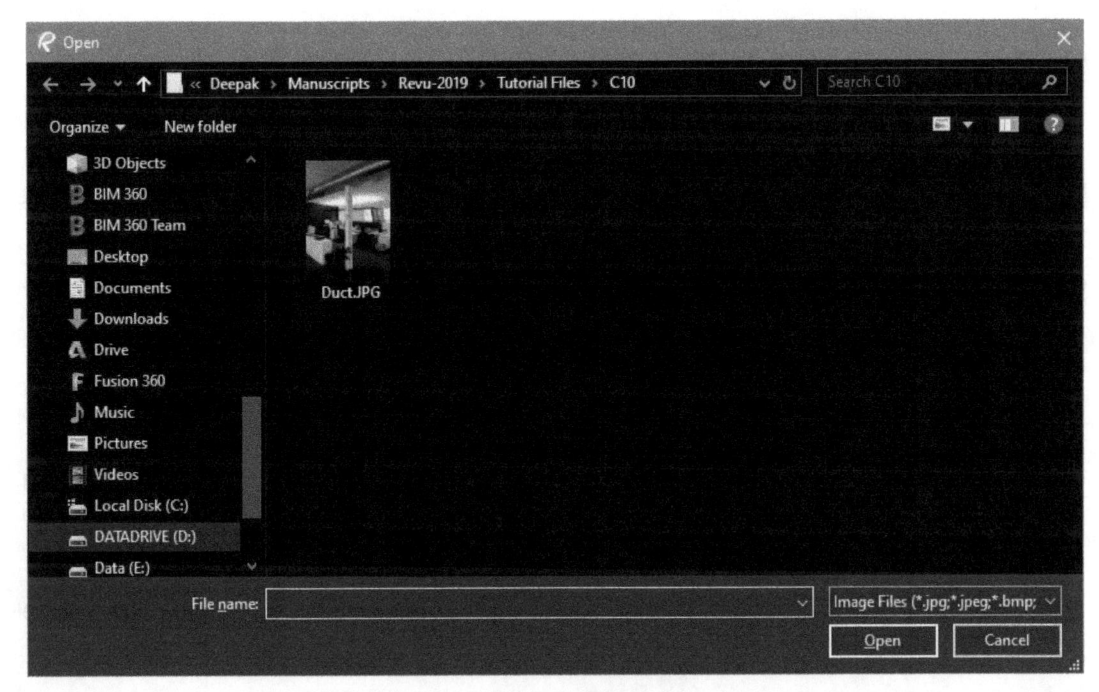

*Figure 30 The **Open** dialog box to select the image to insert*

4. Browse to the **C10** folder; the **Duct.JPG** image is available in the folder.

5. Double-click on the **Duct.JPG** file; you are returned to the PDF file and are prompted to select a region to place the image.

6. Drag two opposite corners to the right of the cloud to place the image, as shown in Figure 31.

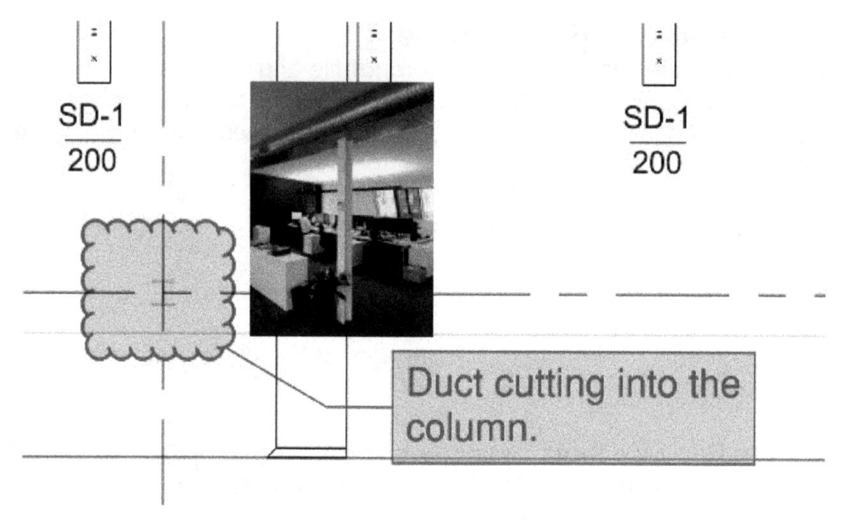

Figure 31 *The image inserted on the right of the cloud*

7. Zoom to the extents of the page and notice that the image looks tiny on the page.

   You will now insert a new blank page after the current page and insert a bigger size of the same image.

8. Using the **Thumbnails** panel, insert a blank **ANSI C** (Imperial) or **Iso_A3** (Metric) sheet with landscape orientation after the current page.

9. Press the **I** key on the keyboard to invoke the **Image From File** tool; the **Open** dialog box is displayed.

10. Select the same image and place a bigger version of it on the new page you inserted, as shown in Figure 32.

    With more space around this image, you can add markups to explain what is going on and how to resolve this issue. You can also add additional images on this page to discuss the problem and its potential solutions.

    To ensure you are navigated to this page from the smaller image, you will now create hyperlinks between this image and the smaller image on the first page of the PDF.

11. Navigate back to the cloud+ markup and the smaller image.

Figure 32 The bigger size image inserted on the new page

12. From the **Tools** menu, invoke the **Hyperlink** tool and click two opposite corners around the image to define the hyperlink region.

13. In the **Action** dialog box, select the **Jump to** radio button.

14. In the **Page** spinner, enter **2** to jump to page 2.

15. Click **OK** in the **Action** dialog box.

16. Press the ESC key to exit the **Hyperlink** tool.

    To ensure the people reviewing this PDF file now there is a hyperlink, you will edit the cloud+ callout.

17. Double-click on the cloud+ callout and add the following text:

    Click on the image to know more.

18. Click on the image and notice that the hyperlink navigates you to page 2.

Now to return back to the location of the smaller image on page 1, you will add a text and then create a hyperlink on it.

19. Using the **Project Manager's Text Box** tool, add a 36 font size text on the right of the image that says **Click here to return to the Original Image**, as shown in Figure 33.

Figure 33 *The text added to the right of the image*

You will now create a hyperlink on this text.

20. From the **Tools** menu, invoke the **Hyperlink** tool.

21. Draw a box around the text to define the region for the hyperlink; the **Action** dialog box is displayed.

    Because you want to return to the location of the original image, you need to create a place using the **Action** dialog box.

22. In the **Action** dialog box, click the **Create** button on the right of the **Place** radio button; the **Place** dialog box is displayed.

23. Enter **Image** as the name of the place.

24. Click the **Get Rectangle** button to define the location of the place; you are returned to the PDF file.

25. Navigate back to the location of the smaller image on page 1.

26. Specify two opposite corners around the image to image and the cloud+ markup to define the region for the place you are creating, as shown in Figure 34.

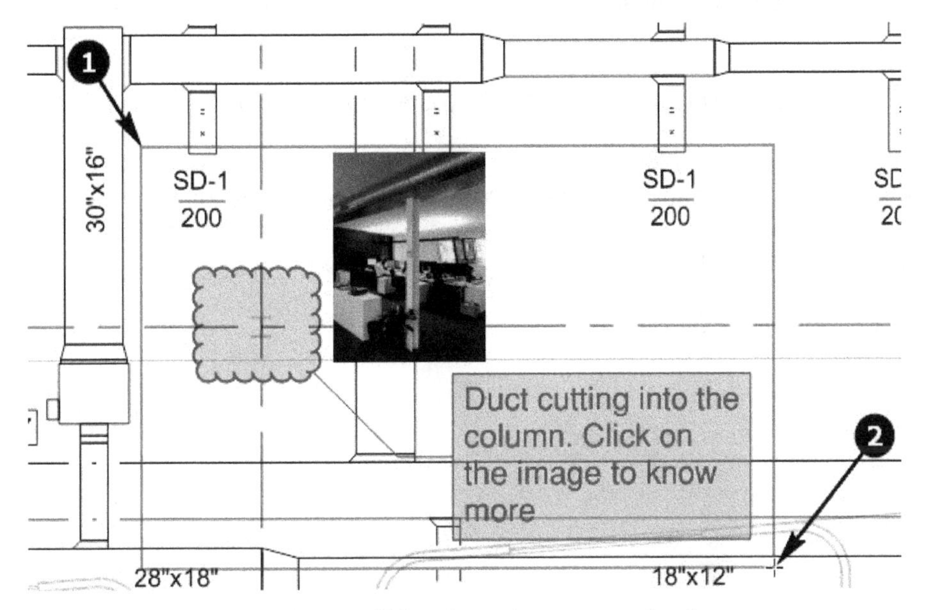

Figure 34 Specifying the region to create the place

You are returned to the **Place** dialog box. This dialog box now shows the information about the place you created, as shown in Figure 35.

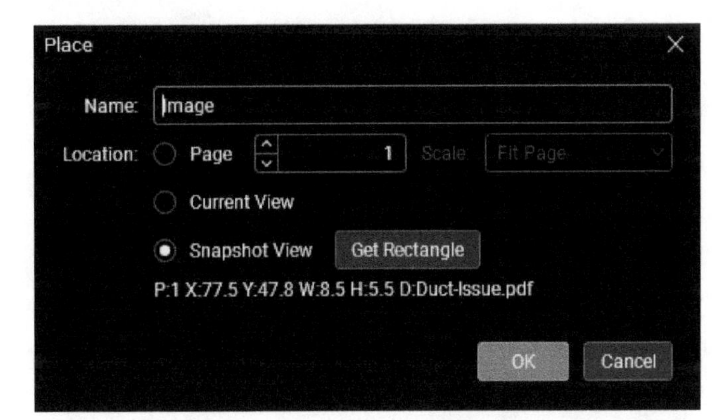

*Figure 35 The **Place** dialog box showing the information about the place you created*

27. Click **OK** in the **Place** dialog box to return to the **Action** dialog box.

28. Click **OK** in the **Action** dialog box.

29. Press the ESC key to exit the **Hyperlink** tool.

    You will now test the hyperlinks.

30. Click on the hyperlink on the smaller image; you are navigated to the bigger image on page 2.

31. Click on the hyperlink on the text on page 2; you are navigated to the smaller image and the cloud+ markup.

    This is an extremely handy workflow to insert images and creating hyperlinks.

32. Zoom to the extents of the PDF file.

33. Save and close the PDF file.

Skill Evaluation

Evaluate your skills to see how many questions you can answer correctly. The answers to these questions are given at the end of the book.

1. In Revu, you can only search for texts. (True/False)

2. You cannot compare two different versions of a PDF file in Revu. (True/False)

3. The **Overlay Pages** tool does not work on the PDF files. (True/False)

4. The images inserted in the PDF are not embedded inside it. (True/False)

5. You can only insert an existing image in Revu. (True/False)

6. Which tool is used to insert an image using the camera of the current device?

   (A) **Image** (B) **From Camera**
   (C) **From File** (D) **From Scanner**

7. Which tool allows you to compare two PDF files?

   (A) **Compare Documents** (B) **Compare Files**
   (C) **Show Changes** (D) **Compare**

8. Which tool allows you to overlay PDF files?

   (A) **Compare** (B) **Overlay Pages**
   (C) **Share** (D) **None**

9. Which option in the **Search** panel allows you to perform visual searches?

   (A) **Text** (B) **Image**
   (C) **Block** (D) **Visual**

10. Which option in the **Overlay Pages** dialog box allows you to manually align points of the PDF files?

    (A) **Align Points** (B) **Link Points**
    (C) **Share Points** (D) **None**

Class Test Questions

Answer the following questions:

1. Explain briefly the process of performing a visual search in a PDF file.
2. Explain briefly how to apply a count measurement to the searched items.
3. Explain the process of manually aligning PDFs while overlaying them.
4. Explain the process of comparing two documents.
5. Explain the process of inserting an existing image.

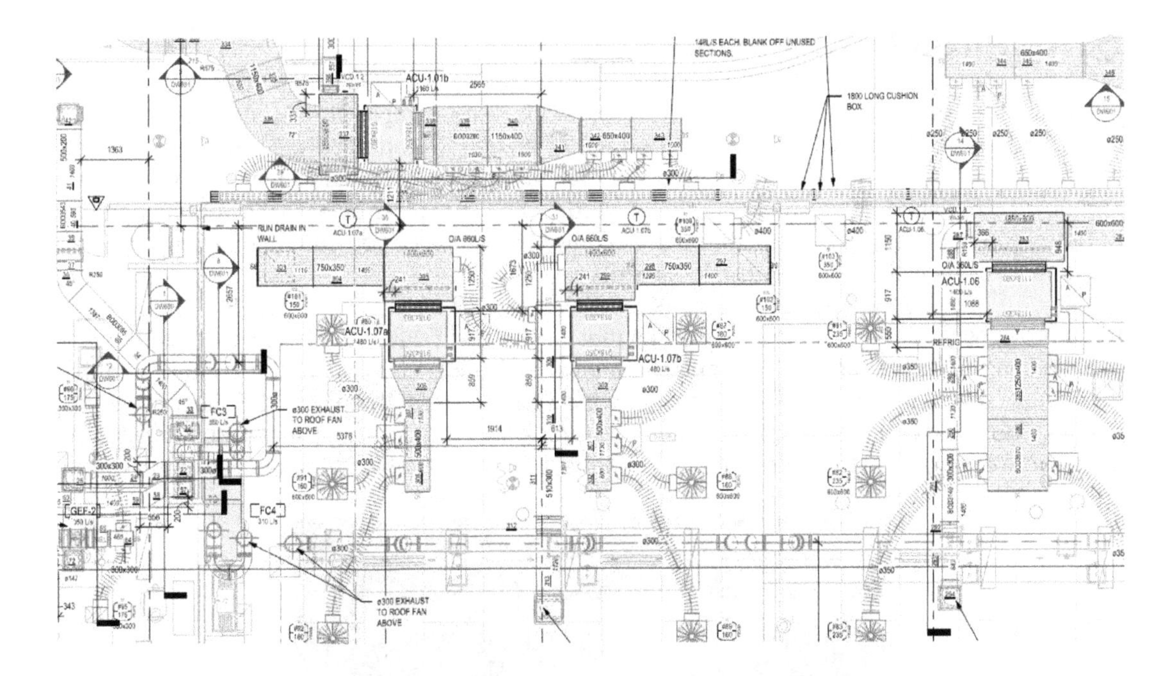

Chapter 11 - Working with Stamps

The objectives of this chapter are to:

√ *Explain the types of stamps available in Revu*
√ *Explain how to place the default stamps*
√ *Explain how to change the blend mode of the stamp*
√ *Explain the process of batch stamping documents*
√ *Explain how to edit the default stamps folder*
√ *Explain the process of creating custom stamps*
√ *Explain how to add dynamic text in the custom stamps*
√ *Explain how to import an existing stamp*

STAMPING DOCUMENTS

Stamping documents is an integral part of any design review process. Once the documents are reviewed, they need to be stamped for the next steps. For example, the documents that meet the specified standards have to be stamped as "Approved". Alternatively, if the documents do not meet the specified standards, they need to be stamped as "Rejected". Similarly, there are a number of other stamp types that need to be used to show the state of the reviewed documents. In Revu, you can stamp a single page of a document at a time or you can batch stamp multiple documents or multiple pages of the documents.

Revu provides you with a number of stamps that can be used out of the box. These stamps are available in the **Tools** menu > **Stamps**, as shown in Figure 1. Notice that the stamps in this menu have extension as .pdf. This is because the stamps are nothing but PDF files with some specific characters that make them behave as stamps. You can create your own custom stamps or import them from other PDF programs using the options available at the bottom of this menu.

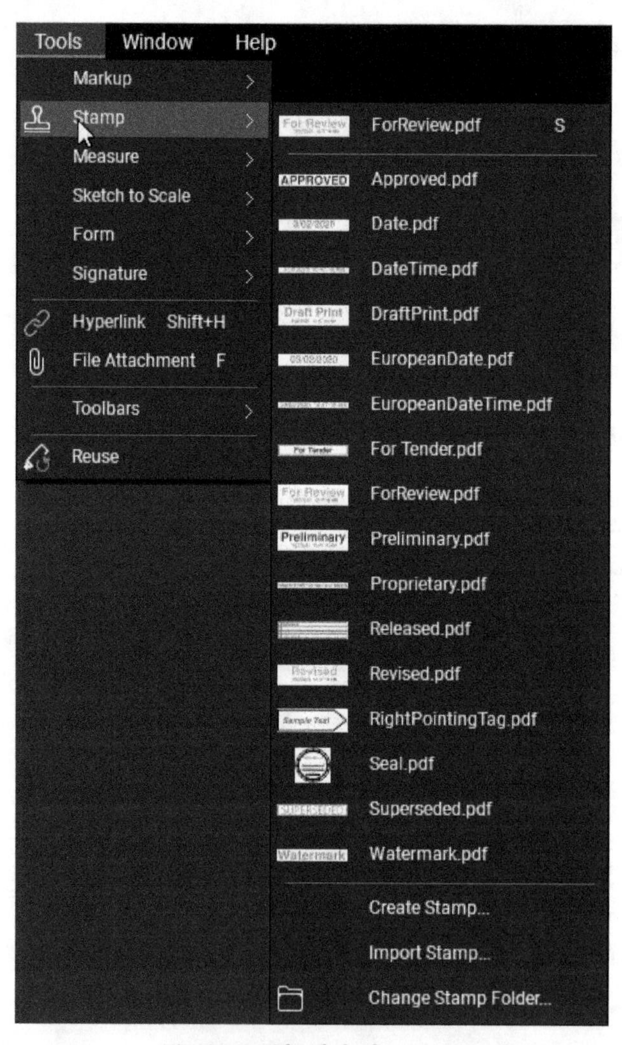

***Figure 1** The default stamps*

In Revu, the stamps are of two types:

Static Stamps
These stamps are placed as static images and their content cannot be modified. For example, the **Approved** stamp. The content in this stamp cannot be changed. However, you can change the size and orientation of this stamp.

Dynamic Stamps
These stamps include dynamic text that is automatically updated during placement. For example, the **DateTime** stamp. The information in this stamp is dynamic and is updated depending on when the stamp is placed.

The Procedures for Working with Stamps

The following are the various procedures for working with the stamps.

The Procedure for Stamping a Single Page of the PDF File

The following is the procedure for stamping a single page of the PDF file.

1. Open the PDF file to be stamped.

2. From the **Menu Bar**, click **Tools > Stamp** and then select the stamp you want to place; the cursor will show the stamp symbol and you will be prompted to select a region to place the stamp.

3. Click on a location to place the stamp with its original size or drag two opposite corners to size the stamp; the stamp will be placed at the specified location on the PDF. Figure 2 shows a static stamp placed to approve the document and Figure 3 shows a dynamic stamp that shows the date and time below the static stamp.

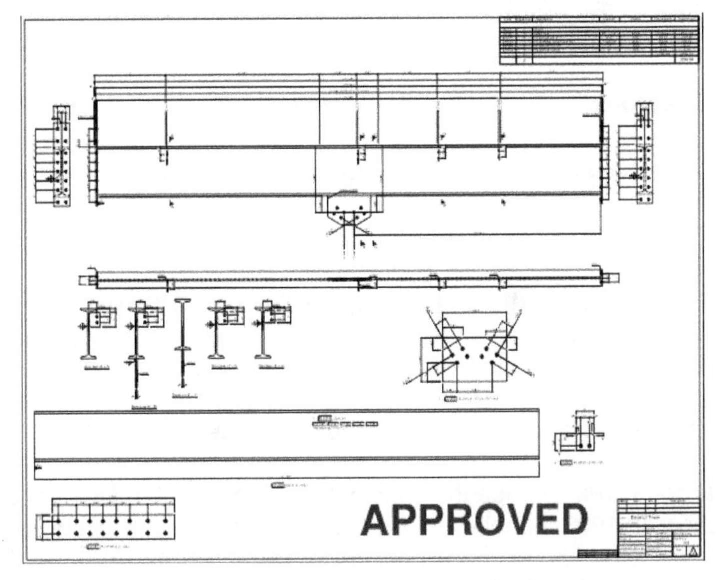

Figure 2 *A static stamp to approve the document*

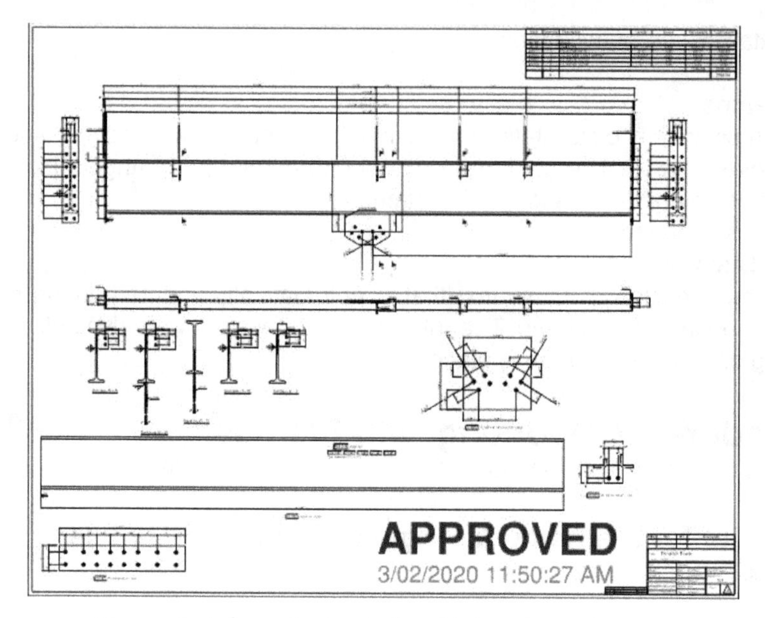

Figure 3 A dynamic stamp placed below the static stamp

The Procedure for Changing the Color of the Stamps

As mentioned earlier, the content of the stamps cannot be edited. However, you can modify some appearance properties, such as the color, opacity and blend mode. Additionally, you can change the scale and rotation of the stamps. The following is the procedure for changing the color of the stamps.

1. Click the stamp that is placed and activate the **Properties** panel.

2. In the **Appearance** area, click the **Change Colors** option; the **Color Processing** dialog box will be displayed. This dialog box is similar to the one discussed in the earlier chapters.

3. Click the **Source Color** button and select the source color of the stamp you want to change.

4. Click the **To** button and select the color you want to change the stamp to.

5. Click **OK** to close the dialog box; the color change will be reflected in the PDF file.

The Procedure for Changing the Opacity of the Stamps

By default, the stamps are inserted with 100% opacity. However, this can be changed in the **Properties** panel. The following is the procedure for doing this.

1. Click the stamp that is placed and activate the **Properties** panel.

2. In the **Appearance** area, change the opacity of the stamp in the **Opacity** spinner. Figure 4 shows a stamp with 100% opacity and Figure 5 shows the same stamp with 40% opacity.

Figure 4 The stamp with 100% opacity

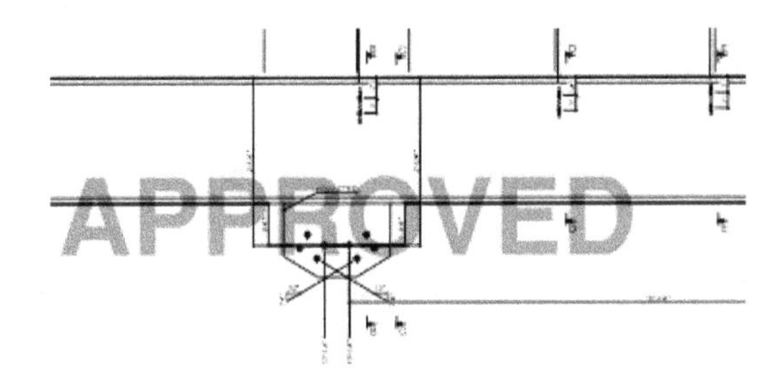

Figure 5 The stamp with 40% opacity

The Procedure for Changing the Blend Mode of the Stamps

By default, the stamps are placed on top of the PDF content. This means that the content behind the stamps is hidden, especially while placing stamps that have images. The blend mode controls how these stamps blend with the PDF content. The following is the procedure for changing the blend mode of the stamps.

1. Click the stamp that is placed and activate the **Properties** panel.

2. In the **Appearance** area, select the required blend mode from the **Blend Mode** drop-down list. Normally, it is recommended to use the **Multiply** blend mode to ensure the stamps blend well with the content of the PDF file. Figure 6 shows a stamp with the **Normal** blend mode and Figure 7 shows the same stamp with the **Multiply** blend mode.

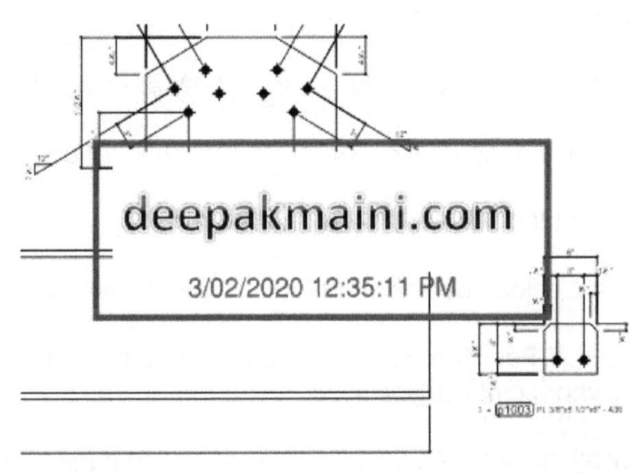

*Figure 6 The stamp with the **Normal** blend mode*

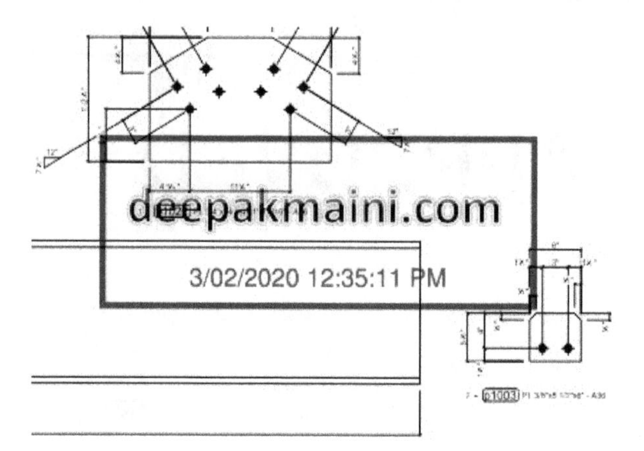

***Figure 7** The stamp with the **Multiply** blend mode*

The Procedure for Batch Stamping Multiple Pages and PDFs

Stamping a large number of pages individually is a tedious and time-consuming process. Therefore, Revu allows you to batch stamp multiple pages or multiple PDFs together. The following is the procedure for doing this.

1. From the **Menu Bar**, click **Batch > Apply Stamp**; the **Apply Stamp** dialog box will be displayed, as shown in Figure 8.

   The currently opened file is listed in the **Files** area.

2. In the **Files** area, click the **Add** button and add additional files or folders to select files.

3. From the drop-down list at the top in the **Stamp** area, select the stamp you want to apply.

4. From the **Blend Mode** drop-down list, select the required blend mode. As mentioned earlier, the stamps with images in them should use **Multiply** as their blend mode.

5. In the **Opacity** spinner, set the opacity value.

6. Select the **Lock** check box to lock the stamps once they are placed.

7. In the **Rotation** spinner, set the rotation for the stamps that will be placed.

8. In the **Scale** spinner, set the scale for the stamps.

9. From the **Anchor** area, select the anchor point to locate the stamp on the page.

10. In the **X Position** and **Y Position** edit boxes, enter the offset values along the X and Y direction from the anchor point you specified.

11. Click the **OK** button; all the selected files and their specified pages will be stamped.

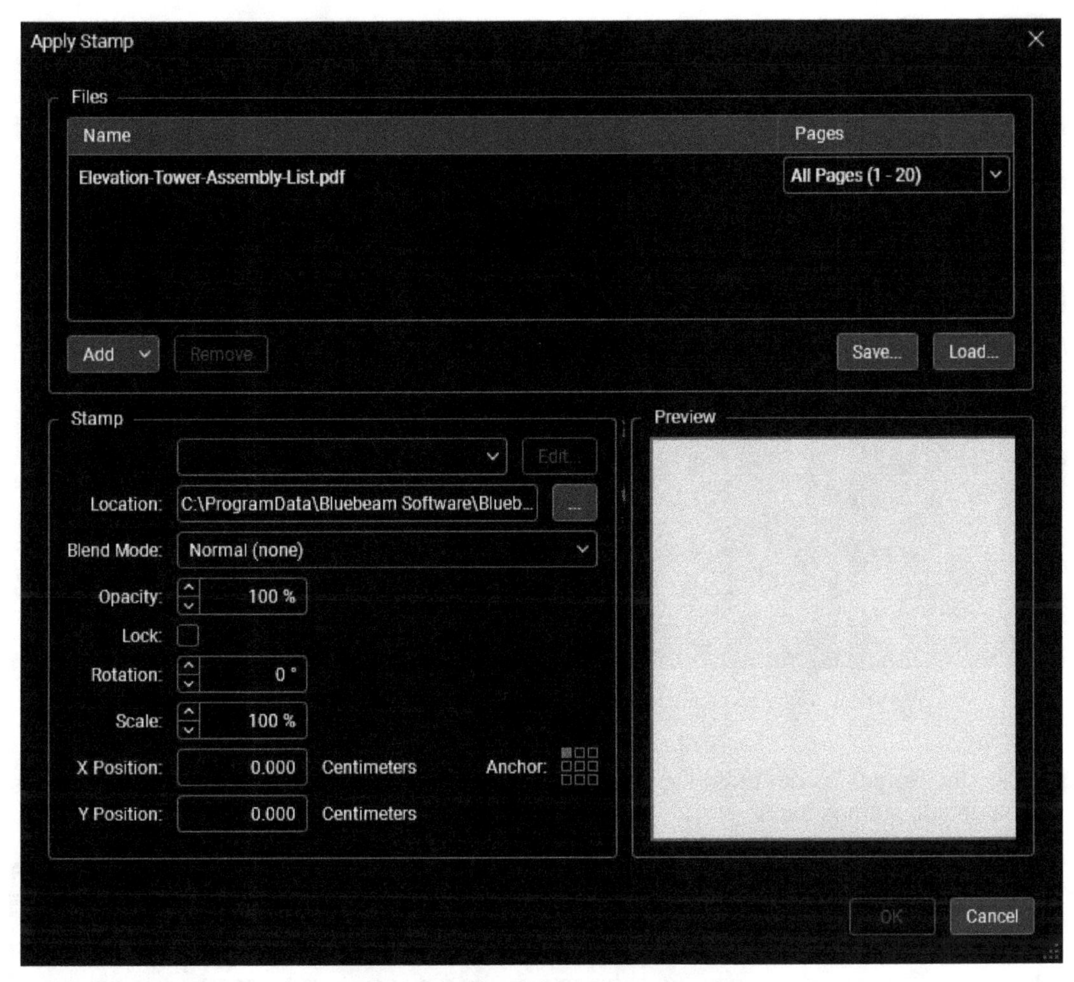

*Figure 8 The **Apply Stamp** dialog box*

The Procedure for Changing the Stamps Folder

By default, Revu sets a local folder to place stamps. However, this folder can be changed to a central folder on a network drive so everyone in the team has access to the same stamps. The following is the procedure to change the default stamps folder.

1. From the **Menu Bar**, click **Tools > Stamp > Change Stamp Folder**; the **Select Folder** dialog box will be displayed, as shown in Figure 9.

   By default, this folder is set to **C:\ProgramData\Bluebeam Software\Bluebeam Revu\2019\Stamps**. Notice the various localized language folders located inside this folder, as shown in Figure 9. If you want to place stamps in localized languages, such as Spanish, French, Italian, and so on, you can select one of those localized language folders.

2. Browse and select the folder where you have saved your stamps.

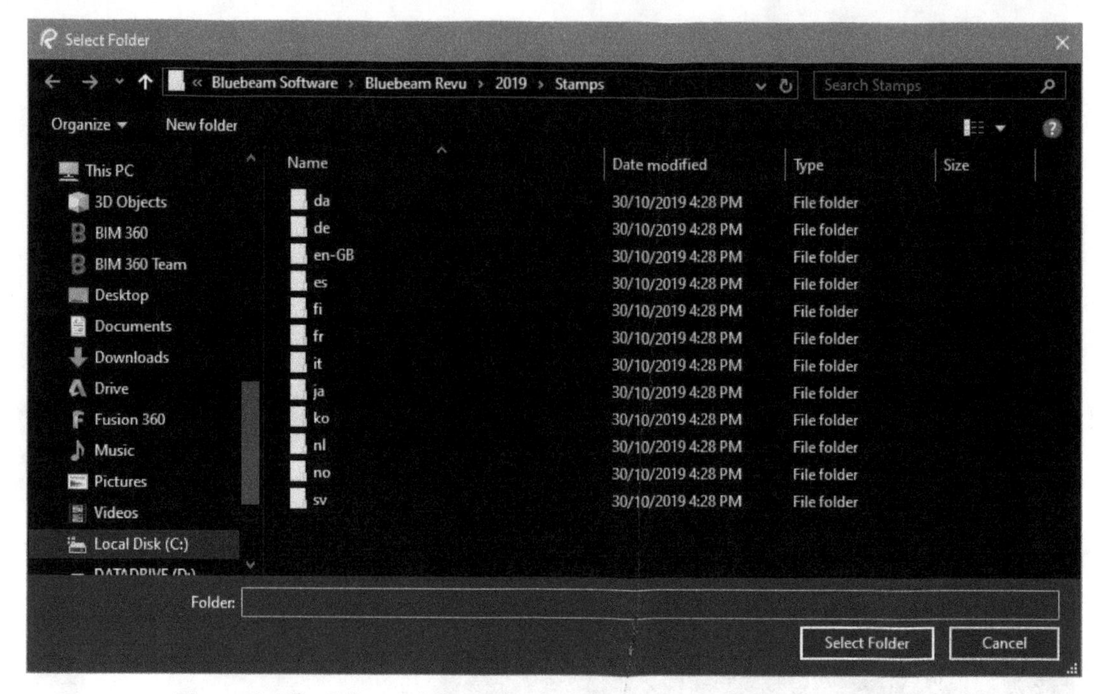

*Figure 9 The **Select Folder** dialog box to change the default stamps folder*

3. Click the **Select Folder** button in the **Select Folder** dialog box; the new folder will be set as the default stamps folder.

> **What I do**
>
> *I strongly recommend not to set the default stamp folder to the network location if you create custom stamps with the image of your signature in it. This is because you do not want the stamp with your signature to be saved on a network drive to for others to access it to stamp documents.*

The Procedure for Creating Custom Stamps

In most cases, you will require a stamp customized to your company and your own individual requirements. Revu provides you with really simple but powerful tools to create custom stamps. The following is the procedure for creating a custom stamp.

1. From the **Menu Bar**, click **Tools > Stamp > Create Stamp**; the **Create Stamp** dialog box will be displayed, as shown in Figure 10.

2. In the **Subject** field, enter the name of the stamp. This will be the name of the PDF file of the stamp that will be created. Also, this is the name that will be listed in the **Stamp** menu. Also, this will be the text that will appear by default in the new custom stamp PDF file.

3. The **Author** field is automatically populated based on the author's information in Revu preferences. If required, you can change the text in this field.

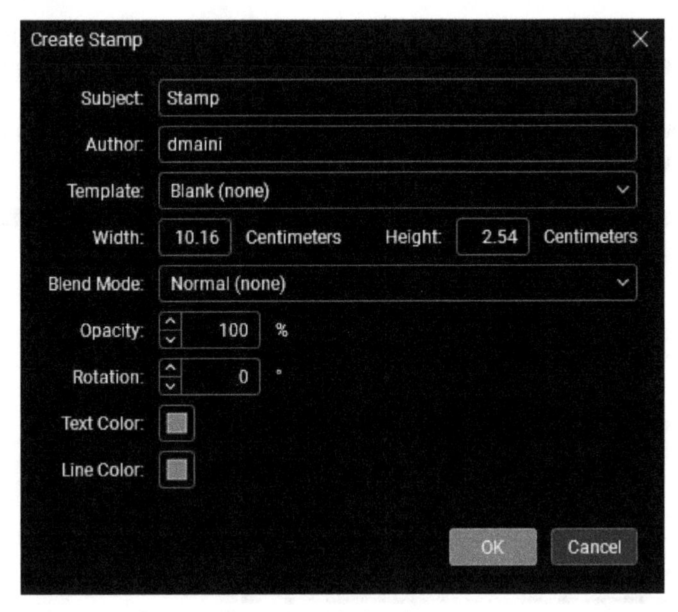

***Figure 10** The **Create Stamp** dialog box to create a custom stamp*

4. From the **Template** drop-down list, select the required template.

> **_What I do_**
>
> *I normally use the **Text with Date and Border** as the template from the **Template** drop-down list while create a custom stamp. This is because this template starts the PDF file of custom stamp with a border and also some default text in it. This makes it easier to create a new stamp instead of writing all the text from scratch.*

5. In the **Width** and **Height** edit boxes, enter the width and height of the custom stamp that will be created.

6. From the **Blend Mode** drop-down list, select the default blend mode for the stamp. It is recommended to create all stamps with **Multiply** as the blend mode so the stamps blend with the content of the PDF file.

7. In the **Opacity** spinner, set the default opacity of the stamp.

8. In the **Rotation** spinner, set the default value of the stamp rotation.

9. From the **Text Color** swatch, select the default color of the text in the stamp.

10. From the **Line Color** swatch, select the default color of the stamp border, if you selected the template that creates a default border.

11. Click **OK** in the **Create Stamp** dialog box; a new PDF file will be created with the name specified in the **Subject** field of the dialog box and will be opened in the Revu window.

    Figure 11 shows a stamp created with the name **Deepak's Stamp**. The template used for this was **Text with Date and Border** and the size was 10 cm X 5 cm.

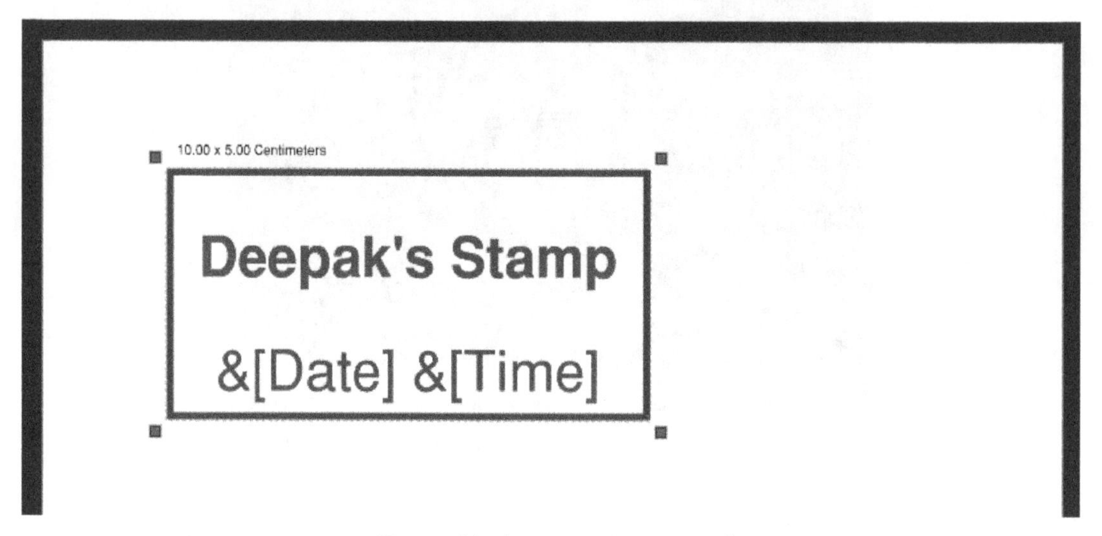

Figure 11 A custom stamp created

12. Select the existing text in the stamp PDF file and modify its properties using the **Properties** panel, if required.

13. Save the PDF file and close it.

14. From the **Menu Bar**, click **Tools > Stamp**; your custom stamp is available to be placed, as shown in Figure 12.

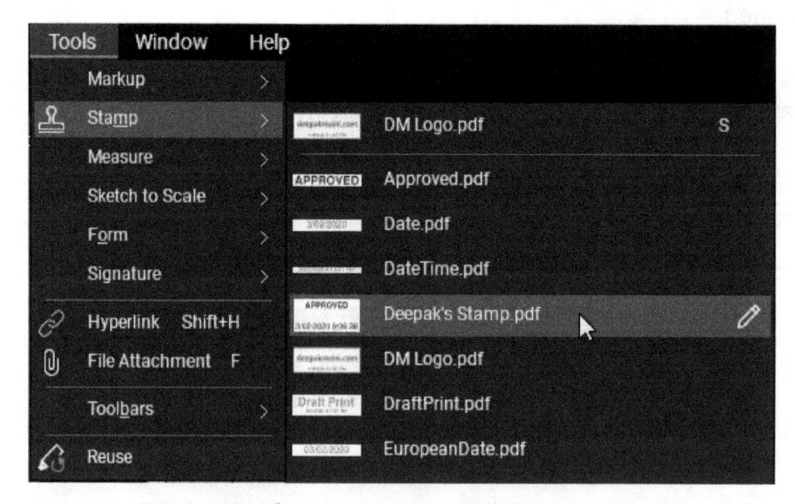

Figure 12 The custom stamp available to be placed

The Procedure for Inserting Dynamic Text in the Custom Stamps

If you created a stamp with the date template, it automatically inserts a dynamic text that updates the date and time when the stamp is inserted. There are a number of other dynamic text parameters you can insert in your custom stamps. The following is the procedure for doing this.

1. In the stamp PDF file, insert a new text box markup or double-click on the existing text; the **Dynamic** drop-down list is displayed below the text, as shown in Figure 13.

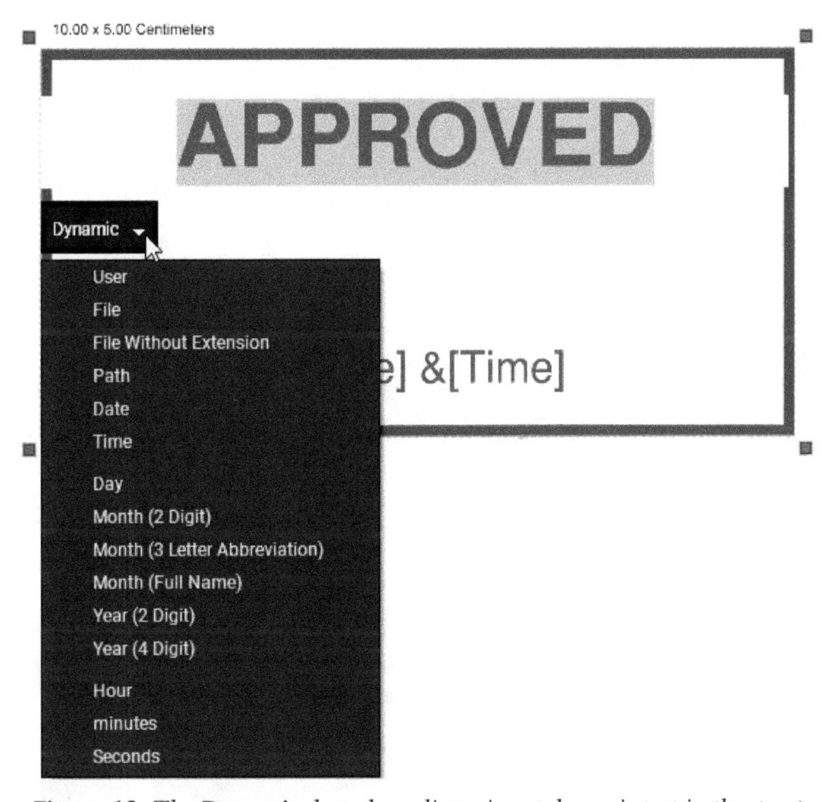

Figure 13 *The* ***Dynamic*** *drop-down list to insert dynamic text in the stamp*

2. Select the dynamic text you want to insert in your stamp.

3. Adjust the appearance and font properties of the dynamic text.

4. Save and close the PDF file of the stamp.

5. Use the **Tools > Stamp** menu to insert the stamp.

Figure 14 shows the PDF file of the custom stamp that automatically inserts a static text **APPROVED**. This stamp also has dynamic text that will extract the user name of the person placing this stamp, filename without extension, and the date and time. These dynamic texts appear with **&** as the prefix and are inside the square brackets [].

Figure 14 *The custom stamp PDF file with dynamic text*

Figure 15 shows this stamp inserted. Notice the user name, the file name, and the date and time are automatically extracted based on when and who places this stamp and on which file.

Figure 15 *The custom stamp placed in the* ***Elevation-Tower-Assembly-List*** *file*

The Procedure for Inserting Images in the Custom Stamps

Sometimes you need to insert images, such as your company logo or the scanned image of our signature in your custom stamp. It is important to remember that the stamps that will have images inserted in them should have their blend mode set to **Multiply**. The following is the procedure for inserting an image in the stamp.

1. In the stamp PDF file, from the **Menu Bar**, click **Tools > Markup > Image > From File**; the **Open** dialog box will be displayed.

2. Browse and select the image file to be used.

3. Drag two opposite corners to insert the image.

4. Save the stamp PDF file.

5. Insert the stamp with the image. Make sure you change the blend mode is set to **Multiply**.

   Figure 16 shows a stamp with an image and dynamic text placed in the PDF file.

APPROVED
deepakmaini.com
Elevation-Tower-Assembly-List
3/02/2020 9:50:28 PM

Figure 16 *The custom stamp with an image and dynamic text*

The Procedure for Editing a Custom Stamp

The content of a stamp once placed cannot be edited. However, if required, you can edit the stamp and place it again to ensure the changes are reflected in the stamp. The following is the procedure for editing a stamp.

1. If the stamp to be edited is already placed in the PDF file, select it and delete it.

2. From the **Menu Bar**, click **Tools > Stamp**; the available stamps are displayed in the cascading menu.

3. Hover the cursor over the stamp in the cascading menu; a pencil icon is displayed on the right of the name of the stamp PDF file.

4. Click on the pencil icon, as shown in Figure 17.

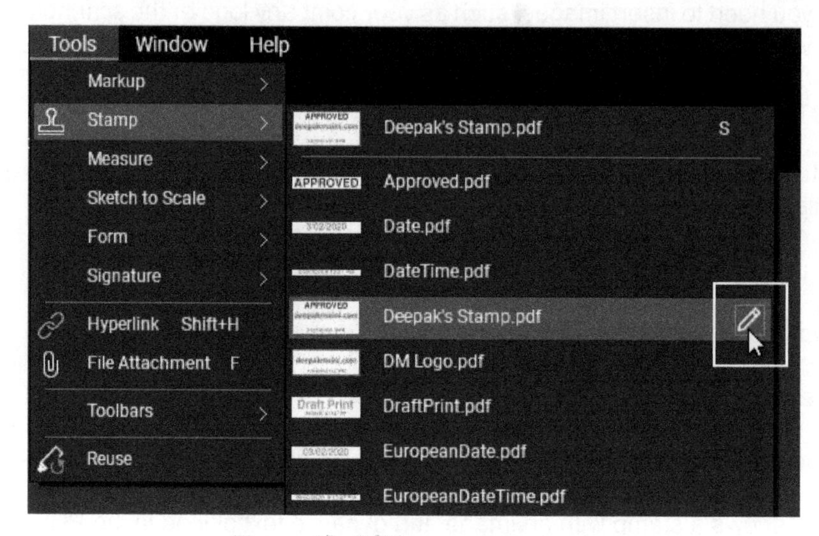

Figure 17 *Editing a custom stamp*

On doing so, the PDF file of the stamp is opened.

5. Make the required changes in the PDF file of the stamp.

6. Save and close the PDF file of the stamp.

7. Place the stamp again and you will see the changes you made.

The Procedure for Importing a Custom Stamp

If your company has a custom stamp already created in Revu or another PDF program, you can directly import it. The following is the procedure for importing a stamp.

1. From the **Menu Bar**, click **Tools > Stamp > Import Stamp**; the **Open** dialog box will be displayed, as shown in Figure 18.

2. Browse and select the PDF file of the stamp you want to import.

3. Click **Open** in the **Open** dialog box; the stamp will be imported and will be available in the **Stamp** menu for you to place.

Note: *Depending on how the stamps were created and in which program, some dynamic text in the stamps may not work. In that case, you can edit the stamp and insert the dynamic text using the* ***Dynamic*** *text drop-down list in Revu.*

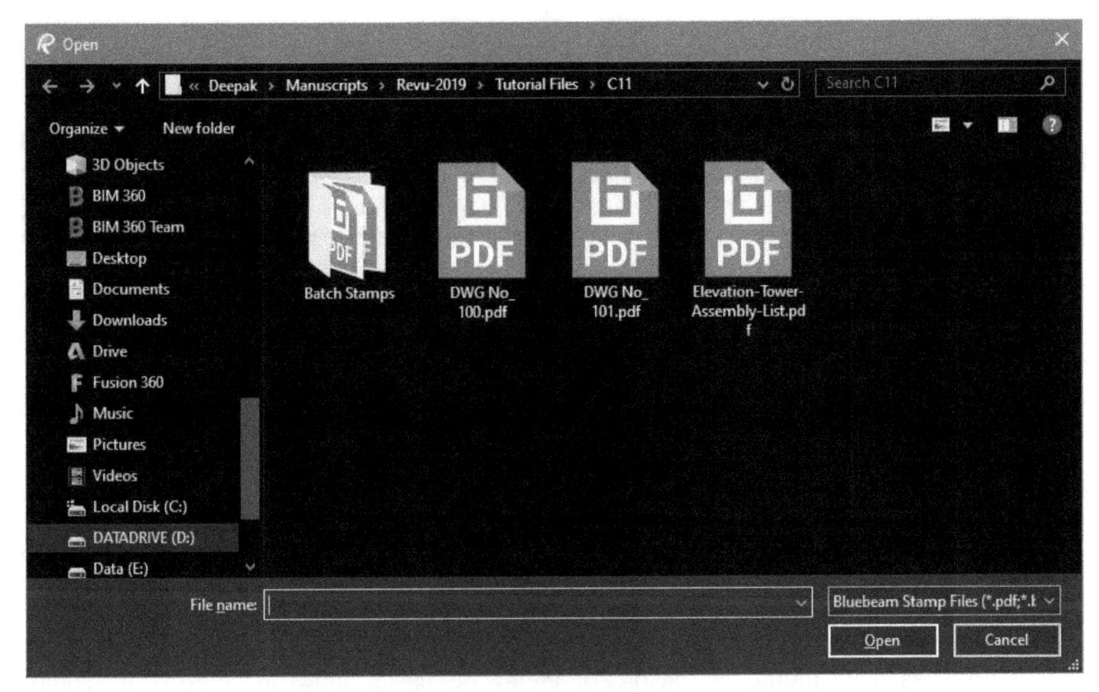

***Figure 18** The **Open** dialog box to import a stamp*

Hands-on Tutorial

In this tutorial, you will complete the following tasks:
1. *Stamp a document using one of the default stamps.*
2. *Change the properties and scale of the stamp.*
3. *Create a custom stamp with dynamic text and image.*
4. *Stamp a single document with the custom stamp.*
5. *Batch stamp multiple documents and pages.*

Section 1: Stamp a Document Using a Default Stamp

In this section, you will open the **DWG No_100.pdf** file from the **C11** folder. You will then use the **ForReview** stamp to stamp this document.

1. From the **C11** folder, open the **DWG No_100.pdf** file.

2. From the **Menu Bar**, click **Tools > Stamp**; the default stamps are shown in a cascading menu.

   You will now place the **ForReview** stamp, which is one of the default stamps that are delivered with Revu.

3. From the cascading menu, click on the **ForReview.pdf** file, as shown in Figure 19; the cursor shows a stamp icon and you are prompted to select a region to place the stamp.

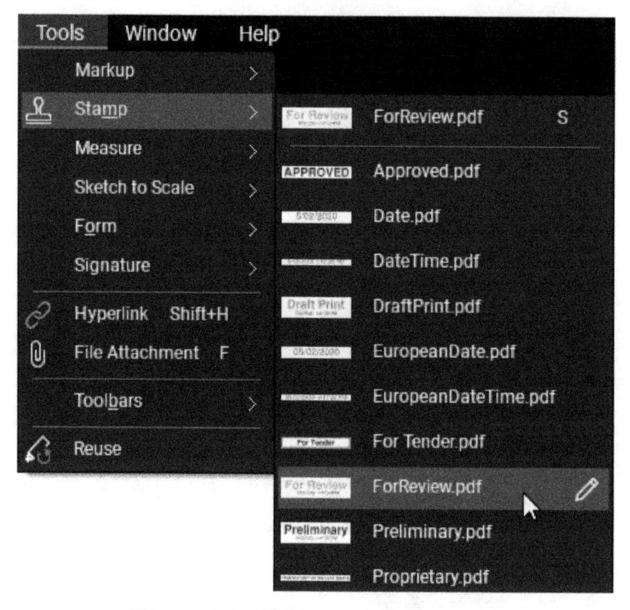

***Figure 19** Placing a default stamp*

You will place the stamp on the lower right corner, above the revision table.

4. Click above the revision table on the lower right corner; the stamp is placed, as shown in Figure 20.

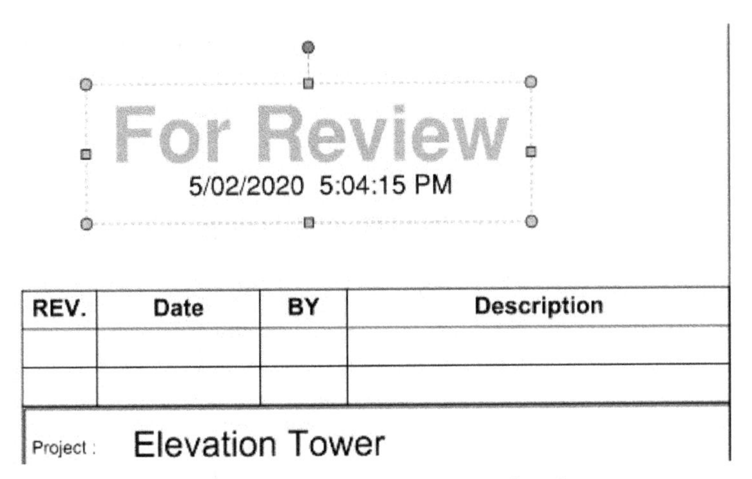

Figure 20 The default stamp placed

Notice that this stamp has a static text **For Review** and dynamic texts that show the date and time when this stamp was placed.

Section 2: Changing the Properties and Scale of the Stamp

In this section, you will change the color of the **For Review** text to Red. You will also change the scale of this stamp.

1. With the **For Review** stamp still selected above the revision table, display the **Properties** panel.

2. In the **Appearance** area, click the **Change Colors** button; the **Color Processing** dialog box is displayed.

3. Click the swatch on the right of the **Source Color** button and select the Green color, as shown in Figure 21.

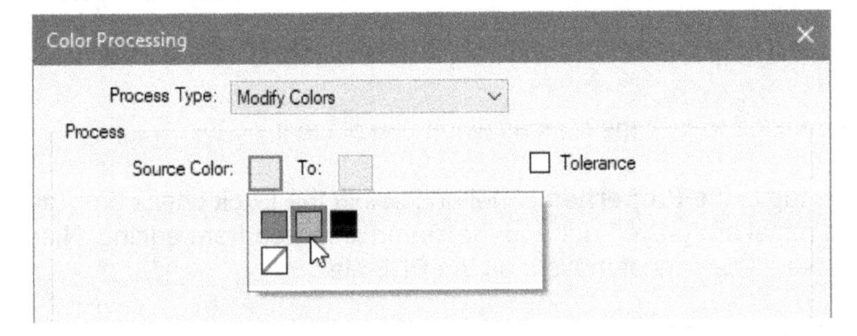

Figure 21 Selecting the source color to change

4. Click on the swatch on the right of the **To** button and select Red color, as shown in Figure 22; the preview in the **After** area of the **Color Processing** dialog box is updated.

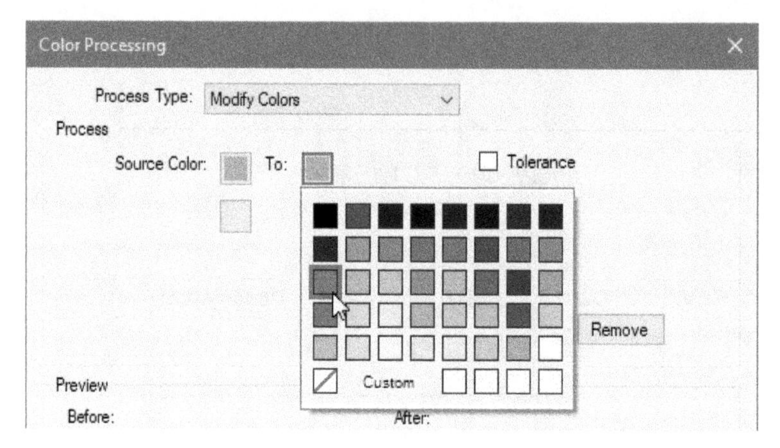

Figure 22 Selecting the required color for the stamp

5. Click **OK** in the dialog box; the Green color in the stamp is updated to Red.

6. Scroll down in the **Properties** panel to the **Layout** area.

7. Change the **Scale** value to **300**.

8. Move the stamp to the left of the titleblock, as shown in Figure 23.

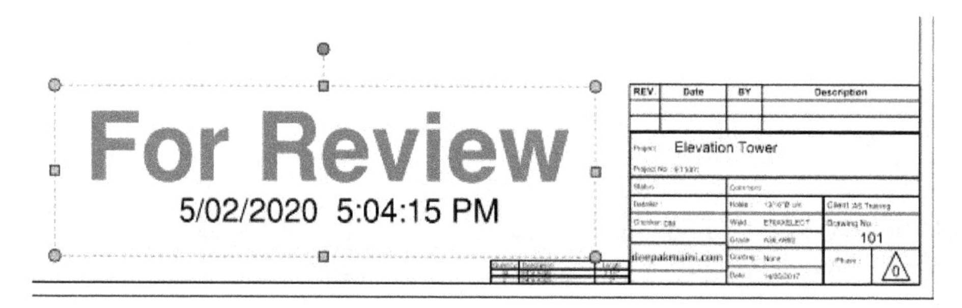

Figure 23 The stamp scaled and moved to the left of the titleblock

It is recommended to lock the stamps when you place it.

9. Scroll to the top in the **Properties** panel and select the **Lock** check box; the options in the **Properties** panel are grayed out and the stamp is locked from editing. Note that once the stamp is locked, you cannot move it on the PDF file.

10. Save and close the PDF file.

Section 3: Creating a Custom Stamp with Dynamic Text and an Image

In this section, you will create a custom stamp with dynamic text and an image. You will use the **Text with Date and Border** template and will use the 4 in X 2 in (Imperial) or 10 cm X 5 cm (Metric) as the size of the stamp.

1. From the **C11** folder, open the **DWG No_101.pdf** file.

2. From the **Menu Bar**, click **Tools > Stamp > Create Stamp**, as shown in Figure 24.

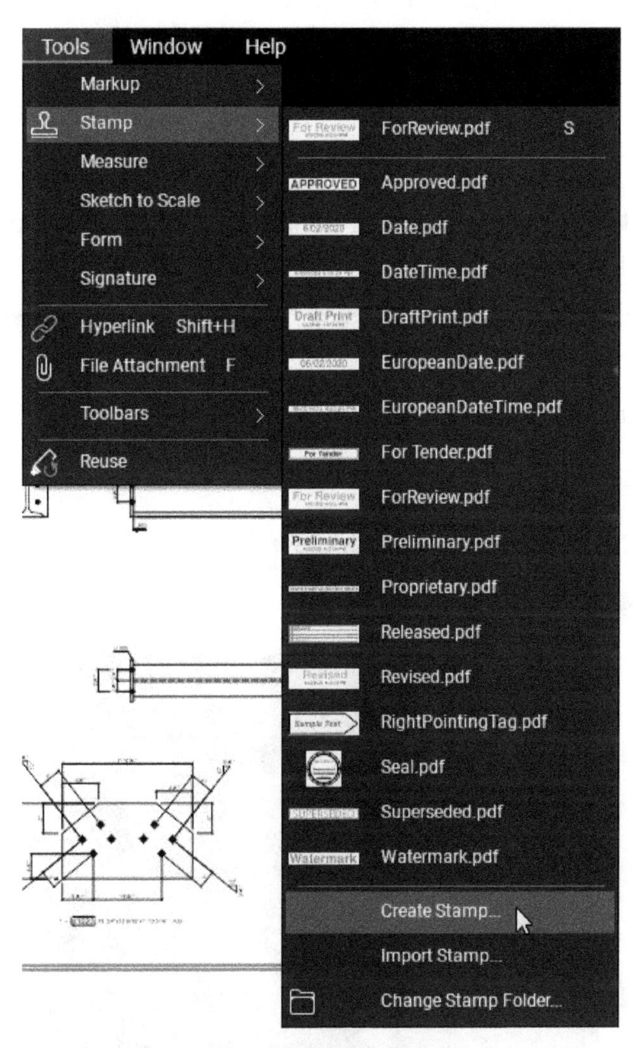

Figure 24 Creating a custom stamp

On doing so, the **Create Stamp** dialog box is displayed.

3. In the **Subject** field, enter **Training Stamp** as the name of the stamp. Note that this is the name of the stamp PDF file that will be created. Also, this will be the default main text in the stamp.

4. From the **Template** drop-down list, select **Text with Date and Border**. This will ensure the stamp PDF file has a default border and the date and time as the text below the main text.

5. In the **Width** edit box, enter **4** for Imperial units or **10** for the Metric units. This will ensure the width of the stamp is 4 inches or 10 cm.

6. In the **Height** edit box, enter **2** for Imperial units or **5** for the Metric units. This will ensure the height of the stamp is 2 inches or 5 cm.

   It is important to set the right blend mode for the stamps while creating them. This ensures that you do not need to change the blend mode when you insert the stamp. In this case, you will use the **Multiply** blend mode. As mentioned earlier in this chapter, this blend mode ensures the images in the stamps do not hide the content of the PDF behind the stamps.

7. From the **Blend Mode** drop-down list, select **Multiply** as the blend mode for the custom stamp.

   You will use the default opacity of 100% and the default rotation of 0-degree for this stamp.

8. From the **Text Color** swatch, select the Blue color.

9. From the **Line Color** swatch, select Dark Green.

   Figure 25 shows the **Create Stamp** dialog box will all these parameters.

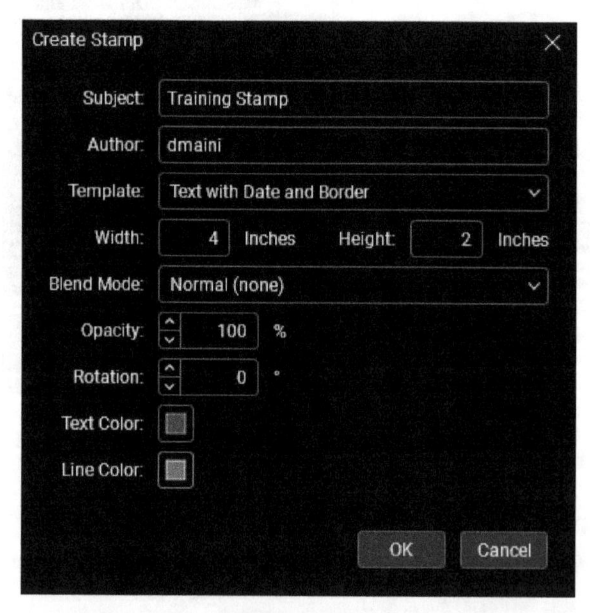

*Figure 25 The **Create Stamp** dialog box*

10. Click **OK** in the **Create Stamp** dialog box; the new stamp PDF file is created and a stamp with the size you specified is displayed on the screen. Figure 26 shows a partial view of this stamp. As evident at the top, this PDF file has the same name as the one you specified in the **Subject** field in the **Create Stamp** dialog box.

Figure 26 *The custom stamp PDF file*

Important Note: Before proceeding any further, you need to know about the four Blue handles displayed around the stamp border. These handles represent the boundary of the stamp that will be placed. If any part of the stamp is outside the boundary represented by these handles, that part of the stamp will not be displayed when the stamp is placed.

You will now edit the text in this stamp and add more dynamic text. To help line up the text with the other content of the stamp, you will turn on the **Snap to Markup** option from the **Status Bar**.

11. From the **Status Bar**, turn on the **Snap to Markup** option.
12. Double-click on the **Training Stamp** text and change it to **APPROVED**.
13. Move the **APPROVED** just below the top edge of the stamp, refer to Figure 27.
14. Change the font size of the **&[Date] &[Time]** text to **12**.

Next, you will increase the width of the text box of the **&[Date] &[Time]** text so the date and time when placed in the stamp fit in one line.

15. Drag the text box of the **&[Date] &[Time]** text so it spans across the width of the stamp.

16. Move the **&[Date] &[Time]** text just above the bottom edge of the stamp, refer to Figure 27.

Next, you will insert a dynamic text that will insert the author's name in the stamp and also the name of the file without the extension. You can insert a new text box markup, but it is easier to copy and paste the **&[Date] &[Time]** text.

17. Copy and paste the **&[Date] &[Time]** text above the previous instance of this text.

18. Double-click on the new text; the **Dynamic** drop-down list is displayed.

19. Highlight the text and from the **Dynamic** drop-down list, select **File Without Extension**, as shown in Figure 27.

Figure 27 Selecting the required dynamic text

On doing so, the **&[Date] &[Time]** text is replaced with the **&[FileNoExt]** text. You now need to ensure you widen the text box of this text so it spans across the width of the stamp.

20. Increase the width of the **&[FileNoExt]** text so it matches the width of the text below it.

You will now copy the **&[FileNoExt]** text and place it above it. You will then replace this text with the dynamic text that will display the user name when the stamp is placed.

21. Copy and paste the **&[FileNoExt]** text above it, refer to Figure 28.

22. Double-click on the **&[FileNoExt]** text; the **Dynamic** drop-down list is displayed.

23. Highlight the text and from the **Dynamic** drop-down list, select **User**, as shown in Figure 28.

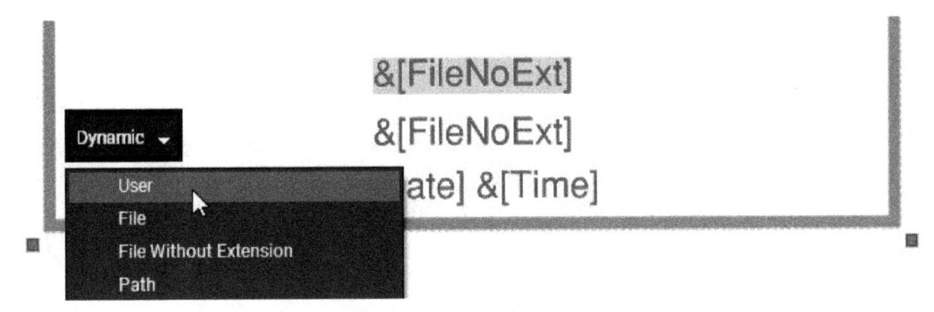

Figure 28 *Selecting the **User** dynamic text*

On doing so, the **&[FileNoExt]** text is replaced with the **&[User]** text. You now need to ensure you widen the text box of this text so it spans across the width of the stamp.

24. Increase the width of the **&[User]** text so it matches the width of the text below it.

Next, you will align the three dynamic texts so they are equally spaced.

25. Hold down the SHIFT key and select the three dynamic texts.

26. Right-click on the bottom dynamic text and from the shortcut menu, select **Alignment > Distribute Vertically**. Figure 29 shows the three dynamic texts aligned correctly.

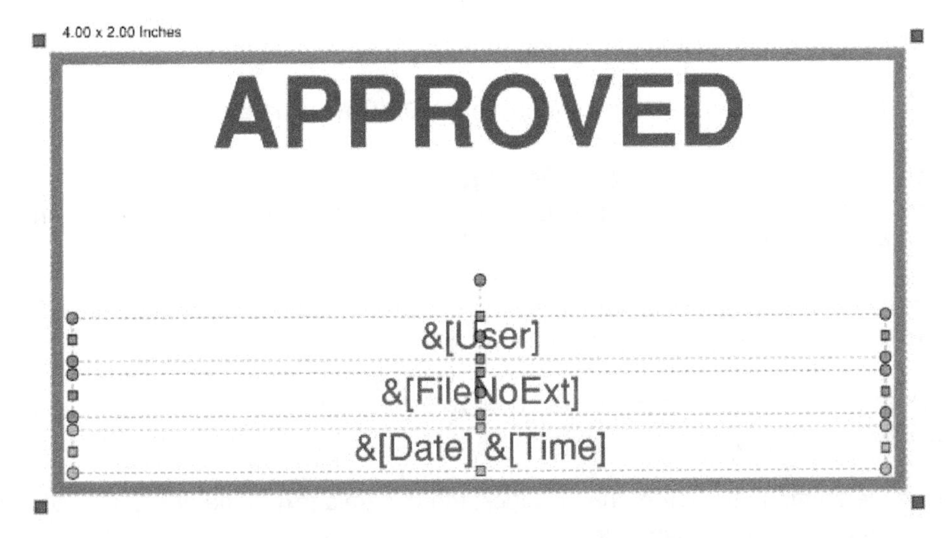

Figure 29 *The dynamic texts aligned together*

Next, you will insert an image in this stamp. The image to be used is available in the **C11** folder, Alternatively, you can use your own company logo as the image for this stamp.

27. Press the **I** key on the keyboard; the **Open** dialog box is displayed.

28. Browse to the **C11** folder and select the **DM-LOGO.PNG** image file.

29. Place the image, as shown in Figure 30.

Figure 30 The image inserted in the stamp

This completes the process of creating this stamp. You will now save and close the PDF file of this stamp.

30. From the **Menu Bar**, click **File > Save**; the PDF file of the stamp is saved.

31. Close the PDF file of the stamp to return to the **DWG No_101.pdf** file.

Section 4: Placing the Custom Stamp

In this section, you will place the custom stamp you created on the lower right corner of the **DWG No_101.pdf** file.

1. From the **Menu Bar**, click **Tools > Stamp**; the custom stamp you created is listed near the bottom of the cascading menu, as shown in Figure 31.

2. Click on the **Training Stamp.pdf** stamp from the cascading menu; you are returned to the PDF file.

3. Place the stamp to the left of the titleblock; the dynamic text in the stamp is automatically updated, as shown in Figure 32.

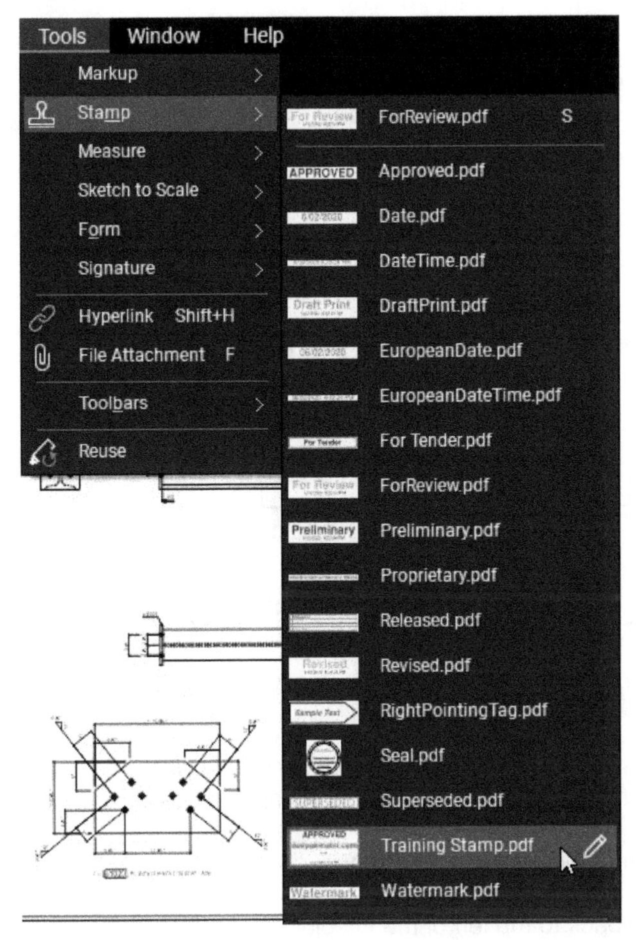

Figure 31 The custom stamp listed in the cascading menu

Figure 32 The dynamic text in the stamp updated on placing

Next, you will test the blend mode of the stamp and notice how the image in the stamp blends with the content of the PDF file.

4. Drag the stamp on top of the content of the PDF file; the stamp blends with the content and does not hide the content of the PDF file, as shown in Figure 33.

Figure 33 The stamp blending with the content of the PDF file

5. With the stamp still selected, change its blend mode to **Normal (none)** from the **Properties** panel; the content behind the image in the stamp is hidden.

6. Change the blend mode back to **Multiply**.

7. Move the stamp back to the left of the titleblock.

8. In the **Properties** panel, select the **Lock** check box to lock the stamp.

9. Save and close the PDF file.

Section 5: Batch Stamping Multiple PDF Files and Sheets Using the Custom Stamp

In this section, you will batch stamp multiple PDF files and sheets using the custom stamp that you created in the previous section.

1. From the **C11** folder, open the **Elevation-Tower-Assembly-List.pdf** file. Notice that this file has 20 pages.

   You will now batch stamp all pages of this PDF file and also some other PDF files that you will open during the batch stamping process.

2. From the **Menu Bar**, click **Batch > Apply Stamp**; the **Apply Stamp** dialog box is displayed with the current file listed in the **Files** area at the top of this dialog box, as shown in Figure 34. Notice that this area shows all 20 pages of the current PDF selected to be stamped.

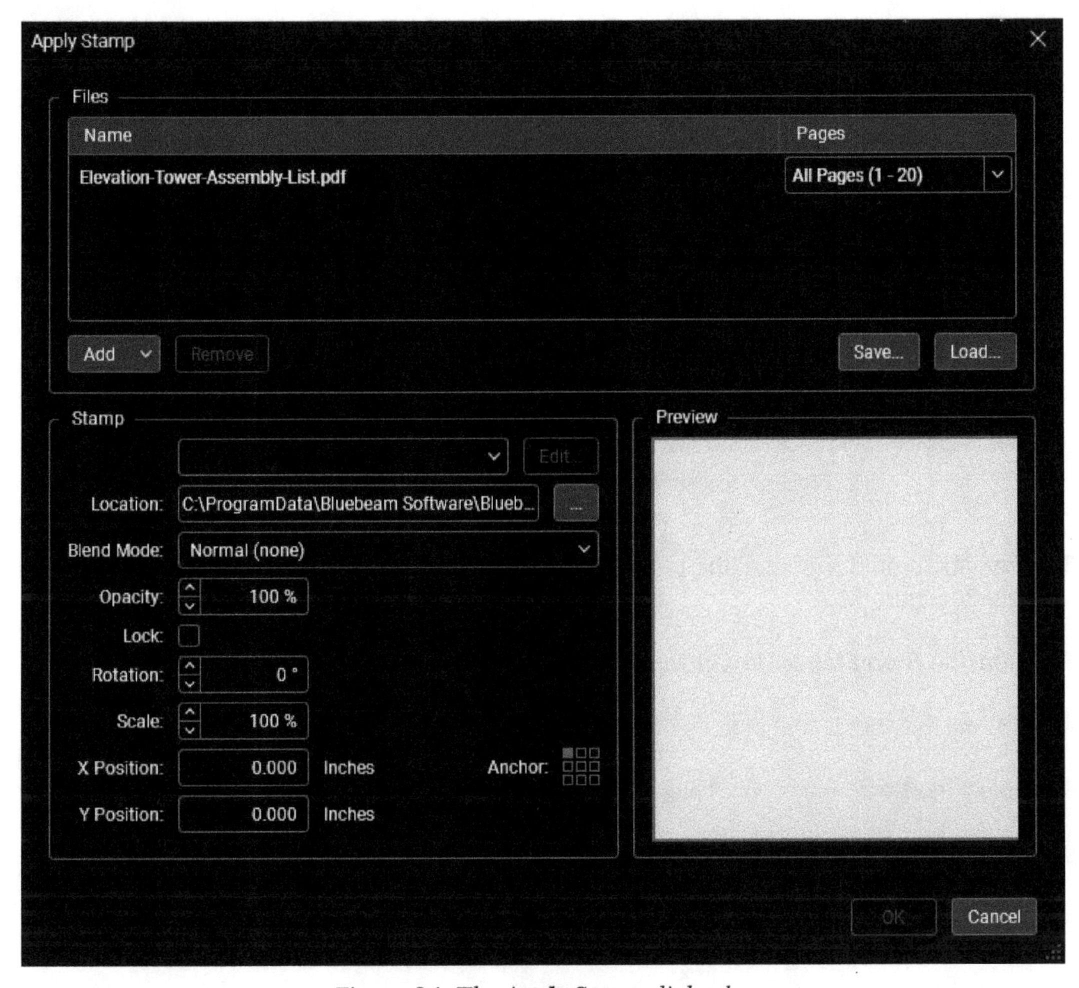

*Figure 34 The **Apply Stamp** dialog box*

You will now add a folder and all its included files to batch stamp along with the current file.

3. From the bottom left of the **Files** area, click **Add > Folder**; the **Add Folder** dialog box is displayed.

4. Browse to the **C11 > Batch Stamps** folder.

5. Click the **Select Folder** button in the **Add Folder** dialog box; all the files in this folder are not added in the **Files** area, as shown in Figure 35.

Figure 35 *All files from the folder added to apply stamp*

6. From the drop-down list at the top in the **Stamps** area, select the **Training Stamp.pdf** file, refer to Figure 36.

7. From the **Blend Mode** drop-down list, select **Multiply**, refer to Figure 36.

   You will use the default opacity of 100%.

8. Select the **Lock** check box. This ensures the stamps are locked once they are placed.

9. In the **Rotation** spinner, enter **-30** as the value.

   This stamp will be placed at the center of all pages. Therefore, you need to select the center anchor point.

10. From the **Anchor** area, select the center option.

    Figure 36 shows the lower part of the **Apply Stamp** dialog box after configuring all these settings.

11. Click **OK** in the **Apply Stamp** dialog box; the **Bluebeam Revu** dialog box is displayed warning you that you are about to make changes in multiple files.

12. Click **OK** in the **Bluebeam Revu** dialog box; the process of stamping multiple PDF files and multiple pages starts.

    Once the process is finished; the current PDF file shows the stamps placed on all pages.

13. Scroll through the pages of the current PDF file and notice the stamp placed at the center of every page. Figure 37 shows the last page of the PDF file with the stamp inserted.

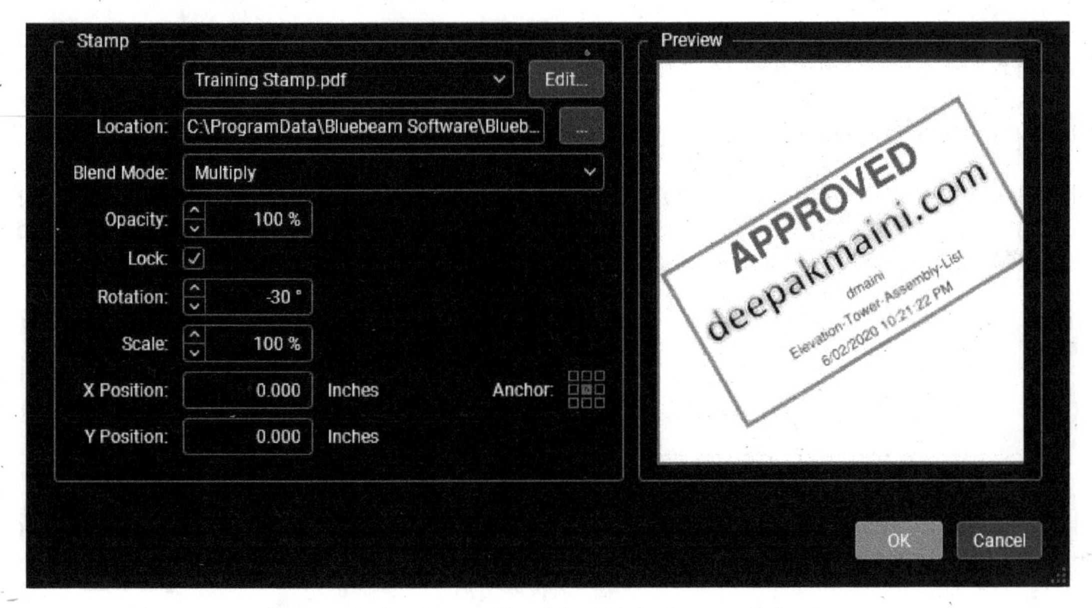

Figure 36 The lower side of the **Apply Stamp** dialog box

Quantity	Mark	Description	Length	Coating	Part weight	Total weight	Part surface	Total surface
			(inch)	Grade	pound	pound	(ft²)	(ft²)
1	**S103**	**C10X30**	**20' 9 5/8"**					
1	S103	C10X30	19' 2 1/8"	A992	575.29	575.29	49.18 ²	49.18 ²
1	s1009	C10X30	1' 1 3/4"	A992	34.42	34.42	2.94 ²	2.94 ²
1	s1011	C10X30	1'	A992	29.97	29.97	2.56 ²	2.56 ²
1	s1001	L3X3X1/4	10 5/16"	A992	4.21	4.21	0.86 ²	0.86 ²
1	s1002	L3X3X1/4	8"	A992	3.27	3.27	0.67 ²	0.67 ²
4		A325 3/4 x 2 1/4	2 1/4"	10.9	0.62	2.47		
36		A325 1/2 x 1 3/4	1 3/4"	10.9	0.21	7.39		
4		A563 Nut M 3/4		10.9	0.19	0.77		
36		A563 Nut M 1/2		10.9	0.07	2.36		
4		Washer F436 - 3/4		[illegible]	0.04	0.18		
36		Washer F436 - 1/2		10.9	0.02	0.72		
					TOTAL	661.05		56.21 ²

TOTAL QUANTITY	127
TOTAL WEIGHT	109,1[illegible].46 pound
TOTAL PAINT AREA	7,33[illegible] ft²

APPROVED
deepakmaini.com
dmaini
Elevation-Tower-Assembly-List
6/02/2020 10:27:12 PM

Return to Page 1

List produced by AUTODESK Advance Steel

Page 19 / 19

Figure 37 The last page of the current PDF file with the stamp placed

Next, you will open PDF files from the **Batch Stamps** folder and review the stamp placed in it.

14. From the **C11 > Batch Stamps** folder, open the four PDF files.

15. Review the PDF files and notice the stamps placed in them. Notice the drawing number in the stamp in each PDF file matches the name of that file. Figure 38 shows the stamped **DWG No_ B1000.pdf** file.

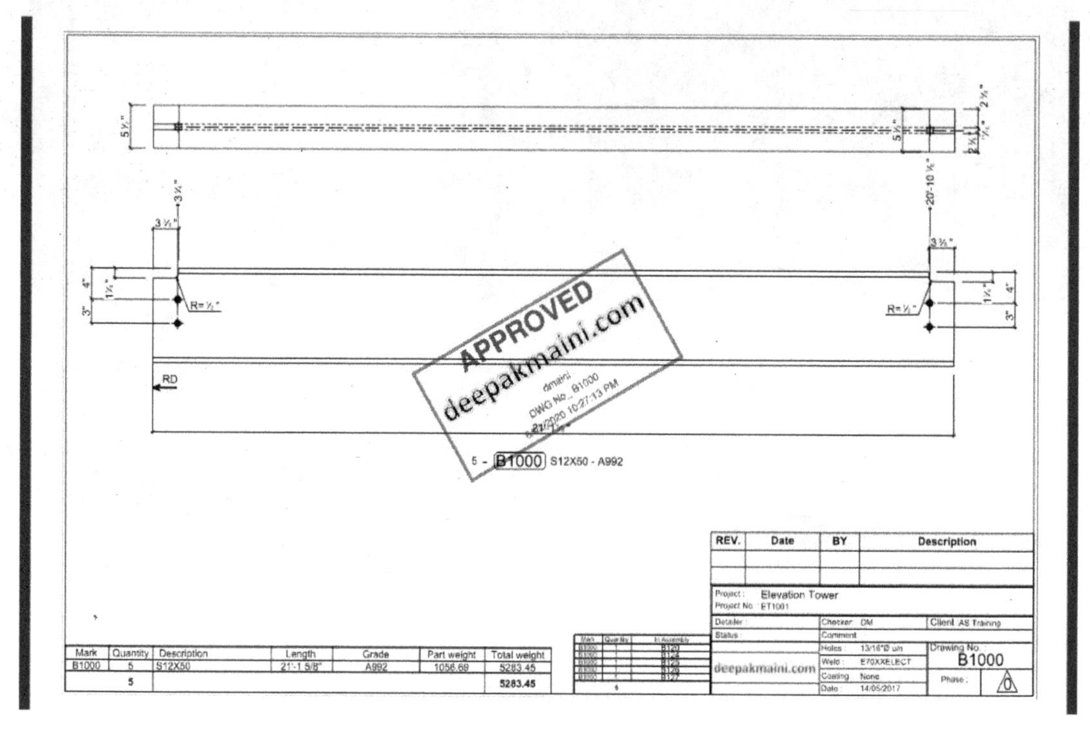

Figure 38 One of the files that was batch stamped

***Note:** Because the stamps were locked during the placement process, you cannot move them. If you need to move any of these stamps, you will have to first clear the **Lock** check box in the **Properties** panel.*

16. Save and close all the files.

Skill Evaluation

Evaluate your skills to see how many questions you can answer correctly. The answers to these questions are given at the end of the book.

1. In Revu, you can stamp single or multiple PDF files. (True/False)

2. You cannot create custom stamps in Revu. (True/False)

3. You cannot modify the color of the text in the stamp once you have placed it. (True/False)

4. You cannot insert images in the stamps. (True/False)

5. The custom stamp once created can be edited. (True/False)

6. Which stamps have their text updated automatically when placed?

   (A) Dynamic (B) Static
   (C) Both (D) None

7. Which menu option allows you to batch stamp multiple PDF files?

   (A) **Batch > Apply Stamp** (B) **Tools > Create Stamp**
   (C) **Markup > Stamp** (D) **Tools > Stamp**

8. Which blend mode allows the images to be blended with the content of the PDF file?

   (A) **Multiply** (B) **Divide**
   (C) **Normal** (D) **None**

9. The stamp files are saved in which format?

   (A) Text (B) Word
   (C) Excel (D) PDF

10. Which option in the **Tools > Stamp** menu allows you to import an existing stamp?

    (A) **Copy Stamp** (B) **Import Stamp**
    (C) **Create Stamp** (D) **None**

Class Test Questions

Answer the following questions:

1. Explain briefly the process of stamping a single page PDF file.
2. Explain briefly how to create a custom stamp.
3. Explain the process of batch stamping multiple files.
4. Explain the process of changing the blend mode of a placed stamp.
5. Explain the process of inserting a dynamic text in the custom stamp.

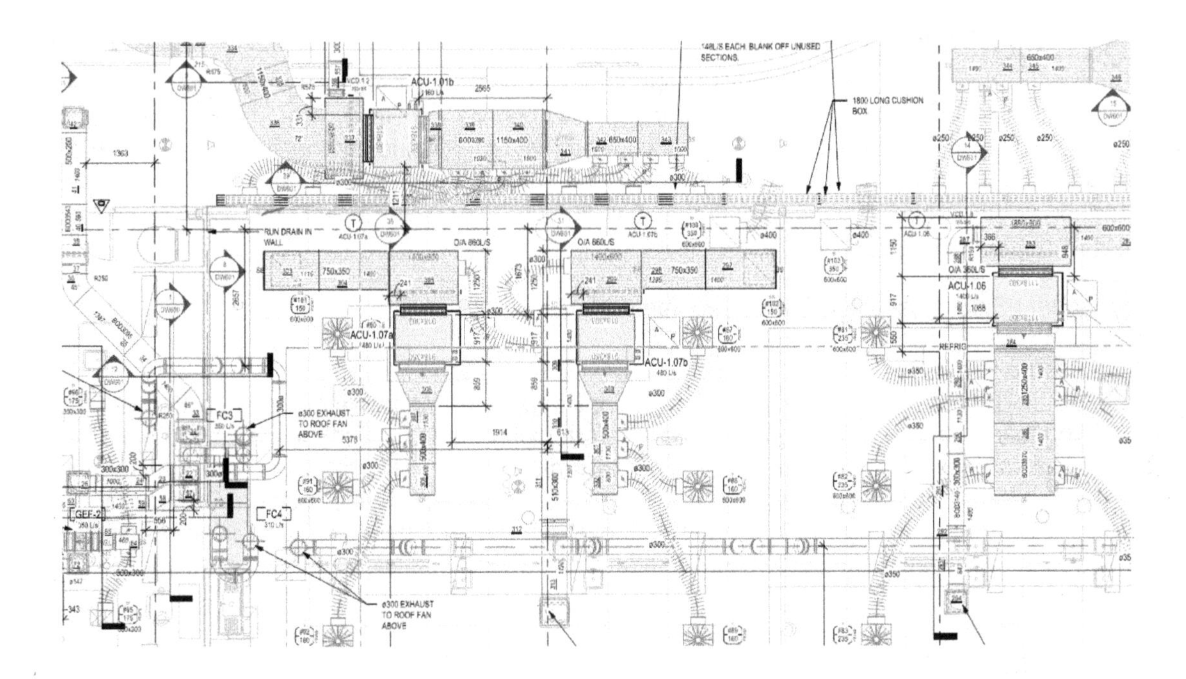

Chapter 12 - Working with Bluebeam Studio

The objectives of this chapter are to:

- √ *Explain what is Bluebeam Studio*
- √ *Explain the components of Bluebeam Studio*
- √ *Explain Studio Projects and Studio Sessions*
- √ *Explain the process of creating a Studio Project and adding files and folders to it*
- √ *Explain how to control the permissions of a Studio Project*
- √ *Explain the process of creating a Studio Session and how to start it*
- √ *Explain how to control the permissions of a Studio Session*

BLUEBEAM STUDIO

Bluebeam Studio is a cloud-based light-weight document management system and an online PDF-based collaboration program. Included in every version of Revu, you can use Bluebeam Studio as a cloud-based repository for all your project data, including non-PDF files. The PDF files can be directly viewed in Revu through Studio and non-PDF files can be downloaded and opened in their native program. Bluebeam Studio can be accessed in offline mode as well at sites where Internet connectivity is an issue.

Bluebeam Studio has the following two components:

Studio Projects

Studio Project is a cloud-based light-weight document management system that allows you to host and access the entire project data using the Revu interface. This can act as a single source of truth for all your project documents. The PDF files in the projects can be checked out and opened for editing and then checked back in after editing. On doing this, Studio Project will bump the revision of the document and maintain the revision history for you to review at a later stage. You can also restore an old revision as the current revision of the PDF file.

Studio Sessions

Studio Sessions allow online design review and markup of PDF files by multiple stakeholders at the same time from within their Revu environment, without the worry of people modifying each other's markups. These sessions can be scheduled to expire after a certain period of time, after which the attendees will not be able to access the data in the sessions. The host of the Studio Session needs a Revu license and can invite an unlimited number of stakeholders to the session. However, a maximum of 500 attendees can mark up the PDF document in a Studio Session at any given time. These attendees can use their own license of Revu to join. The attendees who do not have a license of Revu can also join using Revu in the view mode, which means they do not need a license.

The following table has some key points about Studio Projects and Sessions:

Description	Studio Projects	Studio Sessions
License Required	Host Only	Host Only
Studio Account Required	Yes	Yes
File Support	All Files	PDFs
Number of Attendees at a Given Time	Unlimited	500
Maximum File Size	Unlimited	1 GB each
Maximum Number of Files Allowed	Unlimited	5000
Maximum Space Allowed	Unlimited	Unlimited
Scheduled Expiry	No	Yes

The Procedures for Working with Studio Projects

The following are the various procedures for working with Studio Projects.

The Procedure for Creating a Bluebeam ID to use Bluebeam Studio

Before you start hosting a Studio Project or Session, you first need to sign into Bluebeam Studio. The following is the procedure for creating a Bluebeam ID to use Bluebeam Studio:

1. From the **Panel Access Bar**, invoke the **Studio** panel, as shown in Figure 1.

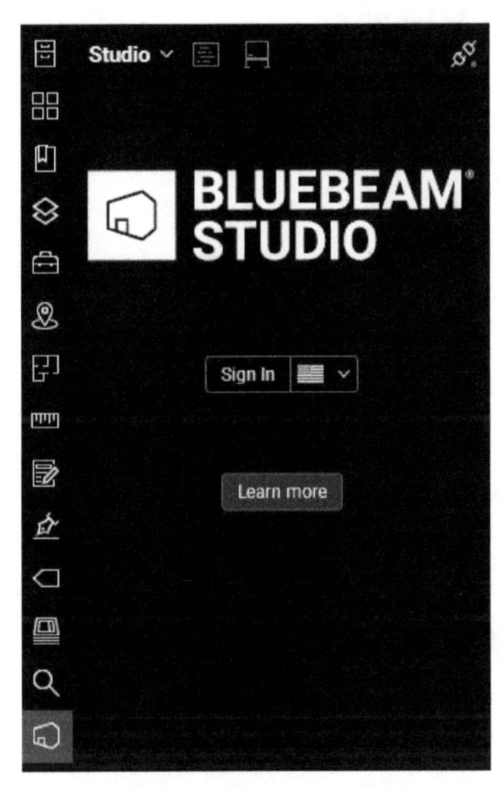

Figure 1 The ***Studio*** *panel*

   Notice that on the right of the **Sign In** button, the American flag appears. This is used to specify whether you want to use the US-based cloud server or UK-based. You can click on the American flag and change the server to the UK-based server. Generally, anyone outside the UK is recommended to use the US-based server.

2. Click the **Sign In** button; the **Bluebeam Studio** dialog box will be displayed prompting you to sign in.

   In this dialog box, there are two tabs. The **SIGN IN** tab is for the users who already have a Studio account created. The **CREATE ACCOUNT** tab is for the users who do not have a Bluebeam ID so they can create it.

3. Click the **CREATE ACCOUNT** tab in the **Bluebeam Studio** dialog box; the options related to creating a Bluebeam ID will be displayed, as shown in Figure 2.

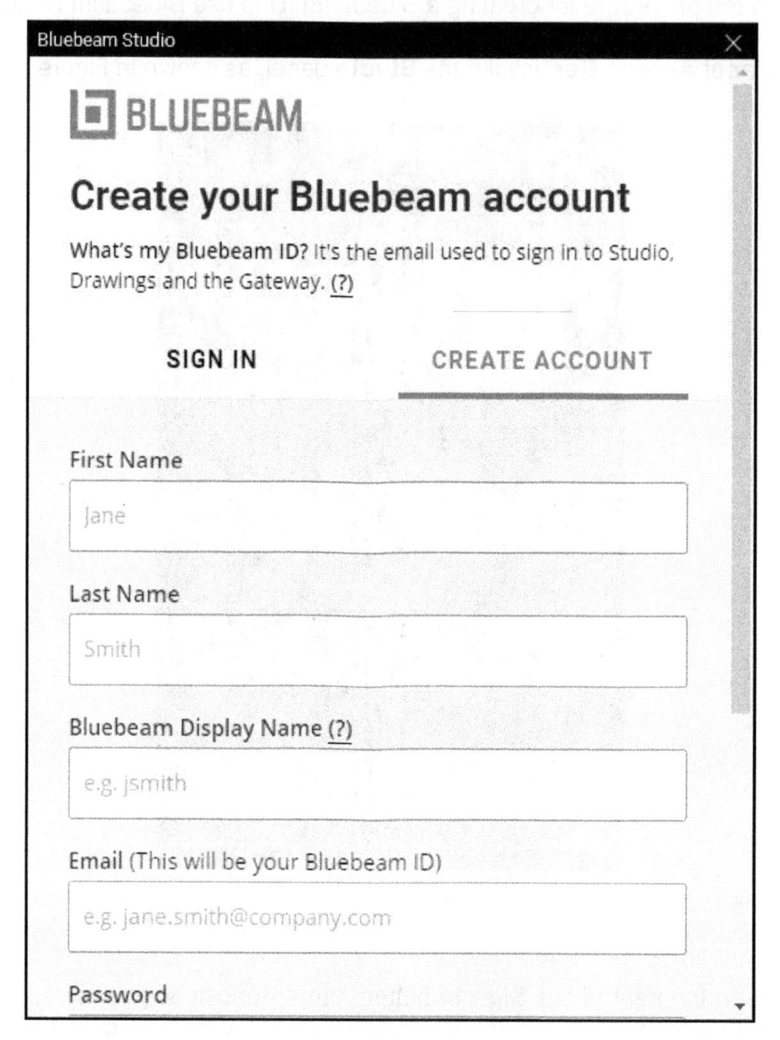

Figure 2 *Creating a Bluebeam ID*

4. Enter the information in this tab.

5. Scroll to the bottom and click **Create Account**; the Bluebeam's General Service and Software Terms of Use window will be displayed for you to review the terms and conditions of using Revu.

6. Once you have reviewed the terms and conditions and are OK with it, click the **I ACCEPT** button; a confirmation Email will be sent to the Email address you used to create the ID.

7. In the confirmation Email, click on the **Confirm My Bluebeam ID** button to confirm the creation of your Bluebeam account that will be used to access Studio Projects and Sessions.

8. Return to the Revu window and now sign in to Bluebeam Studio using the same credentials.

The Procedure for Starting a New Studio Project

The following is the procedure for starting a new Studio Project:

1. From the **Panel Access Bar**, invoke the **Studio** panel.

2. Sign in using your Bluebeam ID. Once you sign in, the **Studio** panel has the **Projects** option selected at the top, labeled as **1** in Figure 3. By default, this window will be blank if you have not hosted any Project and are not invited to join any Project, as shown in Figure 3.

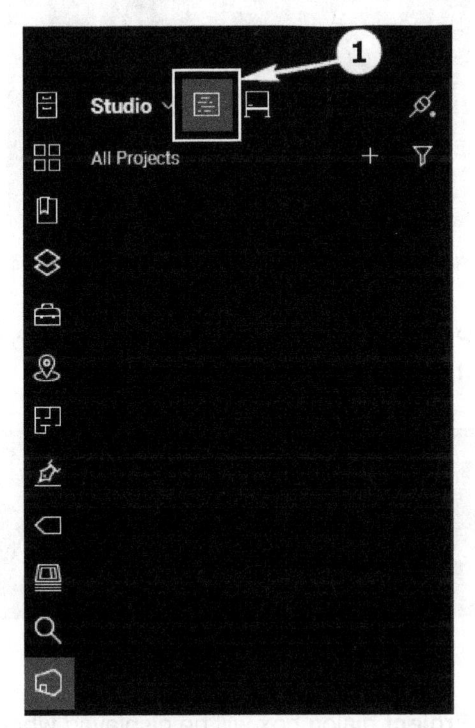

Figure 3 The ***Studio*** *panel after signing in*

However, if you have hosted a Studio Project or were invited to any Project, they will be listed in this window, as shown in Figure 4. Note that in this figure, the IDs of various projects are blurred for privacy purpose.

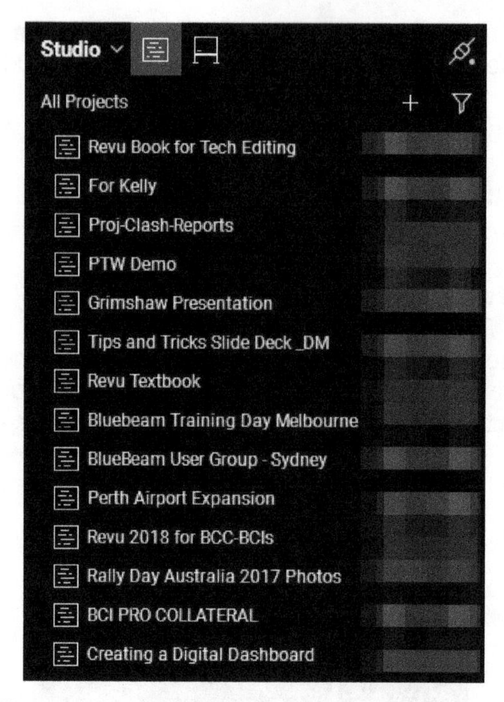

Figure 4 *Studio Projects listed in the* ***Studio*** *panel*

3. To start a new Studio Project, click the **Add** button near the top right in the **Studio** panel and select **New Project**, as shown in Figure 5.

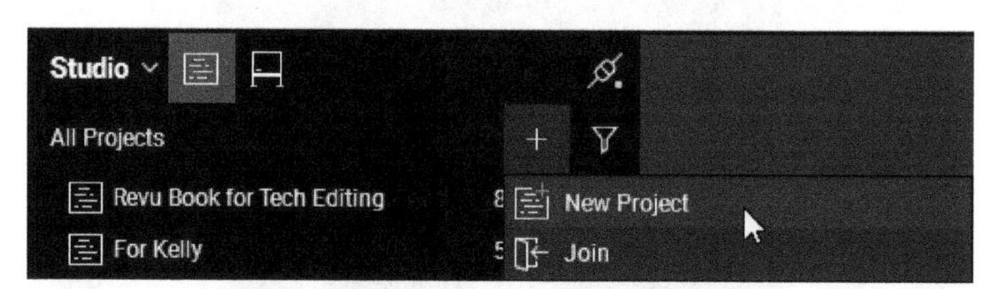

Figure 5 *Adding a new Studio Project*

On doing so, the **New Project** dialog box will be displayed where you can enter the name of the new project, as shown in Figure 6.

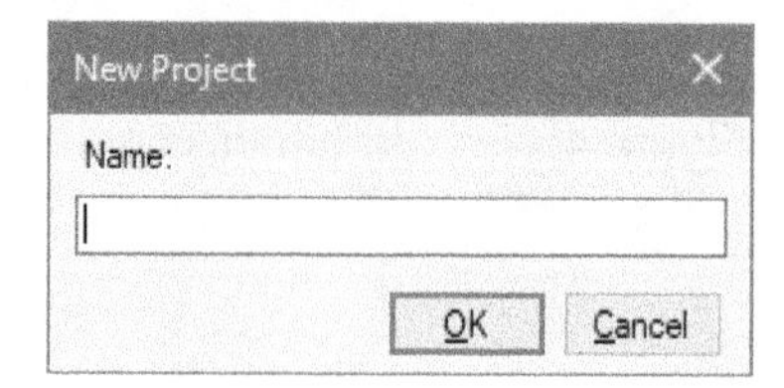

Figure 6 *The* ***New Project*** *dialog box*

4. Enter the name of the new Studio Project and then click **OK** in the **New Project** dialog box; the process of creating the new Studio Project will start.

   Once the Studio Project is created, it will be listed at the top of the Projects list in the **Studio** panel and is also opened in its own tab in the Revu window, as shown in Figure 7.

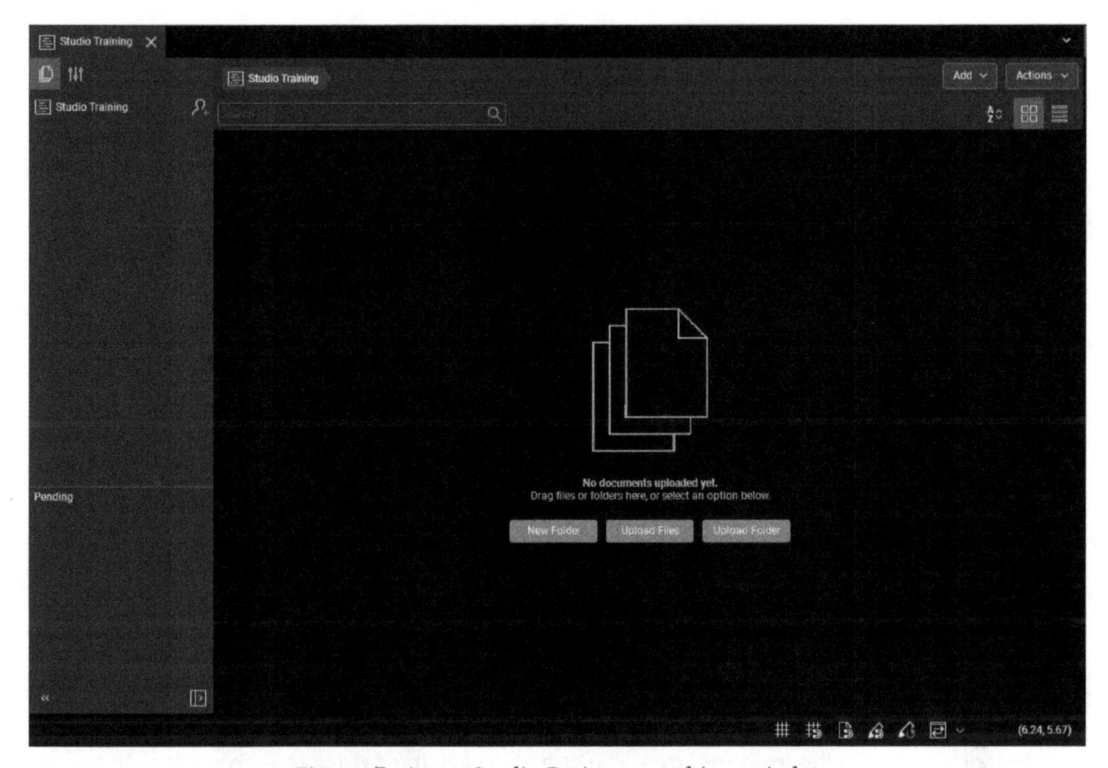

Figure 7 *A new Studio Project opened in a window*

The Procedure for Inviting Members to a Studio Project

The following is the procedure for inviting members to a Studio Project:

1. In the Project tab, click on the **Invite** button on the right of the project name, as shown in Figure 8.

Figure 8 *Inviting members to the Project*

   On doing so, the **Project Invitation** dialog box will be displayed, as shown in Figure 9.

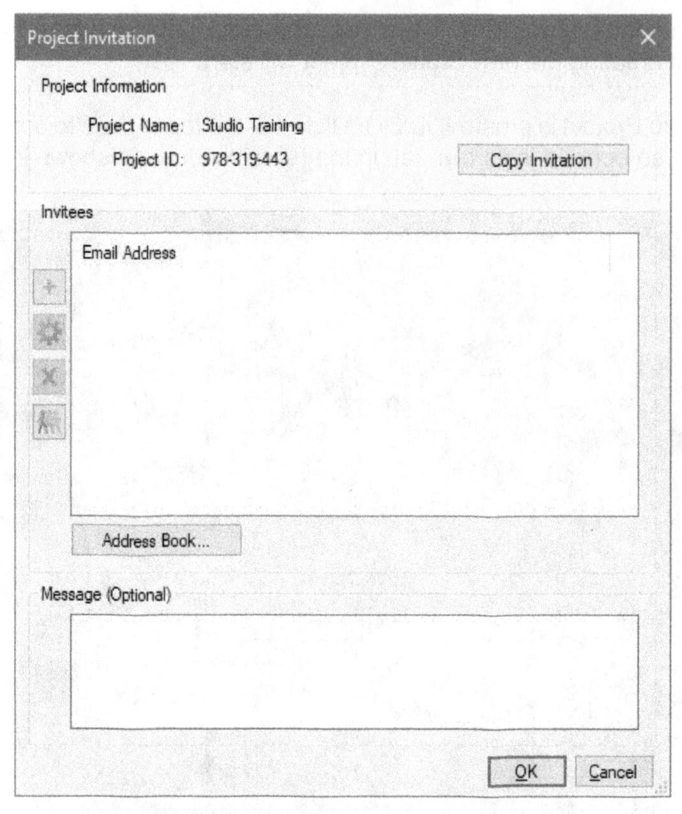

***Figure 9** The **Project Invitation** dialog box*

2. In the **Invitees** area, click the **+** button to enter the email address of the members to invite. Alternatively, if you have a group of users defined, you can add the group of users by clicking the fourth button in the **Invitees** area. You can also click the **Address Book** button to access the address book of your email program such as Outlook to copy the email addresses. Once you add the email addresses, they will be listed in the **Invitees** area, as shown in Figure 10.

***Figure 10** The Emails of the invitees added*

3. If required, type a message in the **Message** area of the **Project Invitation** dialog box.

4. Click **OK** in the dialog box; the **Bluebeam Revu** dialog box will be displayed informing you that the email invitations have been sent to the attendees.

5. Click **OK** in the dialog box to return to the Project tab.

6. The invitees will be able to click on their Project invite and then access the Studio Project.

The Procedure for Configuring Project Settings

Once you have created the Studio Project, it is important to configure its settings so the invitees are managed better in the project. The following is the procedure for configuring the Project settings:

1. In the Project tab, click on the **Project Settings** button, as shown in Figure 11.

Figure 11 *Configuring the Studio Project settings*

On doing so, the **Project Settings** dialog box will be displayed with the **General** tab active, as shown in Figure 12.

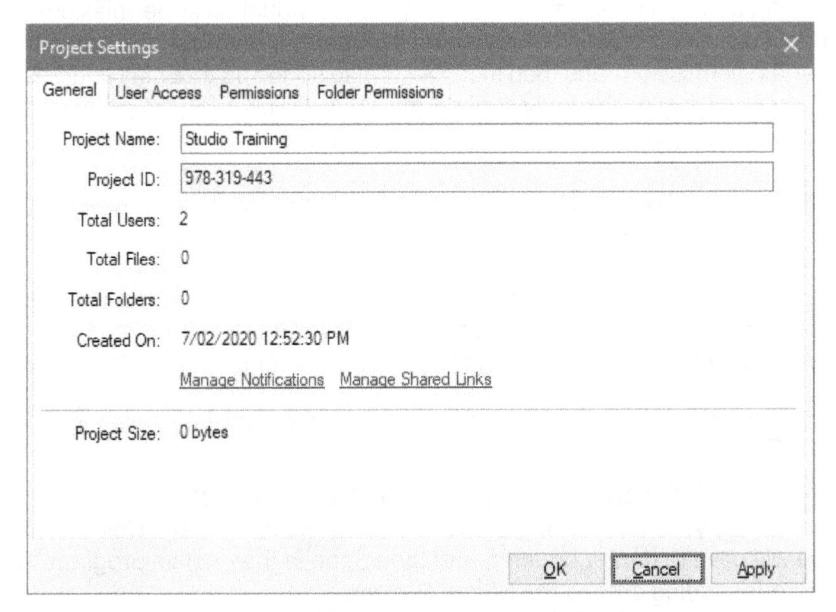

Figure 12 *The* ***Project Settings*** *dialog box*

2. Activate the **User Access** tab; all the users of the current project will be listed.

3. From the bottom of this tab, click the **Restrict User** check box. This ensures that only the users that are added to the Studio Project can access it. Entering the Project ID will not give the users access to the Studio Project.

4. Activate the **Permissions** tab; the default applied permissions are displayed in the **Applied Permissions** area of this tab, as shown in Figure 13.

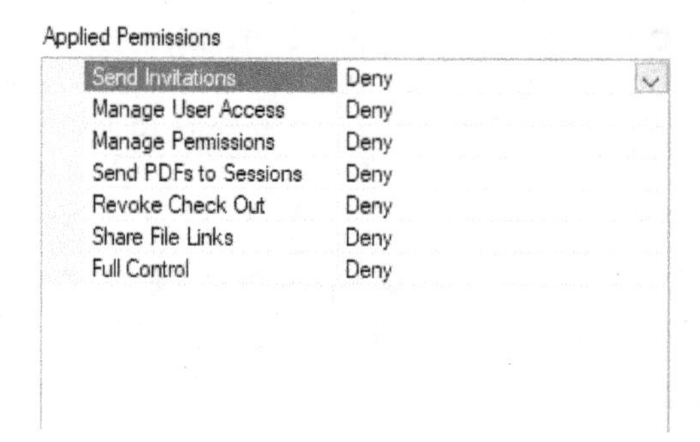

Figure 13 *The permissions applied by default*

5. If required, click the Green + button and then add the users to control each permission type for that user. Alternatively, you can add the user group to control their permissions. You can either apply the **Deny** to **Allow** permission to the selected user or the group.

   Important Note: It is important to note that the individual level permissions override the group permissions of the user. For example, if a user is a member of a group that has been denied certain permissions and then that user is also added as an Email address and allowed certain permissions, then the individual permissions will override the permissions assigned to the group.

6. Activate the **Folder Permissions** tab; all the folders in the Studio Project will be listed in this tab.

7. If required, click the Green + button or the **Group** button and add the users to control the folder permission type for each user or group. For example, the team leads can be given the **Read / Write / Delete** permission to the folders that host files related to their discipline.

8. Click **OK** in the dialog box to finish configuring the Studio Project settings.

The Procedure for Adding Data to the Studio Project

As mentioned earlier, you can add any format file to the Studio Project. However, to open non-PDF files, you will have to download them and then open in their native program. The following is the procedure for adding data to the Studio Project:

1. In the **Main Workspace** of the Project tab, click the **Upload Folder** button, the **Select Folder** dialog box will be displayed.

2. Browse and select the folder to be added to the Studio Project. Note that you can only select one folder at a time. Figure 14 shows the **Select Folder** dialog box with the folder selected to be added.

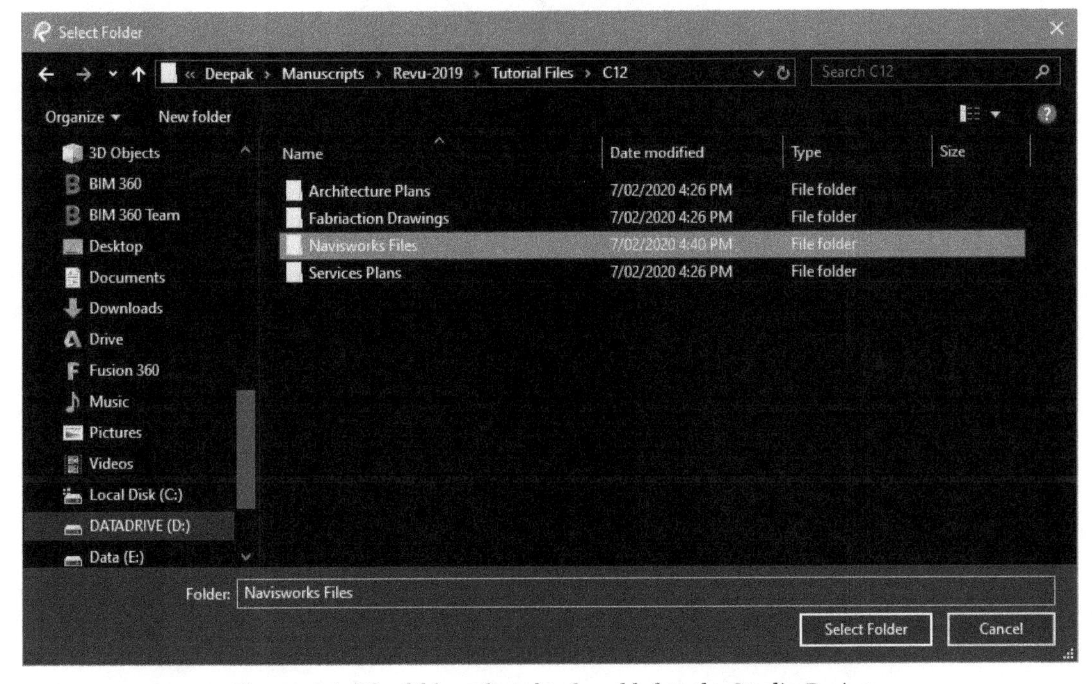

Figure 14 The folder selected to be added to the Studio Project

3. Click the **Select Folder** button; the process of adding the selected folder and its content to the Studio Project will start. Once the process is completed, the folder and its content will be displayed in the Studio Project tab, as shown in Figure 15.

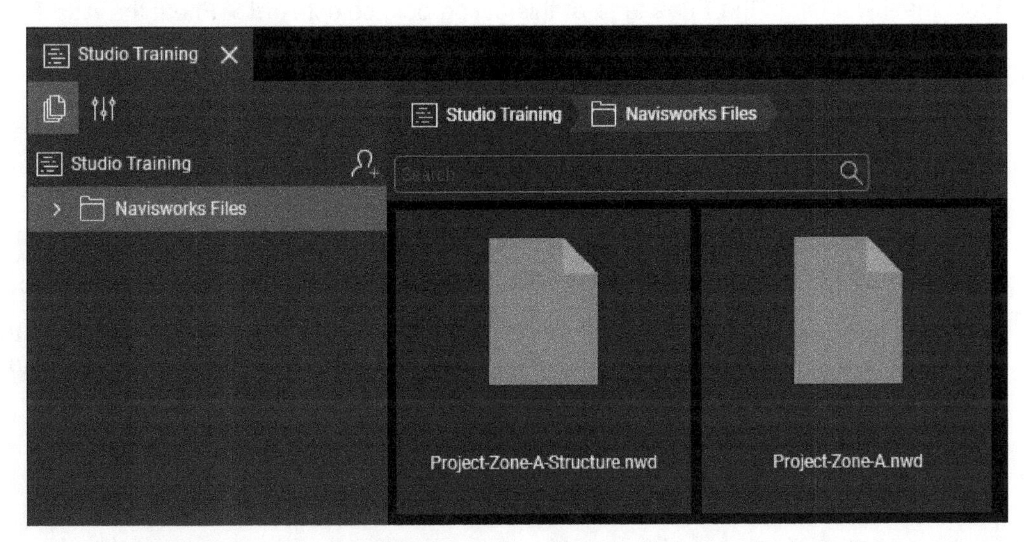

Figure 15 The selected folder and its content added to the Studio Project

4. To add additional folders, make sure you select the Studio Project name in the Project tab. Next, right-click and select **Add Folder** from the shortcut menu, as shown in Figure 16.

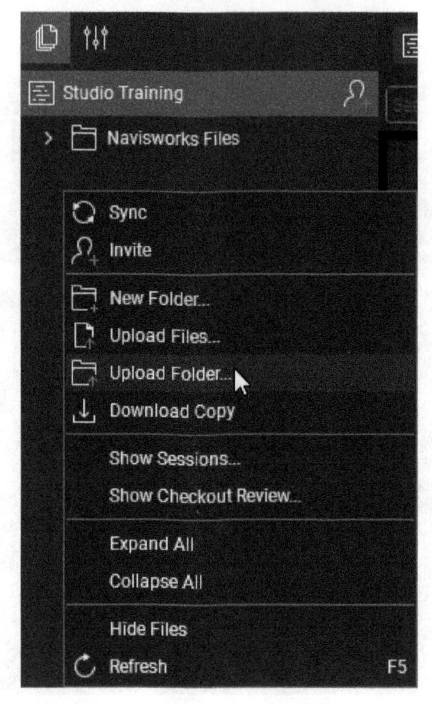

Figure 16 *Adding additional folders to the Project*

5. Similarly, to add files to the Studio Project, right-click and select **Upload Files** from the shortcut menu; the **Upload Files** dialog box will be displayed.

6. From the lower left in the **Files** area of this dialog box, select **Add > Files**; the **Add Files** dialog box will be displayed.

7. Browse to the folder where the files are located.

8. Hold down the SHIFT or CTRL key and select all the files to be uploaded.

9. Click the **Open** button in the **Add Files** dialog box; all the selected files will be added to the **Upload Files** dialog box, as shown in Figure 17.

   Once the files are uploaded, they will be listed in the Studio Project. Figure 18 shows the various folders and files uploaded to the Project. As evident in this figure, if you upload a PDF file to the Project, its preview is shown in the **Main Workspace** of the Project tab.

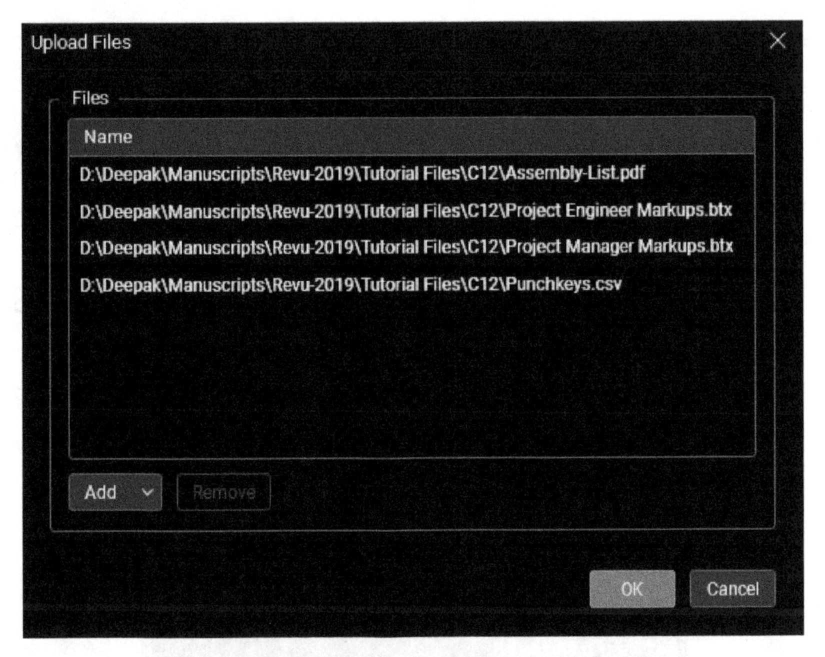

Figure 17 *The* ***Upload Files*** *dialog box*

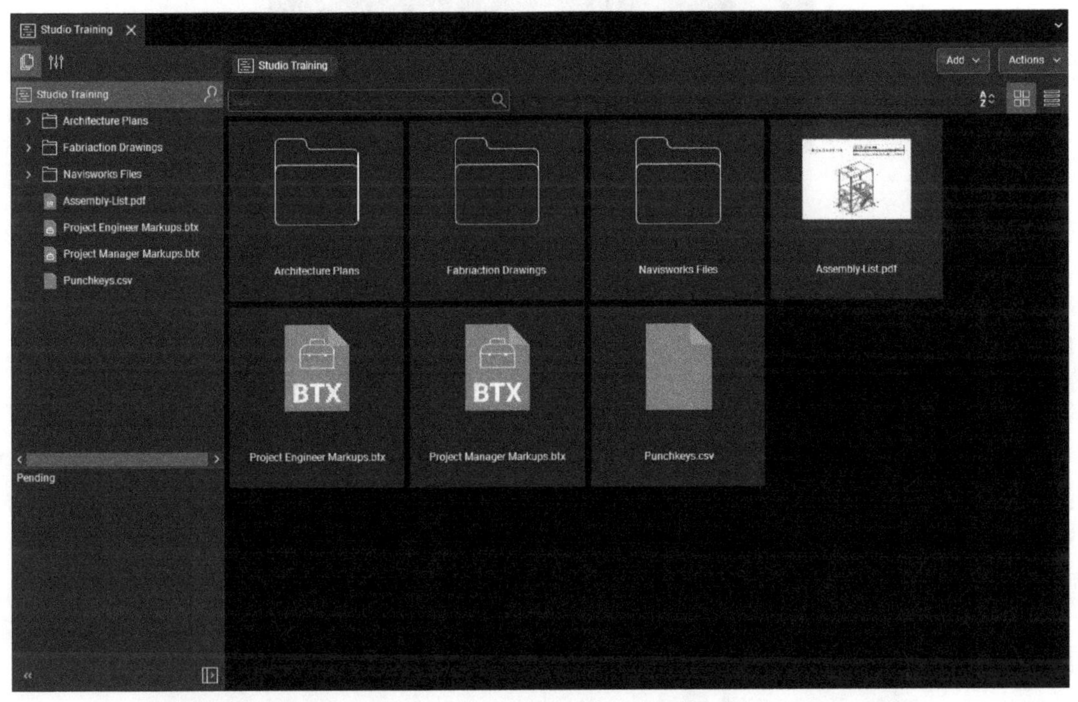

Figure 18 *The Studio Project with the various folders and files uploaded*

The Procedure for Downloading a Local Copy of the Studio Project Folder or File

You can manage the data in the Studio Project by right-clicking on the files or folders. However, it is important to note that the options available in the right-click shortcut menu are different for files and folders. For example, you can right-click on a file and check it out for editing. However, a folder cannot be checked out for editing. The following is the procedure for downloading a local copy of a folder or a file from the Studio Project:

1. To download a local copy of a folder, right-click on it and select **Download Copy** from the shortcut menu, as shown in Figure 19.

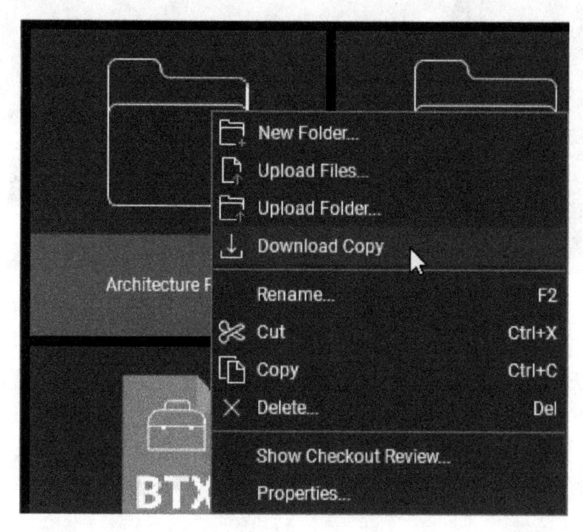

***Figure 19** Download a local copy of the folder*

   On doing so, the **Select Folder** dialog box will be displayed.

2. Browse and select the location where you want to download the folder.

3. Click the **Select Folder** button in the dialog box to start the process of downloading a local copy of the folder and its content.

4. To download a local copy of the file, right-click on it and select **Download Copy** from the shortcut menu, as shown in Figure 20.

   On doing so, the **Save As** dialog box will be displayed.

5. Browse and select the folder in which you want to save the local copy of the file.

6. Click **Save** in the **Save As** dialog box to save the local copy of the Studio Project file.

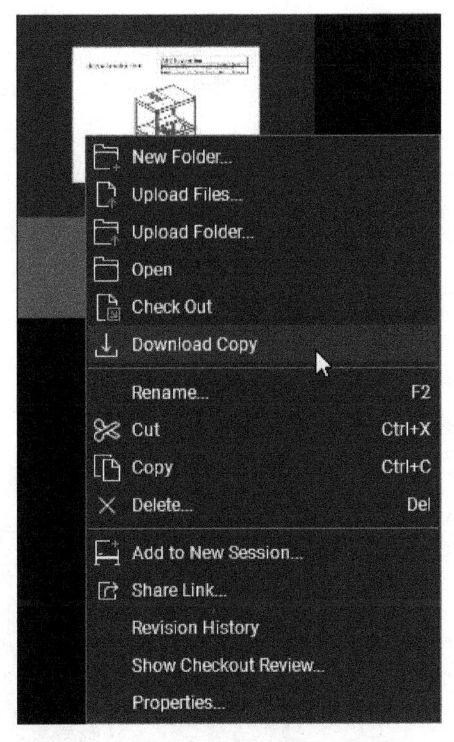

Figure 20 Download a local copy of the selected file

The Procedure for Checking Out a Studio Project File for Editing

Once a PDF or a non-PDF file is uploaded to the Studio Project, you cannot edit it as it is locked from editing. To edit the file, you need to first check it out. The following is the procedure for checking out a file and editing it:

1. Right-click on the PDF file in the Studio Project and select **Check Out** from the shortcut menu, as shown in Figure 21.

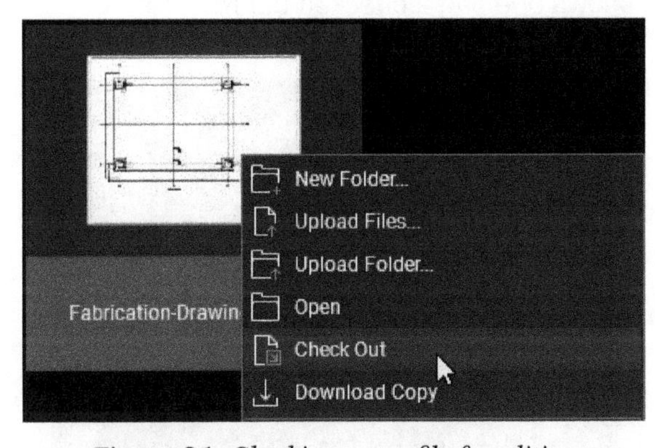

Figure 21 Checking out a file for editing

On doing so, the file will be checked out and will be listed in the **Pending** area below the folder hierarchy in the Project tab, labeled as **1** in Figure 22. Also, the checked out symbol will be displayed on the top left of the file, labeled as **2** in Figure 22.

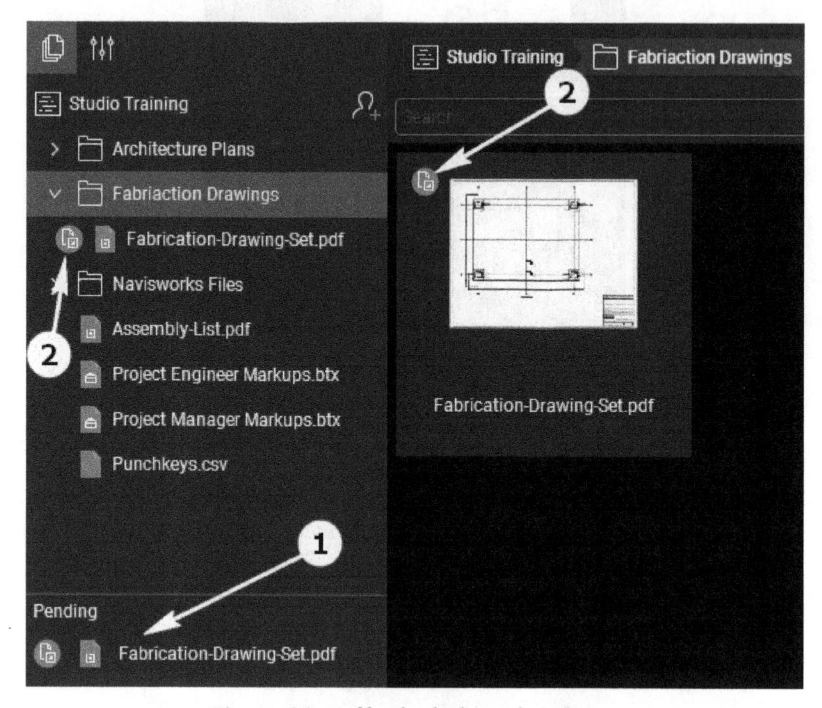

Figure 22 *A file checked out for editing*

***Note**: Only one user can check out a file at any given point of time. If another user needs to edit the file, they need to wait for the first user to check the file back in. However, the Studio Project owner has the right to undo checkout, if required.*

2. Once the file is checked out, right-click on it and select **Open**; the file is opened in its native program. Figure 23 shows opening a checked out PDF file.

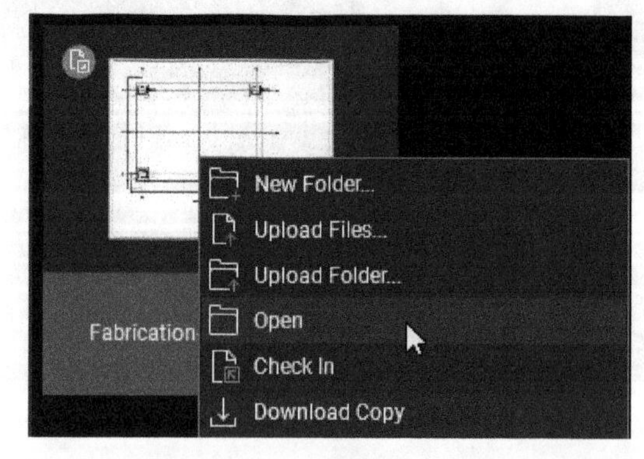

Figure 23 *Opening a checked out file for editing*

3. Make the necessary modifications to the file you have opened and then save the file.

4. If you edited the PDF file in the Revu window, close the file; the **Bluebeam Revu** window will be displayed, as shown in Figure 24.

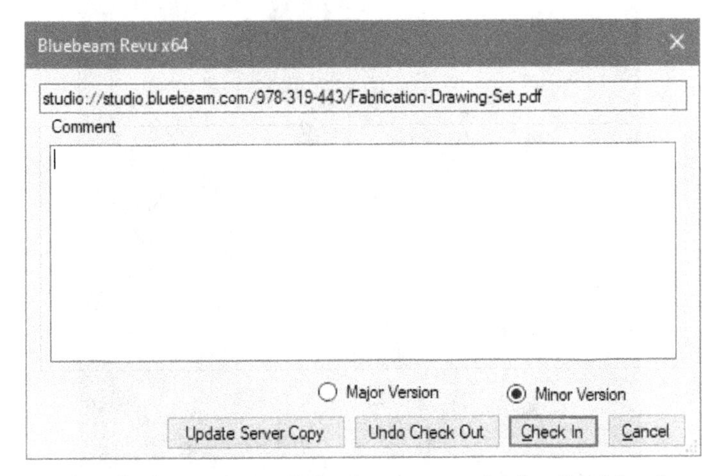

***Figure 24** The **Bluebeam Revu** dialog box to enter the details of the changes made*

***Tip**: It is strongly recommended to enter the details of the changes made in the **Comment** box. This allows you to review the changes made when you check the revision history of the file.*

5. If you want to keep the file checked out for further editing, click **Cancel** in this dialog box.

6. If you are finished with the editing and want to check in the file, enter the details about the changes made in the **Comment** box.

7. Select the **Major Version** or **Minor Version** radio button, depending on whether it was a major change or a minor change.

8. Click the **Check In** button; the file will be checked in and the version updated.

9. For a non-PDF file, save the file after making changes and then close the native program in which the file was opened and return to the Revu window.

10. Return to the Studio Project in the Revu window.

***Tip**: When you create a Studio Project and upload files in it, a local copy of the Project is automatically created and all its files saved in C:\Users\<Username>\AppData\Local\Revu\00\<Project ID>\<Project Folder Name>. You can browse to this folder and review the Project files.*

11. In the Studio Project, right-click on the non-PDF file and select **Check In** from the shortcut menu. Figure 25 shows checking in an Autodesk Navisworks file.

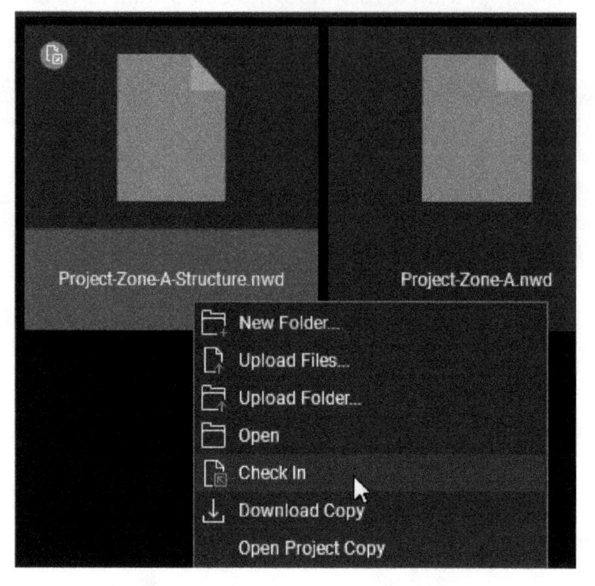

Figure 25 *Checking in a non-PDF file*

On doing so, the **Check In** dialog box will be displayed, as shown in Figure 26.

Figure 26 *The* ***Check In*** *dialog box*

12. Add the details about the changes made in the **Comment** box and then click the **Check In** button; the file will be checked back in and the version of the file will be bumped.

The Procedure for Reviewing the Revision History of a File and Restoring an Old Version of that File

One of the main advantages of Studio Projects is that it keeps the revision history of the files. This way you can review the changes made to the file during various revisions. Also, if required, you can restore an old version of the file as the current version. The following is the procedure for doing this:

1. Right-click on the file in the Studio Project and select **Revision History** from the shortcut menu, as shown in Figure 27.

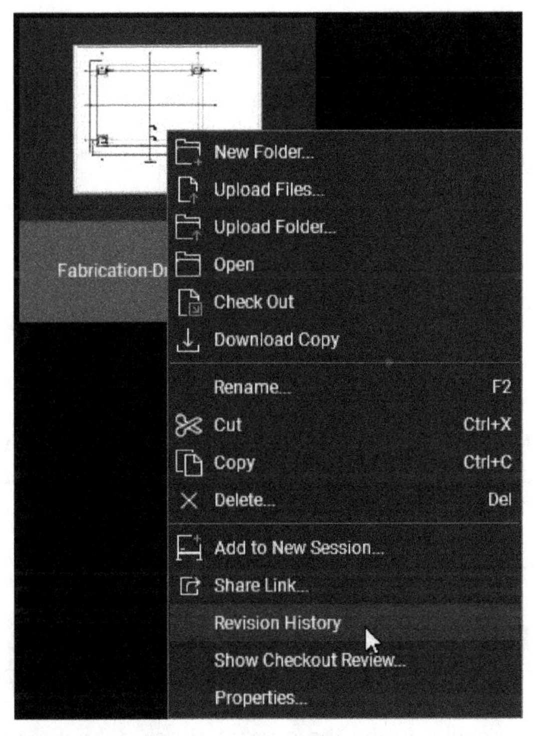

***Figure 27** Reviewing the revision history of a file*

On doing so, the **Revision History** dialog box will be displayed, as shown in Figure 28. Notice that in this dialog box, all versions starting from 1.2 have comments added. These are the comments that were added in the **Comment** box while checking in the file. This makes it a lot easier for you to understand what were the major changes made in that version of the file. However, because there is no comment added for version 1.1, you do not really know what changes were made in that version. Therefore, it is strongly recommended to add the notes in the **Comment** box about the changes made in the file while checking it in.

2. Review the various versions of the current file, the date when the changes were made, and the email address of the users who made the changes, and the revision comments.

3. To restore an old version of the file, click on that version in the list; the **Restore Revision** button is activated near the bottom left of the dialog box.

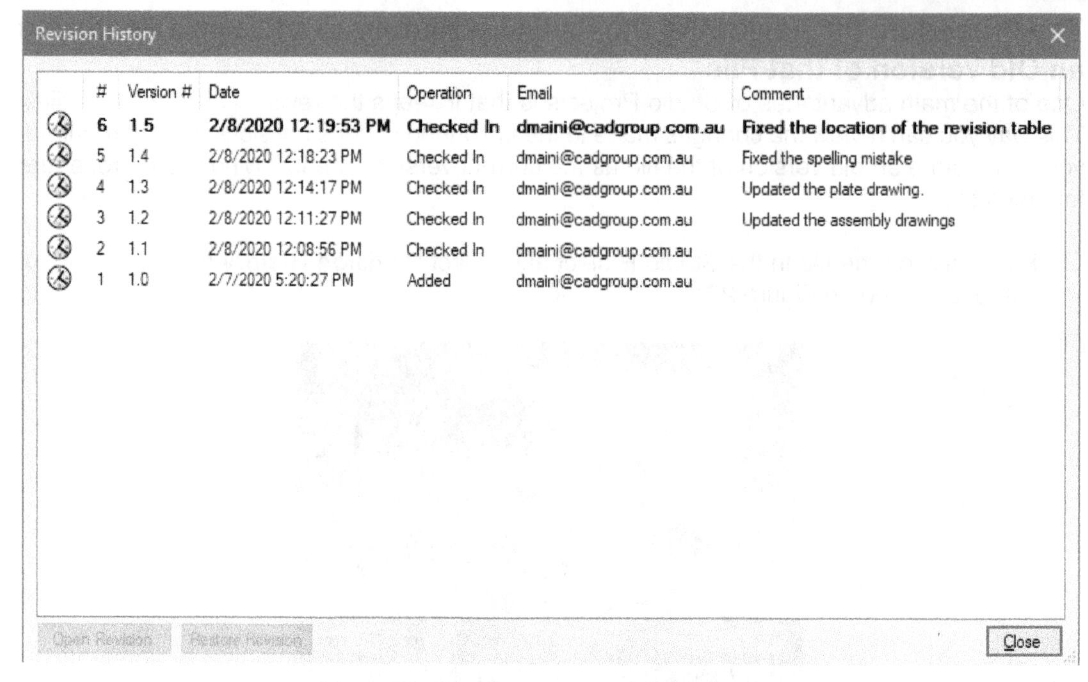

#	Version #	Date	Operation	Email	Comment
6	1.5	2/8/2020 12:19:53 PM	Checked In	dmaini@cadgroup.com.au	Fixed the location of the revision table
5	1.4	2/8/2020 12:18:23 PM	Checked In	dmaini@cadgroup.com.au	Fixed the spelling mistake
4	1.3	2/8/2020 12:14:17 PM	Checked In	dmaini@cadgroup.com.au	Updated the plate drawing.
3	1.2	2/8/2020 12:11:27 PM	Checked In	dmaini@cadgroup.com.au	Updated the assembly drawings
2	1.1	2/8/2020 12:08:56 PM	Checked In	dmaini@cadgroup.com.au	
1	1.0	2/7/2020 5:20:27 PM	Added	dmaini@cadgroup.com.au	

*Figure 28 The **Revision History** dialog box*

4. Click the **Restore Revision** button; the process of restoring the version starts. Once the process is completed, the top version will have a comment **Restores from revision X** where X represents the revision that was restored.

5. Click **Close** in the **Revision History** dialog box.

The Procedure for Replacing a Studio Project File

The Studio Projects allow you to replace the file that was originally uploaded. However, it can only be done if the file is first checked out to you. Also, it is important to remember that all the changes made to the Project file during the current check out will be lost when you replace it with a new file. The following is the procedure for replacing a Studio Project file:

1. Check out the Studio Project file that needs to be replaced.

2. Once the file is checked out to you, right-click on it and select **Replace File** from the shortcut menu, as shown in Figure 29.

   On doing so, the **Replace File** message box will be displayed informing you that your changes will be overwritten by the file you select.

3. Click **OK** in the **Replace File** message box; the **Open** dialog box will be displayed.

4. Browse and select the new file that will replace the Studio Project file.

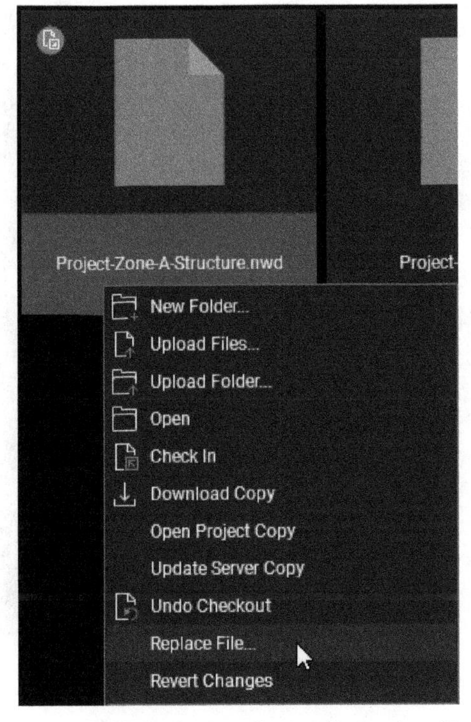

Figure 29 Replacing the Studio Project file

5. Click the **Open** button in the dialog box; the Studio Project file will be replaced by the new file you selected.

6. Checked the file back in.

7. Right-click on the file and select **Revision History** to review the revision history of the file.

The Procedure for Working with the Studio Project in the Offline Mode

Studio Projects allow you to synchronize the files locally so you can access them in the offline mode as well. However, if you want to edit a file in the offline mode, you will have to first check it out so others do not check it out and edit it when you are offline. The following is the procedure for syncing a Studio Project:

1. Right-click on the name of the Studio Project and select **Sync** from the shortcut menu, as shown in Figure 30.

   On doing so, the process of syncing the project settings will start. Once the syncing process is completed, the changes made to the project will be reflected in the Project tab and the files are copied locally on your machine.

2. Check out the file you need to edit in the offline mode.

3. Once you are back online, check the file in with the comments about the changes you made.

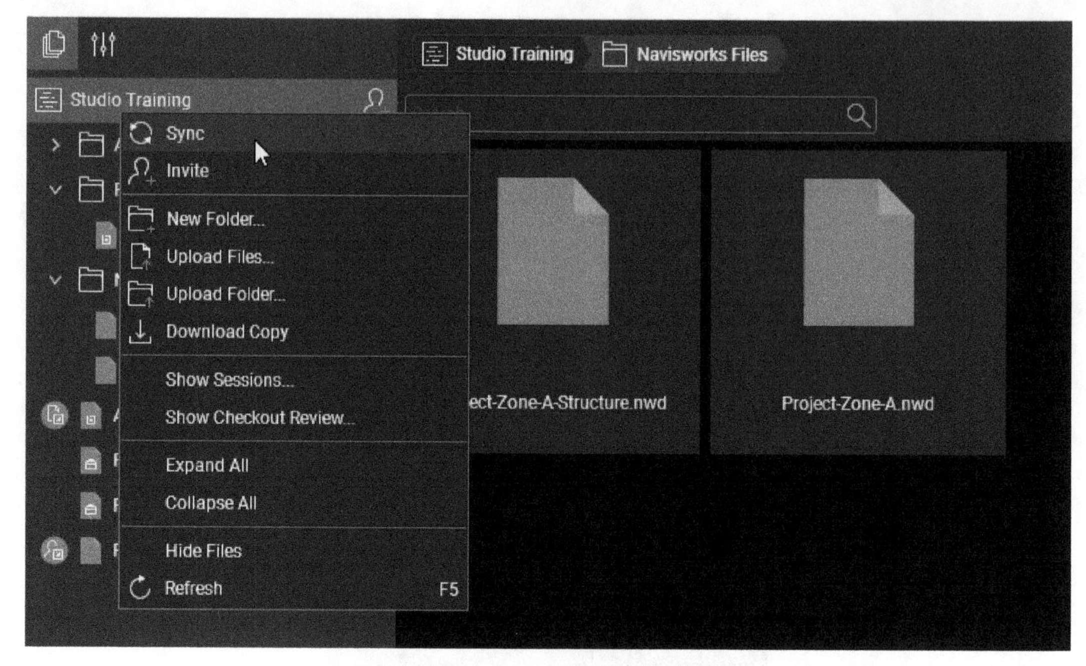

Figure 30 *Syncing a Studio Project*

The Procedures for Working with Studio Sessions

As explained earlier, Studio Sessions allow online design review and markup of PDF files by multiple stakeholders at the same time from within their Revu environment. The host of the Studio Session requires a license of Revu but the attendees can either use a licensed version of Revu or use it in the View Mode which does not require a license. The attendees who use the iPad can purchase the **Bluebeam Revu** iPad app or use **Bluebeam Vu**, which is the free Revu viewer. The following are the various procedures for working with Studio Projects.

The Procedure for Starting a New Studio Session and Adding Files to it From the Local Drive

The following is the procedure for starting a new Studio Session and adding files to it from the local drive:

1. From the **Panel Access Bar**, invoke the **Studio** panel and sign in.

2. From the top of the **Studio** panel, click the **Sessions** button, labeled as **1** in Figure 31.

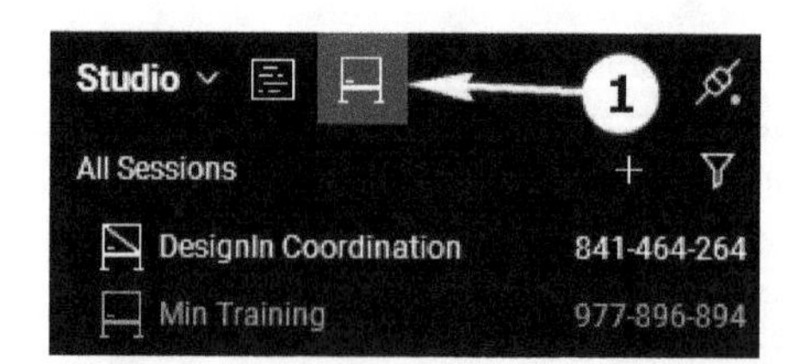

Figure 31 *The **Sessions** button*

On doing so, all the Studio Sessions that you are invited to will be listed in the **Studio** panel.

3. Click the **Add > New Session** on the top right in the **Studio** panel, as shown in Figure 32.

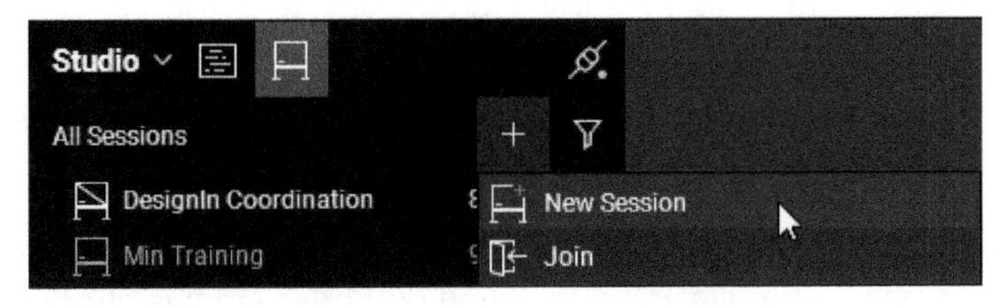

Figure 32 Adding a new Studio Session

On doing so, the **Start Studio Session** dialog box will be displayed, as shown in Figure 33.

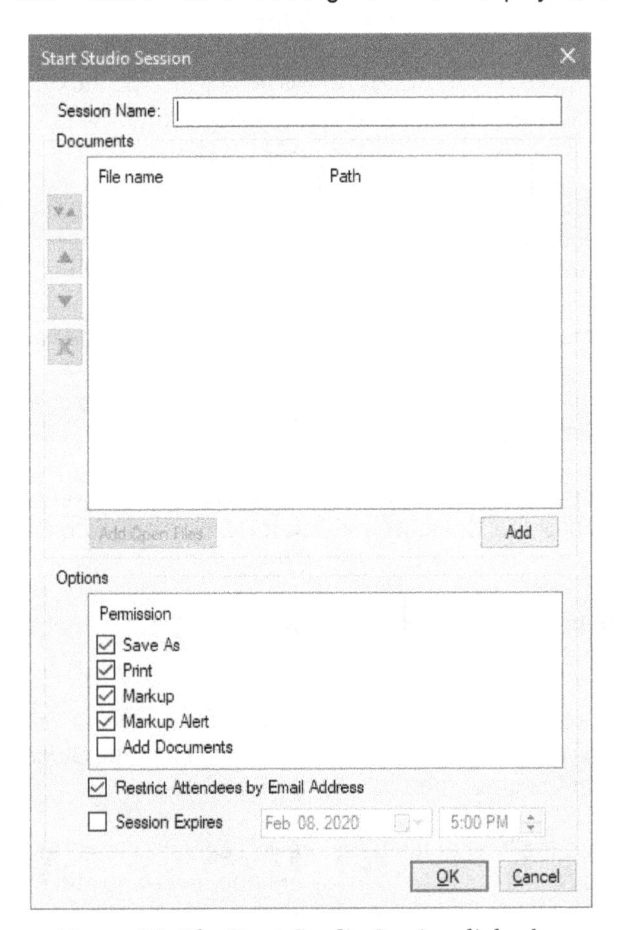

*Figure 33 The **Start Studio Session** dialog box*

4. Enter the Session name in the **Session Name** field.

The files opened in the current Revu session can be added using the **Add Open Files** button. However, if the files you want to add are not currently opened, you can add them using the **Add** button.

5. Using the **Add Open Files** or **Add** button, add the files to the Studio Session.

6. In the **Options > Permission** area, select the check boxes of the permissions you want to assign to the attendees of the session.

7. Select the **Restrict Attendees by Email Address** check box if you to restrict this session only to the users whose Emails you add. If you do not select this check box, any member who has the Session ID will be able to join the Session.

8. To schedule the expiry of this Studio Session, select the **Session Expires** check box and then set the date and time for the session to expire.

9. Click **OK** in the dialog box; the new Session will be started and the **Session Invitation** dialog box will be displayed to invite members to the session. This dialog box is similar to the one discussed while inviting members to the Studio Projects.

   IMPORTANT NOTE: It is extremely important to note that even though you may be the host of the project, but you cannot edit the markups added by the other members in the session. The markups can only be edited by the members who added them.

10. Invite the required members by adding their Email addresses or adding them through groups.

    Remember that similar to the Studio Projects, in Studio Sessions also the individual level permissions override the group level permissions.

    If you have not restricted the session to attendees by Emails, you can also click the **Copy Invitation** at the top of the **Session Invitation** dialog box and paste it in an Email to the invitees.

11. Click **OK** in the **Session Invitation** dialog box; you will be informed that an email is sent to the invitees.

    Once the invitees accept the invite, they will be able to join the Studio Session and start marking up the PDF file. The **Chat** field at the bottom left of the Studio Session window can be used to chat with the other members in the Session.

12. Even though you are the host of the Session, you can still close Revu and the Session will remain open for the rest of the members to continue reviewing and marking up the PDF file.

    Figure 34 shows the Revu window with a Studio Session active. The area labeled as **1** in this figure shows the members invited to the Studio Sessions. Any inactive member will have **Offline** on the right of their name. The area labeled as **2** shows the PDF files added to the session. The area labeled as **3** shows the area where you can review the record of what has been happening in the session and also view any notifications or pending activities.

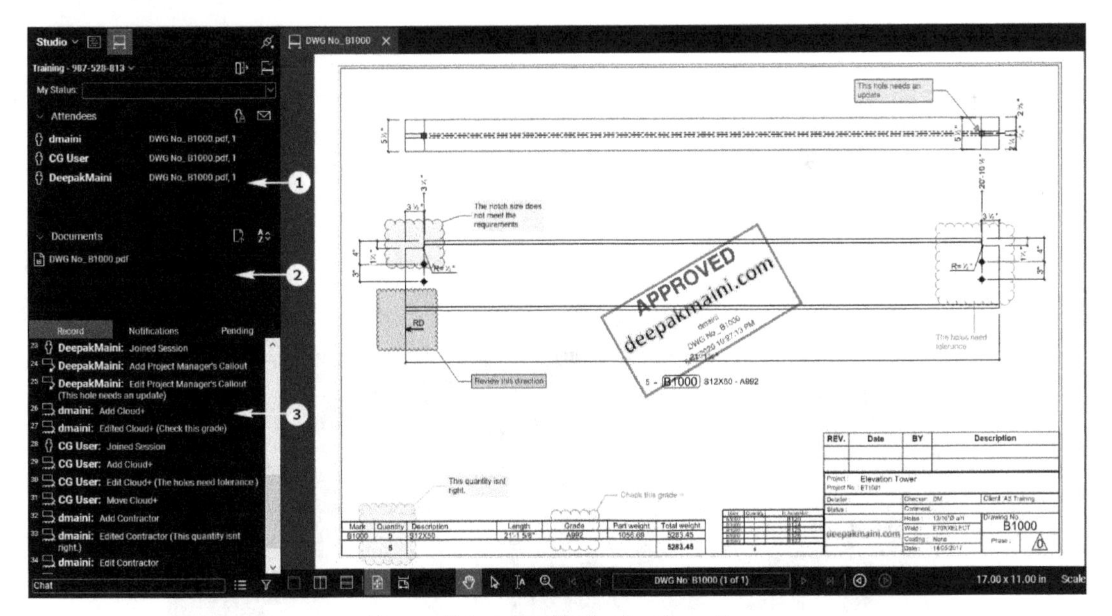

Figure 34 A Studio Session in action

The Procedure for Starting a New Studio Session Using a PDF File from a Studio Project

The following is the procedure for starting a new Studio Session using a PDF file from a Studio Project:

1. From the **Panel Access Bar**, invoke the **Studio** panel and sign in.

2. In the Studio Project, right-click on the PDF file and select **Add to New Session**, as shown in Figure 35.

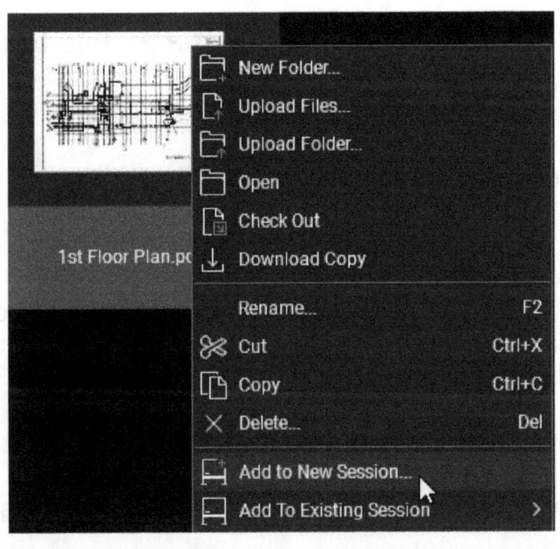

Figure 35 Starting a new Session using a Studio Project PDF File

On doing so, the **Session Name** dialog box will be displayed.

3. Enter the Session name.

4. Click **OK** in the **Session Name** dialog box; a new Studio Session will be started and the PDF file will be added to that session.

   Once a PDF file is added to a Studio Session, the **In Session** icon, labeled as **1** in Figure 36 will be displayed on the file.

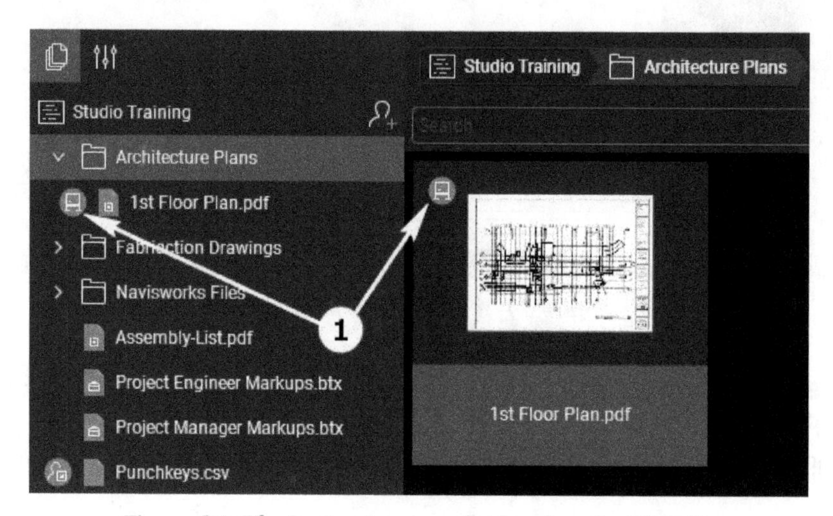

*Figure 36 The **In Session** icon displayed on the PDF file*

5. In the **Studio** panel, click the **Sessions** button to participate in the session started using the PDF file from the Studio Project.

The Procedure for Adding a Studio Project PDF File to an Existing Studio Session

The following is the procedure for adding a Studio Project PDF file to an existing Studio Session:

1. Sign in to the Studio Project.

2. Browse to the PDF file to be added to the existing Studio Project.

3. Right-click on the PDF file and select **Add to Existing Session**; a cascading menu will be displayed showing all the available Studio Sessions that were started using the PDF files from the current Studio Session.

4. Select the Studio Session to which you want to add the PDF file, as shown in Figure 37.

   On doing so, the selected PDF file will be added to the Studio Session.

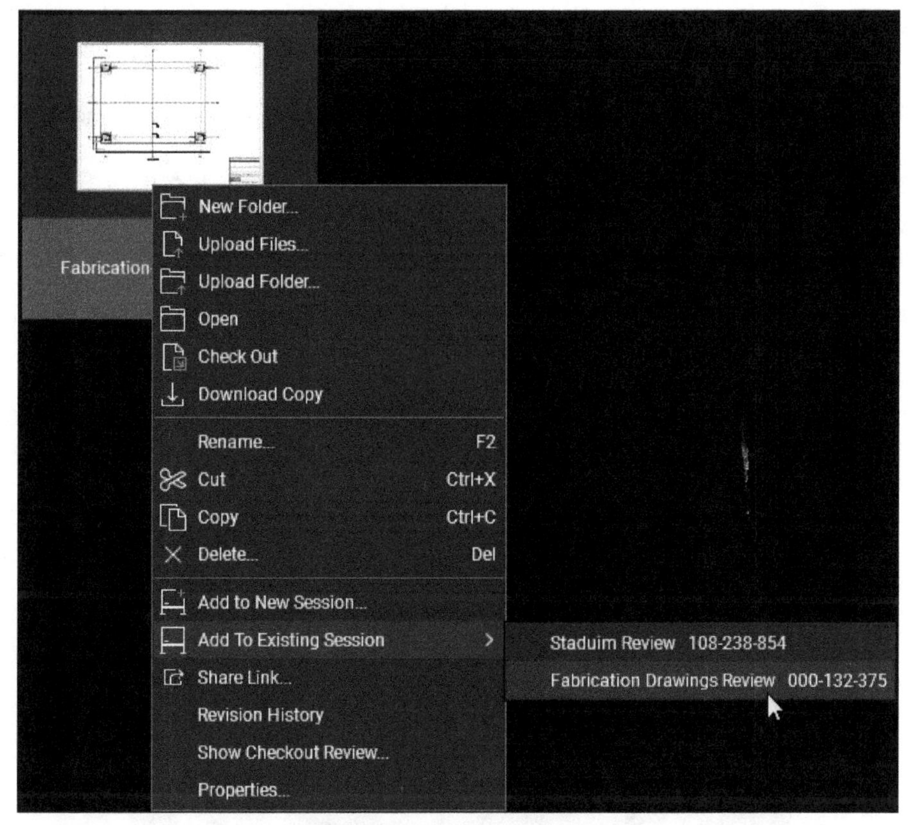

Figure 37 Adding a Studio Project PDF file to an existing Studio Session

The Procedure for Switching Between Different Studio Sessions

The following is the procedure for switching between different Studio Session:

1. In the **Studio** panel > **Sessions** area, click the **Leave Session** button on the top right, as shown in Figure 38.

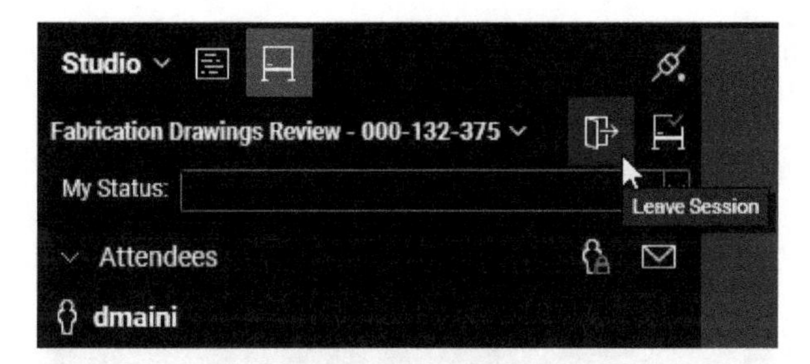

Figure 38 Leaving a Studio Session to join another Session

It is important to note that if you are a Studio Session host, leaving that session will not close it for the other members. They can continue reviewing and marking up the PDF file.

Once you leave a session, the list of the other sessions that you are a member of will be displayed in the **Studio** panel > **Sessions** area.

2. Click on the name of the other Studio Session you want to start reviewing the PDF files in.

The Procedure for Following or Unfollowing an Attendee in a Studio Sessions

If you want to watch what a particular attendee of the Studio Session is doing, you can follow them. This allows you to watch in realtime as they navigate around their PDF file and add markups to the PDF file. Once you are done watching that attendee, you can unfollow them and return to marking up the PDF files. The following is the procedure for doing this:

1. In the **Studio** panel > **Sessions** area > **Attendees** list, hover the cursor over the name of the attendee you want to follow; the **Follow Attendee** icon is displayed on the right of their name, as shown in Figure 39.

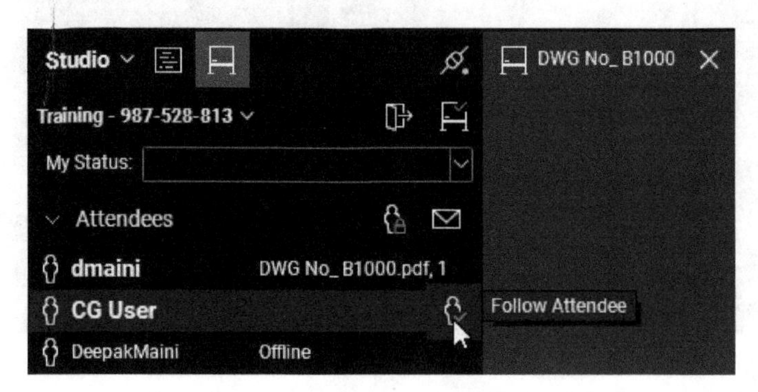

Figure 39 *The* ***Follow Attendee*** *icon displayed to follow an attendee*

2. Click on the **Follow Attendee** icon.

   On doing so, the **Attendee** area will show a note that about you following that attendee. Also, a check mark will be displayed on the attendee's icon on the left of that attendee's name.

3. To unfollow that attendee, zoom or pan on the PDF file. Alternatively, you can click on the icon on the left of their name, as shown in Figure 40.

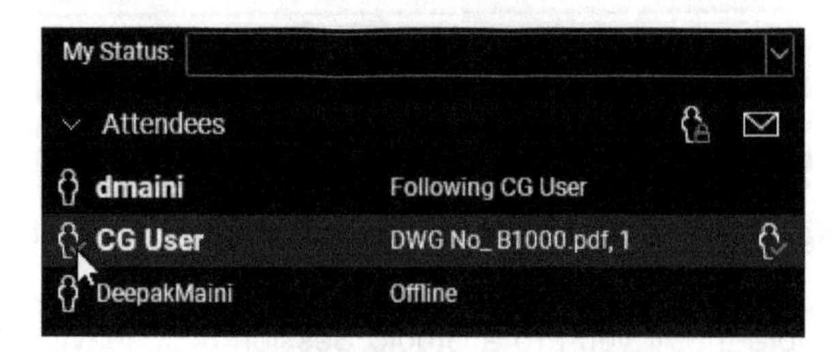

Figure 40 *Unfollowing an attendee*

The Procedure for Finishing a Studio Session and Generating a Report

Once the process of reviewing and marking up the PDF files in a Session is completed, you can finish the Session and generate a report of that session. Note that once you close the session, none of the attendees will be able to markup the PDF file anymore. The following is the procedure for doing this:

1. From the top in the **Studio** panel > **Sessions** area, click the **Finish Session** button, as shown in Figure 41.

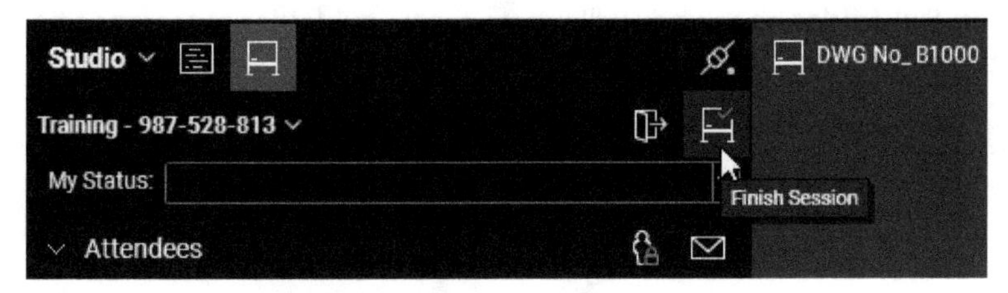

Figure 41 *Finishing a Studio Session*

On doing so, the **Finish Session** dialog box will be displayed, as shown in Figure 42.

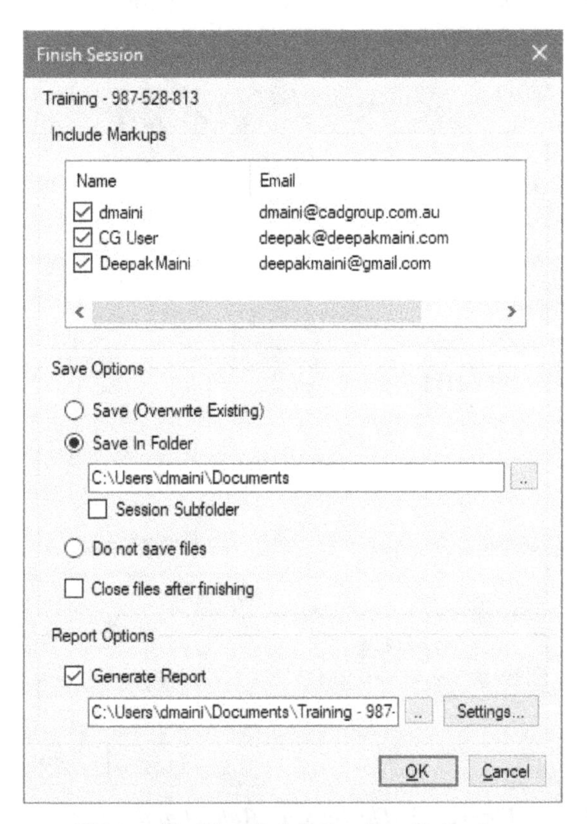

Figure 42 *The* ***Finish Session*** *dialog box*

2. In the **Save Options** area, select the option to save the markups in the Studio Session PDF file. The following are the options you can select:

   Save (Overwrite Existing)
   This option will save all the markups on the original Studio Project PDF file and will overwrite the Project PDF file with this marked up one.

   Save in Folder
   This option will save a new PDF file with all the markups in the specified folder. Selecting the **Session Subfolder** check box will create a new subfolder at the specified location with the ID of the current session.

   Do not save files
   This option will not save any markups added to the PDF file in the Session.

3. Select the **Close files after finishing** to close the PDF file from the session for everyone.

4. In the **Report Options** area, select the **Generate Report** check box and then select the location where you want to save the generated report.

5. Click the **Settings** button on the right of the report location drop-down list; the **Session Report** dialog box will be displayed, as shown in Figure 43.

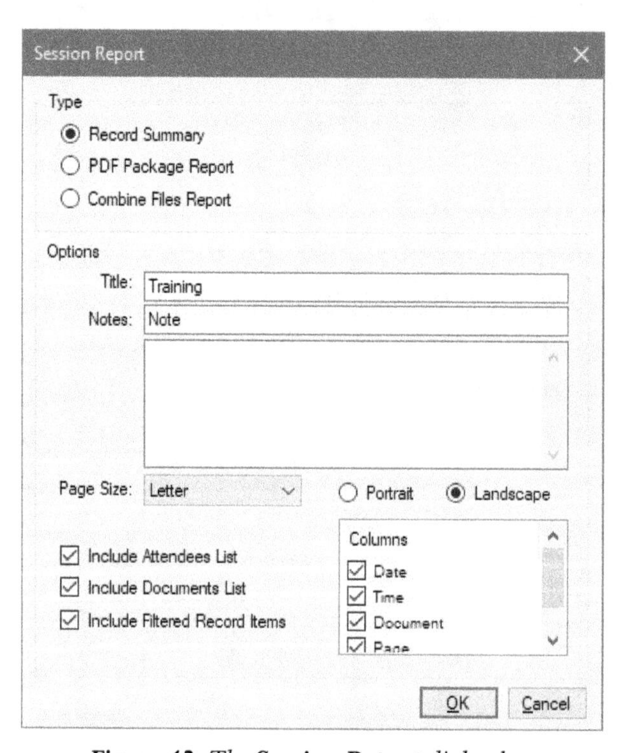

*Figure 43 The **Session Report** dialog box*

6. Configure the report settings in the **Session Report** dialog box and then click **OK** in this dialog box to return to the **Finish Session** dialog box.

7. Click **OK** in the **Finish Session** dialog box; the process of generating the report will start.

   Once the report is generated, the Session will be closed and the **PDF Package** window will be displayed, as shown in Figure 44.

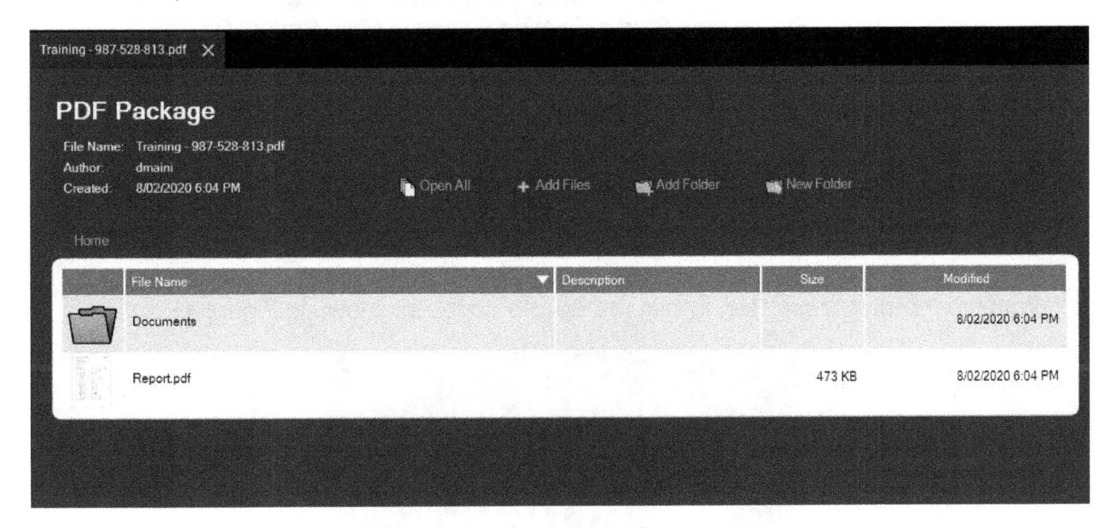

*Figure 44 The **PDF Package** report*

8. If you generated the PDF package report, click on **Open All** to open all the files from that report.

***Tip:** In the report, a hyperlink is automatically created between the markups summary and the original PDF file.*

Hands-on Tutorial	*In this tutorial, you will complete the following tasks:* *1. Create a Bluebeam ID, if you do not have it.* *2. Start a new Bluebeam Studio Project and upload files and folders.* *3. Invite members to the Studio Project and control their permissions.* *4. Checkout a Studio Project file and edit it.* *5. Review revision history of the Studio Project file.* *6. Replace a Studio Project file with a local copy.* *7. Start a new Studio Session using a Studio Project file.* *8. Invite users to the Studio Session and perform live design review.* *9. Finish the Studio Session and generate a report.*

Section 1: Creating Bluebeam ID if you Do Not Have it (Optional)

In this section, you will create a Bluebeam ID if you do not have it already. If you already have a Bluebeam ID, go to Section 2.

1. From the **Panel Access Bar**, invoke the **Studio** panel, as shown in Figure 45.

*Figure 45 The **Studio** panel*

Notice the American flag on the right of the **Sign In** button. This is used to specify whether you want to use the US-based cloud server or UK-based. You can click on the American flag and change the server to the UK-based server. Generally, anyone outside the UK is recommended to use the US-based server.

2. Click the **Sign In** button; the **Bluebeam Studio** dialog box is displayed, prompting you to sign in.

   This dialog box has two tabs: the **SIGN IN** tab is for the users who already have Bluebeam ID and the **CREATE ACCOUNT** tab is to create a Bluebeam ID.

3. Click the **CREATE ACCOUNT** tab in the **Bluebeam Studio** dialog box; the options related to creating a Bluebeam ID are displayed, as shown in Figure 46.

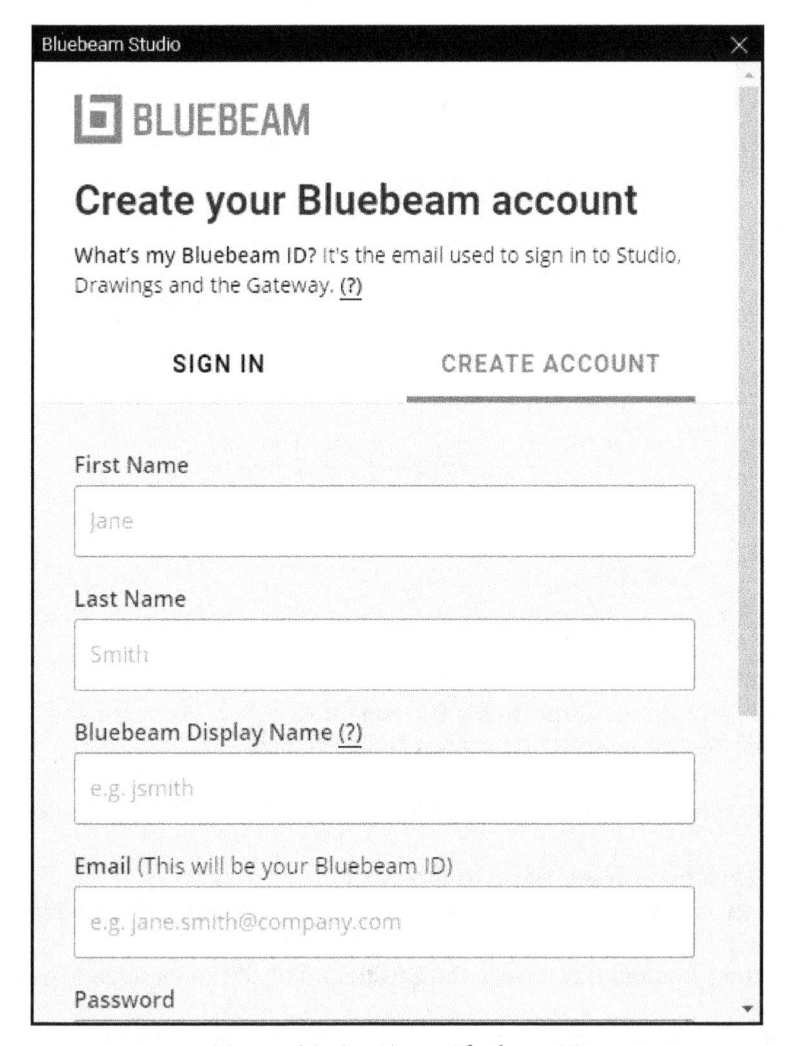

Figure 46 *Creating a Bluebeam ID*

4. Enter the information in this tab.

5. Scroll to the bottom and click **Create Account**; the Bluebeam's General Service and Software Terms of Use window is displayed for you to review terms and conditions of using Revu.

6. Once you have reviewed the terms and conditions and are OK with it, click the **I ACCEPT** button; a confirmation Email is sent to the Email address you used to create the ID. Figure 47 shows a sample of this email.

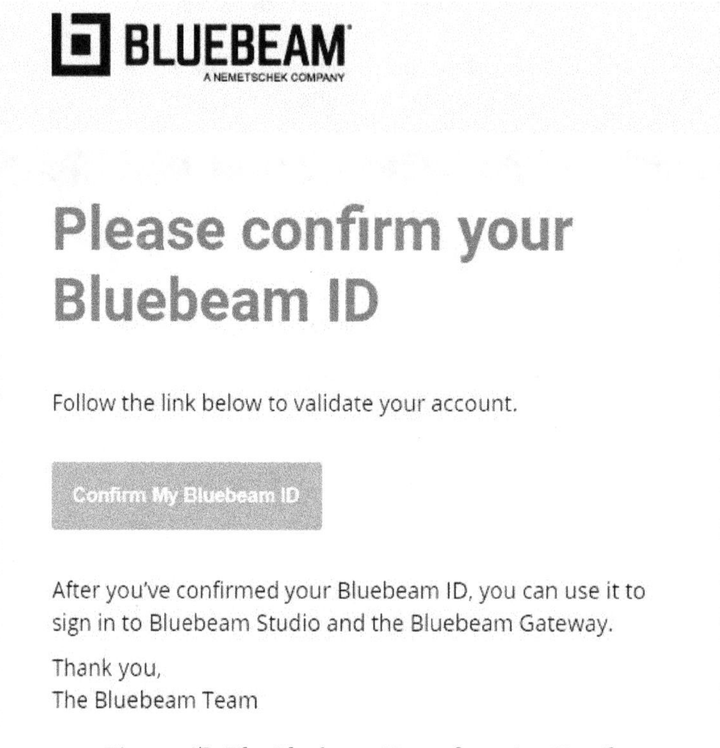

Figure 47 *The Bluebeam ID confirmation Email*

7. In this Email, click on the **Confirm My Bluebeam ID** button to confirm the creation of your Bluebeam ID that will be used to access Studio Projects and Sessions.

8. Return to the Revu window and now sign in to Bluebeam Studio using the same credentials.

Section 2: Starting a New Studio Project

In this section, you will start a new Studio Project with the name **Studio Training**.

1. From the **Panel Access Bar**, invoke the **Studio** panel, if it is not already displayed.

2. Sign in using your Bluebeam ID. Once you sign in, the **Studio** panel has the **Projects** option selected at the top, labeled as **1** in Figure 48. By default, this window will be blank if you have not hosted any Project and are not invited to join any Project, as shown in Figure 48.

   However, if you have hosted a Studio Project or were invited to any Project, they will be listed in this window, as shown in Figure 49. Note that in this figure, the IDs of various projects are blurred for privacy purpose.

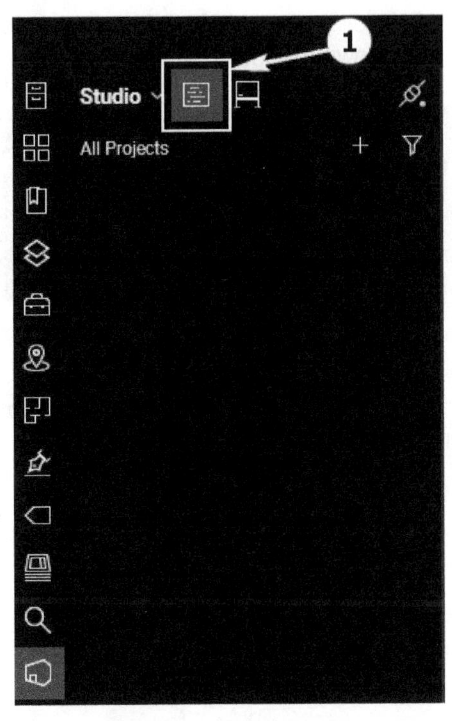

Figure 48 The ***Studio*** *panel >* ***Projects*** *area after signing in*

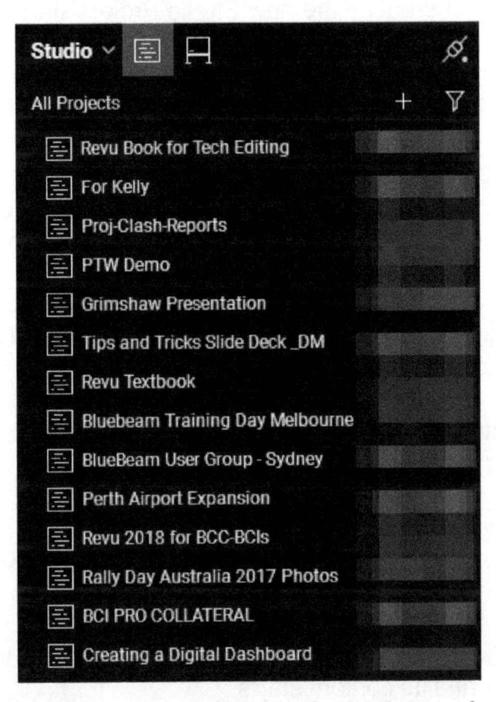

Figure 49 *Various Studio Projects listed in the* ***Studio*** *panel >* ***Projects*** *area*

3. To start a new Studio Project, click the **Add** button near the top right in the **Studio** panel and select **New Project**, as shown in Figure 50.

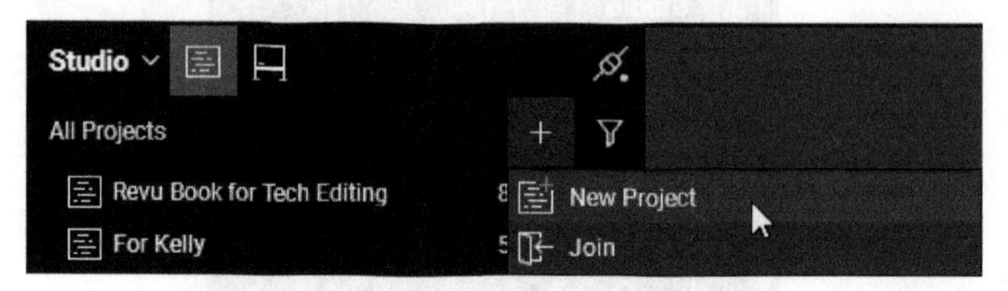

Figure 50 *Adding a new Studio Project*

On doing so, the **New Project** dialog box is displayed where you can enter the name of the new project, as shown in Figure 51.

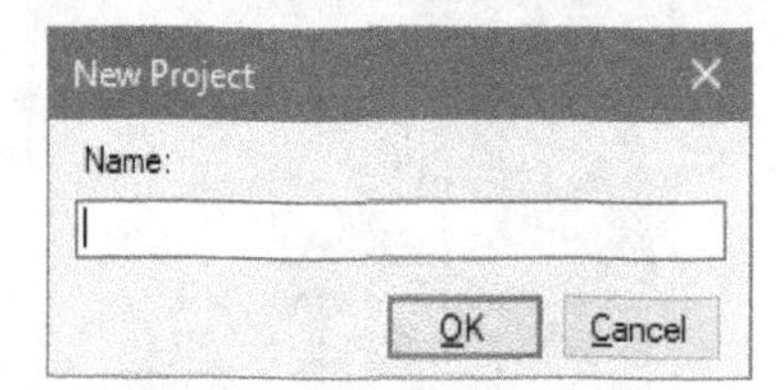

Figure 51 *The **New Project** dialog box*

4. Enter **Studio Training** as the name of the new Project and then click **OK** in the **New Project** dialog box; the process of creating the new Studio Project starts.

   Once the Studio Project is created, it will be listed at the top of the Projects list in the **Studio** panel and is also opened in its own tab in the Revu window, as shown in Figure 52.

Section 3: Uploading Folders and Files to the Studio Project

In this section, you will upload various folders and files to the **Studio Training** Project. You will start with uploading a local folder and will then drag and drop various folders and files to upload.

1. From the **Studio Project** tab in the **Main Workspace**, click the **Upload Folder** button; the **Select Folder** dialog box is displayed.

2. Browse to the **Tutorial Files > C12** folder; the various folders inside the **C12** folder are listed there, as shown in Figure 53.

   You will first upload the **Architecture Plans** folder.

3. Double-click on the **Architecture Plans** folder.

4. In the **Select Folder** dialog box, click the **Select Folder** button on the lower right; the process of uploading the folder and its content starts.

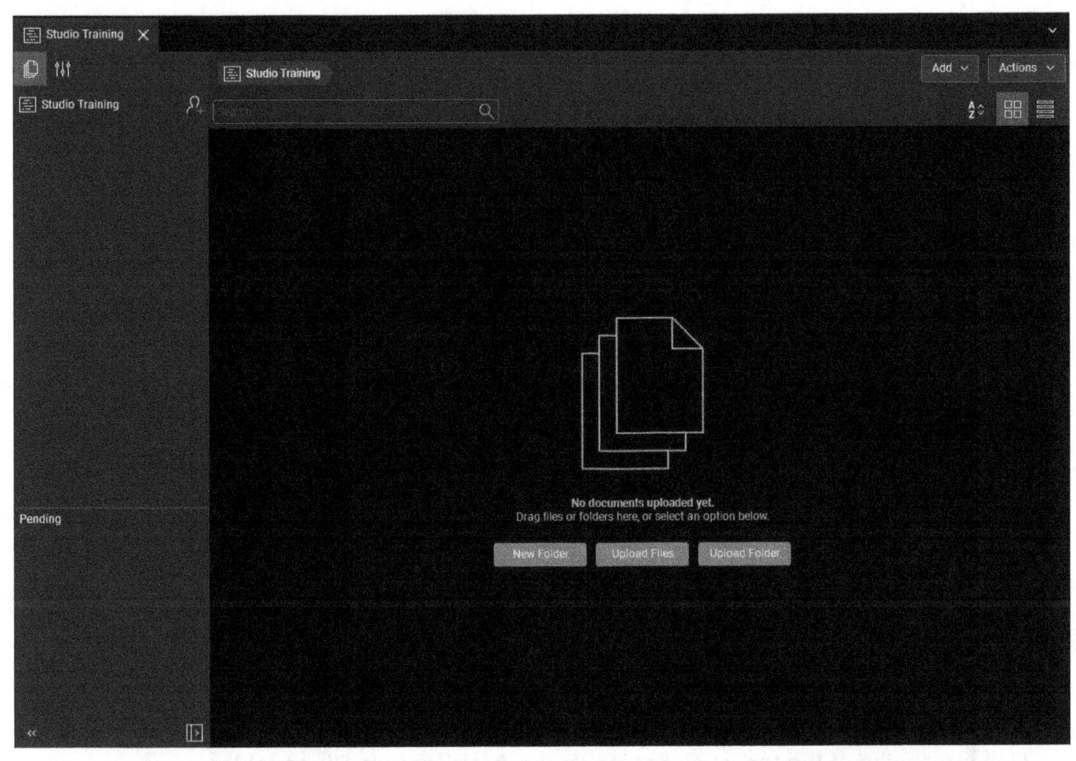

Figure 52 The new Studio Project opened in its own tab in the ***Main Workspace***

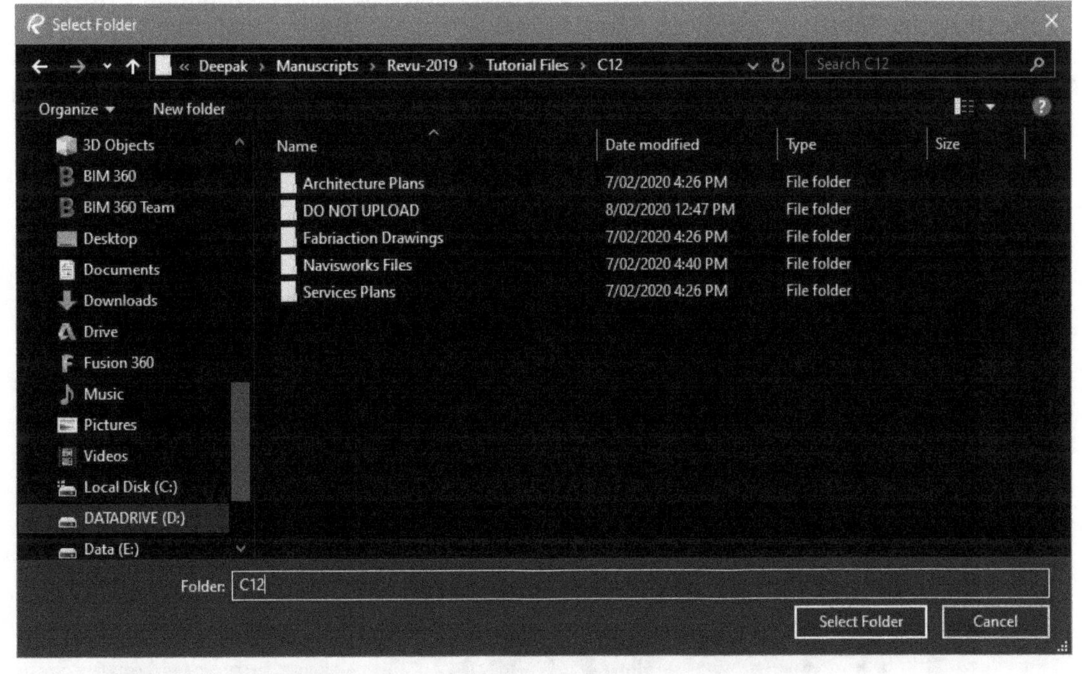

Figure 53 The folders available to upload to the Studio Project

Once the folder and its content are uploaded, click on the **Architecture Plans** folder in the **Studio Training** project tab; the file inside that folder is listed in the **Main Workspace**, as shown in Figure 54.

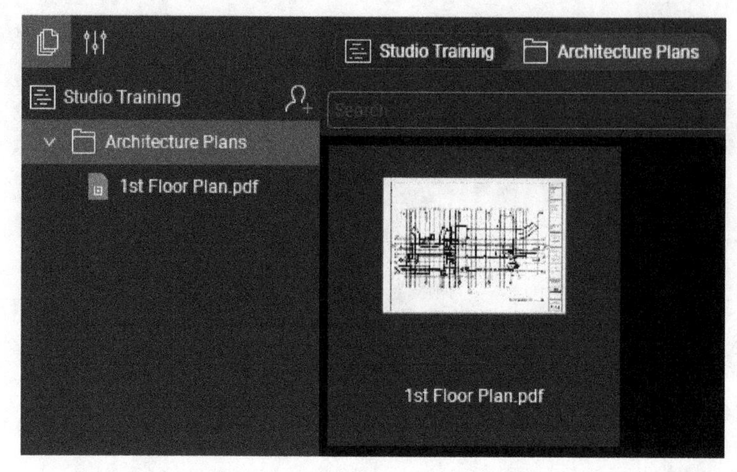

***Figure 54** The uploaded folder and the file inside it listed in the Studio Project*

You will now use Windows Explorer to drag and drop more folders and files into the Project.

5. From the Project tab, click on the **Studio Training** Project name to ensure the folders and files are copied at the top level, not inside the **Architecture Plans** folder.

6. Using Windows Explorer, browse to the **C12** folder and select the **Fabrication Drawings**, **Navisworks Files**, and **Services Plans** folders and all the files in this folder.

7. Drag and drop them in the **Main Workspace** of the **Studio Training** Project, as shown in Figure 55.

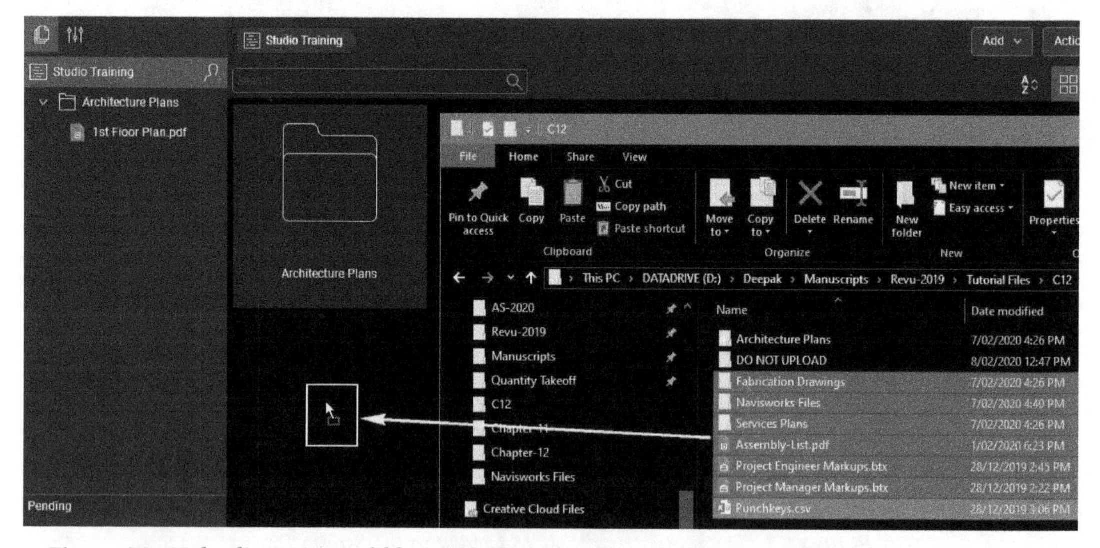

***Figure 55** Uploading various folders and files by ragging and dropping them in the Studio Project*

On doing so, the process of uploading the selected folders and files to the Studio Project starts. Once the folders and files are uploaded, they will be listed in the Project, as shown in Figure 56.

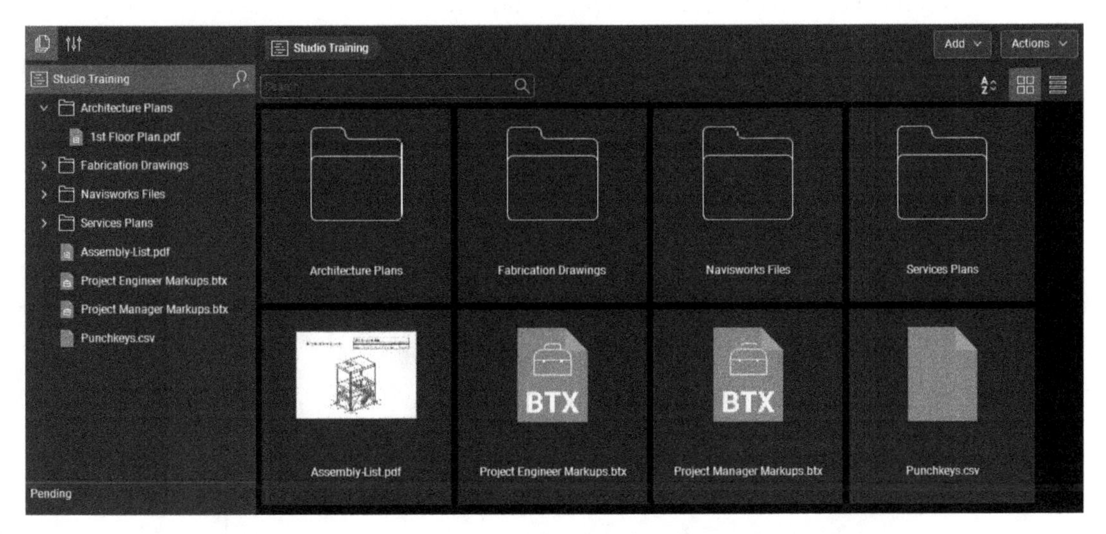

Figure 56 The uploaded folders and files listed in the ***Studio Training*** *Project*

Section 4: Inviting Members to the Studio Project

In this section, you will invite members to this Studio Project.

1. In the Project tab, click on the **Invite** button on the right of the project name, as shown in Figure 57.

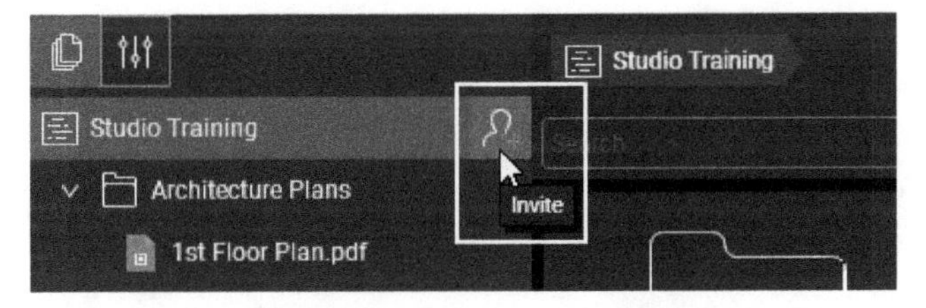

Figure 57 Inviting members to the Studio Project

On doing so, the **Project Invitation** dialog box is displayed. Using this dialog box, you can invite members by typing their Email addresses or selecting them from the Address Book. Alternatively, you can add a group of users, if you already have a group created. These two methods are used to invite members to a Project restricted by Email addresses. When you configure the Project settings later on, you will ensure this is a restricted Project.

2. In the **Invitees** area of the **Project Invitation** dialog box, click the Green + button and enter the Email address of the member to invite. You can invite one of your colleagues to this Project. Alternatively, click the **Address Book** button to access the address book of your Email program, such as Microsoft Outlook, to add the Email addresses.

3. Repeat this process to add members to be invited to this Project. Once you add the email addresses, they will be listed in the **Invitees** area, as shown in Figure 58.

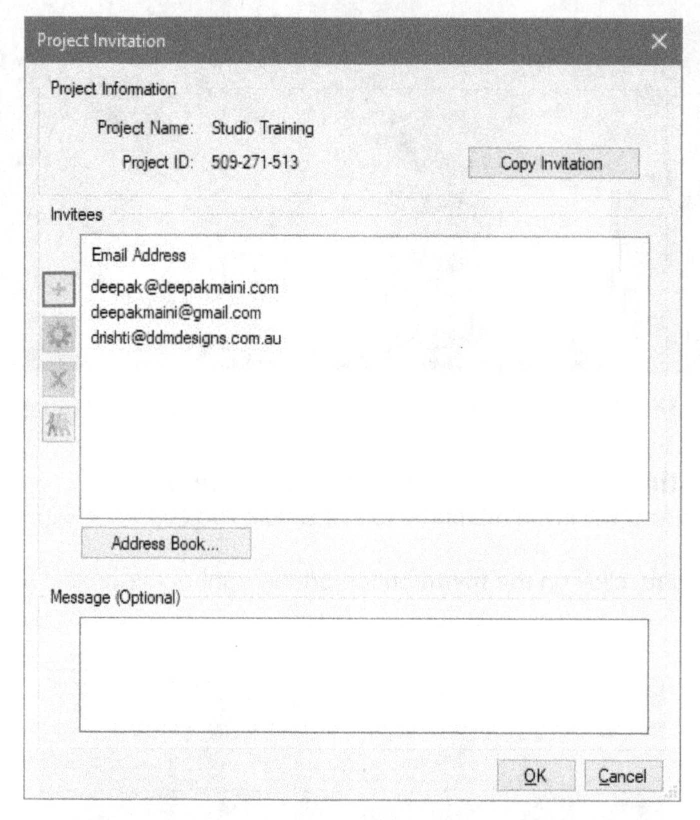

Figure 58 *The Email addresses of the invitees added*

4. In the **Message** area of the **Project Invitation** dialog box, type the following message:

   This is a training Project to explain how Bluebeam Studio Projects and Sessions work.

5. Click **OK** in the dialog box; the **Bluebeam Revu** dialog box is displayed informing you that the Email invitations have been sent to the attendees.

6. Click **OK** in the dialog box to return to the Project tab.

   The invitees will be able to click on their Project invite and then access the Studio Project.

Figure 59 shows the screenshot of accessing the Studio Project using the Bluebeam Revu iPad app.

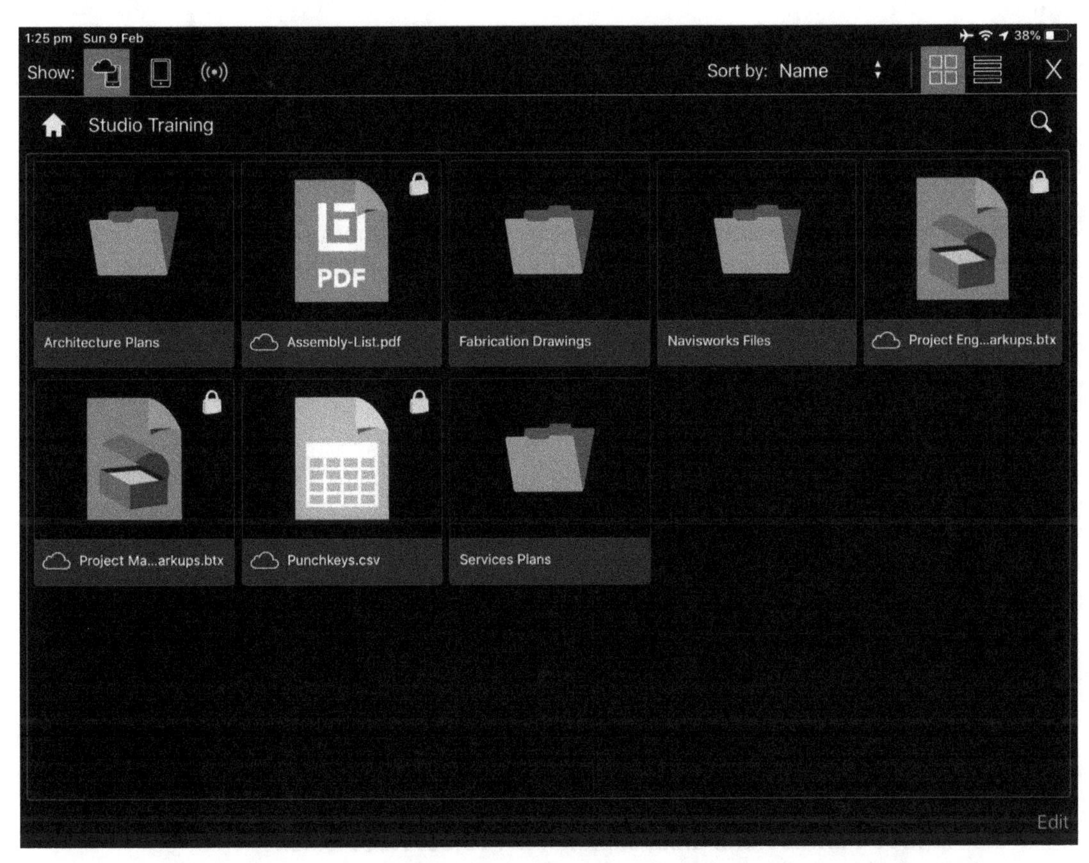

***Figure 59** The screen grab of the Studio Project accessed using the Bluebeam Revu iPad app*

Section 5: Configuring the Studio Project Settings

In this section, you will configure various Project settings and user permissions.

1. In the Project tab, click on the **Project Settings** button, as shown in Figure 60.

***Figure 60** Configuring the Studio Project settings*

On doing so, the **Project Settings** dialog box is displayed with the **General** tab active. You can use this tab to manage the notifications related to all Studio Projects.

2. Activate the **User Access** tab; all the users of the **StudioTraining** Project who have accepted the invite are listed there.

   You will now select the **Restrict Users** check box if it is not already selected. This will ensure that only the users you add to the Project via their Emails are able to access the Project. Any other user with the Project ID will not be able to access this Project.

3. From the bottom left of this tab, make sure the **Restrict Users** check box is selected.

4. Activate the **Permissions** tab. Notice that by default, all the members of the Project are denied from performing any of the activities listed in the **Applied Permissions** area, as shown in Figure 61.

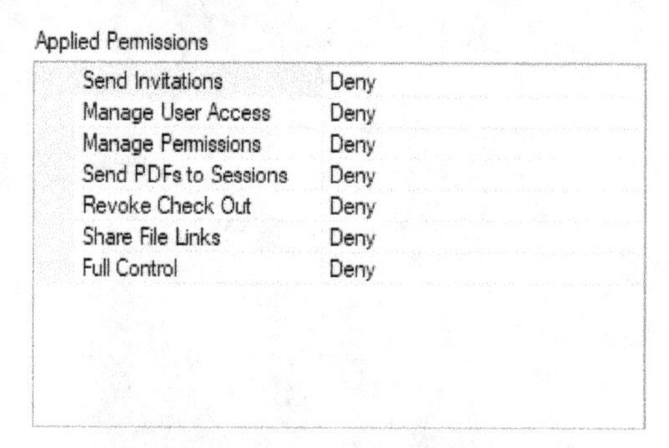

Figure 61 *All the users denied permissions to perform any of these activities*

5. Activate the **Folder Permissions** tab.

6. In the **Applied Permissions** area, expand the **Project Root** folder.

   Notice that in the **Applied Permissions** area, the **Read** permission is assigned to the **Project Root** folder and all its subfolders, as shown in Figure 62. This means that all the users by default are given read-only permission to the Project root folder and all the subfolders.

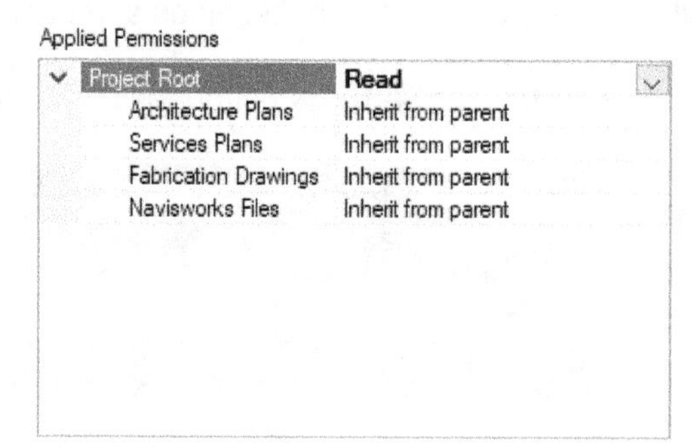

Figure 62 *The default permissions assigned to the Project folders*

You will now give one of the users in the Project Read / Write / Delete access to the **Navisworks Files** folders.

7. From the **User/Groups** area in the **Folder Permissions** tab, click the Green **+** button; the **Add Users/Groups** dialog box is displayed with all the members of the current Project who have accepted the invite.

8. Select one of the members of the Project and then click **OK** in the **Add Users/Group** dialog box; the selected Email address is listed in the **Users/Groups** area.

9. In the **Applied Permissions** area, expand the **Project Root** folder again.

10. Click on the field on the right of the **Navisworks Files** folder and select **Read / Write / Delete** from the drop-down list.

    Figure 63 shows the **Folder Permissions** tab with this permission defined for the selected user. This user will be able to read, write, or delete files from the **Navisworks Files** folder.

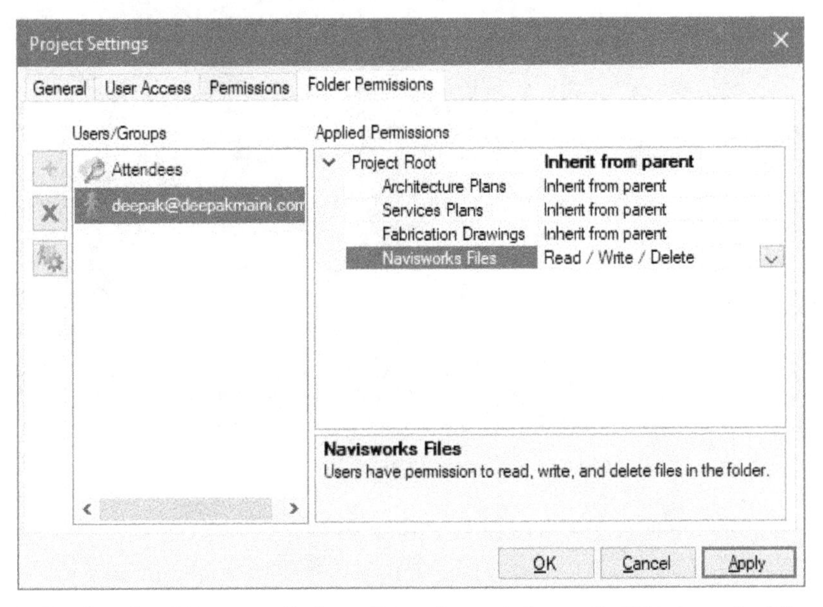

***Figure 63** The selected user applied an elevated permission to the **Navisworks Files** folder*

11. Click **OK** in the **Project Settings** dialog box; you are returned to the **Studio Training** Project tab.

Section 6: Checking Out Studio Project Files to Edit

By default, the Studio Project files are locked from editing. To edit them, you need to first check them out. In this section, you will check out the **Fabrication-Drawing-Set.pdf** file from the **Fabrication Drawings** folder and make changes to it. You will then check this file back in.

1. In the Project tab, click on the **Fabrication Drawings** folder; the **Fabrication-Drawing-Set.pdf** file in this folder is listed in the **Main Workspace**.

2. In the **Main Workspace**, right-click on this file and select **Check Out** from the shortcut menu, as shown in Figure 64.

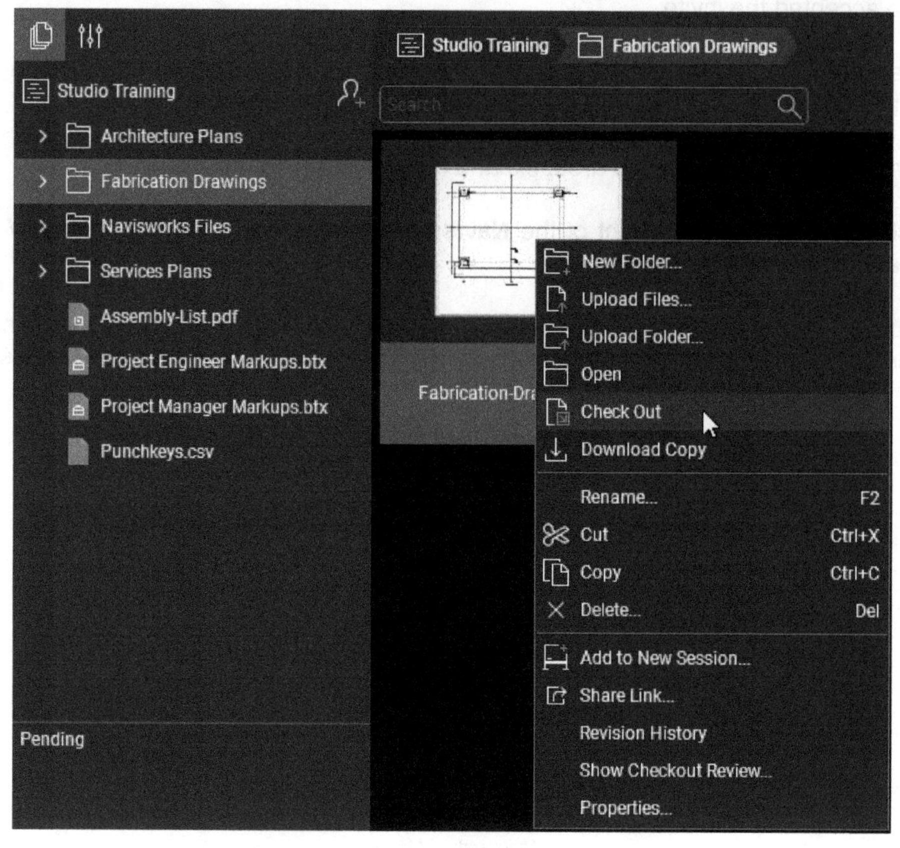

***Figure 64** Checking out a file for editing*

On doing so, the file is checked out the **Checked Out** icon is displayed on its left, labeled as **1** in Figure 65.

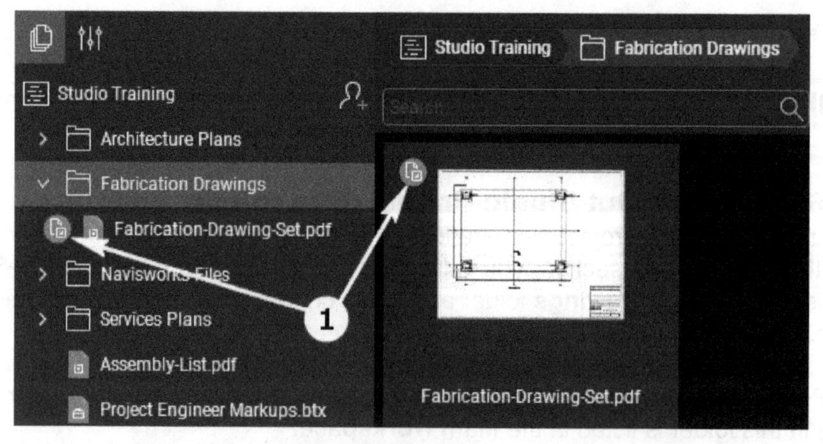

***Figure 65** The **Checked Out** icon displayed on the file*

Tip: Once you check a file out, it is also listed in the ***Pending*** *area below the folder hierarchy of the Project. This is a good place to look at all the files checked out from the Project.*

3. Double-click on the checked out file to open; the **Open** dialog box is displayed prompting you to specify if you want to open this file in the split view.

4. Click **No** in the **Open** dialog box; the checked out file is opened in the Revu window.

5. From the **Menu Bar** click **Edit > PDF Content > Edit Text** and change the drawing number of sheet 4 in the titleblock to **B129**.

6. In the **Thumbnails** panel, double-click on the sheet label and change it to **DWG No: B129**, as shown in Figure 66.

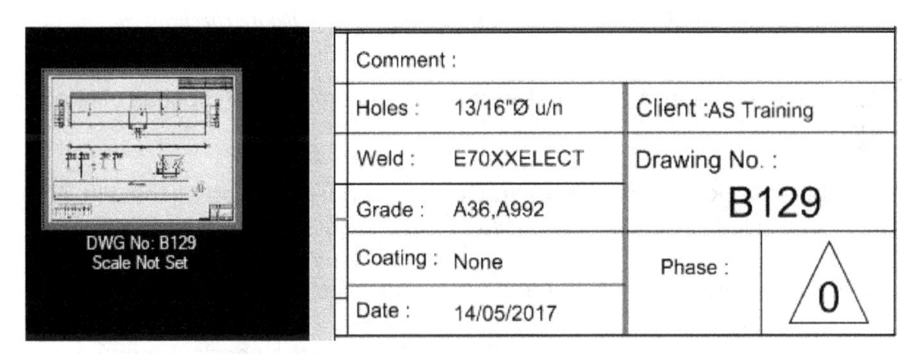

Figure 66 *The drawing number and label edited*

7. Similarly, change the drawing number and sheet label of sheet 5 to **B128**.

8. In the **Thumbnails** panel, drag and drop the **B128** sheet before the **B129** sheet.

9. Zoom to the extents of the sheet.

   You will now save this PDF file and check it back in.

10. Save the PDF file and close it; the **Bluebeam Revu** dialog box is displayed.

    I strongly recommend adding comments about the major changes made to the document before checking it back in. This helps in understanding the changes made to the file while reviewing its revision history.

11. In the **Comment** box, type the following comment:

    Changed the drawing numbers and sheet labels of the two assembly drawings to the assembly numbers 128 and 129.

    Figure 67 shows the comment added before checking the file back in.

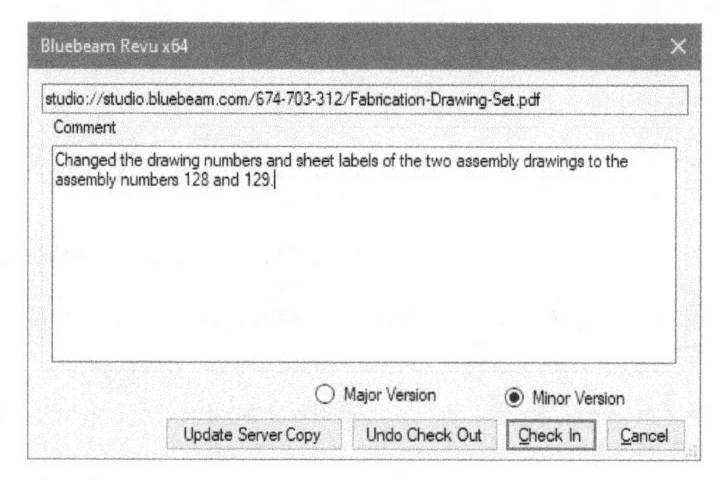

Figure 67 Adding the comment about the changes made before checking the file in

12. Click the **Check In** button in the **Bluebeam Revu** dialog box; the process of checking in the file starts.

    Once the file is checked in, the **Checked Out** icon is not displayed on that file. Also, that file is no more displayed in the **Pending** area of the Project tab.

    You will now check this file back out and will make more modifications to it.

13. Right-click on the same file and select **Check Out** from the shortcut menu to check it out again.

14. Once the **Checked Out** icon is displayed on the file, double-click on it to open it in its own window. Do not open it in the split view.

15. Rename the Drawing 1005 in the titleblock to **P1005**.

16. Also, rename the label of this sheet to **DWG No: P1005**, as shown in Figure 68.

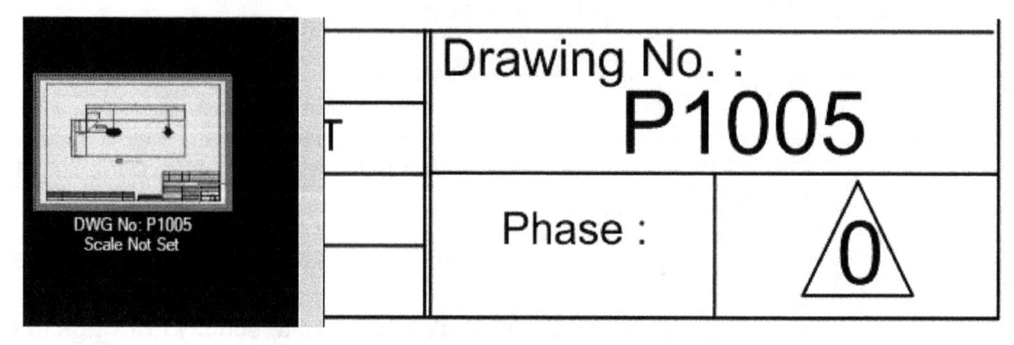

Figure 68 The drawing number and label edited

17. Zoom to the extents of the sheet and then save the PDF file.

18. Close the PDF file; the **Bluebeam Revu** dialog box is displayed.

19. In the **Comment** box, type the following comment:

    Changed the drawing number and sheet label of the plate drawing to P1005.

20. Click the **Check In** button in the **Bluebeam Revu** dialog box; the file is checked back in.

    In the next steps, you will check out the CSV file and edit it. You require Microsoft Excel or a similar program installed on your machine to edit this CSV file. If you do not have that program installed, you can skip to the next section.

21. In the Project folder hierarchy, right-click on the **Punchkeys.csv** file and select **Check Out** from the shortcut menu to check this file out.

22. Now, right-click on the checked out **Punchkeys.csv** file and select **Open** from the shortcut menu; the CSV file is opened in Microsoft Excel or similar program.

23. Change the **WallCloseout** symbol from **CO** to **WO**, as shown in Figure 69.

	A	B	C	D	E
1	Sprinkler	SK	Sprinkler out of place		
2	Slabcut	SC	Slab Cut too close to edge		
3	Switchboard	SB	Switchboard too close to the edge		
4	Firedoor	FD	Fire door non compliant		
5	VAV	VV	VAVs installed incorrectly		
6	Windows	WN	Window panels		
7	Clash	CL	Clash reviewed		
8	WallFinish	WF	Wall finish not to standard		
9	WallCloseout	WO	Wall closeout		

Figure 69 *Editing the CSV file*

24. Save and close the CSV file to return to the Studio Project tab.

25. In the Project tab, right-click on the CSV file and select **Check In** from the shortcut menu, as shown in Figure 70.

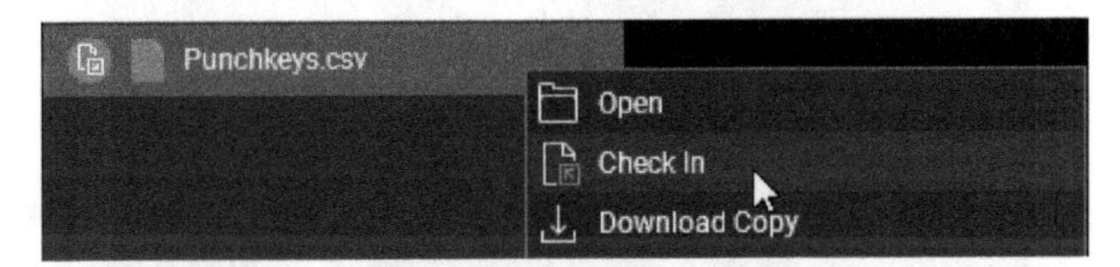

Figure 70 *Checking in the CSV file*

On doing so, the **Check In** dialog box is displayed.

26. In the **Comment** box, type the following comment:

    Changed the Wall Closeout symbol from CO to WO.

27. Click the **Check In** button in the dialog box; the file is checked back in.

Section 7: Replacing the Project File with a Local File

As mentioned earlier, the Studio Projects allow you to replace the file that was originally uploaded. However, it can only be done if the file is first checked out to you. In this section, you will check out the **Fabrication-Drawing-Set.pdf** file again and replace it with a local copy.

1. From the **Fabrication Drawings** folder, check out the **Fabrication-Drawing-Set.pdf** file.

2. Once the file is checked out, right-click on it and select **Replace File** from the shortcut menu, as shown in Figure 71; the **Replace File** dialog box is displayed, as shown in Figure 72.

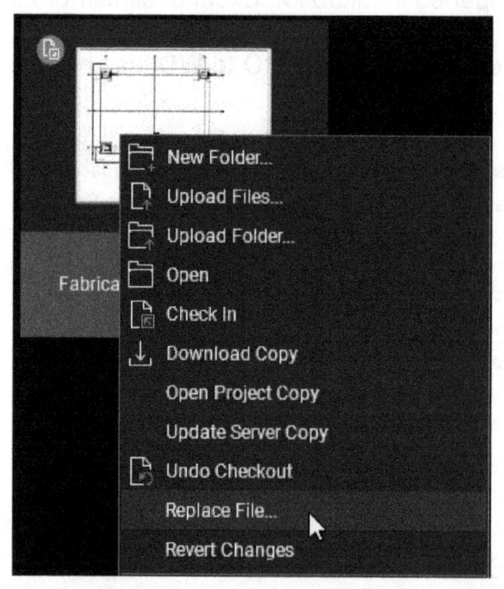

Figure 71 Replacing the Project file with a local copy

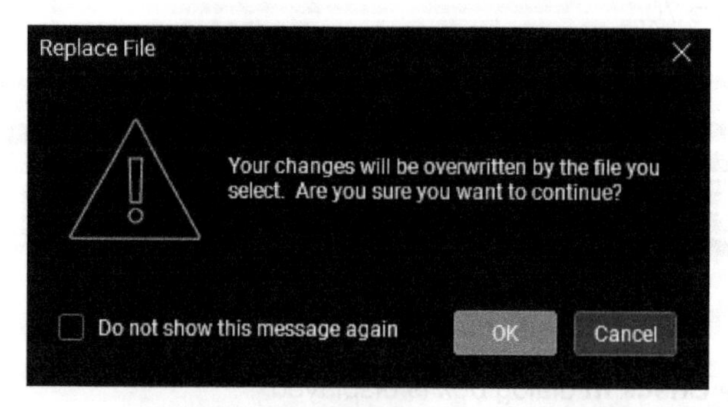

Figure 72 The ***Replace File*** *dialog box*

The **Replace File** dialog box is warning you that by replacing the file, any changes that you made in the current check out will be lost. Because you did not make any changes to the file during this check out, it is OK to replace the file.

3. Click **OK** in the **Replace File** dialog box; the **Open** dialog box is displayed.

4. Browse to the **C12 > DO NOT UPLOAD** folder; a file with the same name as the one you checked out is listed in this folder, as shown in Figure 73.

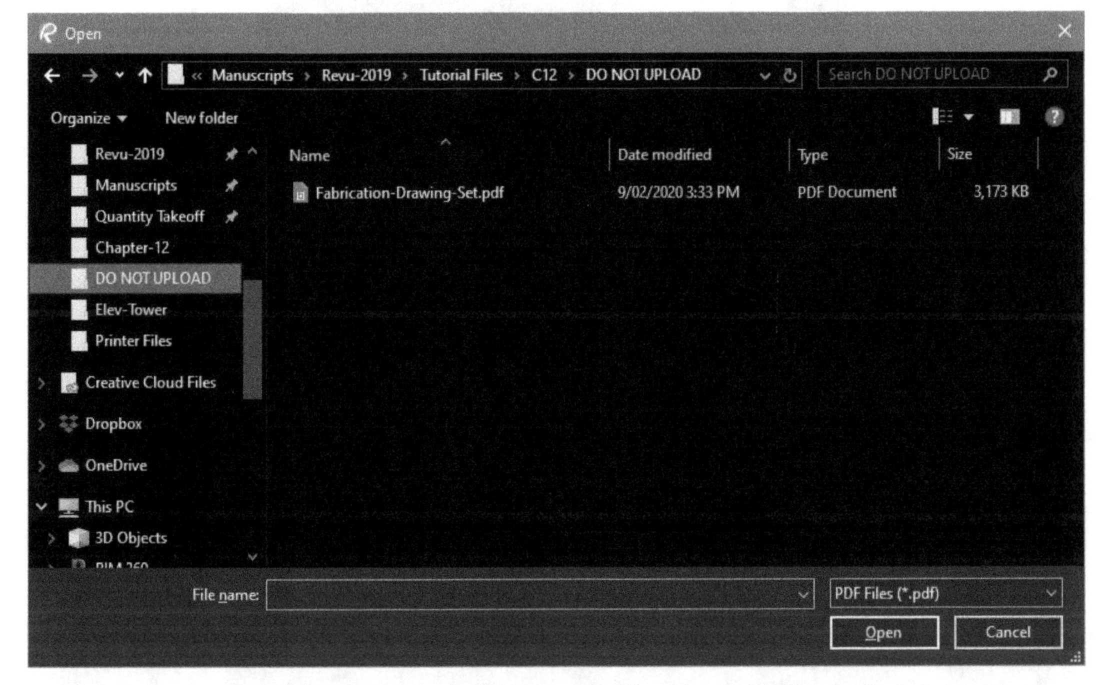

Figure 73 Selecting the local file to replace the Studio Project file

5. Double-click on the **Fabrication-Drawing-Set.pdf** file; the Studio Project file is replaced with this local copy of the same file.

   You are now ready to check this file back in.

6. Check the **Fabrication-Drawing-Set.pdf** file back in with the following comment:

   Replaced the Studio Project file with a local file with additional assembly drawings.

Section 8: Reviewing the Revision History of the Files

In this section, you will review the revision history of the files that you edited in the earlier sections. You will first review the revision history of the **Fabrication-Drawing-Set.pdf** file and then will do the same for the **Punchkeys.csv** file.

1. In the Studio Project, right-click on **Fabrication-Drawing-Set.pdf** file and select **Revision History** from the shortcut menu, as shown in Figure 74.

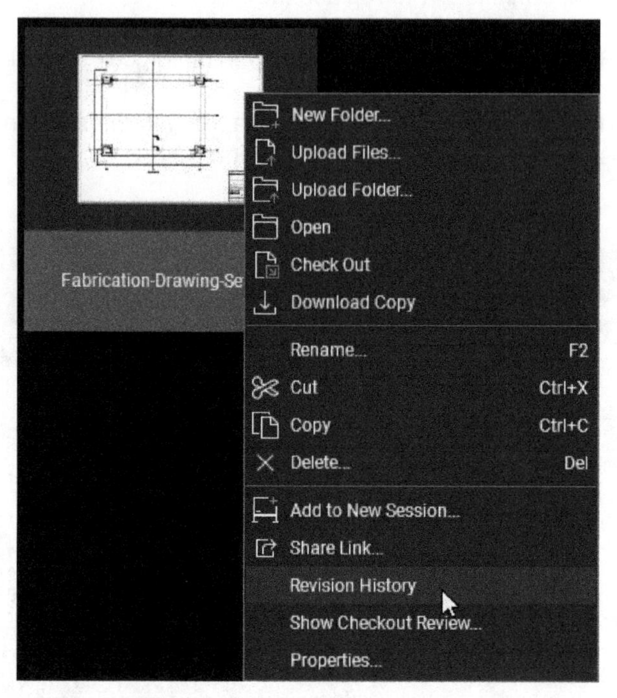

Figure 74 Reviewing the revision history of the file

On doing so, the **Revision History** dialog box is displayed, as shown in Figure 75.

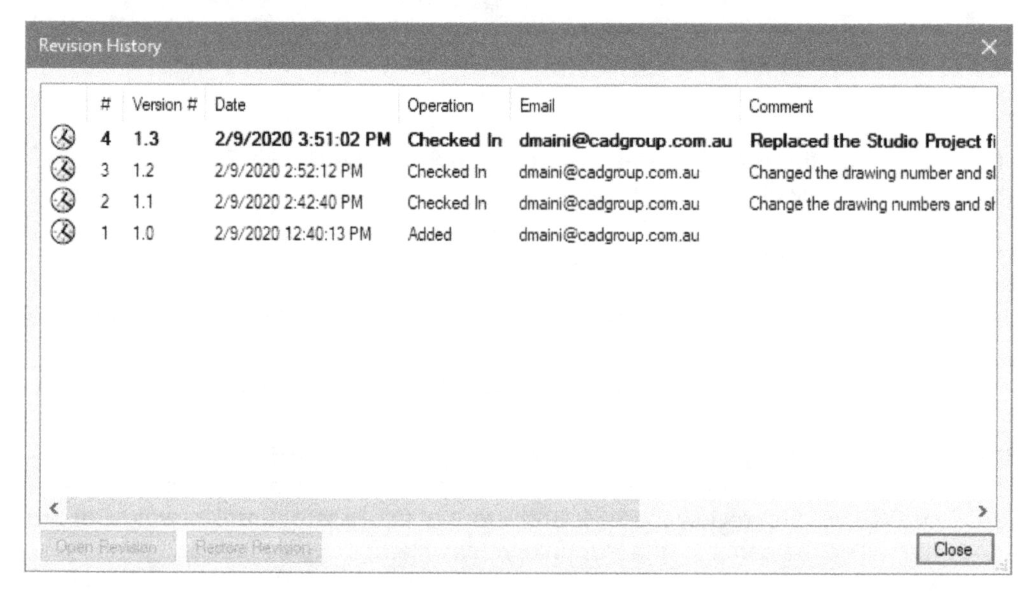

*Figure 75 The **Revision History** dialog box showing the revisions of the selected file*

2. Scroll to the right in the dialog box and review the various revision comments.

3. Click on one of the older versions and notice the **Restore Revision** button is activated.

   This allows you to restore any old version of the file if the changes were made by mistake.

4. Click **Close** in the dialog box.

5. Similarly, review the revision history of the **Punchkeys.csv** file.

Section 9: Starting a New Studio Session Using a Studio Project PDF File

As mentioned earlier, you can start a new Studio Session using a PDF file in the Studio Project. In this section, you will use the **1st Floor Plan.pdf** file to start a new Studio Session. You will then invite members to this Session and perform realtime collaborative markups to this PDF file.

1. In the Studio Project, click on the **Architecture Plans** folder; the **1st Floor Plan.pdf** file in this folder is listed in the **Main Workspace**.

2. Right-click on the **1st Floor Plan.pdf** file and select **Add to New Session** from the shortcut menu, as shown in Figure 76; the **Session Name** dialog box is displayed.

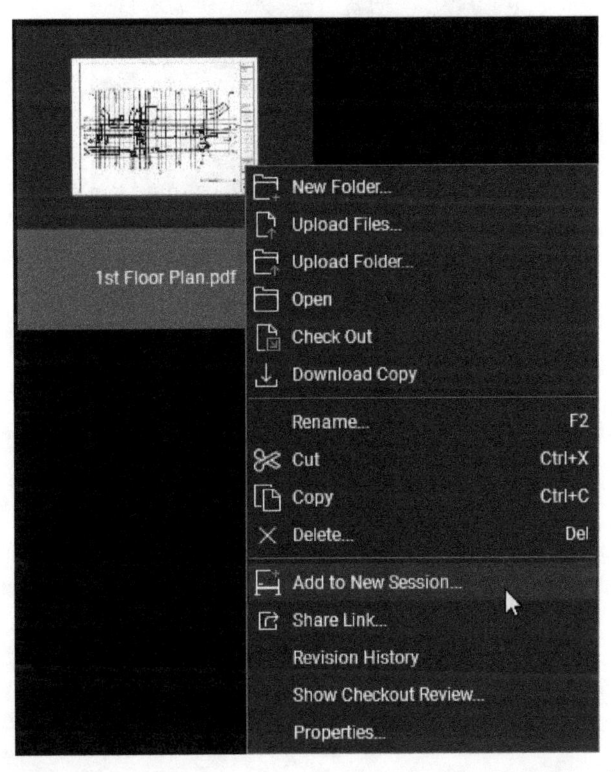

Figure 76 Starting a Studio Session using a Studio Project file

3. In the **Session Name** dialog box, type **SessionTraining** as the name, as shown in Figure 77, and then click **OK**.

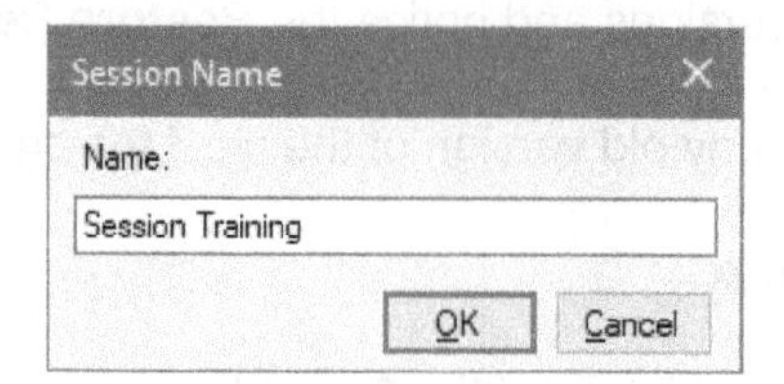

Figure 77 *The* ***Session Name*** *dialog box*

On doing so, a new Studio Session is started and the **Studio** panel has the **Sessions** button selected, labeled as **1** in Figure 78. The **Attendees** area shows only your user name as currently, you are the only member in this Session. The **Documents** area shows the name of the PDF file that was used to start this Session. Also, the **Record** area shows that you joined the Session and added the file to the Session, as shown in Figure 78.

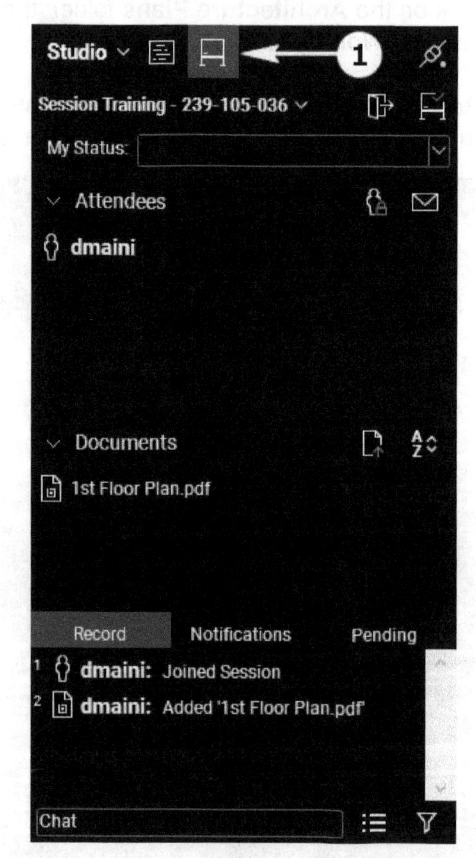

Figure 78 *The* ***Studio*** *panel with the Session*

Also, in the Studio Project tab, an icon is displayed on the **1st Floor Plan.pdf** file that shows this file is being used in Studio Session.

It is recommended to close the Studio Project tab as you will now work with Studio Session.

4. Close the Studio Project tab.

   The Sessions started using the files from a Studio Project are restricted by default by attendees only. This means that only the members you add to the Session can join. Therefore you will now invite members to this Session.

5. From the top right in the **Attendees** area, click the **Invite** button, as shown in Figure 79.

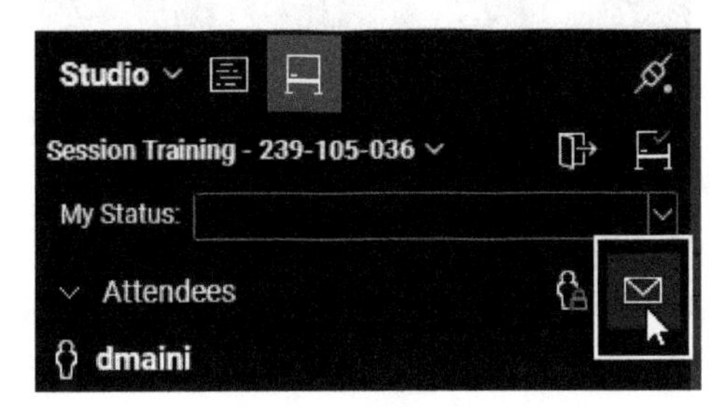

Figure 79 Inviting attendees to the Studio Session

   On doing so, the **Session Invitation** dialog box is displayed that allows you to invite members to this Session.

6. Using the Green + icon or the **Address Book** button, add the Email address of your colleagues to add to the Session. Figure 80 shows three invitees added to this area.

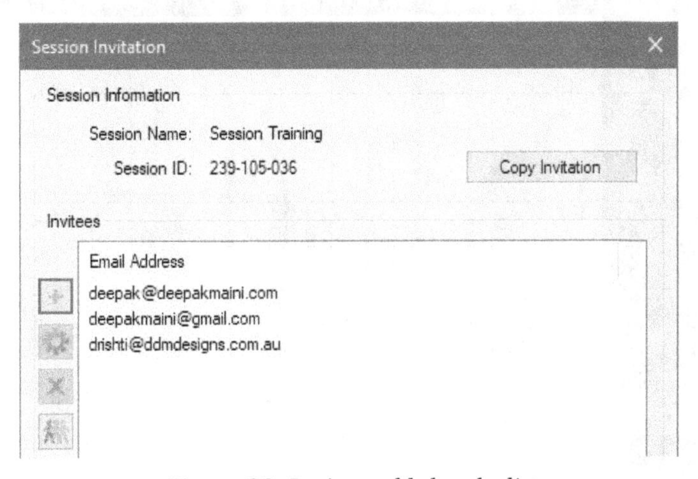

Figure 80 Invitees added to the list

7. In the **Message** area, add the following message:

   This Studio Session will be used to perform realtime collaborative design review.

8. Click **OK** in the **Session Invitation** dialog box; the **Bluebeam Revu** dialog box is displayed informing you that the invitation Emails have been sent to the attendees.

9. Click **OK** in the dialog box.

   Once the members accept the Session invite, they will be listed in the **Attendees** area of the Session. Also, if any of the attendees open the PDF file added to the session, that will also be shown in the right of their user name, as shown in Figure 81.

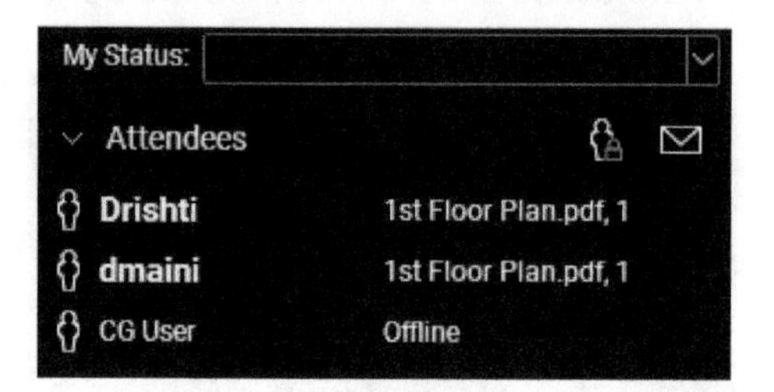

Figure 81 The attendees of the Session and the file they have opened

*Tip: In the **My Status** area above the **Attendees** list, you can enter a status if you are leaving for a meeting and will not be available for some time.*

10. Click on the PDF in the **Documents** area and using the various customized or standard markup tools, markup the PDF file. Figure 82 shows the PDF file with markups from various attendees.

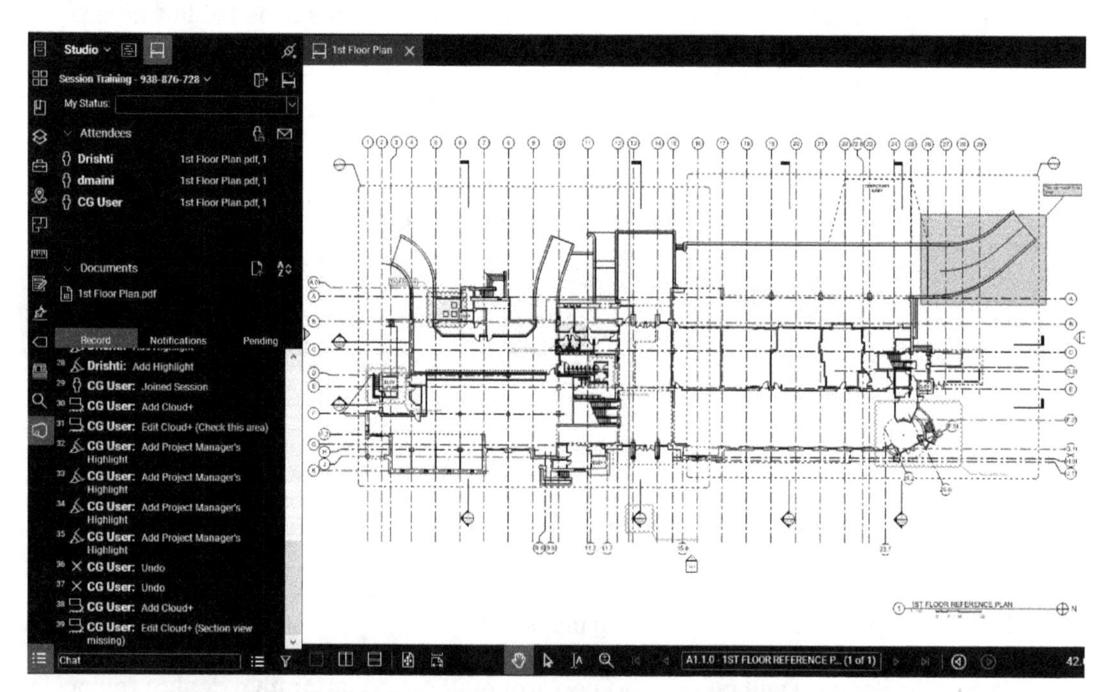

Figure 82 A PDF file in the Studio Session being marked up in realtime

11. If you have a colleague currently in the session, you can hover the cursor over their name in the **Attendee** list and click on **Follow Attendee**, as shown in Figure 83.

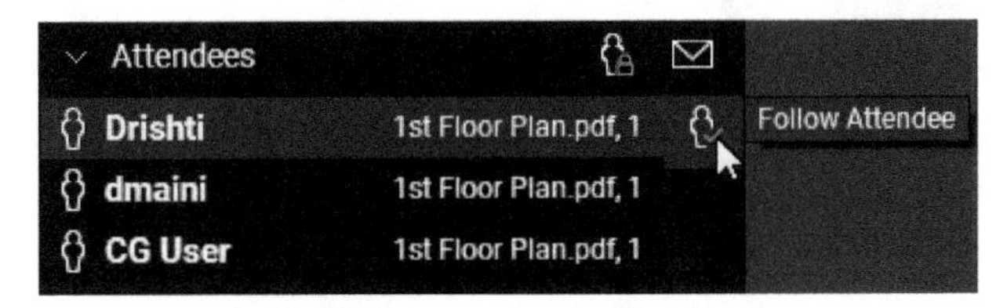

Figure 83 *Following an attendee*

On doing so, you will be able to live view the actions performed by this attendee. The PDF file in your Revu window will update as they navigate around on their screen.

12. To unfollow the attendee, simply zoom or pan in the PDF file in your Revu window.

Section 10: Finishing the Studio Session and Generating the Report

In this section, you will finish the Studio Session and generate the report of the markups performed in the Studio Session.

1. From the top right in the **Studio** panel > **Sessions** area, click **Finish Session**, as shown in Figure 84.

Figure 84 *Finishing the Studio Session*

On doing so, the **Finish Session** dialog box is displayed. You will now configure the settings related to saving this marked up PDF file and also the report that will be generated with all the activities performed in the Session.

2. From the **Save Options** area, make sure the **Save In Folder** radio button is selected.

3. Browse and select the **Tutorial Files > C12** as the folder to save the PDF file.

4. Select the **Session Subfolder** check box. This ensures a subfolder with the Studio Session name and ID is created inside the selected folder.

5. Select the **Close files after finishing** check box. This will ensure the PDF file is closed after the session is finished.

Figure 85 shows the **Finish Session** dialog box with these settings configured.

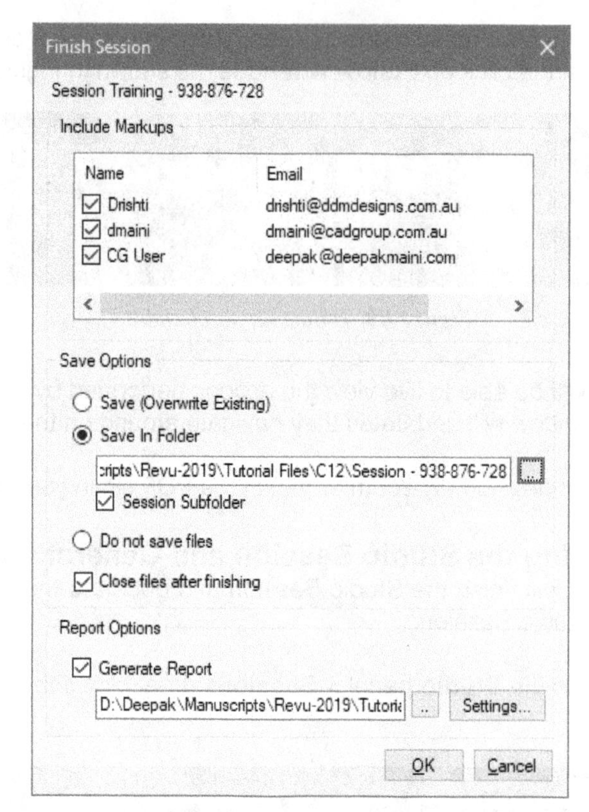

*Figure 85 The **Finish Session** dialog box with the save options configured*

6. In the **Report Options** area, make sure the **Generate Report** check box is selected.

7. Browse and select the **C12** folder as the location to generate the report.

   Next, you will configure the report settings.

8. Click the **Settings** button in the **Report Options** area; the **Session Report** dialog box is displayed.

9. From the **Type** area, select **PDF Package Report**, refer to Figure 86.

10. In the **Options** area > **Title** field, enter **Studio Session Training Report**.

11. From the **Page Size** drop-down list, select **Letter** (Imperial) or **A4** (Metric).

12. Select the **Portrait** radio button.

    Figure 86 shows the **Session Report** dialog box with these settings configured.

13. Click **OK** in the **Session Report** dialog box to return to the **Finish Session** dialog box.

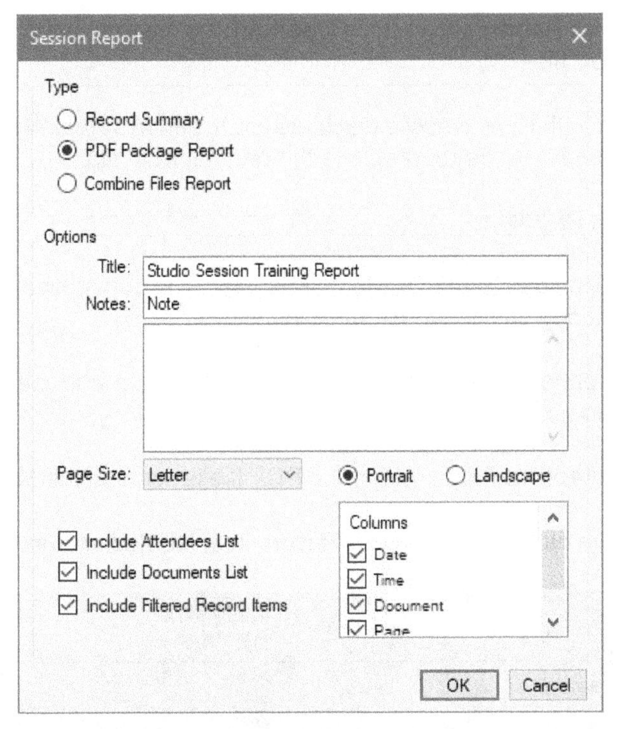

Figure 86 *The* ***Session Report*** *dialog box with the settings configured*

With all the settings configured, you are now ready to close the session and generate the report.

14. Click **OK** in the **Finish Session** dialog box; the Studio Session is closed and the **Bluebeam Revu** dialog box is displayed informing you that the report creation is completed.

15. Click **OK** in the **Bluebeam Revu** dialog box; you are returned to the Revu window and the PDF package report is shown in its own tab, as shown in Figure 87.

Figure 87 *The* ***Session Report*** *dialog box with the settings configured*

16. In the **PDF Package Report** tab, click on **Open All**; the marked up PDF file is opened, along with the report PDF file.

17. Activate the report PDF file and notice that there are hyperlinks created automatically between the markup thumbnails in the report file to the original PDF file.

18. Close all the open PDF files.

    When you add a Studio Project PDF file to a Session, it is automatically checked out to you. Therefore, you need to check that file back in.

19. From the **Studio** panel, click the **Projects** button at the top and then click on the **Studio Training** Project to activate it.

20. Click on the **Architecture Plans** folder, the PDF file is listed in the **Main Workspace**.

    Notice this PDF file has the **Checked Out** icon on it, labeled as **1** in Figure 88.

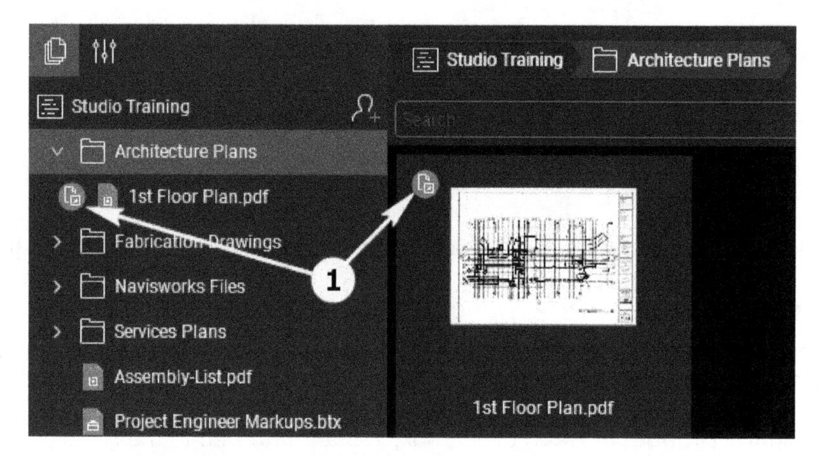

*Figure 88 The **Checked Out** icon displayed on the file*

21. Right-click on this file and click **Undo Checkout** so the file is checked back in without any revision change.

22. Close the **Studio Training** tab.

Skill Evaluation

Evaluate your skills to see how many questions you can answer correctly. The answers to these questions are given at the end of the book.

1. To access Studio Sessions or Projects, you need to first create the Bluebeam ID. (True/False)
2. You cannot add non-PDF files to the Studio Projects. (True/False)
3. Files once added to the Studio Projects cannot be replaced. (True/False)
4. You can start a new Studio Session using a file in the Studio Project. (True/False)
5. You can schedule for a Studio Session to be automatically finished on a specified date. (True/False)
6. Before editing a Studio Project file, what needs to be done to the file?

   (A) Create Local　　(B) Check out
   (C) Add to the Studio Session　　(D) Check in

7. Which file format can be used to start a new session?

   (A) PDF　　(B) DOC
   (C) XLS　　(D) DWG

8. What is the default permission assigned to the invitees of a Studio Project folders?

   (A) Read/Write　　(B) Read
   (C) Write　　(D) Read/Write/Delete

9. Which file format can be uploaded in a Studio Project?

   (A) PDF　　(B) Word
   (C) Excel　　(D) All of them

10. Which option is used to check the file back in without saving the checkout changes?

    (A) **Undo Checkout**　　(B) **Check In**
    (C) **Delete Checkout**　　(D) **None**

Class Test Questions

Answer the following questions:

1. Explain briefly the process of uploading multiple folders to a Studio Project.
2. Explain briefly how to start a new Studio Session using a Studio Project file.
3. Explain the process of editing a Studio Project file.
4. Explain the process of reviewing the revision history of a Studio Project file.
5. Explain the process of replacing a Studio Project file.

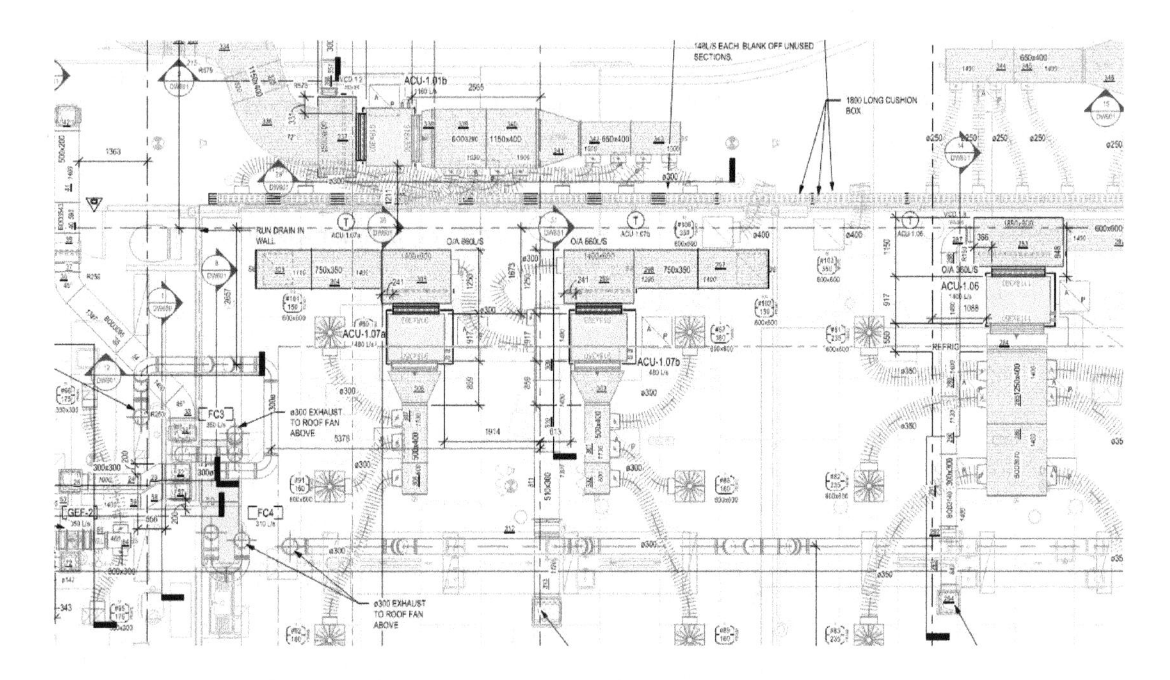

Project 1 - Quantity Takeoff from a HVAC PDF File

In this project, you will do the following:

√ *Create a new profile to take off quantities form the HVAC PDF file*
√ *Calibrate the PDF file*
√ *Import the Mechanical QTO tool set*
√ *Import the Mechanical QTO custom columns*
√ *Takeoff quantities from the HVAC file*
√ *Add a legend showing the quantities taken off*
√ *Export the quantities into a CSV file*

Section 1: Creating a New Profile

In this section, you will open the HVAC file to take off the quantities. You will then create a new profile called **HVAC QTO**. This profile will be based on the **Training QTO** profile you created in Chapter 6 of this book.

1. From the **Tutorial Files > P01** folder, open the **MECHANICAL-LAYOUT.pdf** file; the file is opened in the Revu window, as shown in Figure 1. **This file is a sample PDF file provided by Bluebeam for the users to learn the tools in Revu.**

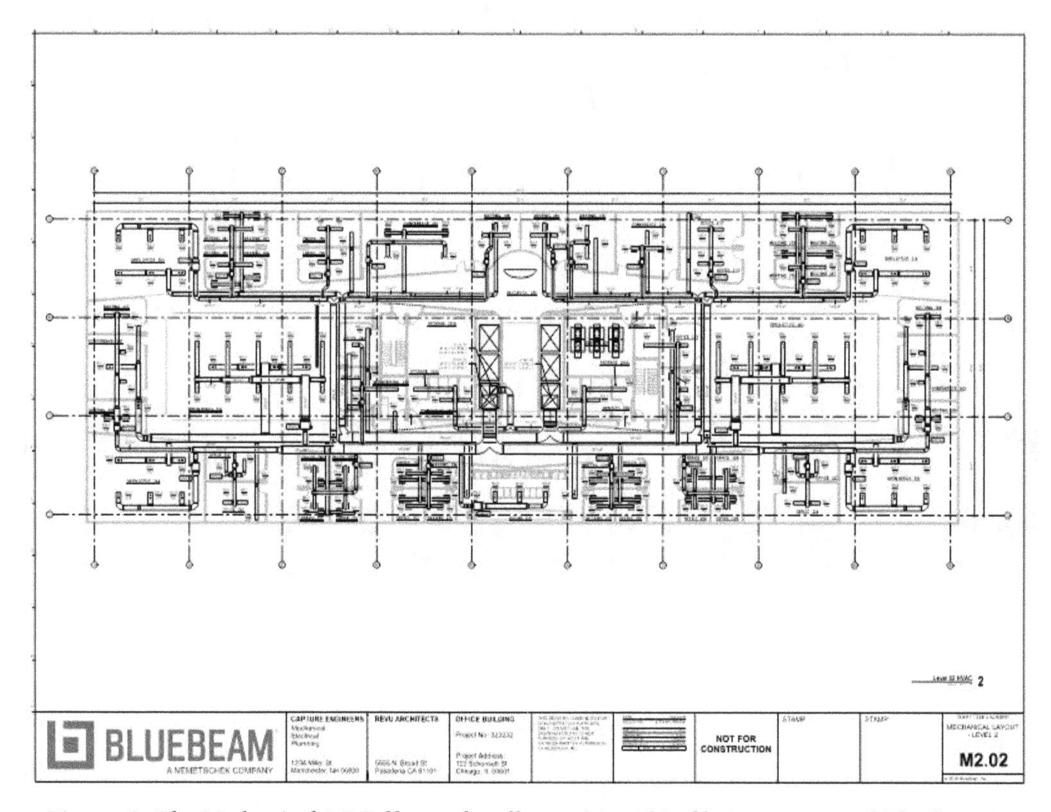

Figure 1 *The Mechanical PDF file to take off quantities (**this file is courtesy of Bluebeam**)*

2. Navigate around this PDF file and notice that the plan shows various types of ducts and VAVs.

3. Double-click the Wheel Mouse Button to zoom to the extents of the PDF file.

   You will now restore the **Training QTO** profile you created in Chapter 6 of this book.

4. From the top left of the Revu window, click **Revu > Profiles > Training QTO**, as shown in Figure 2; the **Training QTO** profile is restored.

5. Hide any panel displayed on the left, right, or below the **Main Workspace**.

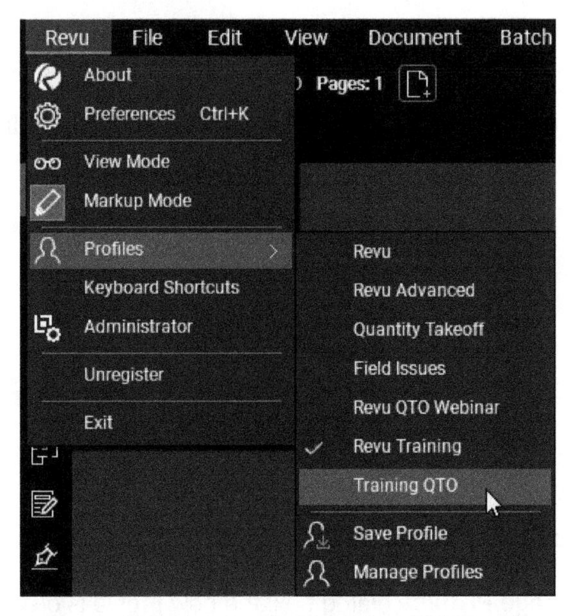

*Figure 2 Activating the **Training QTO** profile*

You will now create a new profile called **HVAC QTO**. The reason you are creating a new profile is so that you can save the custom tool sets and custom columns to this profile.

6. From the top left of the Revu window, click **Revu > Profiles > Manage Profiles**; the **Manage Profiles** dialog box is displayed, as shown in Figure 3.

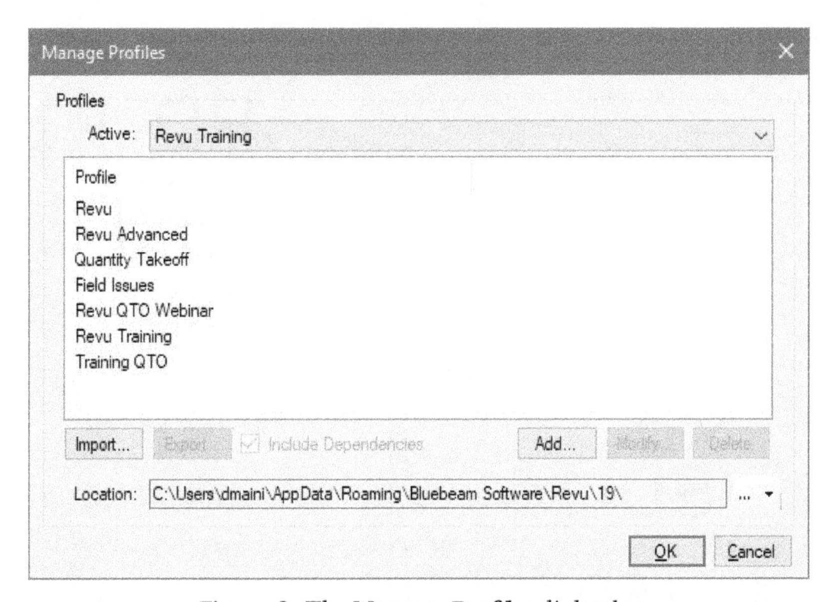

*Figure 3 The **Manage Profiles** dialog box*

7. Near the bottom right in the **Manage Profiles** dialog box, click **Add**; the **Add Profile** dialog box is displayed.

8. Enter **HVAC QTO** as the name of the profile in the **Add Profile** dialog box, as shown in Figure 4.

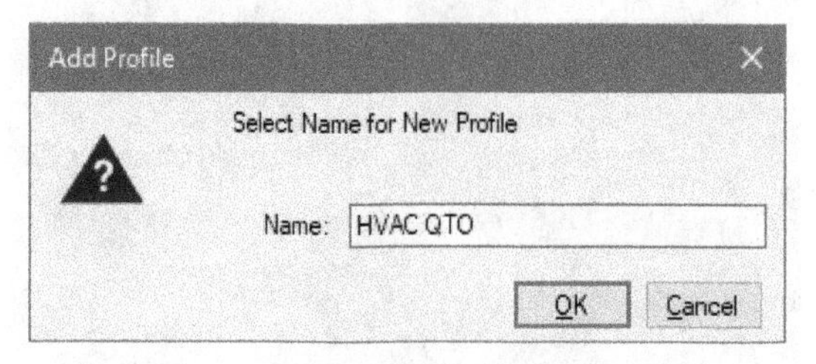

Figure 4 The ***Add Profile*** *dialog box*

9. Click **OK** in the **Add Profile** dialog box; you are returned to the **Manage Profiles** dialog box.

10. Click **OK** in the **Manage Profiles** dialog box; the new profile is created and activated.

Section 2: Calibrating the Sheet

In this section, you will calibrate the sheet to the Imperial or Metric scale, depending on the units you prefer. You will use the door in the **OFFICE 243** room located near the bottom left of the plan to calibrate as the doors are generally standard sizes.

1. Zoom close to the bottom left of the plan where the **OFFICE 243** is located.

2. Now, pan to the door on the right side of this room, as shown in Figure 5.

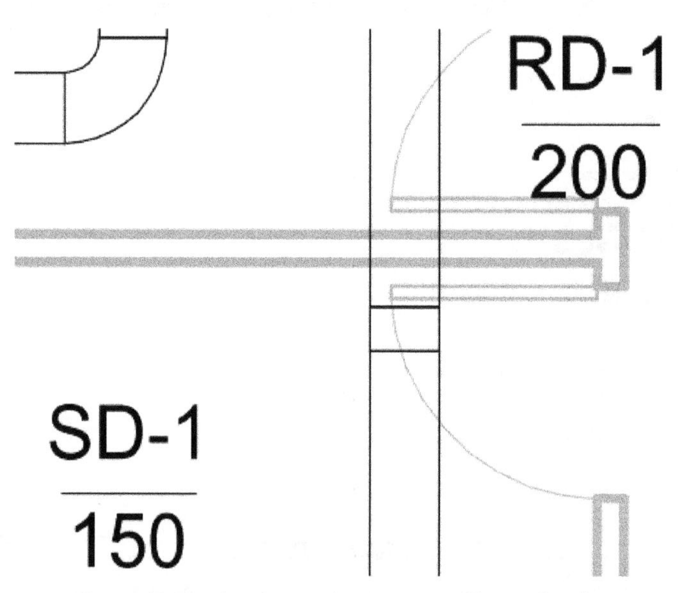

Figure 5 Navigating to the area to calibrate the sheet

Because the current PDF file does not have any scale defined, the lower right corner of the **Navigation Bar** shows the text **Scale Not Set**. You will now click on this text to start the process of calibrating the sheet.

3. From the right side of the **Navigation Bar**, click on the text **Scale Not Set**, labeled as **1** in Figure 6.

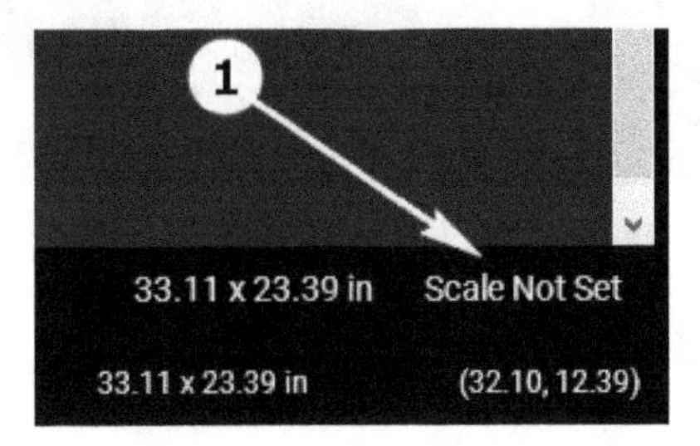

Figure 6 Starting the process of calibrating the sheet

4. On doing so, the **Set Scale** dialog box is displayed, as shown in Figure 7.

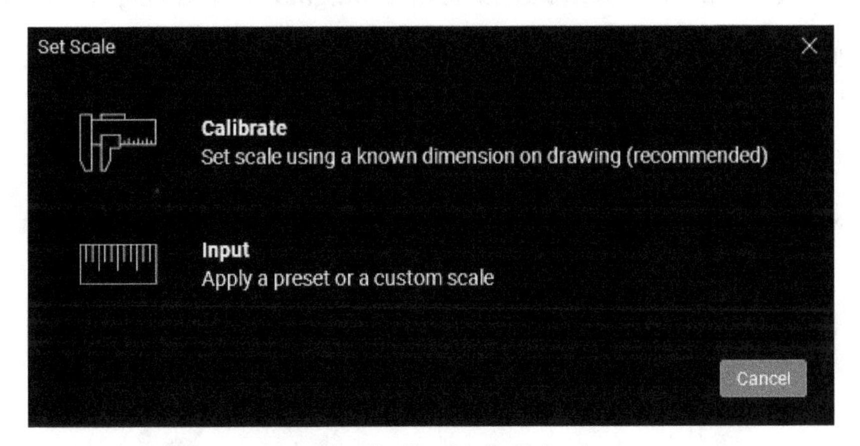

*Figure 7 The **Set Scale** dialog box*

As you do not really know if this PDF was printed to the right scale, you cannot set a standard scale for the measurements. Therefore, you will use the **Calibrate** method from this dialog box.

5. From the **Set Scale** dialog box, click **Calibrate**; the **Calibrate** dialog box is displayed, as shown in Figure 8.

   This dialog box informs you that you need to select two points of a known dimension to calibrate the measurement tools.

***Note**: In the **Calibrate** dialog box, if the **Do not show this message again** check box was selected at a previous stage, then this dialog box is not displayed when you click **Calibrate** in the **Set Scale** dialog box.*

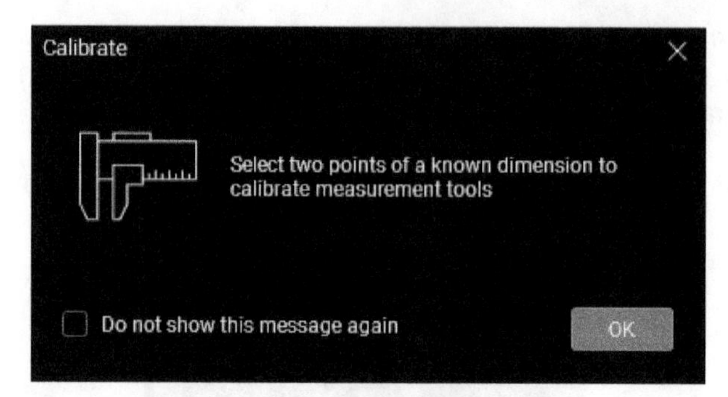

Figure 8 *The **Calibrate** dialog box*

6. Click **OK** in the **Calibrate** dialog box; you are returned to the PDF file.

   Before you pick points to calibrate the sheet, it is recommended to turn off the line weights so the lines in the PDF do not appear thick.

7. From the **Menu Bar**, click **View > Disable Line Weights**; the line weights are turned off.

   You will now use the two points on the door to calibrate the sheet. Remember that you need to hold down the SHIFT key to ensure you pick the correct points.

8. Snap to the point labeled as **1** in Figure 9 to specify the first point of calibration.

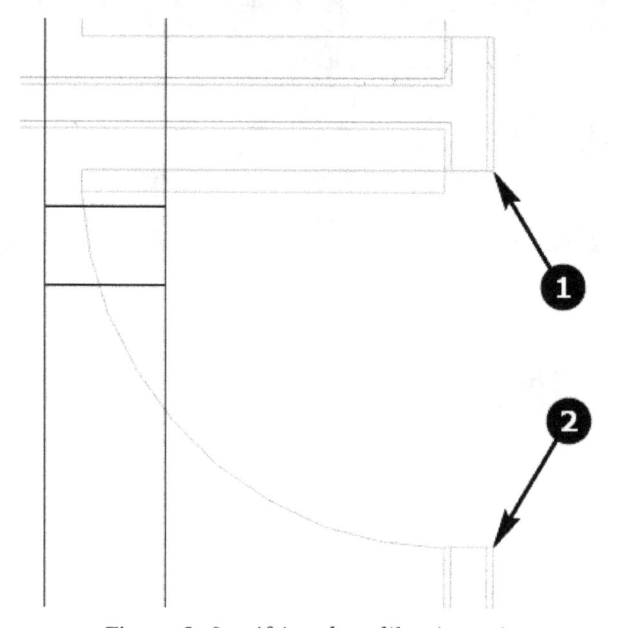

Figure 9 *Specifying the calibration point*

9. Hold down the SHIFT key and snap to the point labeled as **2** in Figure 9 to specify the second point of calibration.

On doing so, the **Calibrate** dialog box is redisplayed, but this time with additional options. The values that you enter on the right side of the = symbol will depend on whether you prefer using the Imperial units or the Metric units.

10. In the field on the right of the = symbol below the **Custom** radio button, enter the appropriate value:

    For the Imperial units, type **3**
    For the Metric units, type **0.92**

11. From the **Units** field on the right of the value you typed, select the appropriate units:

    For the Imperial units, select **ft' in"**
    For the Metric units, select **m**

12. From the **Precision** drop-down list, select **1** for the Imperial units and **0.1** for the Metric units. Figure 10 shows the **Calibrate** dialog box for the Imperial units and Figure 11 shows this dialog box for the Metric units.

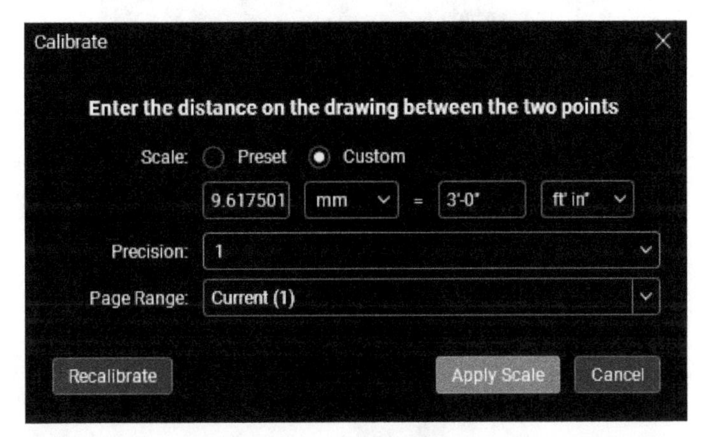

Figure 10 The ***Calibrate*** *dialog box configured for the Imperial units*

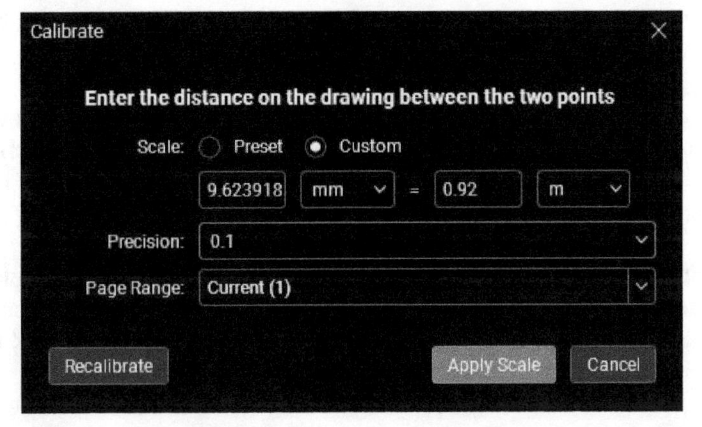

Figure 11 The ***Calibrate*** *dialog box configured for the Metric units*

13. Click the **Apply Scale** button in the **Calibrate** dialog box; the scale is applied and is listed on the right side of the **Navigation Bar**.

Section 3: Importing Custom Tool Set

In this section, you will import the custom tool set that is provided to you to take off the quantities. The detailed process of how to create custom tools to take off quantities was discussed in Chapter 7 of this book.

1. From the **Panel Access Bar**, click the **Tool Chest** button to display various tool sets.

2. From the top of the **Tool Chest** panel, click **Tool Chest > Manage Tool Sets**, as shown in Figure 12.

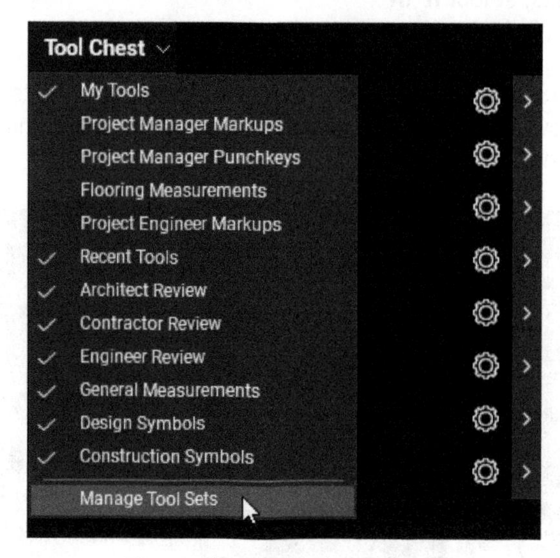

Figure 12 *Managing tool sets*

On doing so, the **Manage Tool Sets** dialog box is displayed.

3. From the lower left in the **Manage Tool Sets** dialog box, click the **Import** button; the **Open** dialog box is displayed.

   There are two different custom tool sets provided to you, depending on whether you use the Imperial units or the Metric units. These are saved in the **P01 > Mechanical Package Imperial** or the **P01 > Mechanical Package Metric** folder. You will now browse to one of these folders, depending on the units you prefer to use.

4. Depending on your preferred units, browse to the **P01 > Mechanical Package Imperial** or the **P01 > Mechanical Package Metric** folder; the BTX format file of the tool set is displayed in the folder. Figure 13 shows the **Open** dialog box with the Imperial BTX file.

5. Double-click on the BTX file of the tool set; you are returned to the **Manage Tool Sets** dialog box and the tool set you imported is listed at the bottom.

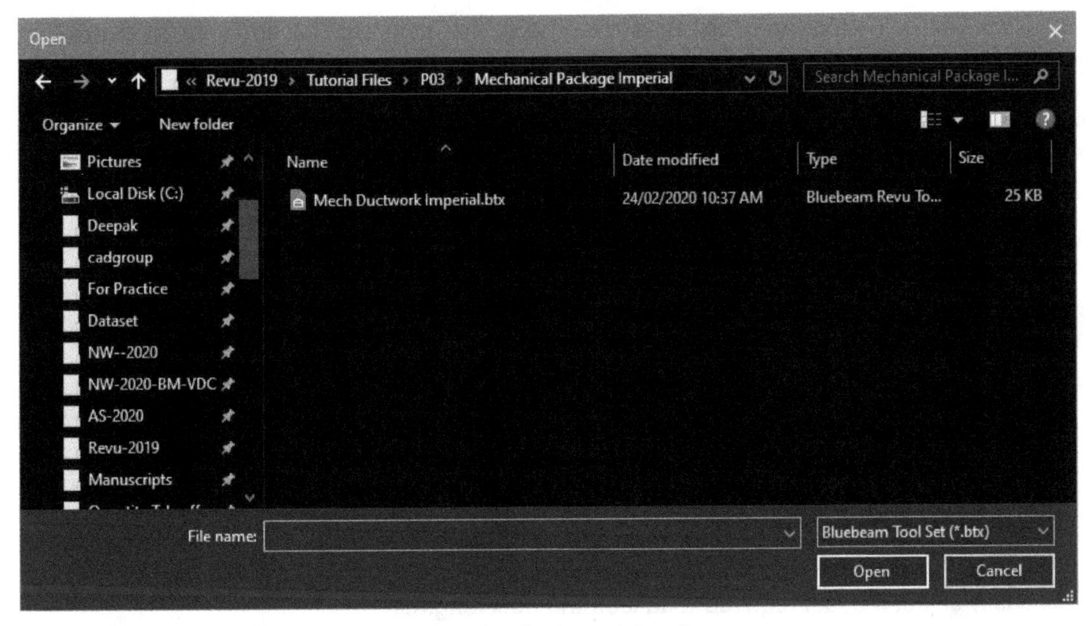

Figure 13 The Imperial tool set

You will now select this tool set and move it up on the list.

6. In the **Manage Tool Sets** dialog box, select the tool set you imported.

7. Use the **Up** arrow key on the left side of the dialog box and reorder this tool set so it is located below the **My Tools** tool set. Figure 14 shows the **Manage Tool Sets** dialog box with the **Mech Ductwork Imperial** tool set listed as the second tool set.

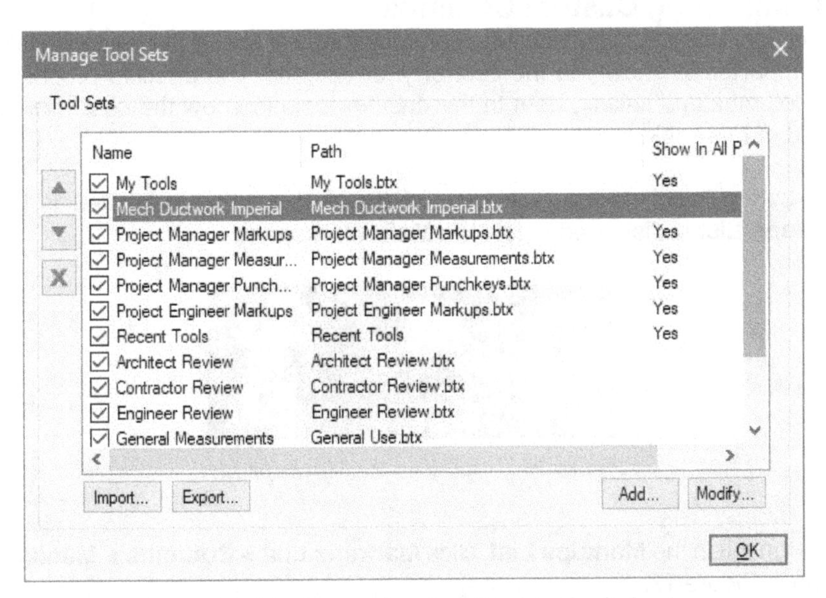

Figure 14 The imported tool set listed in the dialog box

8. Click **OK** in the **Manage Tool Sets** dialog box; the imported tool set is listed in the **Tool Chest** panel, as shown in Figure 15.

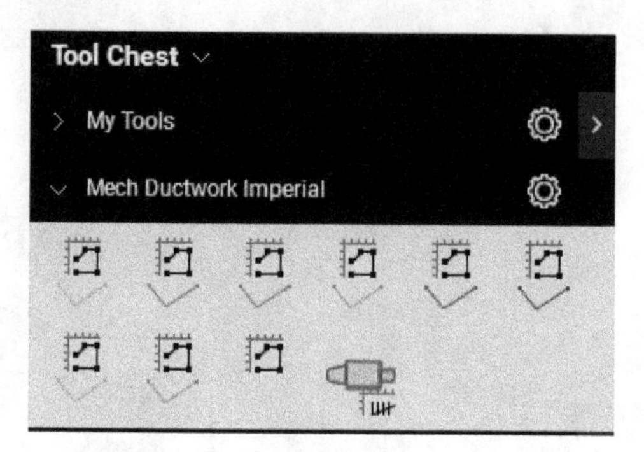

*Figure 15 The imported tool set listed in the **Tool Chest** panel*

For the ease of selecting the custom tools, it is recommended to change the display of this tool set to **Detail**. In this view, the names of the tools are displayed that will allow you to easily select the tool required to takeoff the duct lengths.

9. Click on the cogwheel on the top right of the **Mech Ductwork Imperial** or **Mech Ductwork Metric** tool set you imported and select **Detail** from the shortcut menu; the tools are displayed in the detail view.

***Tip**: The tool used to create these custom duct types in the imported tool set is the **Polylength** tool. The reason is that the **Polylength** tool allows you to add rise and drop values, which are important while taking off the duct lengths.*

Section 4: Importing Custom Columns

In this section, you will import custom columns that will link to the custom tools in the tool set you imported to show you the cost of your takeoffs. The detailed process of creating custom columns and linking them to the custom tools to show the cost was discussed in Chapter 8 of this book.

1. From the bottom left of the Revu window, click the **Markups** button, labeled as **1** in Figure 16; the **Markups List** is displayed below the **Main Workspace**.

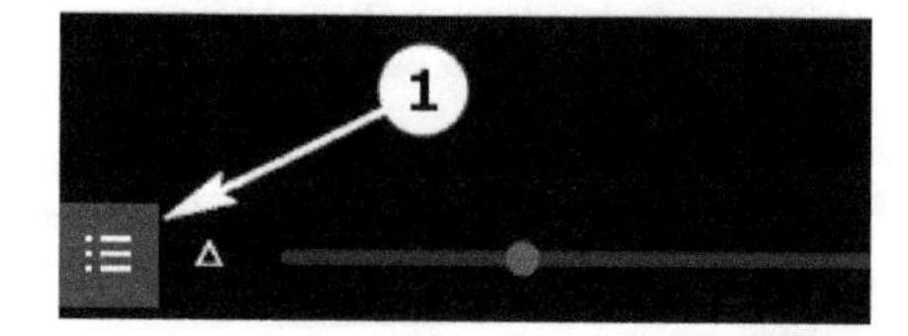

*Figure 16 Displaying the **Markups List***

2. From the top left in the **Markups List**, click **Markups List > Columns > Manage Columns**, as shown in Figure 17.

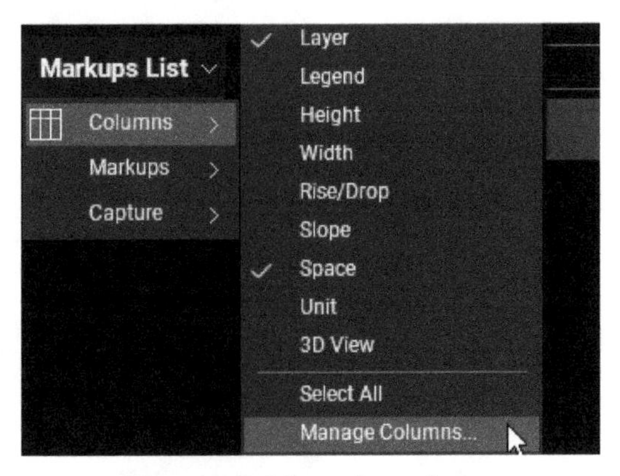

Figure 17 Managing columns

On doing so, the **Manage Columns** dialog box is displayed. You will now activate the **Custom Columns** tab and import the custom columns.

3. In the **Manage Columns** dialog box, activate the **Custom Columns** tab.

4. Click the **Import** button in the **Custom Columns** area; the **Open** dialog box is displayed.

5. If not already in that folder, browse to the **P01 > Mechanical Package Imperial** folder or the **P01 > Mechanical Package Metric** folder, depending on the units you want to use; the custom columns XML file is listed. Figure 18 shows the Imperial custom column file in this dialog box.

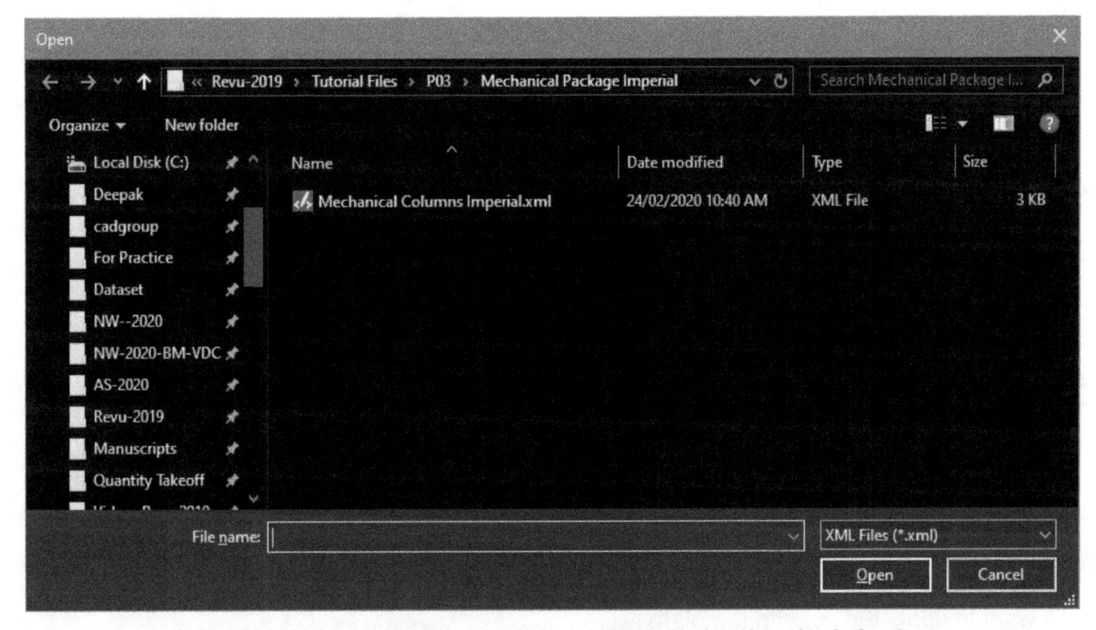

Figure 18 The Imperial custom column XML file listed in the dialog box

6. Double-click on the XML file of the custom columns; the **Import Custom Columns** warning box is displayed, as shown in Figure 19. This warning box informs you that importing these columns will replace any existing columns and will be applied to the open document.

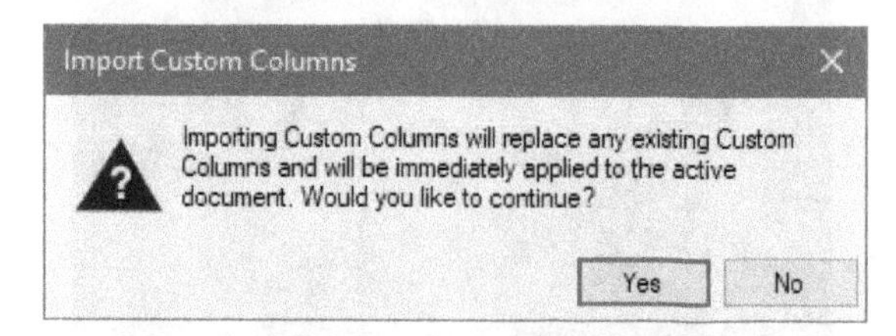

*Figure 19 The **Import Custom Columns** warning box*

7. Click **Yes** in the **Import Custom Columns** warning box; you are returned to the **Manage Columns** dialog box and the two custom columns are listed in this dialog box, as shown in Figure 20.

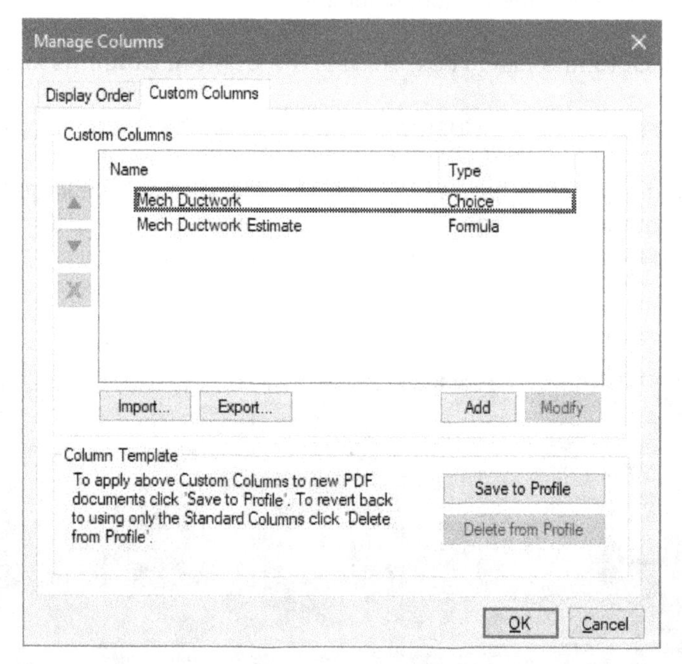

Figure 20 The imported custom columns listed in the dialog box

Before exiting this dialog box, you need to make sure you save these two custom columns to the **HVAC QTO** profile you created at the start of this project.

8. From the lower right in the **Custom Columns** tab of the dialog box, click the **Save to Profile** button; the **Apply Column Template** dialog box is displayed asking you to confirm that you want to apply the column template.

9. Click **OK** in the **Apply Column Template** dialog box; the two custom columns are now saved to the profile.

10. Click **OK** in the **Manage Columns** dialog box; the two custom columns are listed in the **Markups List**; as shown in Figure 21.

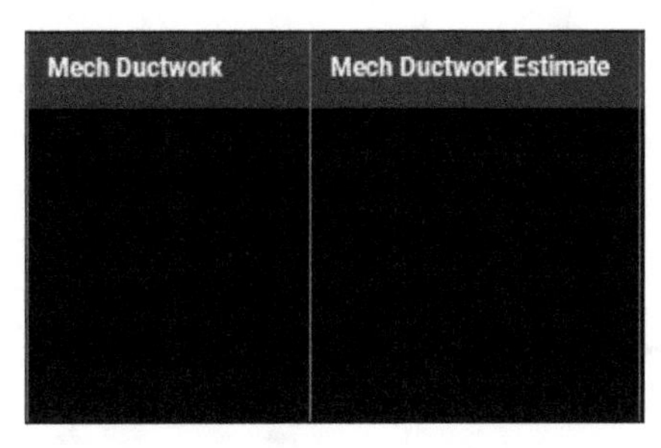

Figure 21 *The imported custom columns listed in the **Markups List***

Because you do not need the **Markups List** displayed at this stage, you will now hide it.

11. Hide the **Markups List**.

Before proceeding any further, you will save the custom profile you created at the start of this project.

12. From the top left in the Revu window, click **Revu > Profiles > Save Profile**; the custom profile is saved.

Section 5: Taking Off Quantities

In this section, you will take off linear quantities of various types of ductwork. As mentioned earlier, the plan has various types of ductwork and VAVs installed. You will start by taking off the duct lengths in the **OPEN OFFICE 200** room on the top left of the plan. The custom tool set you imported has various types of Supply and Return duct tools that you will use to takeoff the quantities.

1. Navigate to the **OPEN OFFICE 200** room a the top left of the plan, as shown in Figure 22.

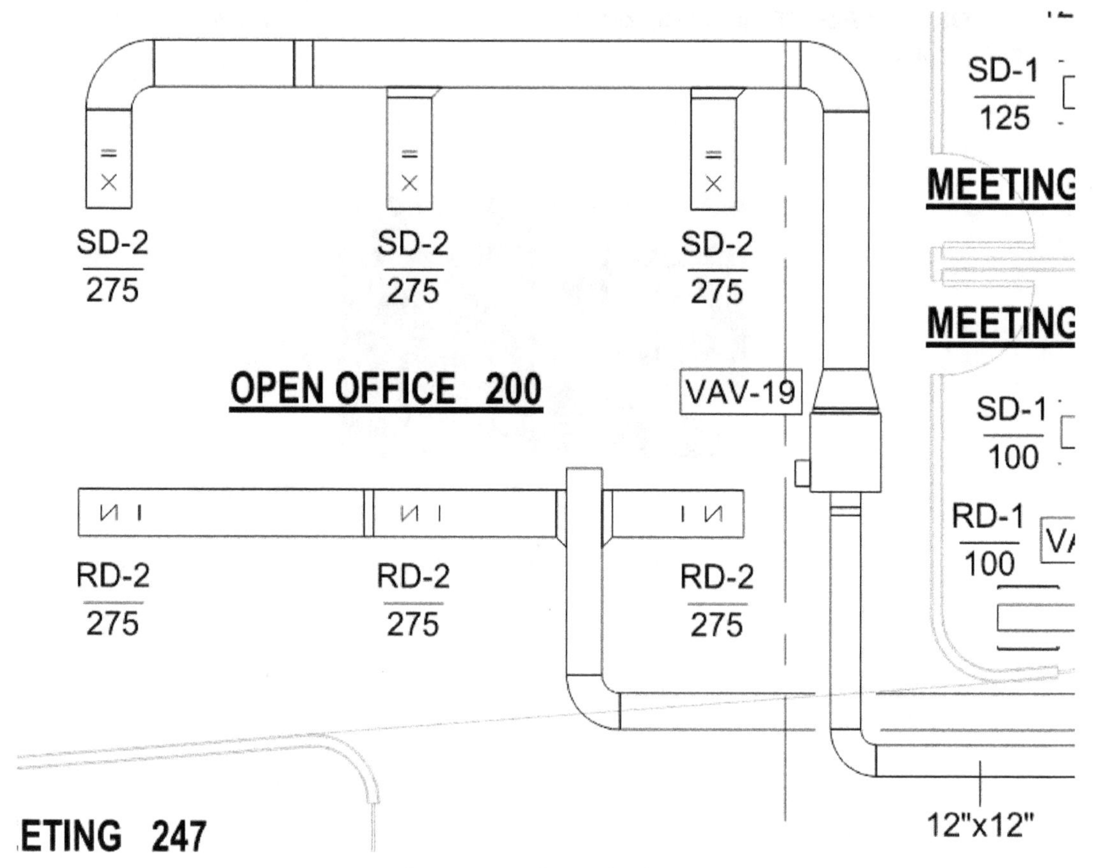

Figure 22 *The room to take off the duct lengths from*

You will start by taking off the lengths of the SD 275 ducts. This tool is listed with the subject **Supply Duct 275** in the custom tool set you imported. Note that you will only takeoff the linear sections of the ducts and not the elbows.

2. From the custom tool set you imported, invoke the **Supply Duct 275** tool; you are prompted to select the vertices of the perimeter to measure.

   IMPORTANT NOTE: If you select an incorrect vertex while taking off the quantities, you can press the BACKSPACE key to deselect that vertex.

3. Snap to the vertex labeled as **1** in Figure 23.

4. Hold down the SHIFT key and snap the vertex labeled as **2** in Figure 23.

5. Press the ENTER key; the measurement is displayed on the PDF file.

   With the **Supply Duct 275** tool still active, you will now takeoff the length of the remaining horizontal and vertical sections. Notice, the second horizontal section has a rise of **2' 5"** or **0.8 m**. Therefore, you need to take off the horizontal length in three sections.

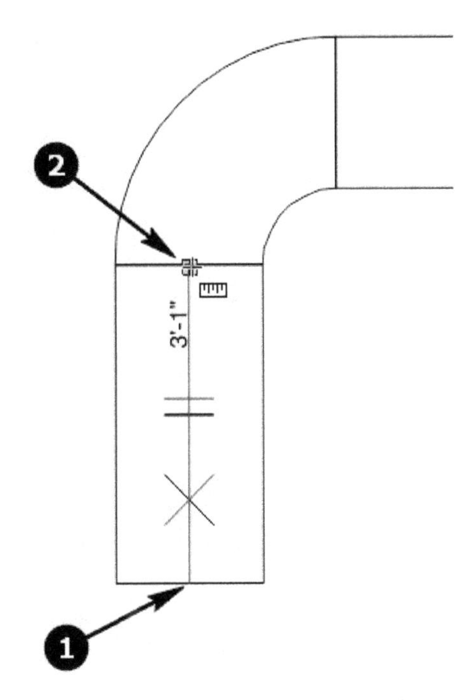

Figure 23 Snapping to the vertices to take off the duct length

6. Hold down the SHIFT key and snap to the vertices labeled as **1** and **2** in Figure 24.

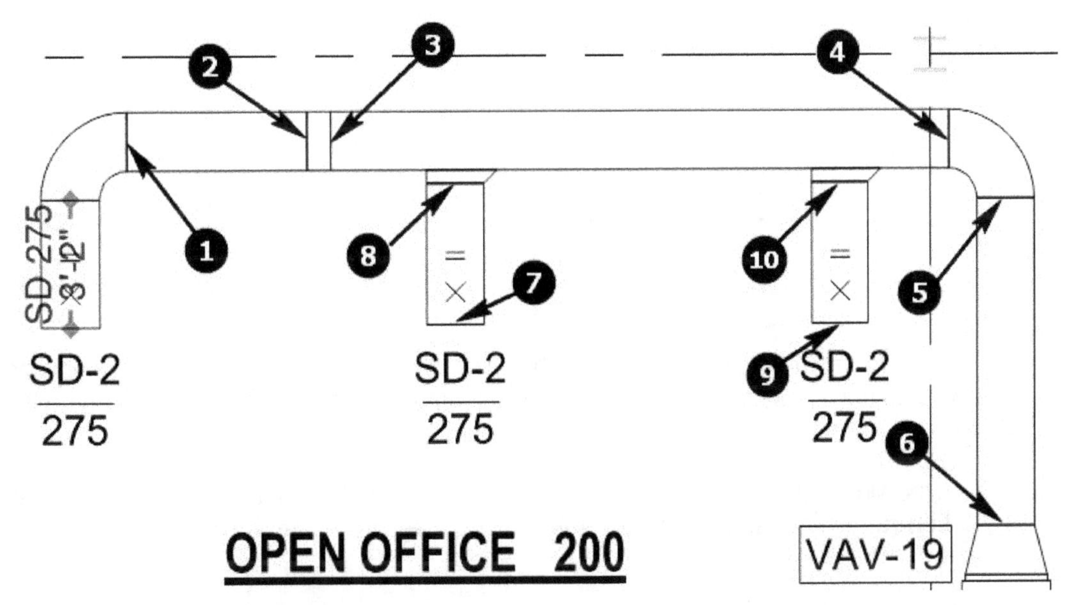

*Figure 24 The vertices to takeoff the **SD 275** duct lengths*

7. Press the ENTER key; the length of this section is displayed.

8. Similarly, takeoff the lengths between vertices **2-3**, **3-4**, **5-6**, **7-8**, and **9-10**.

9. Press the ESC key after taking off the length of the last duct section.

   You will now select the duct section between vertices **2-3** and add a rise value to it.

10. Zoom in and select the duct section between vertices **2-3**.

11. From the right side of the **Main Workspace**, invoke the **Measurements** panel.

12. In the **Measurements** panel, scroll down to the **Polylength Measurement Properties** area.

13. In the **Rise/Drop** field, enter **2'-5"** as the Imperial value or **0.8** as the Metric value.

14. Hide the **Measurements** panel. Figure 25 shows the takeoff in the Imperial file and Figure 26 shows the takeoff in the Metric file. **NOTE THAT YOUR VALUES COULD BE A LITTLE DIFFERENT FROM THE VALUES SHOWN IN THESE FIGURES.**

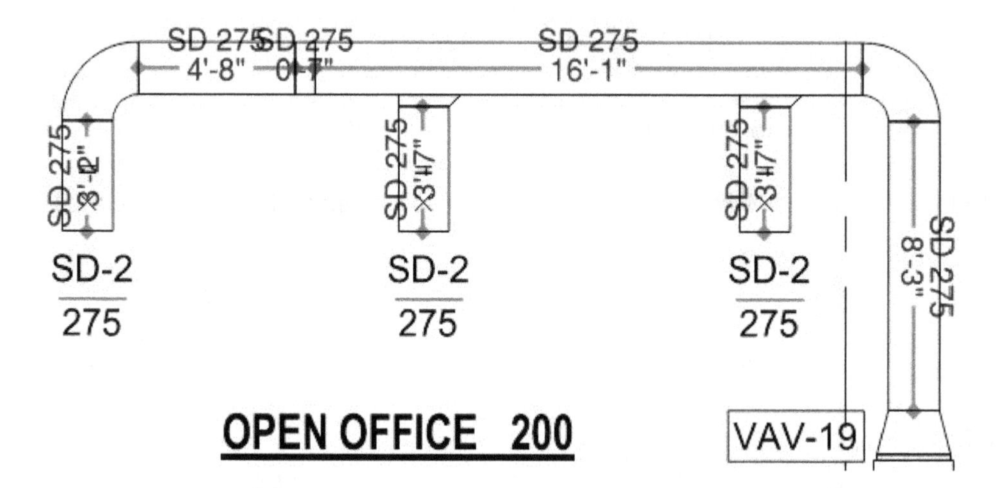

Figure 25 The duct length takeoff in the Imperial file

Before proceeding further, you will check if the cost of this takeoff is displayed in the **Markups List.** Also, to make sure you are able to see the lengths of the takeoffs, you will turn on the **Length** column.

15. Activate the **Markups List** and turn on the **Length** column. Notice the takeoff length displayed in the **Length** column and the cost of takeoff displayed in the **Mech Ductwork Estimate** column, as shown in Figures 27 and 28.

***Tip:** If you want, you can one by one drag the two custom columns and drop them on the right of the **Length** column so the ductwork types and costs are displayed next to each other.*

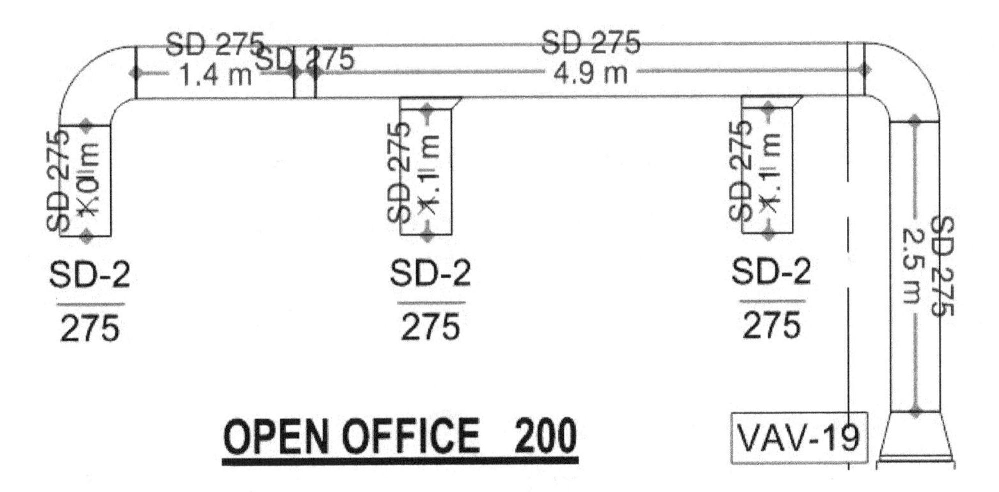

Figure 26 The duct length takeoff in the Metric file

Subject	Page Label	Comments	Length	Mech Ductwork	Mech Ductwork Estimate
Square Duct 275 (7)			42.32 ft' in"		$1,851.06
Square Duct 275	1	SD 275	3.15 ft' in"	SD 275 43.75	$137.81
Square Duct 275	1	SD 275	4.67 ft' in"	SD 275 43.75	$204.31
Square Duct 275	1	SD 275	3.03 ft' in"	SD 275 43.75	$132.56
Square Duct 275	1	SD 275	16.10 ft' in"	SD 275 43.75	$704.38
Square Duct 275	1	SD 275	8.24 ft' in"	SD 275 43.75	$360.50
Square Duct 275	1	SD 275	3.56 ft' in"	SD 275 43.75	$155.75
Square Duct 275	1	SD 275	3.56 ft' in"	SD 275 43.75	$155.75

Figure 27 The duct lengths and costs in the Imperial file

Subject	Page Label	Comments	Length	Mech Ductwork	Mech Ductwork Estimate
Square Duct 275 (7)			12.9 m		$1,673.75
Square Duct 275	1	SD 275	1.0 m	SD 275 128.75	$128.75
Square Duct 275	1	SD 275	1.4 m	SD 275 128.75	$180.25
Square Duct 275	1	SD 275	1.0 m	SD 275 128.75	$128.75
Square Duct 275	1	SD 275	4.9 m	SD 275 128.75	$630.88
Square Duct 275	1	SD 275	2.5 m	SD 275 128.75	$321.88
Square Duct 275	1	SD 275	1.1 m	SD 275 128.75	$141.62
Square Duct 275	1	SD 275	1.1 m	SD 275 128.75	$141.62

Figure 28 The duct lengths and costs in the Metric file

***Note**: The values that you see on your PDF file may be a little different from what you see in this book. This could be due to the vertices selected for calibration and also the vertices selected to take off the quantities.*

Next, you will takeoff the **Return Duct 275** lengths in the same area.

16. From the custom tool set you imported, invoke the **Return Duct 275** tool; you are prompted to select the vertices of the perimeter to measure.

17. One by one, snap to the vertices labeled as **1-2**, **2-3**, **3-4**, and **5-6** in Figure 29 to takeoff the duct lengths.

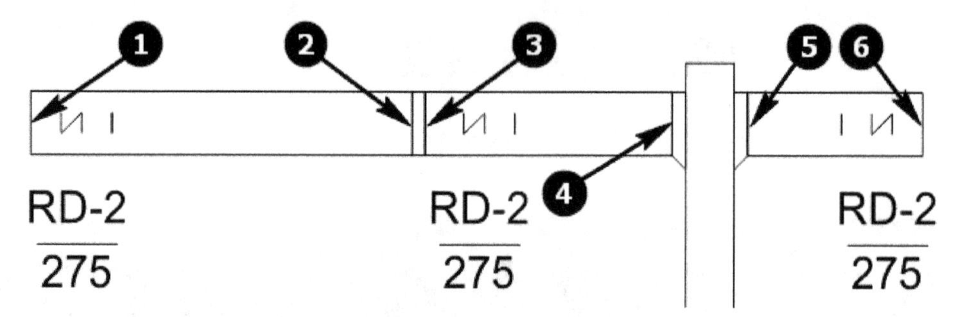

***Figure 29** The vertices to takeoff the **RD 275** duct lengths*

18. Select the duct section between vertices **2-3** and add a **2'-5"** or **0.8 M** as the rise value in the **Measurements** panel.

Figure 30 shows the Imperial takeoff of these ducts and Figure 31 shows the Metric takeoff.

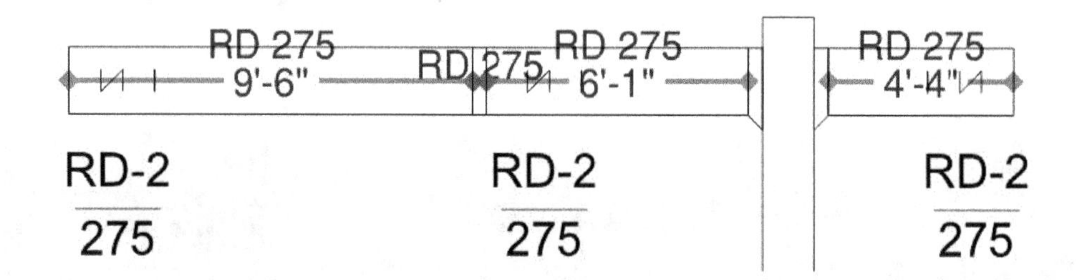

***Figure 30** The duct length takeoff in the Imperial file*

Similarly, takeoff the other duct types available in the custom tool set.

IMPORTANT NOTE: There are some duct types in the PDF file for which the tools are not available in the custom tool set. The idea is for you to try creating these custom tools yourself and then add the custom column choices to link them together. Refer to **Sections 6** and **7** of the tutorial in **Chapter 8** of this book to learn more about how to do this.

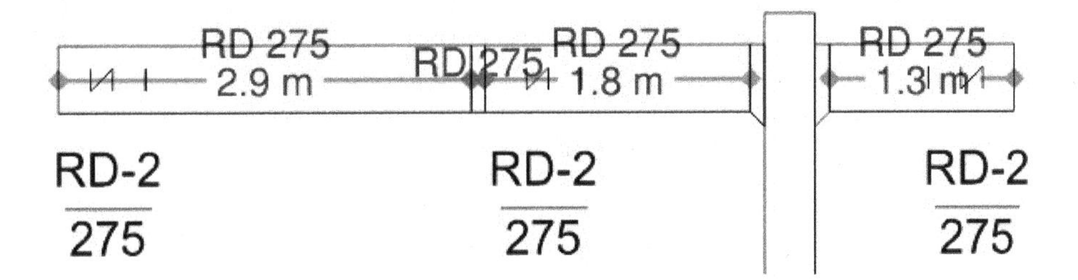

Figure 31 The duct length takeoff in the Metric file

Section 6: Taking Off the VAV Quantities

As mentioned earlier, this PDF file has a number of VAVs. In this section, you will search for the text **VAV** and then add the custom count measurement from your custom tool set to the searched text.

1. From the **Panel Access Bar**, show the **Search** panel.

   The **Search** panel is by default set to search for text. This is because the **Text** button is turned on at the top in this panel. You will now search for the text **VAV**.

2. In the text search field at the top in the **Search** panel, type **VAV**.

3. From the drop-down list below the search field, select **Current Document**.

4. In the **Options** area, select the **Whole Words Only** check box, as shown in Figure 32.

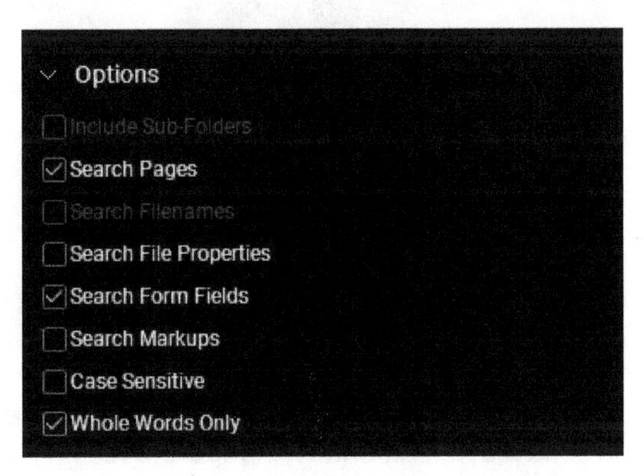

Figure 32 Configuring the search options

5. Make sure the **Case Sensitive** check box is not selected, as shown in Figure 32.

6. Click the **Search** button from the top right in the **Search** panel; the search is performed in the document and the bottom of the **Results** area shows 26 results found.

You will now review these three results. But before you do that, it is recommended to navigate to one of the corners of the PDF so when you click on any of the results, they view zooms to the location of the result in the PDF.

7. Navigate to the top left corner of the PDF file.

8. From the **Search** panel > **Results** area, click on the first result; the view zooms to the search result in the PDF file and the **VAV** text is highlighted near the lower right corner of the **Main Workspace**.

9. Similarly, click on the remaining results and notice how the text is highlighted.

   You will now apply the **VAV** custom count measurement to these results.

10. From the toolbar at the top in the **Results** area, select the **Select All** check box, labeled as **1** in Figure 33; all the search results are selected and have a check mark on their left.

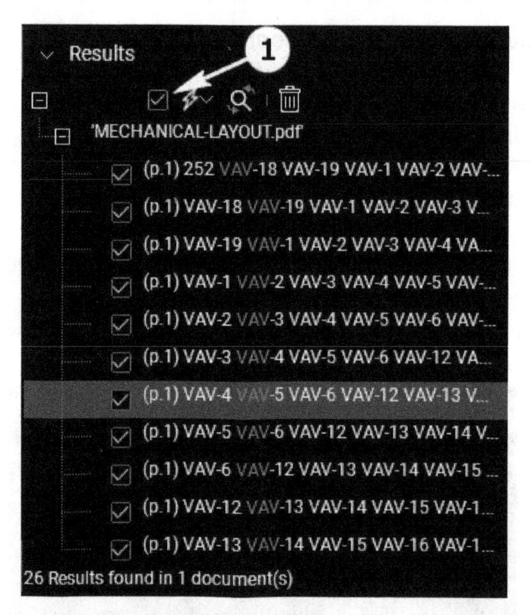

Figure 33 Selecting all the search results

11. Click on the **Check Options** flyout located on the right of the **Select All** check box; the **Check Options** menu is displayed, as shown in Figure 34.

12. Hover the cursor over the **Apply Count Measurement to Checked** option, a cascading menu is displayed showing the **VAV** custom count measurements in the custom tool set you imported, as shown in Figure 35.

13. Select the **VAV** count measurement from the cascading menu, as shown in Figure 35; the custom count symbols are placed on the search results.

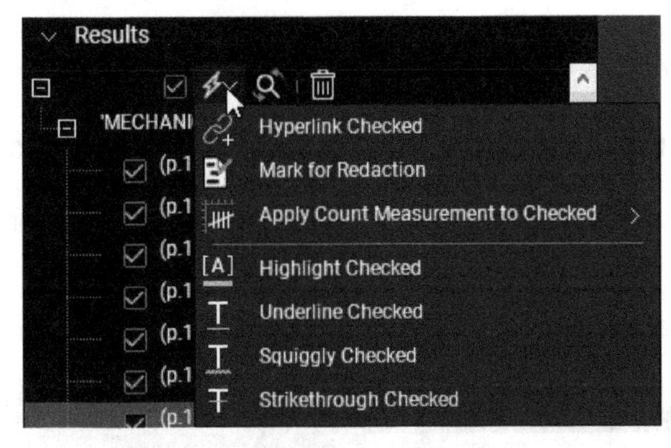

Figure 34 The ***Checked Options*** *menu*

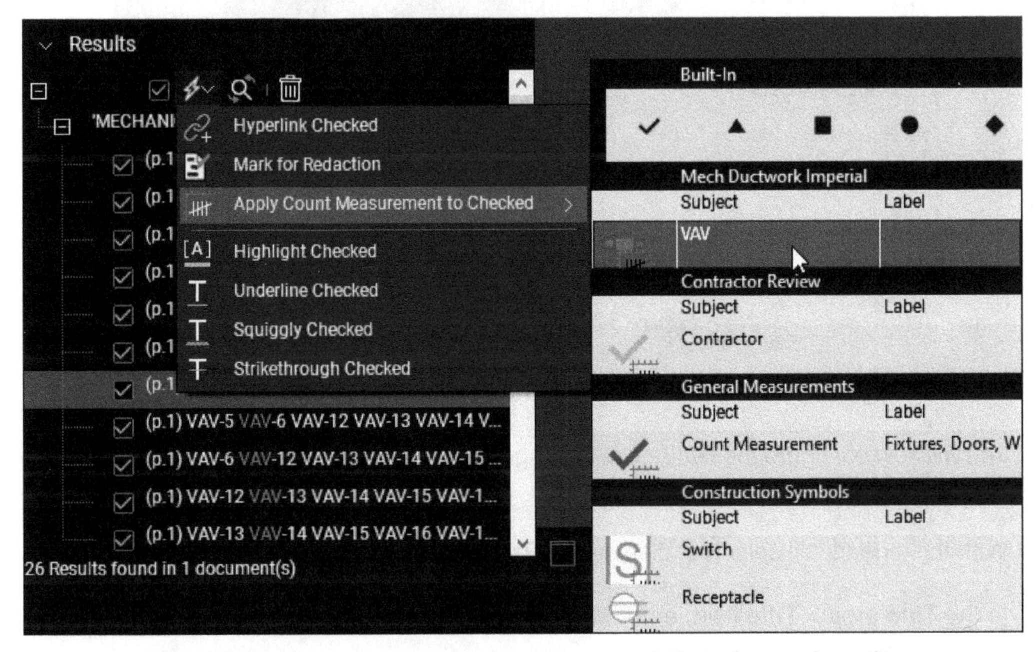

Figure 35 The custom count measurements to apply to the search results

Figure 36 shows the **Markups List** with the 26 VAV count measurements placed and the cost associated with them in the custom column.

Subject	Page Label	Comments	Length	Mech Ductwork	Mech Ductwork Estimate
Supply Duct 350	1	SD 350	17.77 ft' in"	SD 350 48.50	$861.84
Supply Duct 350	1	SD 350	13.52 ft' in"	SD 350 48.50	$655.72
Supply Duct 350	1	SD 350	13.29 ft' in"	SD 350 48.50	$644.56
VAV (1)					$100,750.00
VAV	1	26		VAV 3,875.00	$100,750.00

Figure 36 The ***Markups List*** *shows the custom count added*

*Tip: You can also search for texts such as **150** or **200** to select these duct sizes and then takeoff their lengths using their respective tools in the custom tool set.*

Section 7: Creating a Legend with the Quantities Taken off

In this section, you will create a legend to show the quantities you have taken off from the PDF file. You will then customize this legend and add it to your custom column.

1. From the top right of the custom tool set, click the cogwheel icon and select **Legend > Create New Legend**, as shown in Figure 37; a new legend is attached to the cursor.

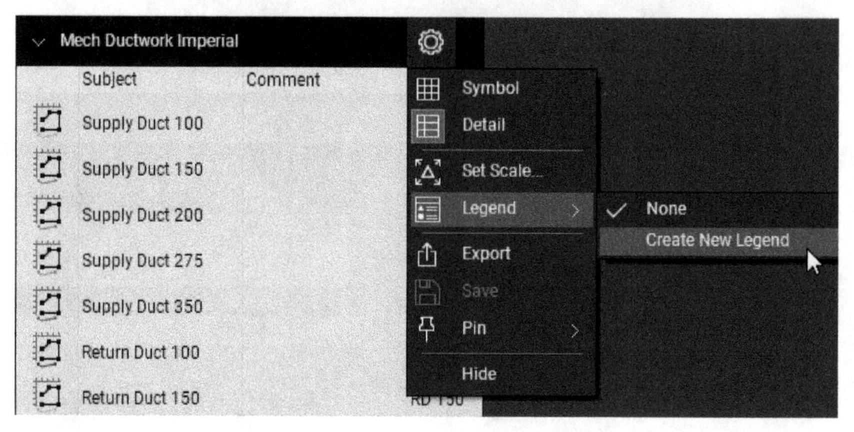

Figure 37 Creating a new legend

2. Place the legend near the bottom right of the plan, above the titleblock.

   You will now configure the settings of this legend and include the custom column that shows the cost of the measurement type.

3. With the legend still selected, display the **Properties** panel.

4. In the **Title** area > **Title** field, enter **MECHANICAL TAKEOFF** as the title of the legend.

5. In the **Columns** area, click the **Edit Columns** button; the **Edit Legend Columns** dialog box is displayed.

6. Scroll down in the dialog box and select the **Mech Ductwork Estimate** column.

7. Click **OK** in the dialog box; the legend is updated and you can now see the cost of each flooring type.

8. In the **Table** area, change the color to Blue.

9. From the **Table Style** drop-down list, select **Gridlines**; the grid lines are added in the legend.

10. Change the **Symbol Size** spinner to **150**; the sizes of the duct type symbols in the legend are increased.

This completes the editing of the legend. Figure 38 shows the legend in the Imperial file and Figure 39 shows the legend in the Metric file. **Note that the values in your file may be a little different from these values.**

MECHANICAL TAKEOFF				
	Description	Quantity	Unit	Mech Ductwork Estimate
	Return Duct 100	78.94	ft	$2,467.18
	Return Duct 150	156.45	ft	$5,608.37
	Return Duct 275	132.34	ft	$6,266.79
	Return Duct 500	64.12	ft	$3,767.04
	Supply Duct 100	179.85	ft	$5,170.13
	Supply Duct 150	146.05	ft	$4,417.73
	Supply Duct 200	92.05	ft	$3,452.62
	Supply Duct 275	81.42	ft	$3,563.00
	Supply Duct 350	7.08	ft	$343.38
	VAV	26	Count	$100,750.00

Figure 38 *The legend showing the quantification values in the Imperial file*

MECHANICAL TAKEOFF				
	Description	Quantity	Unit	Mech Ductwork Estimate
	Return Duct 100	24.2	m	$2,373.85
	Return Duct 150	48.0	m	$5,551.18
	Return Duct 275	43.9	m	$6,525.70
	Return Duct 500	19.7	m	$3,425.10
	Supply Duct 100	55.1	m	$5,080.22
	Supply Duct 150	44.8	m	$4,603.21
	Supply Duct 200	28.2	m	$3,290.14
	Supply Duct 275	26.6	m	$3,450.50
	Supply Duct 350	2.2	m	$325.28
	VAV	26	Count	$100,750.00

Figure 39 *The legend showing the quantification values in the Metric file*

You will now add this legend to the custom tool set you have imported. This way this customized legend can be added to any other file as well where you takeoff the quantities.

11. Right-click on the custom legend and add it to your custom tool set. Figure 40 shows the two custom tool sets in the **Symbol** mode with the legends added.

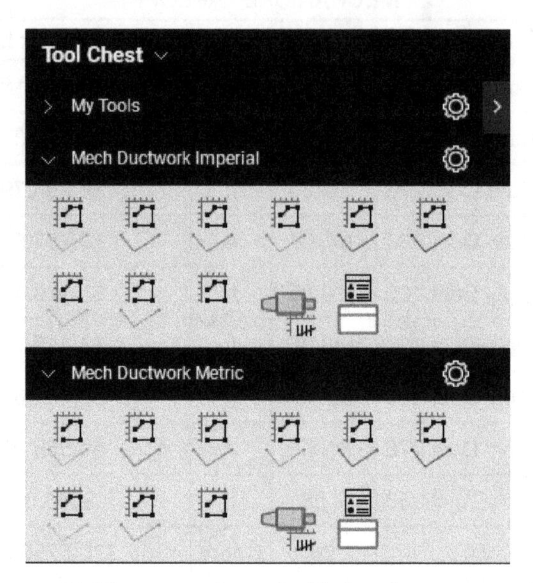

Figure 40 *The custom legend added to the custom tool set*

Section 8: Exporting Quantities as a CSV File

In this section, you will export the quantities you have taken off as a CSV file so the values could be used outside Revu.

1. From the toolbar at the top in the **Markups List**, click the **Summary** flyout, labeled as **1** in Figure 41, and then select **CSV Summary**, as shown in the same figure.

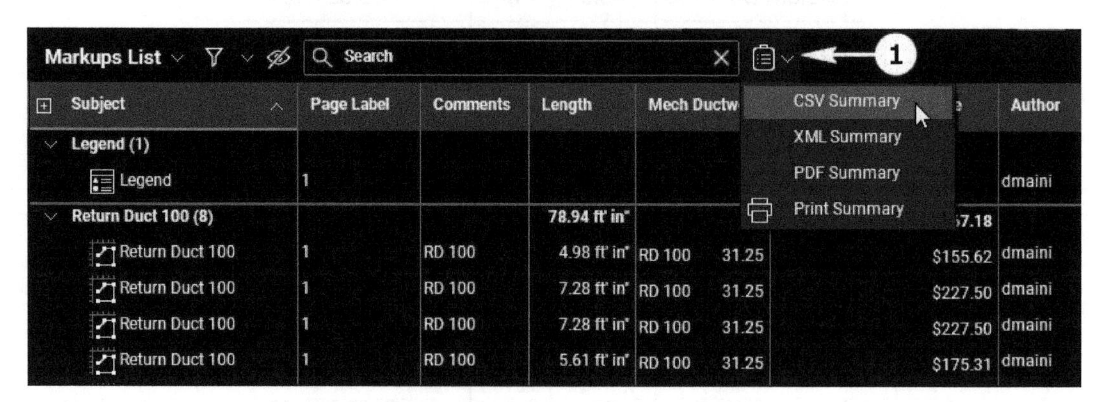

Figure 41 *Exporting the quantities*

On doing so, the **Markup Summary** dialog box is displayed.

2. Activate the **Filter and Sort** tab.

3. Click on the **Subject** drop-down list and deselect **Legend**. Make sure all the flooring types are selected.

4. In the rest of the drop-down lists, make sure **All** is selected, as shown in Figure 42.

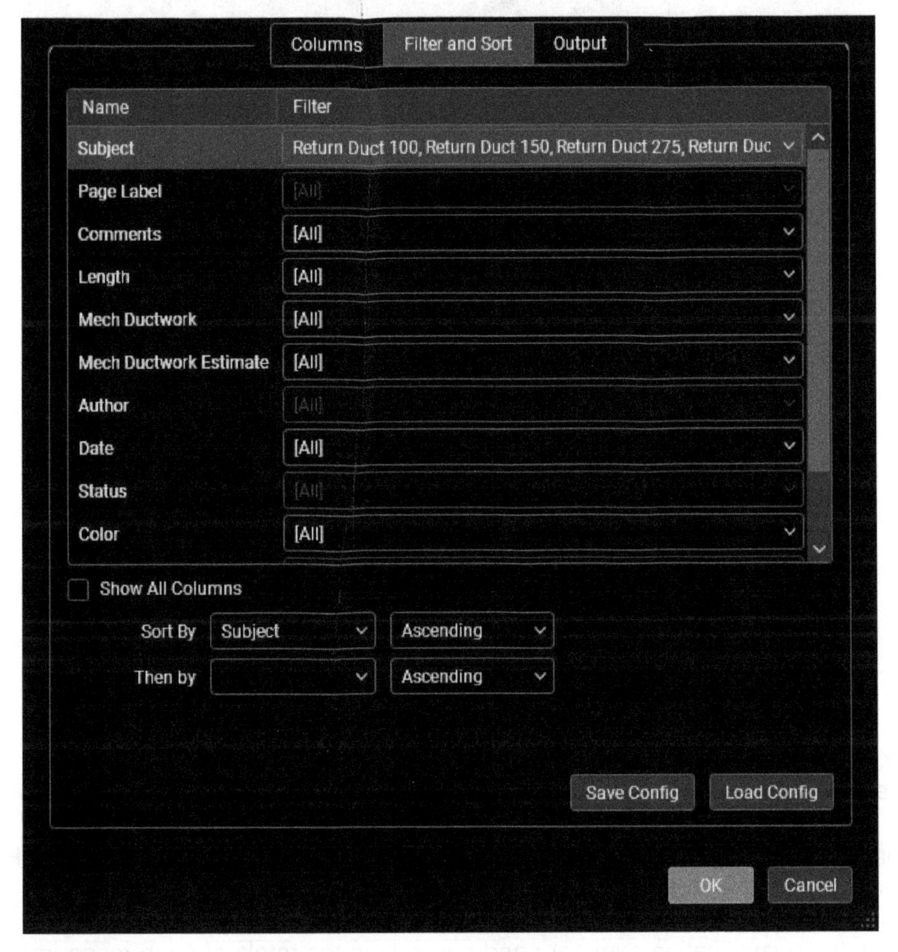

***Figure 42** Configuring the export filters*

5. Activate the **Output** tab.

6. Set the **Export to** folder to the **Tutorial Files > P01** folder.

7. From the **Include** area, select the **Markups & Totals** radio button, as shown in Figure 43.

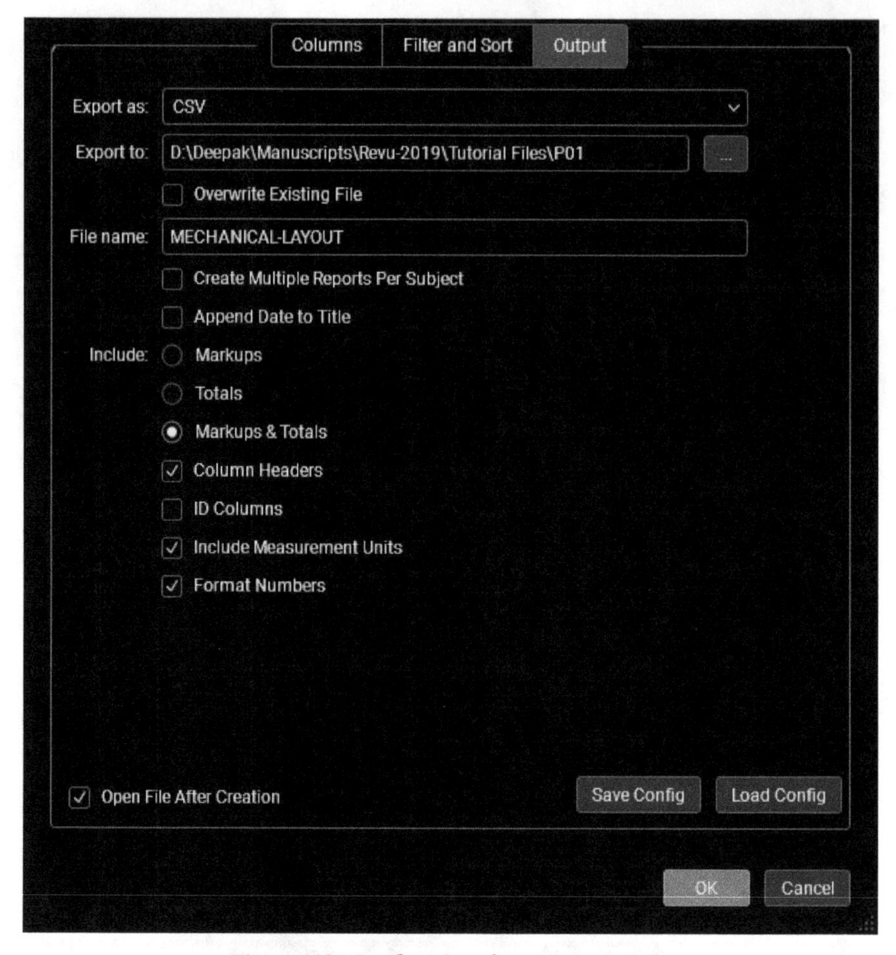

Figure 43 *Configuring the output settings*

8. Click the **OK** button in the **Markup Summary** dialog box; the summary of the quantification is created and the CSV file is opened.

   Figure 44 shows a part of this CSV file with the Imperial quantification summary and Figure 45 shows a part of the Metric quantification summary.

9. Review the quantification summary and notice how the totals of each measurement type are displayed at the top in that column.

	A	B	C	D	E	F	G	H	I	J	K
1	Subject	Page Label	Comments	Length	Length Unit	Mech Ductwork	Mech Ductwork Estimate	Author	Date	Status	Color
2	Return Duct 100 (8)			78.94	ft' in"		$2,467.18				
3	Return Duct 100	1	RD 100	7.28	ft' in"	RD 100	$227.50	dmaini	14/03/2020 16:32		#800040
4	Return Duct 100	1	RD 100	7.28	ft' in"	RD 100	$227.50	dmaini	14/03/2020 16:32		#800040
5	Return Duct 100	1	RD 100	4.98	ft' in"	RD 100	$155.62	dmaini	14/03/2020 16:32		#800040
6	Return Duct 100	1	RD 100	6.27	ft' in"	RD 100	$195.94	dmaini	14/03/2020 16:32		#800040
7	Return Duct 100	1	RD 100	15.91	ft' in"	RD 100	$497.19	dmaini	14/03/2020 16:32		#800040
8	Return Duct 100	1	RD 100	15.85	ft' in"	RD 100	$495.31	dmaini	14/03/2020 16:32		#800040
9	Return Duct 100	1	RD 100	15.77	ft' in"	RD 100	$492.81	dmaini	14/03/2020 16:32		#800040
10	Return Duct 100	1	RD 100	5.61	ft' in"	RD 100	$175.31	dmaini	14/03/2020 16:32		#800040
11	Return Duct 150 (9)			156.45	ft' in"		$5,608.37				
12	Return Duct 150	1	RD 150	9.07	ft' in"	RD 150	$325.16	dmaini	14/03/2020 16:33		#FF80FF
13	Return Duct 150	1	RD 150	15.92	ft' in"	RD 150	$570.73	dmaini	14/03/2020 16:33		#FF80FF
14	Return Duct 150	1	RD 150	30.75	ft' in"	RD 150	$1,102.39	dmaini	14/03/2020 16:33		#FF80FF
15	Return Duct 150	1	RD 150	15.85	ft' in"	RD 150	$568.22	dmaini	14/03/2020 16:33		#FF80FF
16	Return Duct 150	1	RD 150	5.98	ft' in"	RD 150	$214.38	dmaini	14/03/2020 16:33		#FF80FF
17	Return Duct 150	1	RD 150	31.2	ft' in"	RD 150	$1,118.52	dmaini	14/03/2020 16:33		#FF80FF
18	Return Duct 150	1	RD 150	7.92	ft' in"	RD 150	$283.93	dmaini	14/03/2020 16:33		#FF80FF
19	Return Duct 150	1	RD 150	15.85	ft' in"	RD 150	$568.22	dmaini	14/03/2020 16:33		#FF80FF
20	Return Duct 150	1	RD 150	23.9	ft' in"	RD 150	$856.82	dmaini	14/03/2020 16:33		#FF80FF
21	Return Duct 275 (11)			132.34	ft' in"		$6,266.79				
22	Return Duct 275	1	RD 275	6.54	ft' in"	RD 275	$309.67	dmaini	14/03/2020 16:33		#8080FF
23	Return Duct 275	1	RD 275	2.73	ft' in"	RD 275	$129.27	dmaini	14/03/2020 16:33		#8080FF
24	Return Duct 275	1	RD 275	4.27	ft' in"	RD 275	$202.18	dmaini	14/03/2020 16:33		#8080FF
25	Return Duct 275	1	RD 275	21.13	ft' in"	RD 275	$1,000.51	dmaini	14/03/2020 16:33		#8080FF
26	Return Duct 275	1	RD 275	9.39	ft' in"	RD 275	$444.62	dmaini	14/03/2020 16:33		#8080FF
27	Return Duct 275	1	RD 275	6	ft' in"	RD 275	$284.10	dmaini	14/03/2020 16:33		#8080FF
28	Return Duct 275	1	RD 275	12.08	ft' in"	RD 275	$571.99	dmaini	14/03/2020 16:33		#8080FF
29	Return Duct 275	1	RD 275	9.39	ft' in"	RD 275	$444.62	dmaini	14/03/2020 16:33		#8080FF

MECHANICAL-LAYOUT

Figure 44 *The Imperial CSV quantification summary*

	A	B	C	D	E	F	G	H	I	J	K
1	Subject	Page Label	Comments	Length	Length Unit	Mech Ductwork	Mech Ductwork Estimate	Author	Date	Status	Color
2	Return Duct 100 (8)			24.2	m		$2,373.85				
3	Return Duct 100	1	RD 100	2.2	m	RD 100	$216.70	dmaini	14/03/2020 16:37		#800040
4	Return Duct 100	1	RD 100	2.2	m	RD 100	$216.70	dmaini	14/03/2020 16:37		#800040
5	Return Duct 100	1	RD 100	1.5	m	RD 100	$147.75	dmaini	14/03/2020 16:37		#800040
6	Return Duct 100	1	RD 100	1.9	m	RD 100	$187.15	dmaini	14/03/2020 16:37		#800040
7	Return Duct 100	1	RD 100	4.9	m	RD 100	$482.65	dmaini	14/03/2020 16:37		#800040
8	Return Duct 100	1	RD 100	4.9	m	RD 100	$482.65	dmaini	14/03/2020 16:37		#800040
9	Return Duct 100	1	RD 100	4.8	m	RD 100	$472.80	dmaini	14/03/2020 16:37		#800040
10	Return Duct 100	1	RD 100	1.7	m	RD 100	$167.45	dmaini	14/03/2020 16:37		#800040
11	Return Duct 150 (9)			48	m		$5,551.18				
12	Return Duct 150	1	RD 150	2.8	m	RD 150	$323.82	dmaini	14/03/2020 16:37		#FF80FF
13	Return Duct 150	1	RD 150	4.9	m	RD 150	$566.68	dmaini	14/03/2020 16:37		#FF80FF
14	Return Duct 150	1	RD 150	9.4	m	RD 150	$1,087.11	dmaini	14/03/2020 16:37		#FF80FF
15	Return Duct 150	1	RD 150	4.9	m	RD 150	$566.68	dmaini	14/03/2020 16:37		#FF80FF
16	Return Duct 150	1	RD 150	1.8	m	RD 150	$208.17	dmaini	14/03/2020 16:37		#FF80FF
17	Return Duct 150	1	RD 150	9.6	m	RD 150	$1,110.24	dmaini	14/03/2020 16:37		#FF80FF
18	Return Duct 150	1	RD 150	2.4	m	RD 150	$277.56	dmaini	14/03/2020 16:37		#FF80FF
19	Return Duct 150	1	RD 150	4.9	m	RD 150	$566.68	dmaini	14/03/2020 16:37		#FF80FF
20	Return Duct 150	1	RD 150	7.3	m	RD 150	$844.24	dmaini	14/03/2020 16:37		#FF80FF
21	Return Duct 275 (11)			43.9	m		$6,525.70				
22	Return Duct 275	1	RD 275	2	m	RD 275	$297.30	dmaini	14/03/2020 16:37		#8080FF
23	Return Duct 275	1	RD 275	2.5	m	RD 275	$371.62	dmaini	14/03/2020 16:37		#8080FF
24	Return Duct 275	1	RD 275	1.3	m	RD 275	$193.24	dmaini	14/03/2020 16:37		#8080FF
25	Return Duct 275	1	RD 275	2.9	m	RD 275	$431.08	dmaini	14/03/2020 16:37		#8080FF
26	Return Duct 275	1	RD 275	1.8	m	RD 275	$267.57	dmaini	14/03/2020 16:37		#8080FF
27	Return Duct 275	1	RD 275	3.7	m	RD 275	$550.00	dmaini	14/03/2020 16:37		#8080FF
28	Return Duct 275	1	RD 275	2.9	m	RD 275	$431.08	dmaini	14/03/2020 16:37		#8080FF
29	Return Duct 275	1	RD 275	2.5	m	RD 275	$371.62	dmaini	14/03/2020 16:37		#8080FF

MECHANICAL-LAYOUT - M

Figure 45 *The Metric CSV quantification summary*

10. Return to the Revu window.
11. Save the PDF file.
12. Close the PDF file.

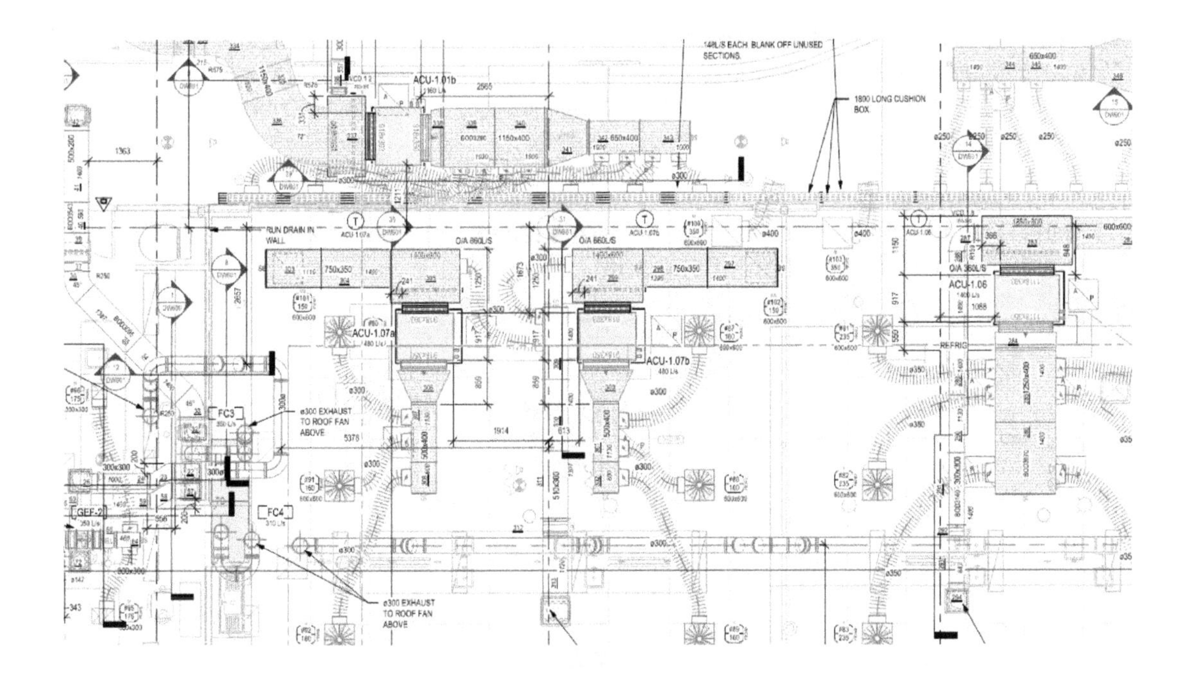

Project 2 - Quantity Takeoff from an Electrical PDF File

In this project, you will do the following:

- √ *Create a new profile to take off quantities form an Electrical PDF file*
- √ *Import the Electrical QTO tool set*
- √ *Import the Electrical QTO custom columns*
- √ *Takeoff quantities from the Electrical file*
- √ *Add a legend showing the quantities taken off*
- √ *Export the quantities into a CSV file*

Section 1: Creating a New Profile

In this section, you will open the Electrical file to take off the quantities. You will then create a new profile called **Electrical QTO**. This profile will be based on the **Training QTO** profile you created in Chapter 6 of this book.

1. From the **Tutorial Files > P02** folder, open the **Electrical_QTO.pdf** file; the file is opened in the Revu window, as shown in Figure 1. **This file is courtesy of Karl De Wet, Director, Draftech Developments (MEP Solutions)**.

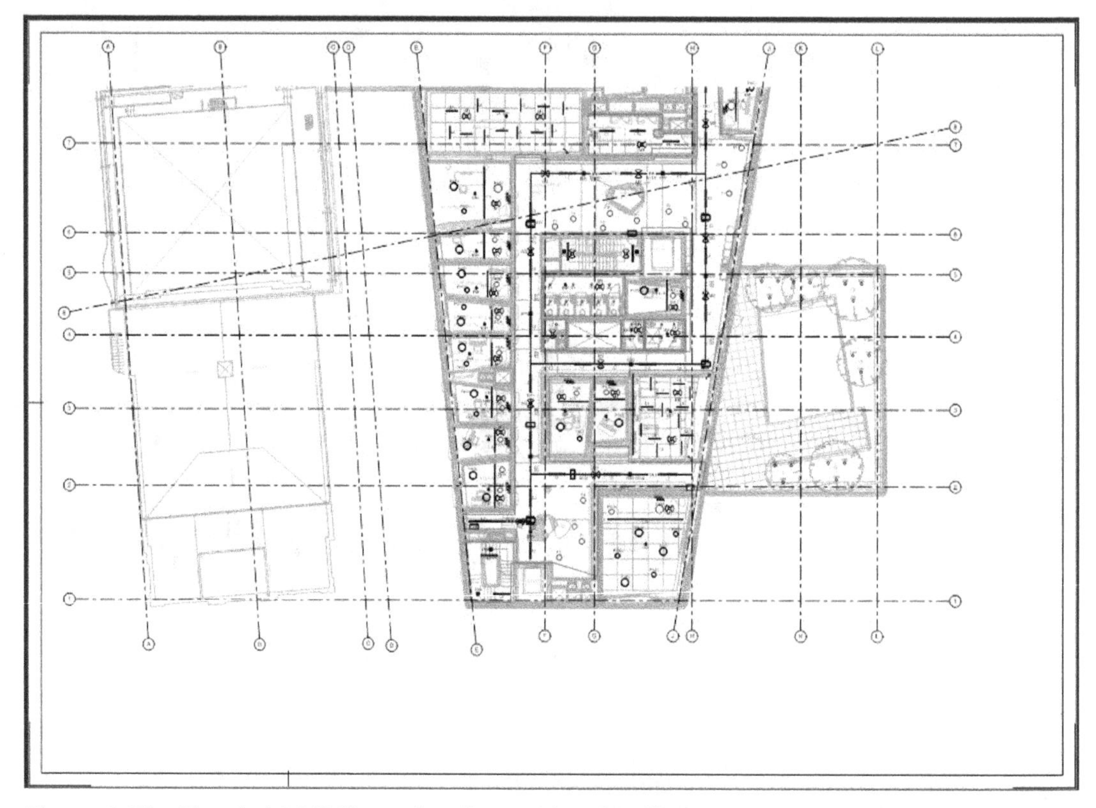

*Figure 1 The Electrical PDF file to take off quantities (**this file is courtesy of Karl De Wet, Director, Draftech Developments (MEP Solutions)**)*

2. Navigate around this PDF file and notice that it has various electrical symbols and labels.

3. Double-click the Wheel Mouse Button to zoom to the extents of the PDF file.

   You will now restore the **Training QTO** profile you created in Chapter 6 of this book.

4. From the top left of the Revu window, click **Revu > Profiles > Training QTO**, as shown in Figure 2; the **Training QTO** profile is restored.

5. Hide any panel displayed on the left, right, or below the **Main Workspace**.

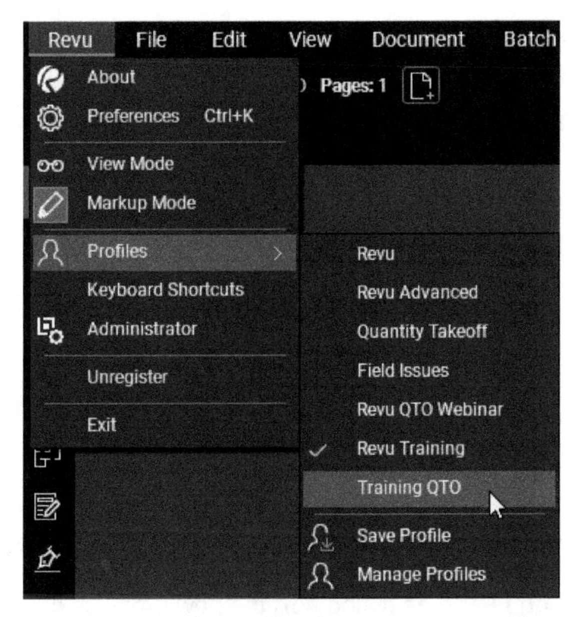

*Figure 2 Activating the **Training QTO** profile*

You will now create a new profile called **Electrical QTO**. The reason you are creating a new profile is so that you can save the custom tool sets and custom columns to this profile.

6. From the top left of the Revu window, click **Revu > Profiles > Manage Profiles**; the **Manage Profiles** dialog box is displayed, as shown in Figure 3.

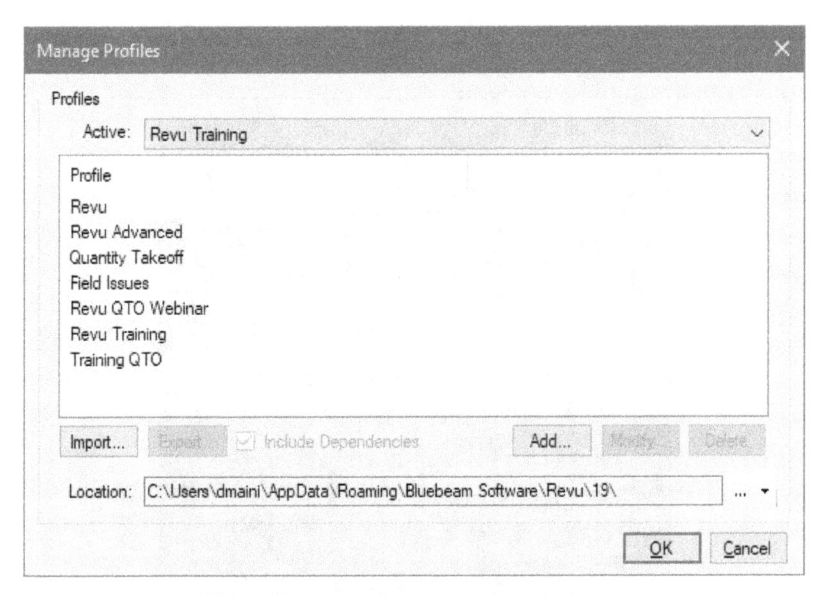

*Figure 3 The **Manage Profiles** dialog box*

7. Near the bottom right in the **Manage Profiles** dialog box, click **Add**; the **Add Profile** dialog box is displayed.

8. Enter **Electrical QTO** as the name of the profile in the **Add Profile** dialog box, as shown in Figure 4.

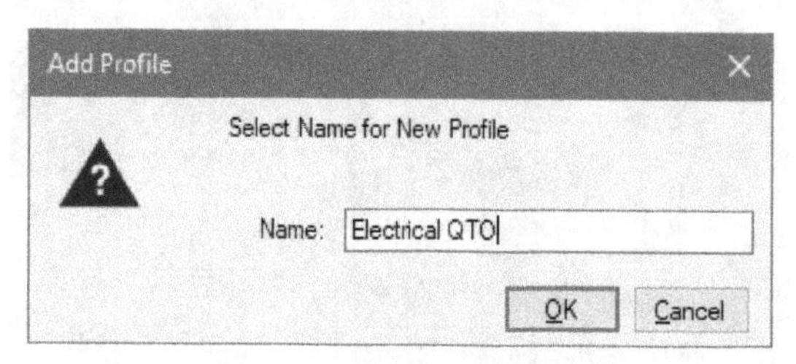

*Figure 4 The **Add Profile** dialog box*

9. Click **OK** in the **Add Profile** dialog box; you are returned to the **Manage Profiles** dialog box.

10. Click **OK** in the **Manage Profiles** dialog box; the new profile is created and activated.

Section 2: Importing Custom Tool Set

In this section, you will import the custom tool set that is provided to you to take off the quantities. The detailed process of how to create custom tools to take off quantities was discussed in Chapter 7 of this book.

1. From the **Panel Access Bar**, click the **Tool Chest** button to display various tool sets.

2. From the top of the **Tool Chest** panel, click **Tool Chest > Manage Tool Sets**, as shown in Figure 5.

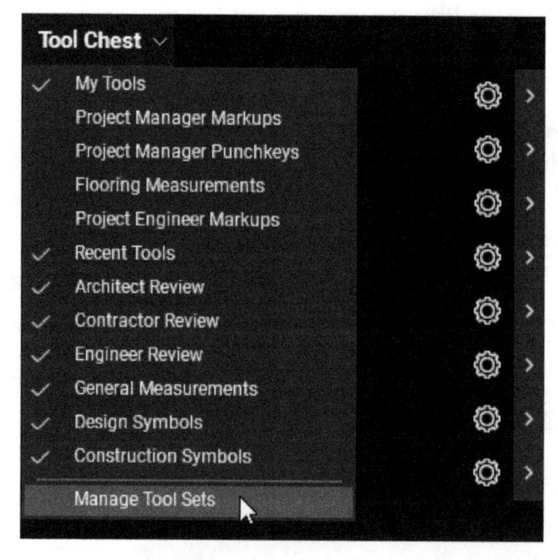

Figure 5 Managing tool sets

On doing so, the **Manage Tool Sets** dialog box is displayed.

3. From the lower left in the **Manage Tool Sets** dialog box, click the **Import** button; the **Open** dialog box is displayed.

   The custom tool set with the Electrical quantification tools is saved in the **Tutorial Files > P02** folder. You will now browse to this folder to import the tool set.

4. Browse to the **Tutorial Files > P02** folder; the BTX format file of the tool set is displayed in the folder, as shown in Figure 6.

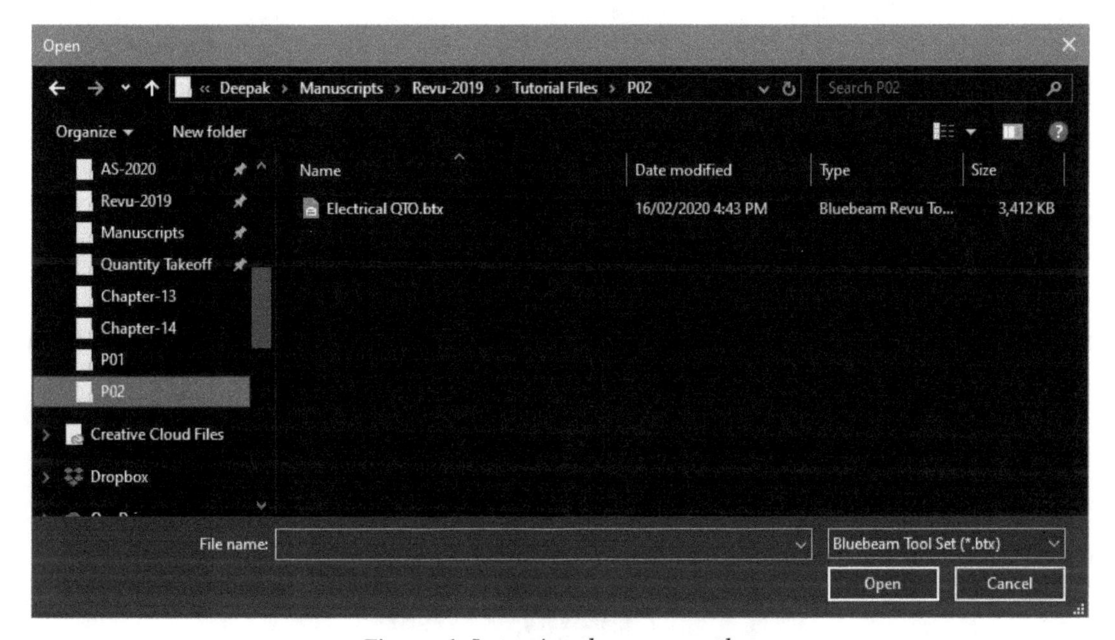

Figure 6 *Importing the custom tool set*

5. Double-click on the BTX file of the tool set; you are returned to the **Manage Tool Sets** dialog box and the tool set you imported is listed at the bottom. Also, the tool set is displayed in the **Tool Chest** panel in the Revu window.

   You will now select this tool set and move it up on the list.

6. In the **Manage Tool Sets** dialog box, scroll down and select the tool set you imported.

7. Use the **Up** arrow key on the left side of the dialog box and reorder this tool set so it is located below the **My Tools** tool set. Figure 7 shows the **Manage Tool Sets** dialog box with the **Electrical QTO** tool set listed as the second tool set.

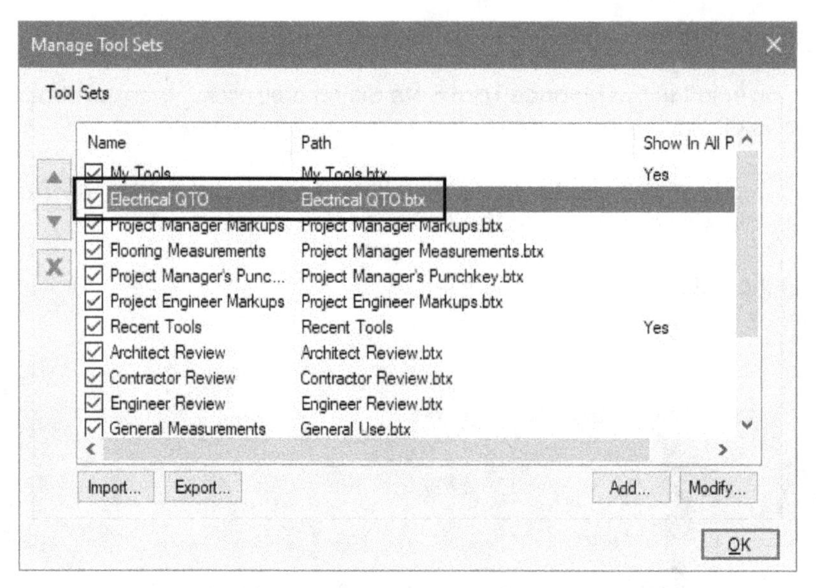

Figure 7 The imported tool set listed in the dialog box

8. Click **OK** in the **Manage Tool Sets** dialog box; the imported tool set is listed in the **Tool Chest** panel, as shown in Figure 8.

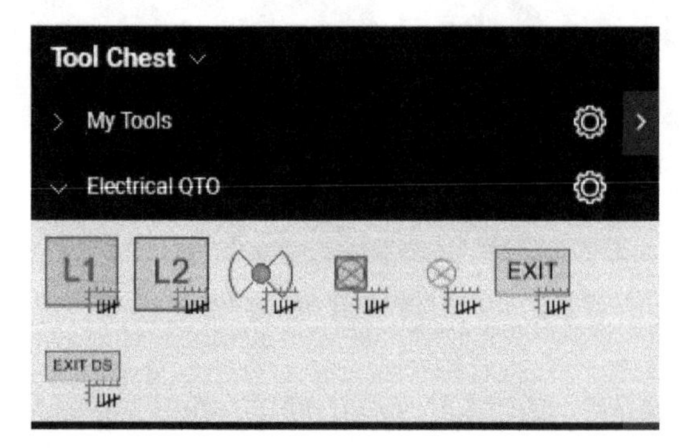

*Figure 8 The imported tool set listed in the **Tool Chest** panel*

Section 3: Importing Custom Columns

In this section, you will import custom columns that will link to the custom tools in the tool set you imported to show you the cost of your takeoffs. The detailed process of creating custom columns and linking them to the custom tools to show the cost was discussed in Chapter 8 of this book.

1. From the bottom left of the Revu window, click the **Markups** button, labeled as **1** in Figure 9; the **Markups List** is displayed below the **Main Workspace**.

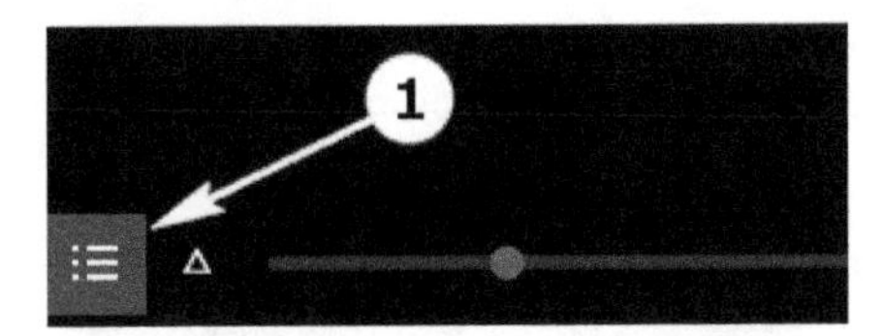

*Figure 9 Displaying the **Markups List***

2. From the top left in the **Markups List**, click **Markups List > Columns > Manage Columns**, as shown in Figure 10.

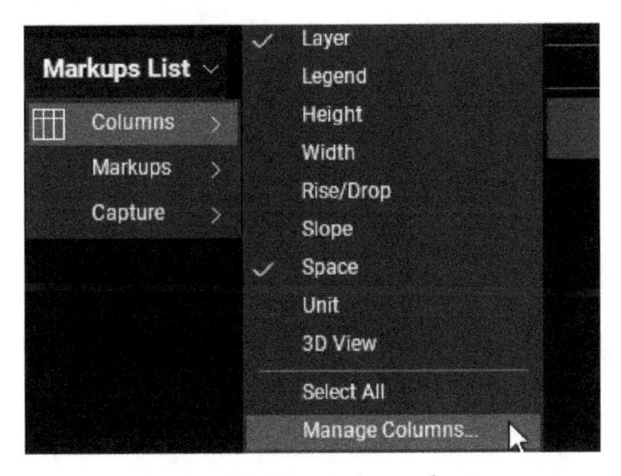

Figure 10 Managing columns

On doing so, the **Manage Columns** dialog box is displayed. You will now activate the **Custom Columns** tab and import the custom columns.

3. In the **Manage Columns** dialog box, activate the **Custom Columns** tab.

4. Click the **Import** button in the **Custom Columns** area; the **Open** dialog box is displayed.

5. If not already in that folder, browse to the **Tutorial Files > P02** folder; the XML file of the custom columns is listed there.

6. Double-click on the XML file of the custom columns; the **Import Custom Columns** warning box is displayed, as shown in Figure 11. This warning box informs you that importing these columns will replace any existing columns and will be applied to the open document.

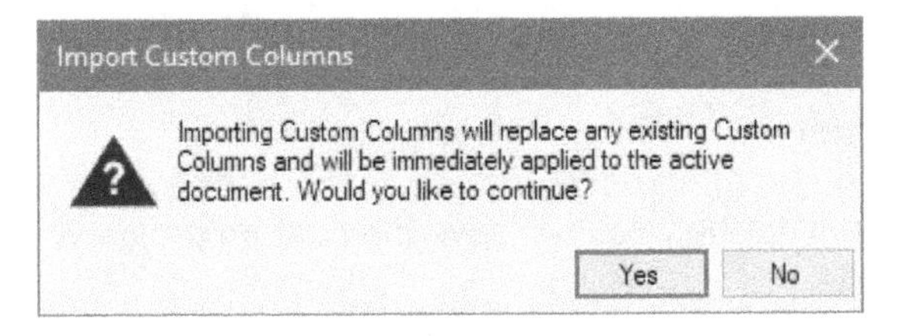

*Figure 11 The **Import Custom Columns** warning box*

7. Click **Yes** in the **Import Custom Columns** warning box; you are returned to the **Manage Columns** dialog box and the two custom columns are listed in this dialog box, as shown in Figure 12.

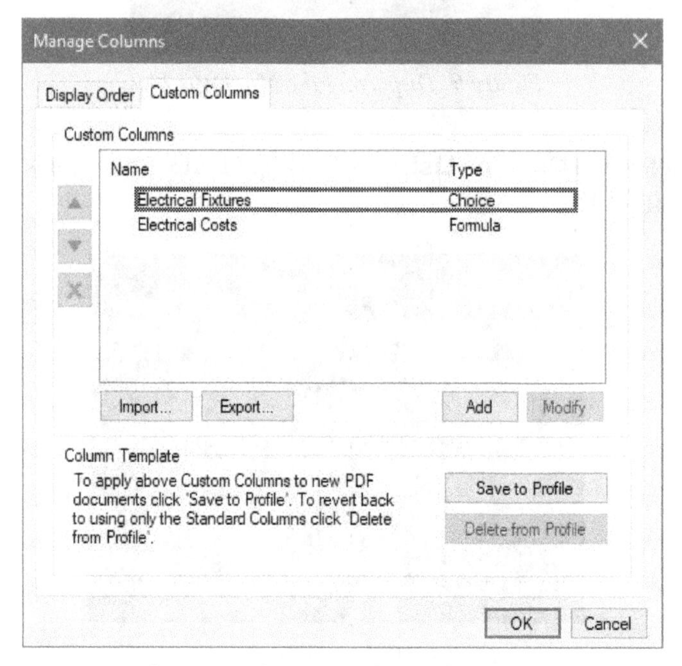

Figure 12 *The imported custom columns listed in the dialog box*

Before exiting this dialog box, you need to make sure you save these two custom columns to the **Electrical QTO** profile you created at the start of this project.

8. From the lower right in the **Custom Columns** tab of the dialog box, click the **Save to Profile** button; the **Apply Column Template** dialog box is displayed asking you to confirm that you want to apply the column template.

9. Click **OK** in the **Apply Column Template** dialog box; the two custom columns are now saved to the profile.

10. Click **OK** in the **Manage Columns** dialog box; the two custom columns are listed in the **Markups List**; as shown in Figure 13.

    Because you do not need the **Markups List** displayed at this stage, you will now hide it.

11. Hide the **Markups List**.

    Before proceeding any further, you will save the custom profile you created at the start of this project.

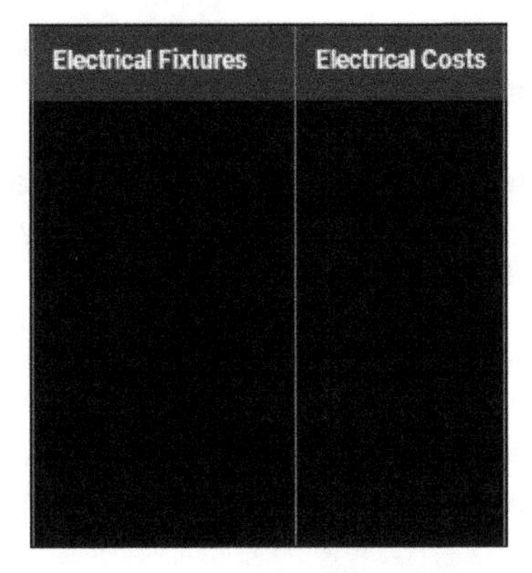

Figure 13 *The imported custom columns listed in the* ***Markups List***

12. From the top left in the Revu window, click **Revu > Profiles > Save Profile**; the custom profile is saved.

Section 4: Taking Off Quantities

For this Electrical takeoff, you will search for text and symbols and will not measure any areas or distances. As a result, you do not need to calibrate the sheet. The search will be performed using the **Search** panel. Therefore, you first need to display this panel.

1. From the **Panel Access Bar**, show the **Search** panel.

   The **Search** panel is by default set to search for text. This is because the **Text** button is turned on at the top in this panel. You will now search for the text **L1** that represent surface mounted/suspended single linear fluorescent lights.

2. In the text search field at the top in the **Search** panel, type **L1**.

3. From the drop-down list below the search field, select **Current Document**.

4. In the **Options** area, select the **Whole Words Only** check box, as shown in Figure 14.

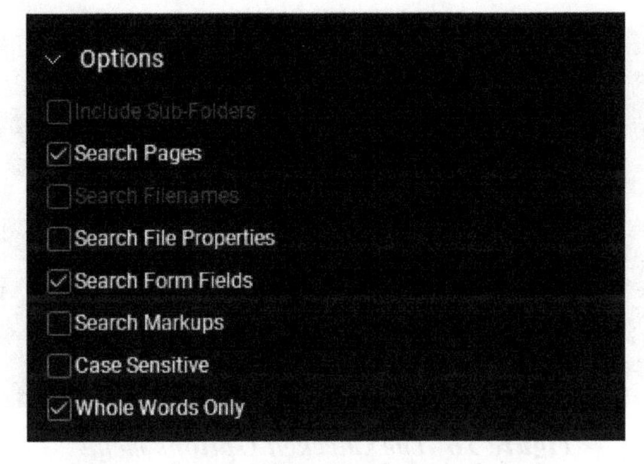

Figure 14 *Configuring the search options*

5. Click the **Search** button from the top right in the **Search** panel; the search is performed in the document and the bottom of the **Results** area shows 3 results found.

   You will now review these three results. But before you do that, it is recommended to navigate to one of the corners of the PDF so when you click on any of the results, they view zooms to the location of the result in the PDF.

6. Navigate to the top left corner of the PDF file.

7. From the **Search** panel > **Results** area, click on the first result; the view zooms to the search result in the PDF file and the **L1** text is highlighted near the lower right corner of the **Main Workspace**.

8. Similarly, click on the remaining results and notice how the text is highlighted.

   You will now apply a custom count measurement to these three results.

9. From the toolbar at the top in the **Results** area, select the **Select All** check box, labeled as **1** in Figure 15; all the search results are selected and have a check mark on their left.

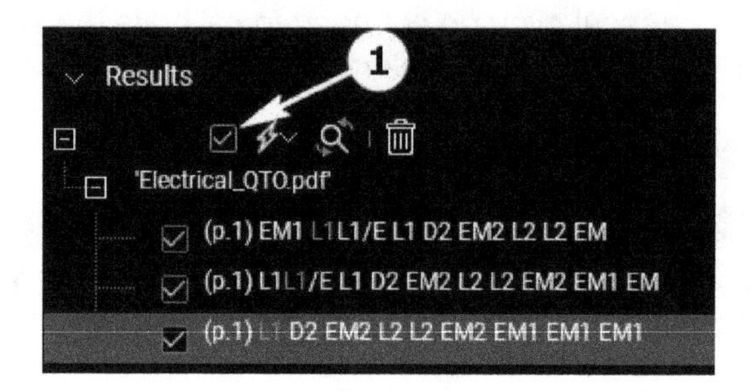

Figure 15 Selecting all the search results

10. Click on the **Check Options** flyout located on the right of the **Select All** check box; the **Check Options** menu is displayed, as shown in Figure 16.

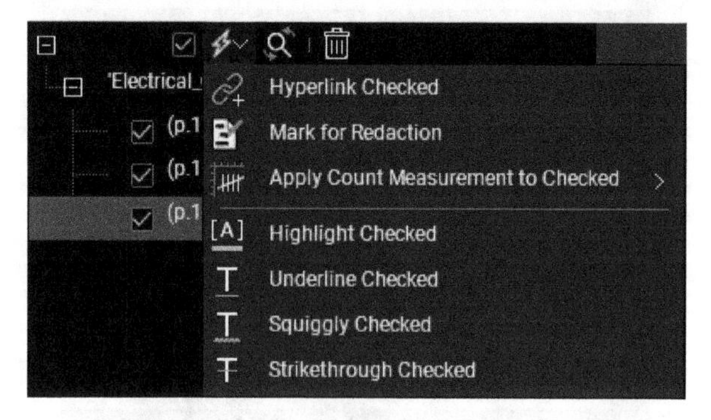

*Figure 16 The **Checked Options** menu*

11. Hover the cursor over the **Apply Count Measurement to Checked** option, a cascading menu is displayed showing various custom count measurements in the custom tool set you imported, as shown in Figure 17.

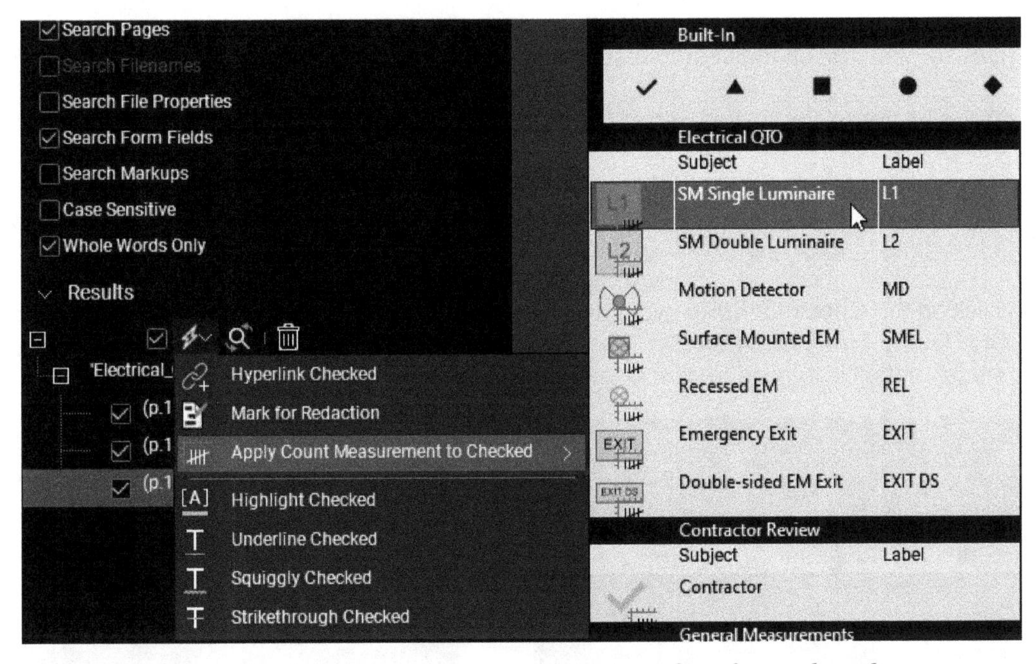

Figure 17 The custom count measurements to apply to the search results

12. Select **SM Single Luminaire** from the cascading menu, as shown in Figure 17; the custom count symbols are placed on the three search results, as shown in Figure 18.

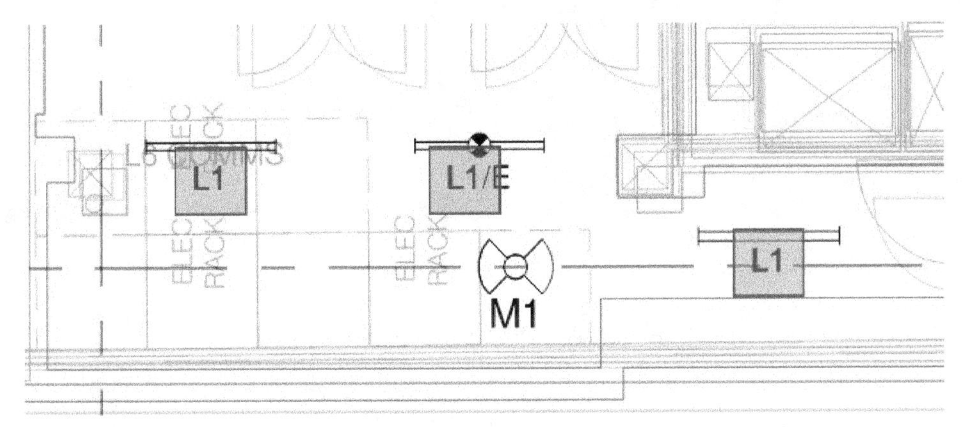

Figure 18 The custom count measurement symbols placed on the search results

Next, you will search for the text **L2** that represents surface mounted/suspended double linear fluorescent lights.

13. In the **Search** panel type **L2** as the text to search.

14. Click the **Search** button; the search is performed and the bottom of the **Results** area shows 23 results found.

15. One by one click on the search results and notice how they are highlighted in the PDF file.

    You will now select all these search results and apply the custom count measurement to them.

16. From the toolbar at the top in the **Results** area, select the **Select All** check box, labeled as **1** in Figure 15 earlier; all the search results have a check mark on their left.

17. Click on the **Check Options** flyout and select **Apply Count Measurement to Checked > SM Double Luminaire** from the cascading menu, as shown in Figure 19; the custom count measurement is applied to all the selected search results.

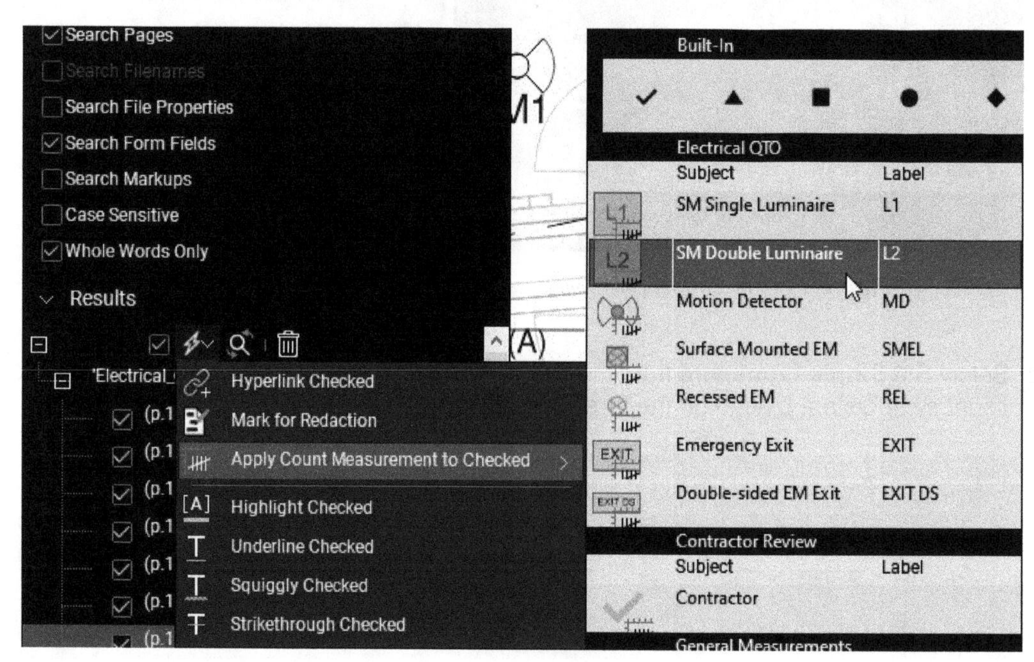

Figure 19 *Applying the custom count measurements to the search results*

Next, you will search for the motion detectors. They are labeled as **M1** and **M2** in the PDF file. You will first search for the text **M1** and apply the count measurement to it.

18. In the **Search** panel, search for the text **M1**; you will find 20 results.

19. Select all the 20 results and apply the **Motion Detector** count measurement to them, as shown in Figure 20.

20. Similarly, search for the text **M2**; you will find 10 results.

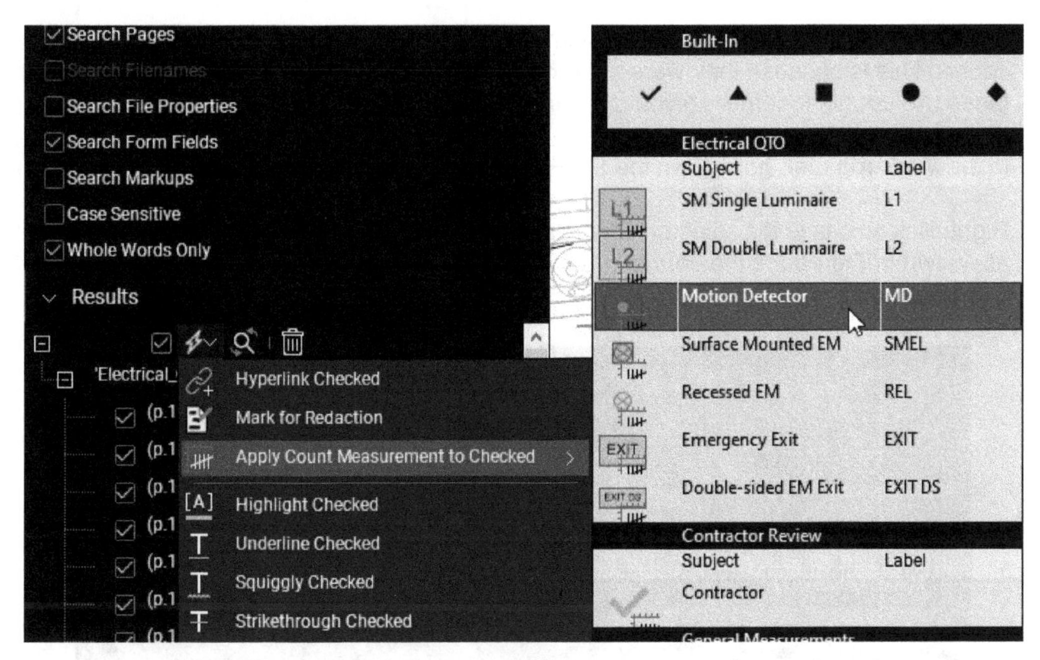

Figure 20 Applying the custom count measurements to the search results

21. Apply the same **Motion Detector** count measurement to the 10 search results.

    Before searching for the other items, you will look at the **Markups List** to review how the costs associated with these searches appear.

22. From the **Panel Access Bar**, click the **Markups** button to display the **Markups List**.

    Figure 21 shows the **Markups List** with the custom counts applied to the search results. Notice the counts are displayed in the **Comments** column, the individual costs are displayed in the **Electrical Fixtures** column, and the total cost is displayed in the **Electrical Costs** column.

Markups List

Subject	Date	Comments	Electrical Fixtures		Electrical Costs
Motion Detector (2)					**$2,820.00**
Motion Detector	17/02/2020 10:44:3...	20	MD	94.00	$1,880.00
Motion Detector	17/02/2020 10:44:4...	10	MD	94.00	$940.00
SM Double Luminaire (1)					**$1,506.50**
SM Double Luminaire	17/02/2020 10:30:4...	23	L2	65.50	$1,506.50
SM Single Luminaire (1)					**$168.30**
SM Single Luminaire	17/02/2020 10:18:1...	3	L1	56.10	$168.30

*Figure 21 The **Markups List** shows the costs associated with the takeoff*

Notice in Figure 21, the **Markups List** shows the **Motion Detectors** listed in two separate counts. This is because they were searched separately as **M1** and **M2**. You will now merge these two counts together so you get one single count for these.

23. In the **Markups List**, hold down the SHIFT key and select the two **Motion Detector** counts.

24. Right-click on one of the selected counts and select **Merge Counts** from the shortcut menu, as shown in Figure 22; the counts are merged and a single value of 30 is displayed in the **Comments** column.

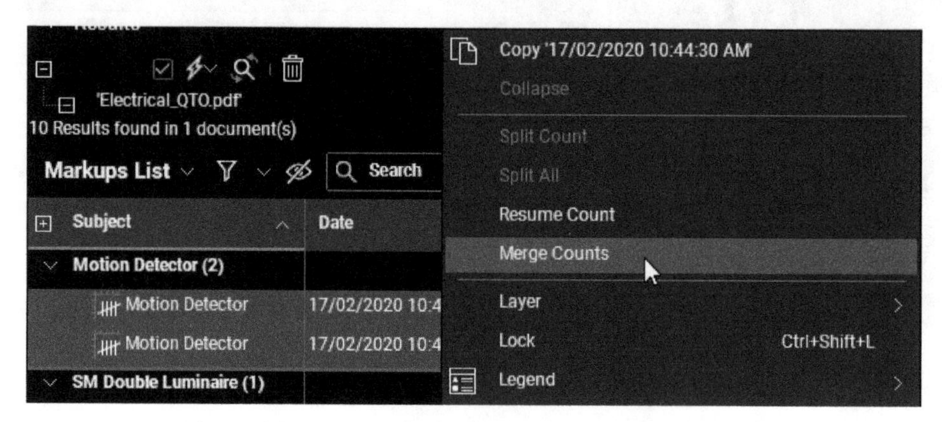

*Figure 22 Merging the **Motion Detector** custom counts*

25. Hide the **Markups List** as you do not need it for the time being.

Next, you will perform a visual search to search for a symbol that represents the Emergency Luminaire. This PDF file has three types of these luminaires and as a result, it is better to use visual search rather than the text search.

26. From the top in the **Search** panel, select the **Visual** button, labeled as **1** in Figure 23; the options in the **Search** panel are modified.

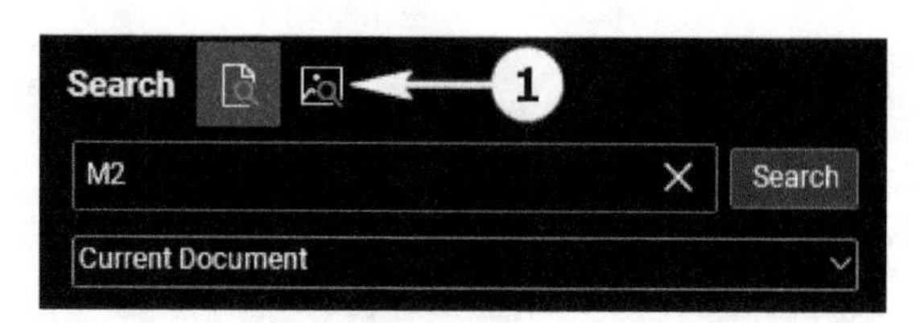

Figure 23 Selecting the option to perform visual search

27. Navigate to the top center of the PDF file, between the **E** and **F** grid lines where the text **EM1** and its symbol appears, refer to Figure 24.

You will now use the **Get Rectangle** tool and drag a rectangle around the EM1 symbol.

28. From the top in the **Search** panel, click the **Get Rectangle** button; the cursor changes to the crosshairs and you are prompted to select the rectangle.

29. Drag the two opposite corners labeled as **1** and **2** in Figure 24 to define the shape you want to search; the specified shape is now shown in the preview box at the top in the **Search** panel.

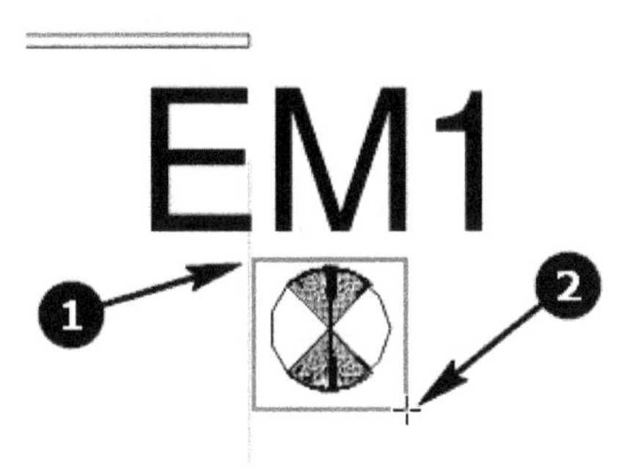

Figure 24 *Dragging the box around the shape to search*

30. In the **Options** area, drag the **Sensitivity** slider all the way to the left.

31. Select the **Search Multiple Rotations** check box, as shown in Figure 25.

32. Clear the **Search Markups** check box, as shown in Figure 25.

33. Clear the **Search for Fine Details** check box. Figure 25 shows the **Options** area with the settings configured for this search.

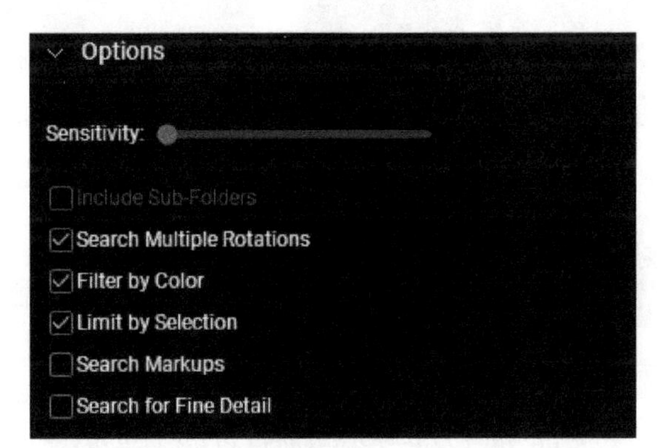

Figure 25 *The visual search options configured*

34. From the top in the **Search** panel, click the **Search** button; the process of performing visual search starts.

    Note that the visual search takes longer than the text search as the program needs to look for anything matching the symbol you specified.

Once the search process is completed, the bottom of the **Results** area shows 30 results found. Therefore, you will now scroll through the results to review what is found in the PDF.

35. One by one scroll through the results and review various items found.

    As mentioned earlier, there are three types of Emergency Luminaires in this PDF file. As you scroll through the results, you will notice that the program has found 18 X EM1s, 11 X EM2s (symbol enclosed in a box), and 1 X EM5. The EM1s and EM5 will be applied the same custom count. However, EM2 will be applied a different custom count.

36. From the top in the **Results** area, select the **Select All** check box; all the 30 search results are selected.

37. Scroll through the search results and deselect the 11 EM2s that have the symbol enclosed in a box.

38. To the remaining 19 selected elements, apply the **Recessed EM** custom count, as shown in Figure 26.

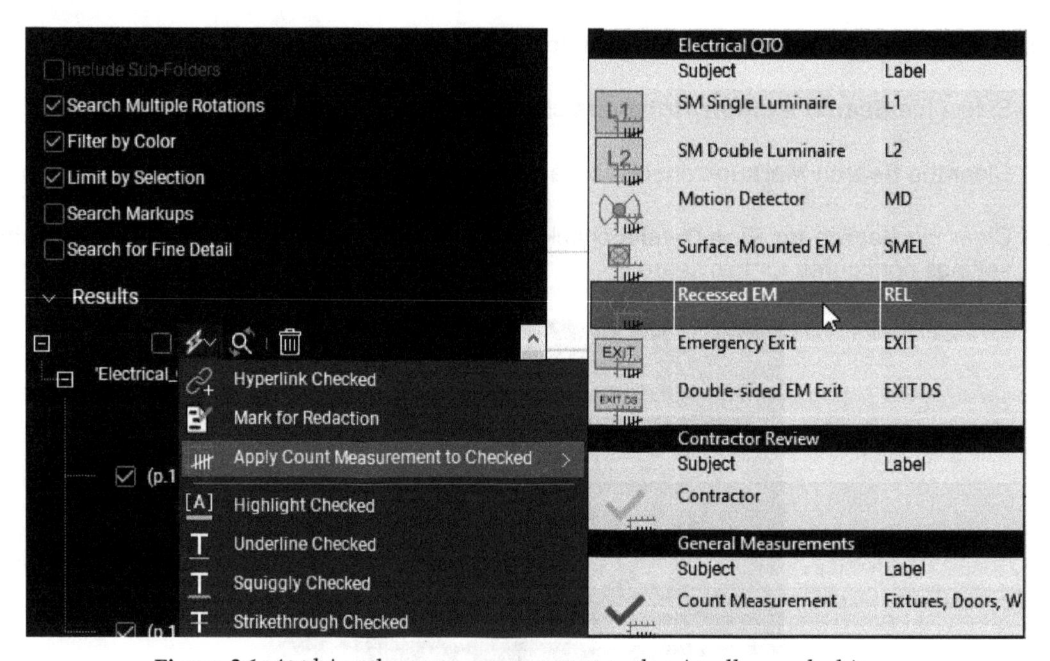

Figure 26 *Applying the count measurement to the visually searched items*

39. At the top in the **Results** area, select the **Select All** check box to select all the results.

40. Now, clear the **Select All** check box to deselect all the results.

    You will now select the 11 counts of EM2s and apply the **Surface Mounted EM** count measurement to them.

41. From the search results, select the 11 counts of the EM2 results.

42. Apply the **Surface Mounted EM** count measurement to the selected search results.

43. Similarly, search for the text **EXIT**. You will find 8 results.

44. To the 4 counts of EXIT with double-sided arrows, apply the **Double-sided EM Exit** count measurement.

45. To the 4 counts of EXIT, apply the **Emergency Exit** count measurement.

46. Save the PDF file.

Section 5: Creating a Legend with the Quantities Taken off

In this section, you will create a legend to show the quantities you have taken off from the PDF file. You will then customize this legend and add it to your custom column to show the cost of the takeoff.

1. From the top right of the custom tool set, click the cogwheel icon and select **Legend > Create New Legend**, as shown in Figure 27.

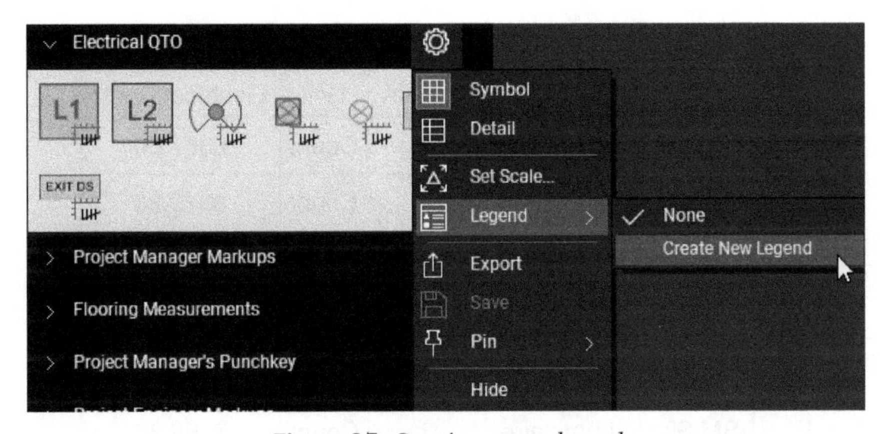

Figure 27 Creating a new legend

On doing so, a new legend is attached to the cursor.

2. Place the legend near the bottom right of the PDF file.

You will now configure the settings of this legend and include the custom column that shows the cost of the takeoff type.

3. With the legend still selected, display the **Properties** panel.

4. In the **Title** area > **Title** field, enter **ELECTRICAL TAKEOFF** as the title of the legend.

5. In the **Columns** area, click the **Edit Columns** button; the **Edit Legend Columns** dialog box is displayed.

6. Scroll down in the dialog box and select the **Electrical Costs** column.

7. Click **OK** in the dialog box; the legend is updated and you can now see the cost of each electrical fixture type.

8. In the **Table** area, change the color to Blue.

9. From the **Table Style** drop-down list, select **Gridlines**; the grid lines are added in the legend.

10. Change the **Symbol Size** spinner to **150**; the sizes of the flooring type symbols in the legend are increased.

    This completes the editing of the legend.

11. If need be, drag the legend so it is located inside the border of the PDF file.

12. Click somewhere outside to deselect the legend. Figure 28 shows the legend for this takeoff.

ELECTRICAL TAKOFF				
	Description	Quantity	Unit	Electrical Costs
EXIT DS	Double-sided EM Exit	4	Count	$367.00
EXIT	Emergency Exit	4	Count	$319.80
	Motion Detector	30	Count	$2,820.00
	Recessed EM	19	Count	$1,805.00
L2	SM Double Luminaire	23	Count	$1,506.50
L1	SM Single Luminaire	3	Count	$168.30
	Surface Mounted EM	11	Count	$1,319.45

Figure 28 *The legend showing the electrical takeoff counts and costs*

You will now add this legend to the custom tool set you have imported. This way this customized legend can be added to any other file as well where you takeoff the quantities.

13. Right-click on the custom legend and add it to your custom tool set.

14. Zoom to the extents of the PDF file and then save the file.

Section 6: Exporting Quantities as a CSV File

In this section, you will export the quantities you have taken off as a CSV file so the values could be used outside Revu.

1. From the toolbar at the top in the **Markups List**, click the **Summary** flyout, labeled as **1** in Figure 29, and then select **CSV Summary**, as shown in the same figure.

Markups List | Search | 1

CSV Summary
XML Summary
PDF Summary
Print Summary

Subject	Date	Comments	Electrical Fixtures		E	
Recessed EM (1)						
Recessed EM	17/02/2020 11:42:3...	19	REL	95.00		
SM Double Luminaire (1)						
SM Double Luminaire	17/02/2020 10:30:4...	23	L2	65.50	$1,506.50	1
SM Single Luminaire (1)					$168.30	
SM Single Luminaire	17/02/2020 10:18:1...	3	L1	56.10	$168.30	1
Surface Mounted EM (1)					$1,319.45	

Figure 29 *Exporting the quantities*

On doing so, the **Markup Summary** dialog box is displayed.

2. Activate the **Filter and Sort** tab.

3. Click on the **Subject** drop-down list and deselect **Legend**. Make sure all the counts are selected.

4. In the rest of the drop-down lists, make sure **All** is selected, as shown in Figure 30.

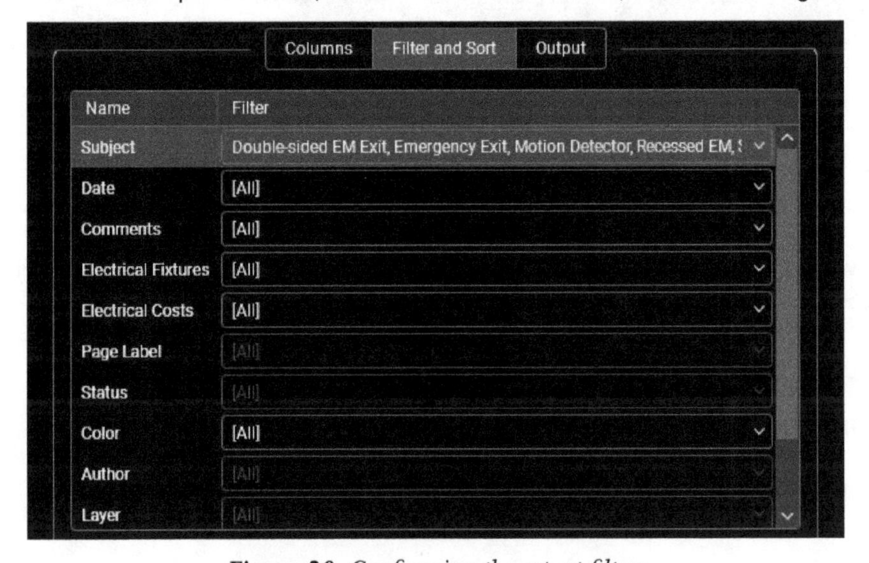

Figure 30 *Configuring the export filters*

5. Activate the **Output** tab.

6. Set the **Export to** folder to the **Tutorial Files > P02** folder.

7. From the **Include** area, select the **Markups & Totals** radio button, as shown in Figure 31.

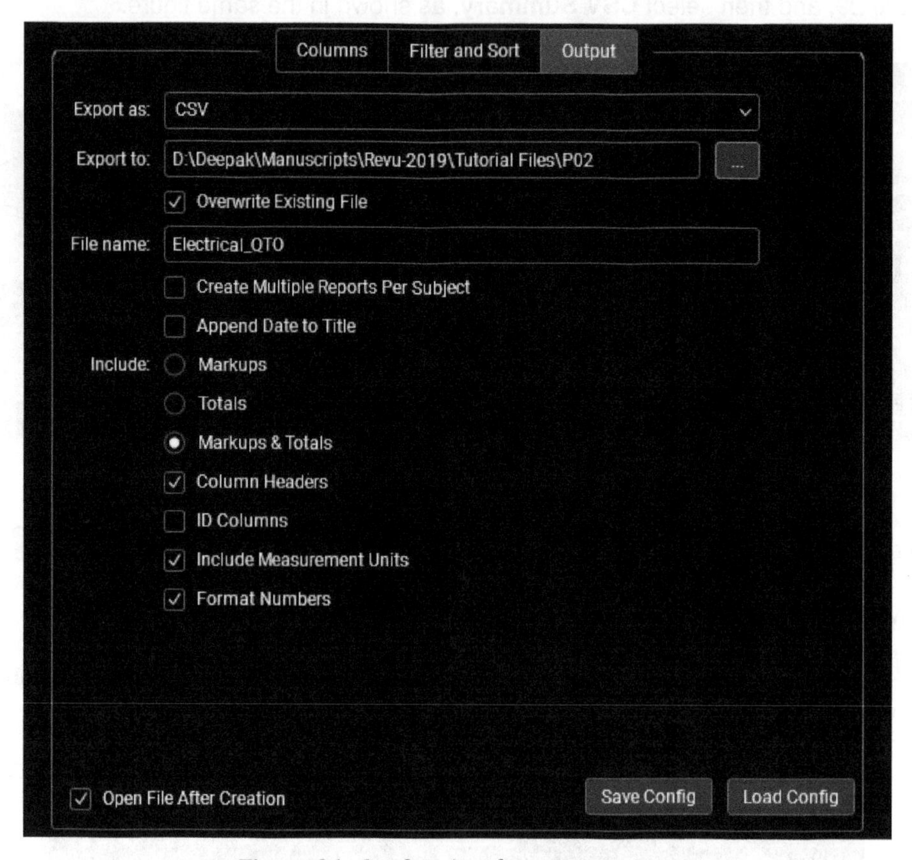

Figure 31 *Configuring the output settings*

8. Click the **OK** button in the **Markup Summary** dialog box; the summary of the quantification is created and the CSV file is opened, as shown in Figure 32.

9. Review the quantification summary and notice how each measurement type is displayed in the CSV summary.

10. Return to the Revu window.

11. Save the PDF file.

12. Close the PDF file.

	A	B	C	D	E	F	G	H	I
1	Subject	Date	Comments	Electrical Fixtures	Electrical Costs	Page Label	Status	Color	Author
2	Double-sided EM Exit (1)				$367.00				
3	Double-sided EM Exit	17/02/2020 12:00	4	EXIT DS	$367.00	1		#8080FF	dmaini
4	Emergency Exit (1)				$319.80				
5	Emergency Exit	17/02/2020 11:49	4	EXIT	$319.80	1		#8080FF	dmaini
6	Motion Detector (1)				$2,820.00				
7	Motion Detector	17/02/2020 11:00	30	MD	$2,820.00	1		#0000FF	dmaini
8	Recessed EM (1)				$1,805.00				
9	Recessed EM	17/02/2020 11:42	19	REL	$1,805.00	1		#FF0000	dmaini
10	SM Double Luminaire (1)				$1,506.50				
11	SM Double Luminaire	17/02/2020 10:30	23	L2	$1,506.50	1		#FF8000	dmaini
12	SM Single Luminaire (1)				$168.30				
13	SM Single Luminaire	17/02/2020 10:18	3	L1	$168.30	1		#8080FF	dmaini
14	Surface Mounted EM (1)				$1,319.45				
15	Surface Mounted EM	17/02/2020 11:45	11	SMEL	$1,319.45	1		#FF0000	dmaini

Figure 32 *The CSV file with the takeoff summary*

Index

Symbols

A

B

C

D

E

F

G

H

I

Answers to Skill Evaluation

Chapter 1
1. F
2. T
3. F
4. T
5. F
6. (D) **File Access**
7. (A) .DWG
8. (A) Profile
9. (C) CTRL
10. (A) Revu

Chapter 2
1. F
2. F
3. T
4. T
5. T
6. (A) **Callout**
7. (B) **Note**
8. (B) **Match Properties**
9. (D) **Recent Tools**
10. (A) **Properties** and (D) **Drawing**

Chapter 3
1. T
2. F
3. F
4. T
5. F
6. (D) **Dimension**
7. (A) **Polyline**
8. (D) **ALT**
9. (D) **Style**
10. (B) **Arrow**

Chapter 4
1. F
2. T
3. F
4. F
5. F
6. (A) **ALT**
7. (D) **Cloud+**
8. (A) **You can import AutoCAD hatches**
   (D) **You can create your custom hatches**
9. (B) **Hatch**
10. (C) **K**

Chapter 5
1. F
2. T
3. T
4. T
5. F
6. (C) BTX
7. (A) CSV
8. (A) **My Tools**
9. (B) **Symbol**
10. (A) **Subject**, (D) **Comments**

Chapter 6
1. F
2. T
3. F
4. F
5. T
6. (B) **Length**
7. (A) Viewports
8. (B) **Status Bar**, (D) **Navigation Bar**
9. (C) **View > Disable Line Weights**
10. (B) Calibrate

Chapter 7
1. F
2. T
3. F
4. T
5. F
6. (D) Second
7. (B) **Dynamic Fill**
8. (A) Polygonal, (B) Elliptical
9. (A) **Add Boundary**
10. (B) Rectangle

Chapter 8

1. T
2. F
3. T
4. T
5. F
6. (A) Word
7. (B) PDF
8. (D) **Hide Markups**
9. (A) Notes
10. (D) Choice

Chapter 9

1. T
2. T
3. F
4. F
5. T
6. (C) **Hyperlink**
7. (A) **Place**
8. (A) **Snapshot**
9. (B) **Create Page Labels**
10. (B) **Open**

Chapter 10

1. F
2. F
3. F
4. F
5. F
6. (B) **From Camera**
7. (A) **Compare Documents**
8. (B) **Overlay Pages**
9. (D) **Visual**
10. (A) **Align Points**

Chapter 11

1. T
2. F
3. F
4. F
5. T
6. (A) Dynamic
7. (A) **Batch > Apply Stamp**
8. (A) **Multiply**
9. (D) PDF
10. (B) **Import Stamp**

Chapter 12

1. T
2. F
3. F
4. T
5. T
6. (B) Check out
7. (A) PDF
8. (B) Read
9. (D) All of them
10. (A) **Undo Checkout**

Printed in the USA
CPSIA information can be obtained
at www.ICGtesting.com
LVHW021041070823
754531LV00012B/502